Vol. 31. **Photometric Organic Analysis** (*in t*

Vol. 32. **Determination of Organic Compou** Weiss

Vol. 33. **Masking and Demasking of Chemic**

Vol. 34. **Neutron Activation Analysis.** By D.

Vol. 35. **Laser Raman Spectroscopy.** By Ma

Vol. 36. **Emission Spectrochemical Analysis.**

Vol. 37. **Analytical Chemistry of Phosphorus Compounds.** Edited by M. Halmann

Vol. 38. **Luminescence Spectrometry in Analytical Chemistry.** By J. D. Winefordner, S. G. Schulman and T. C. O'Haver

Vol. 39. **Activation Analysis with Neutron Generators.** By Sam S. Nargolwalla and Edwin P. Przybylowicz

Vol. 40. **Determination of Gaseous Elements in Metals.** Edited by Lynn L. Lewis, Laben M. Melnick, and Ben D. Holt

Vol. 41. **Analysis of Silicones.** Edited by A. Lee Smith

Vol. 42. **Foundations of Ultracentrifugal Analysis.** By H. Fujita

Vol. 43. **Chemical Infrared Fourier Transform Spectroscopy.** By Peter R. Griffiths

Vol. 44. **Microscale Manipulations in Chemistry.** By T. S. Ma and V. Horak

Vol. 45. **Thermometric Titrations.** By J. Barthel

Vol. 46. **Trace Analysis: Spectroscopic Methods for Elements.** Edited by J. D. Winefordner

Vol. 47. **Contamination Control in Trace Element Analysis.** By Morris Zief and James W. Mitchell

Vol. 48. **Analytical Applications of NMR.** By D. E. Leyden and R. H. Cox

Vol. 49. **Measurement of Dissolved Oxygen.** By Michael L. Hitchman

Vol. 50. **Analytical Laser Spectroscopy.** Edited by Nicolo Omenetto

Vol. 51. **Trace Element Analysis of Geological Materials.** By Roger D. Reeves and Robert R. Brooks

Vol. 52. **Chemical Analysis by Microwave Rotational Spectroscopy.** By Ravi Varma and Lawrence W. Hrubesh

Vol. 53. **Information Theory As Applied to Chemical Analysis.** By Karel Eckschlager and Vladimir Štěpánek

Vol. 54. **Applied Infrared Spectroscopy: Fundamentals, Techniques, and Analytical Problem-solving.** By A. Lee Smith

Vol. 55. **Archaeological Chemistry.** By Zvi Goffer

Vol. 56. **Immobilized Enzymes in Analytical and Clinical Chemistry.** By P. W. Carr and L. D. Bowers

Vol. 57. **Photoacoustics and Photoacoustic Spectroscopy.** By Allan Rosencwaig

Vol. 58. **Analysis of Pesticide Residues.** Edited by H. Anson Moye

Vol. 59. **Affinity Chromatography.** By William H. Scouten

Vol. 60. **Quality Control in Analytical Chemistry.** By G. Kateman and F. W. Pijpers

Vol. 61. **Direct Characterization of Fineparticles.** By Brian H. Kaye

Vol. 62. **Flow Injection Analysis.** *Second Edition*. By J. Ruzicka and E. H. Hansen

Vol. 63. **Applied Electron Spectroscopy for Chemical Analysis.** Edited by Hassan Windawi and Floyd Ho

Vol. 64. **Analytical Aspects of Environmental Chemistry.** Edited by David F. S. Natusch and Philip K. Hopke

Vol. 65. **The Interpretation of Analytical Chemical Data by the Use of Cluster Analysis.** By D. Luc Massart and Leonard Kaufman

(*continued on back*)

Flow Injection Analysis

SECOND EDITION

CHEMICAL ANALYSIS

A SERIES OF MONOGRAPHS ON
ANALYTICAL CHEMISTRY AND ITS APPLICATIONS

VOLUME 62

A WILEY-INTERSCIENCE PUBLICATION

JOHN WILEY & SONS

New York / Chichester / Brisbane / Toronto / Singapore

Flow Injection Analysis

SECOND EDITION

JAROMÍR RŮŽIČKA

Department of Chemistry
University of Washington
Seattle, Washington

and

Chemistry Department A
The Technical University of Denmark
Lyngby, Denmark

ELO HARALD HANSEN

Chemistry Department A
The Technical University of Denmark
Lyngby, Denmark

A WILEY-INTERSCIENCE PUBLICATION

JOHN WILEY & SONS

New York / Chichester / Brisbane / Toronto / Singapore

Library of Congress Cataloging in Publication Data:

Růžička, Jaromír.
Flow injection analysis / Jaromír Růžička, Elo Harald Hansen.—
2nd ed.

p. cm. — (Chemical analysis : v. 62)
"A Wiley-Interscience publication."
Bibliography: p.
Includes indexes.
ISBN 0-471-81355-9
1. Instrumental analysis. I. Hansen, Elo Harald. II. Title.
III. Series.
QD79.I5R89 1988 87-18772
543'.08—dc19

Printed in the United States of America

10 9 8 7 6 5 4 3 2

To Andula and Else

PREFACE TO THE SECOND EDITION

Every revolutionary idea in Science, Politics, Art or Whatever evokes three stages of reaction. They may be summed up by the three phrases:

1. *"It is impossible, do not waste my time."*
2. *"It is possible, but it is not worth doing."*
3. *"I said it was a good idea all along."*

CLARK'S LAW OF REVOLUTIONARY IDEAS

Over 1300 flow injection analysis (FIA) papers have been published since the first edition of *Flow Injection Analysis* [153]* was completed in 1981. Also, two flow analysis conferences have been held (one in Sweden and the other in England), and several national and international symposia on FIA have been organized and given in Scandinavia, the United States, Japan, and China. In Japan The Association for Promotion of FIA has been founded and has begun publishing the first periodical entirely devoted to FIA, *Journal of Flow Injection Analysis*. The number of practicing "FIAists" is beyond estimate.

FIA has passed the Clark test with flying colors, or perhaps surpassed it, since a fourth stage of reaction has been recorded in the literature: "We discovered FIA a long time ago" [1352].

In view of these developments, a mere reprinting of the first edition, which in the interim has been published in Japanese [663] and Chinese [1223], would not have been satisfactory. Yet, when we reviewed the novel approaches and wealth of contributions that our colleagues throughout the world have made, we could not hope realistically to describe the amazing development of FIA comprehensively. Thus, we merely at-

* In view of this avalanche of publications, reference to published FIA papers are throughout this monograph consistently cited with numerals in brackets, e.g. [153], and are gathered together chronologically at the end of the book in the "List of FIA References." References that pertain particularly to one chapter only are cited, e.g. [1.2], where 1 refers to the chapter and 2 to the reference number. The latter type references are listed at the end of each chapter.

tempted to capture the state of the art as we perceive it ourselves, an approach that the reader may justly view as egocentric. In our defense we wish to say that we tried our best to review the technique we were so fortunate to have discovered and named [1].

For other viewpoints on FIA applications, theory, history, or future trends the reader may wish to consult the monograph written by Ueno and Kina [664] (in Japanese), dealing with FIA principles and experimental techniques, the monograph of Valcarcel Cases and Castro [665] (in Spanish), which comprehensively deals with FIA methods, theory, techniques, history, and applications and which recently was translated into English, and the most recently published book of J. Möller (Springer Verlag), available in German and English, dealing with FIA techniques and laboratory applications.

All those acknowledged in the preface to the First Edition supported us most kindly during the preparation of this second edition. Once again we wish to express our gratitude to Ove Broo Sørensen for his outstanding artwork and endless patience (however, to keep up with old tradition, all the "FIAgrams" in this monograph are photographic reproductions of the original chart recordings), to Tage Frederiksen of the mechanical workshop of the Chemistry Department A for his ability to turn our scribbles on odd pieces of paper into workable prototypes, and to Alison Macdonald for prompting and encouraging us in the work on this edition. Finally, we are indebted to Bruce Kowalski of the Center for Process Analytical Chemistry (CPAC) in Seattle, Washington, for opening our eyes to the wonders of word processing by Macintosh, without which the completion of the second edition would have been not only appalling but quite clearly close to impossible.

To the Danish Council for Industrial and Scientific Research and to Pernovo, Sweden, we express our sincere appreciation for their pecuniary support of several aspects of our research and to Gary D. Christian for friendly support.

This edition was completed and sent to the publisher in April 1987.

JAROMÍR RŮŽIČKA
ELO HARALD HANSEN

Copenhagen, Denmark
November 1987

PREFACE TO THE FIRST EDITION

I would never read a book if it were possible to talk half an hour with the man who wrote it.

WOODROW WILSON, ADVICE TO HIS STUDENTS AT PRINCETON

Flow injection analysis should not be explained. It ought to be demonstrated.

BO KARLBERG

To read about flow injection analysis is, unfortunately, the worst of the three possible choices noted above. Therefore, we have included herein some exercises that should allow patient readers to turn to the lab bench and make several simple experiments themselves, thereby experiencing the surprising features of this new analytical technique in much the same way as our students do during their course in instrumental analysis. This may be done using the FIA analyzer that is commercially available, or if one is unwilling or unable to make such an investment, using a home-made machine, which will do amazingly well provided that it is assembled from the proper parts. It is not enough, however, to dash into the children's room and plunder their toy box of Lego building blocks, although admittedly without them your FIA analyzer will not be as colorful and will certainly be less tidy. Some more time has to be invested in finding, making, or borrowing the injection port, the flow-through detectors, and other components and in assembling them according to the rules given herein. By following this more tortuous path, one will certainly be rewarded by a better understanding of what is actually going on when a sample is injected and forced to flow.

The authors made such an experiment for the first time in the early spring of 1974, and it took them six months to find that the idea was worth pursuing and that the method deserved to be given a name. Following this, it took another two years before anyone else took much notice of flow injection analysis, which would still be regarded as an obscure eccentricity had not several of our colleagues become interested in our work. Fortunately, they were often more creative than we were ourselves, and they developed their own experimental approaches which confirmed the versatility of the new technique. We greatly appreciate the research cooperation that has been extended to us by Henrique Bergamin, Elias Zagatto, and F. (Chico) J. Krug in Brazil, Bo Karlberg and Torbjörn Anfält in Sweden, Alison Macdonald and Jack Betteridge in England, Kent Stewart in the United States, and Jens Mindegaard, Søren S. Jørgensen, and Arne Jensen in Denmark. We wish to thank them for becoming our friends in the finest possible way—while disputing and comparing views and ideas.

We also wish to express our gratitude to the many people who have shown interest in our work and have found time to visit our laboratory. We regret that we cannot mention them all. W. E. van der Linden, H. Poppe, and J. M. Reijn were among them, and we are indebted to them for inspiring discussions and for organizing the Flow Analysis Conference in Amsterdam in fall, 1979.

We also must not forget the assistance of those who consistently rejected the concept of FIA, because their scepticism made us work harder. We do not think, however, that they would like to be specifically mentioned at this stage.

No instrumental method has become widely accepted before a commercial instrument has become available. Being aware of this, we tried hard to persuade several manufacturers of analytical instruments to consider giving FIA a try. After having failed miserably and given up entirely, a single person, without funds but with great vision, turned up and took a chance: Rune Lundin, from BIFOK AB in Sweden. We can never thank him enough.

No research can succeed without freedom and a proper spiritual and economic environment. We wish to express our gratitude to Niels Hofman-Bang and Carl Göran Lamm for implanting us into Chemistry Department A of the Technical University of Denmark, and to our colleagues and the members of the staff for nourishing a scientifically stimulating and friendly climate. The skillful technical assistance of Inge Marie Johansen, Eva Thale, and Georg Møller is greatly appreciated as is the electronic wizardry of Carl Erik Foverskov. We are indebted to the Danish Council for Scientific and Industrial Research, to the Danish Natural Science Research Council, and to Pernovo, Sweden, for supporting several aspects of our research.

It was kind of Alison Macdonald and Anders Ramsing to read the manuscript and offer their constructive comments. All the illustrations in the book except the response curves, which were reproduced photographically from recorder charts, are the careful work of Ove Broo Sørensen. Finally, this work was completed and sent to the publisher in May 1980.

Jaromír Růžička
Elo Harald Hansen

Copenhagen, Denmark
January 1981

CONTENTS

LIST OF SYMBOLS

A	absorbance; peak area
a	lag phase
a_m	radial mass transfer rate ($= 1/\bar{T}_i$)
C	concentration (mol/liter)
C^0	concentration at $t = 0$ (prior to injection)
C_{A0}	concentration of species A in a mixing chamber at $t = 0$
C^{max}	concentration at peak maximum
C^{maxSS}	concentration at steady state
D	dispersion coefficient ($= C^0/C$)
D_m	molecular diffusion coefficient (cm^2/s)
De	Dean number
D_f	axial dispersion coefficient
D_r	radial dispersion coefficient
d	internal tube diameter, $d = 2R$
d_p	particle diameter
F	mean linear flow velocity ($Q/\pi R^2$)(cm/s)
f	interstitial flow velocity (cm/s)
f	accommodation factor
h	fraction of suspended matter; peak height, steady-state analysis
H	peak height, FIA
H'	vertical readout
H_{0m}	height of a mixing stage in a straight tube
H_m	height of a mixing stage
k	proportionality constant; reaction rate constant
L	reactor length (cm)
N	number of mixing stages (tanks)
n	molar ratio of reacting components
Pe	Peclet number
p	pressure (atm)
Q	volumetric flow rate (mL/min)
R	internal tube radius
Re	Reynolds number
r_{cyc}	reagent consumption per cycle

Sc	Schmidt number
S_{max}	maximum sampling frequency
S_v	injected sample volume (μL)
$S_{1/2}$	sample volume necessary to reach 50% of the steady-state value, corresponding to $D = 2$
T	peak maximum time
$\bar{T}$	mean residence time
$\bar{T}_i$	mean residence time in a single tank
t	time(s)
t_a	peak arrival time
t_b	baseline to baseline interval
t_{cyc}	measuring cycle time
$t_{1/2}$	time in which any C decreases by one-half
u_r	recorder speed
V	volume
V_m	individual mixing stage volume ($= V_i$)
V_r	reactor volume
W	peak width at the base
x	distance traveled
α	injection parameter (S_v/V_r)
$\beta_{1/2}$	dispersion factor ($S_{1/2}/V_r$)
γ	tortuosity factor
η	viscosity
θ	injected plug width
κ	velocity profile factor
λ	aspect ratio (R/R_c)
ν	kinematic viscosity
ρ	density
σ	axial dispersion
σ^2	variance
τ	normalized residence time
ϕ	angle of coiling, knitting, or bending
χ	physical constant of a tank ($= Qt/V_m$)

Flow Injection Analysis

SECOND EDITION

CHAPTER

1

INTRODUCTION

You press the button and we'll do the rest.

KODAK ADVERTISEMENT 1888

I have no use of automation; I have only a few samples to analyze.

ANALYST'S COMMENT

1.1 AUTOMATION IN A CHEMICAL LABORATORY

Any measurement in a chemical laboratory involving liquid materials comprises the following operations: solution handling, analyte detection, data collection, and computation of results.

There is no shortage of computers and sophisticated detectors to aid chemists in performing the latter two tasks, but solution handling requires an arsenal of skills, which a practicing chemist has to master, since mixing, decanting, pipetting, and other volumetric operations are still performed manually, even in the most advanced laboratories, using tools that were designed more than 200 years ago (Fig. 1.1). It might seem that robots would be suitable tools for automation of such manual tasks; but, it is likely that their impact will remain limited to repetitive operations like weighing of pulverized materials, mechanization of sample injection into chromatographic columns, handling of radioactive materials, or sample preparation. Because manual handling using robots requires extensive programming and active feedback control, the use of robots is justified only if a large series of repetitive operations is to be handled over prolonged periods.

Truly, there seems to be no way of resolving the problem of automated solution handling other than by manual operations, as long as we think in terms of *batch operations*, a concept in which generations of chemists have been trained. Therefore, in freshmen courses, as well as in advanced research laboratories, beakers, flasks, and volumetric glassware are still

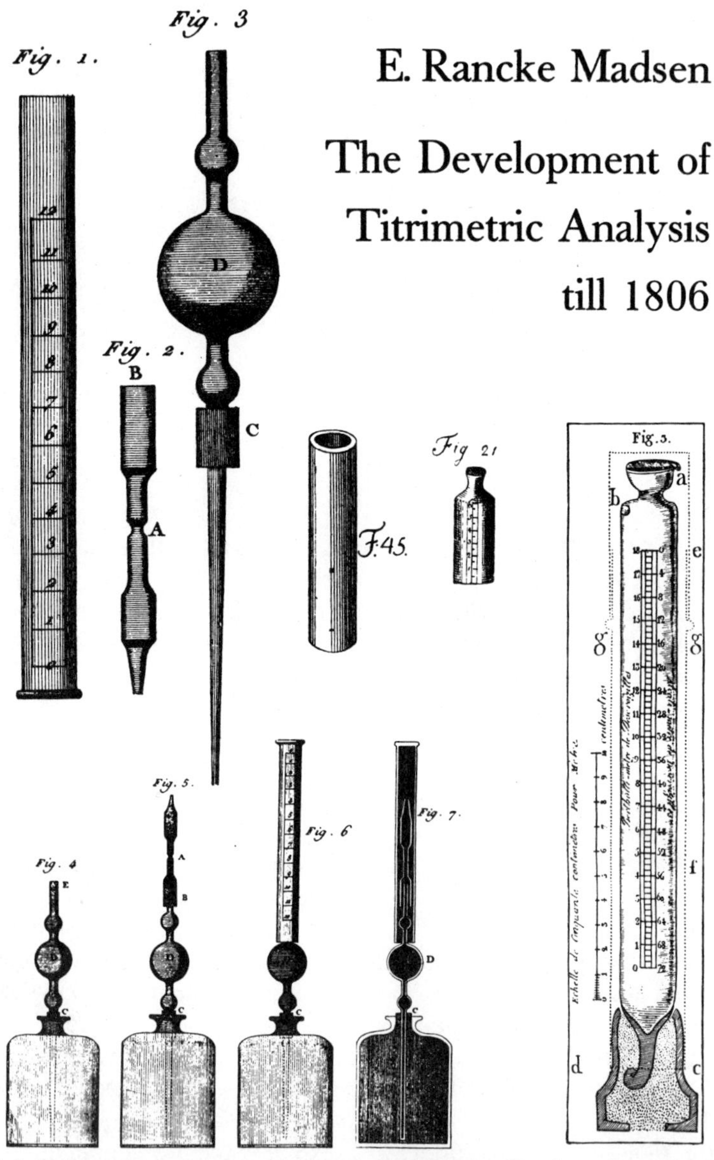

Figure 1.1. Volumetric glassware and concepts of solution handling have changed very little during the last 200 years. The cover of the thesis of the Danish historian of analytical chemistry, E. Rancke Madsen. G. E. C. Gad Publishers, Copenhagen.

the standard tools of the trade, coexisting with the electronics of advanced detectors and computers.

Flow operations are much easier to automate, since they replace the mechanical handling of oddly shaped (and often fragile) containers by sequential movements of liquids in tubes. Flow operations are much easier to miniaturize by using small bore tubing, and the microvolumes are conveniently manipulated and metered by pumping devices, which (unlike pipettes) are not affected by surface tension (or by shaking hands). Flow operations are much easier to control in space and time, since using closed tubing avoids evaporation of liquids, provides exactly repeatable path(s) through which measured solutions move, and provides an environment for a highly reproducible mixing of components and formation of reaction products. Flow operations are very versatile, since flows can be mixed, stopped, restarted, reversed, split, recombined, and sampled, while contact times with selected sections of reactive or sensing surfaces can be precisely controlled. Finally, flow operations allow most detectors and sensors to be used in a more reproducible manner than when used in batch operations and by hand—as is obvious to anyone who has used both conventional and flowthrough cuvettes.

While many of the advantages of flow operations have been exploited in chromatography, why has the batch approach not yet been replaced by flow systems in *all* areas of laboratory practice? The reason must be tradition, and the fact that most chemists are used to thinking in terms of batch operations, where *homogeneous mixing* is thought to be the only reproducible way to bring reactants together and where the homogeneously mixed solution is regarded as the only suitable form in which a reproducible measurement can be taken.

The concept of flow injection analysis (FIA) is changing this prevailing attitude. Originally designed to automate serial assays [1], FIA has emerged during the last decade as a general solution-handling and data-gathering technique [1055]. A present survey of the chemical literature reveals that FIA is finding increasing applications in analytical routine and research, teaching of analytical chemistry, monitoring of chemical processes, sensor testing and development, and enhancing the performance of various instruments. FIA is also used to measure fundamental values, such as diffusion coefficients, reaction rates, stability constants, composition of complexes, and extraction constants and solubility products.

Most FIA applications published so far have been concerned with applications to serial assays, since FIA was originally designed for that purpose. It is hoped, however, that this monograph, written primarily for the analytical community, will contribute to the further exploitation of

FIA in the above-mentioned areas, along the lines discussed in Chapter 8.

1.2 AUTOMATION OF SERIAL ASSAYS

The number of samples analyzed by means of automated instrumentation is large: millions of tests are run annually in clinical chemistry laboratories alone. Much significant research is done and considerable effort is expended in developing and manufacturing automatic analyzers, especially in the United States and Japan. Often, the wrong path is chosen and money is wasted in a futile attempt to produce a "black box" with a magic button to press. (For example, in 1971 Xerox terminated a 5-year program on which $70 million had been spent unsuccessfully [1.1].) The few successful corporations produce reliable, but often very complex, instruments, which cost for the simplest, single-channel analyzer, $35,000; for a computer-based parallel fast (centrifugal) analyzer, about $80,000; and for a multichannel analyzer from about $100,000 to well over $200,000. It is not hard to guess where this leaves analysts, such as those from small laboratories or universities and the often forgotten majority of chemists working in developing countries, who have a limited number of samples to analyze or limited funds.

Thus, today, we have a large number of samples being automatically analyzed by a few people, while the majority of chemists use manual methods because of lack of funds or lack of interest. The latter is not a negligible factor, and the teaching of analytical chemistry is much to blame for this situation. Textbooks on instrumental analysis treat the subject in a limited way. The latest editions of the best and most frequently used books [1.2–1.4] devote only a small portion of the test to automation; only part of this coverage deals with the analysis of discrete samples, the remainder covers continuous monitoring for process control. It is, however, hard to blame the authors, most of whom are university teachers who know that only few students will use automated instruments in their laboratory courses or will use such instruments later in their careers. Therefore, most chemistry graduates have never heard a lecture on automation and know next to nothing about the subject.

Thus, the majority of chemical assays in university courses, research, and small laboratories are performed manually with the aid of traditional tools (Fig. 1.1). To draw an analogy, we find ourselves in a situation similar to that in which computation technology was 15 years ago: large computers were available as were pocket-size electronic calculators, while minicomputers were well beyond the horizon, and menu-driven personal computers with laser printers were unimaginable.

Until the advent of FIA, all automated methods of analysis emulated not only the sequence, but also the underlying concept of individual operations performed in the course of a manual procedure (Fig. 1.2). A simple chemical assay (Fig. 1.2*a*) requires mixing of a precisely metered volume of sample solution with a precisely metered volume of reagent,

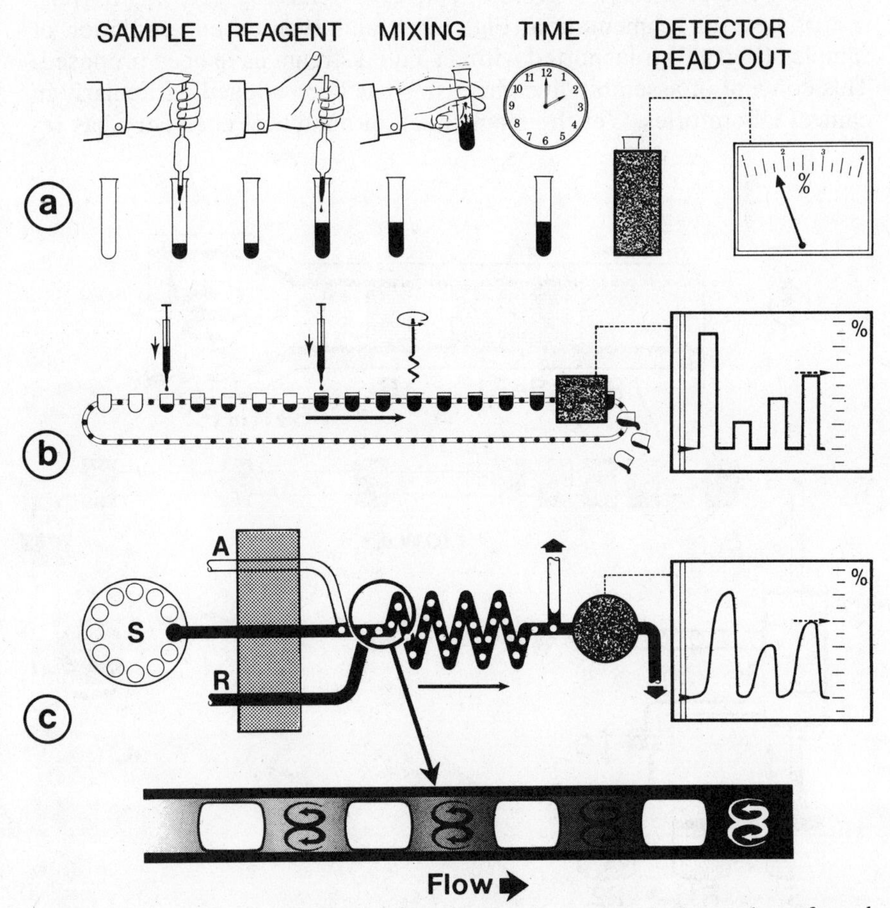

Figure 1.2. The parallel between manual and automated operations that have to be performed in the course of a typical colorimetric assay: (*a*) Manual handling; (*b*) a discrete belt-type analyzer; (*c*) a continuous-flow air-segmented analyzer, with a detail of the air and liquid segments, showing a mixing pattern that leads to homogenization of individual liquid segments. Note that all assays aim to perform the measurements at a steady state (stable readout, that is, "flat top").

followed by the time interval necessary to form the species to be measured. Obviously, these sequential operations must be precisely repeated, otherwise the conditions under which the standards and unknowns have been processed would not match, and the assays will yield erroneous results. Automated instruments can perform these mundane operations of sample metering, diluting, reagent addition, mixing, incubation, heating, separation, and signal monitoring in simple, or complex, sequences, tirelessly and precisely. Therefore, conveyor belts (Fig. 1.2*b*), carousels, centrifuges, air segmentation (Fig. 1.2*c*), and even oil encapsulation of samples (Fig. 1.3) transported within a liquid stream have been proposed. This concept of assembly-line chemistry has been applied particularly in clinical laboratories. Yet the mainstream of analytical chemistry has re-

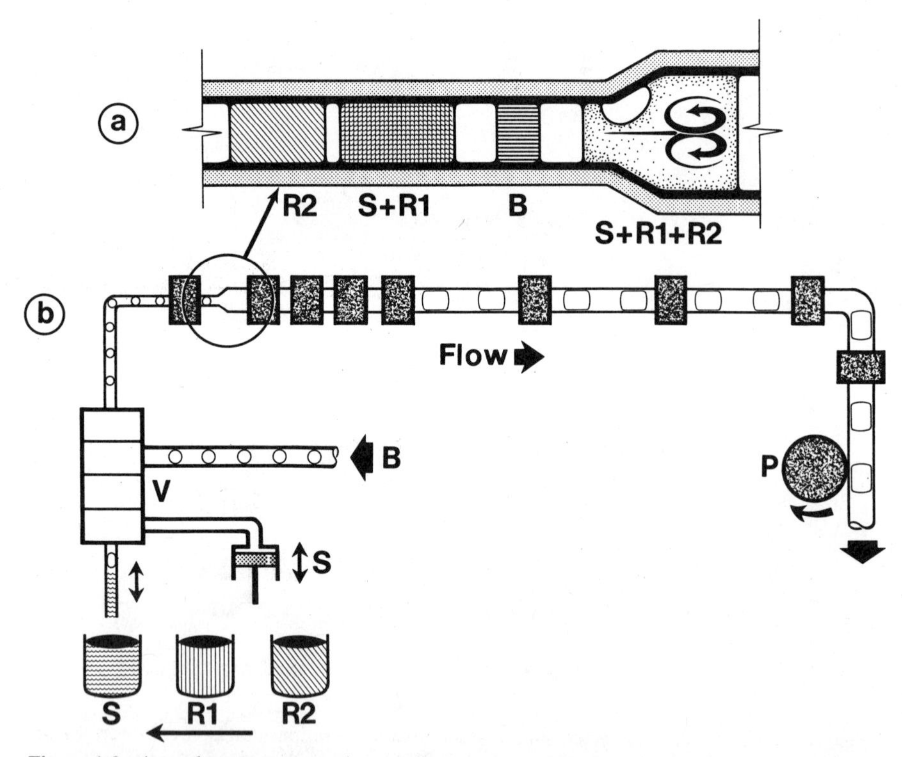

Figure 1.3. A random-access continuous-flow analyzer (Chem 1–Technicon, USA), which aspirates sample (S) and necessary reagents (R1, R2) sequentially from a two carousels (not shown). The resulting stream is air segmented and oil encapsulated (oil is shown as solid lines on top of liquids in the cups and on the walls of tubing to minimize the carryover. Detection is by a series of optical detector stations that discretely record gradual development of color during the flow transport. (P is a peristaltic pump.)

mained untouched by this development, which has been centered on the mechanical and logistic problems of the handling of large series of discrete samples, with the sole aim to increase the sample throughput. It is, however, useful to review briefly the batch and continuous-flow analyzers designed so far, since this will allow us to identify their advantages and shortcomings and to propose what can be viewed as the ultimate approach to solution handling and data gathering in chemical research and technology.

1.3 DISCRETE ANALYZERS

In a batch analyzer, each discrete sample is assigned a container (sample cup or sample bag), within which it is held through all the steps necessary to perform the analysis. The advantage of this approach is that cross-contamination between samples is not possible; the samples preserve their identity and cannot be mismatched, since each container carries an identification label. The disadvantage of all batch analyzers is that they are mechanically very complex, have many moving parts that may become worn, and include components that must be precisely machined; therefore, they have limited lifetimes. Three types of batch analyzers have been suggested [1.2–1.4].

The first type of analyzer, the sequential discrete analyzer with sample cups, is a system in which individual samples are transported through a number of stations where manual operations, such as pipetting, diluting, reagent addition, mixing, heating, incubation, and measurement (usually that of color, turbidity, or electrode potential) are mechanized, thus simulating a manual procedure. The more complex instruments of this kind perform preparations of solid samples (including grinding) as well as titrations.

The use of prepacked reagents in individual sample bags [1.2–1.4] is the basis of the second type of batch analyzer, originally marketed by DuPont. The additional advantage of DuPont's Automated Clinical Analyzer is that the same machine is capable of performing various assays on identical samples sequentially, because each bag may contain different chemicals (in the form of pills), while the various stations through which the bags are carried (suspended on chains) crush the pills, mix the sample with reagent, and finally mold a plastic lens, containing the reacted mixture, for spectrophotometric measurement. This machine is well suited for performing single emergency assays during night duty in a clinical laboratory and fulfills the promise of the 1888 Kodak advertisement (although at a high price).

An ingenious extension of this technology is the introduction of solid-state reagents incorporated into a slide in which they are layered in strategic sequence corresponding to the operations to be performed on a given serum sample. Such a slide, produced by layered coating technology, has a spreading layer; intermediate layers that filter and convert an analyte to a quantifiable material; and a reflectance layer, within which the optical density of the product formed is measured by reflectance spectrophotometry. The layers are backed by a transparent support and sandwiched in a plastmic mount that fits within the optical sphere of a reflectance spectrophotometer. This tool, developed by Eastman Kodak Company [1.5–1.6], is the most advanced approach to reagent prepacking. It is, however, strictly limited to clinical applications where its utility yet has to be tested extensively.

The third batch analyzer, the parallel centrifugal analyzer [1.2–1.4, 1.7], is designed to perform a number of spectrophotometric determinations simultaneously. This instrument contains a flat rotor with a number of radially arranged cavities connected by radial channels to a corresponding number of spectrophotometer cuvettes arranged on the rotor's circumference. Each cavity is separated into two compartments by a low barrier. The reagent solution is pipetted into the compartment closest to the rotor axis, and the individual samples are pipetted into the outer compartments. When a disk is rotated, the reagent solutions are moved by centrifugal force across the barrier into the sample compartments, and these mixtures are further forced through the channels into the appropriate cuvettes. Effective mixing is achieved by rapidly changing the speed of rotation of the rotor. Sample transmittance is measured by means of a light beam that scans all the cuvettes in turn as they pass the stationary light source. The display of absorbances in individual cuvettes is obtained on an oscilloscope, and the machine has a built-in computer to sort and process the signals.

The drawbacks of discrete analyzers are their mechanical complexity and high cost of operation. Sample cups, disposable cuvettes, rotors, and prepacked reagents increase the cost of individual assays above the acceptable limit for the strained budgets of most clinical laboratories. In addition, these machines are seldom used outside the clinical laboratory, because they are designed to handle three dozen of the most frequently required clinical tests. The advantages of the discrete approach are the ability of some of these instruments to perform assays via random access—which allows sequential assay of diverse analytes at will—and the capability of stat operation, which yields the analytical readout within 5–10 min after the machine has been switched on and a sample has been inserted by a technician.

1.4 CONTINUOUS-FLOW ANALYZERS

In its broadest context the term continuous-flow analysis refers to any process in which the concentration of analyte is measured uninterruptedly in a stream of liquid (or gas). For serial assays, successive samples are introduced into the same path (length of tubing) in the analyzer, with the reagents being added at strategic points and the mixing and incubation taking place while the sample solution is on its way toward a flowthrough cuvette, where the signal is monitored continuously and recorded. The greatest difficulty to overcome is to prevent intermixing of adjacent samples during their passage through the analyzer conduits. To minimize such carryover, Skeggs [1.8] introduced air segmentation, thus dividing the flowing stream into a number of compartments separated by air bubbles (see Fig. 1.2*c*). The first system was developed for the determination of urea and glucose in blood 30 years ago; the Technicon AutoAnalyzer originally designed for these methods has since become the most popular automatic analyzer marketed.

The continuous-flow approach is the most flexible way to perform the number of operations necessary in a chemical assay. In addition to sample dispensing, dilution, heating, mixing, and reagent addition—operations equally well executed by a batch analyzer—the continous-flow mode can perform dialysis, distillation, solvent extraction, and other types of separations. Furthermore, a much greater variety of detectors may be applied to a flowing stream than to a centrifugal analyzer, where the cuvette is in a rapid circular motion, or to a batch analyzer, where the sample liquid is encapsulated in a bag. Also, the continuous-flow analyzer is made with fewer moving parts, because it is the liquid that is in motion, and, therefore, the instrument is mechanically much less complex and easier to construct than a batch analyzer.

The greatest drawback of the continuous-flow analyzer, from the viewpoint of a clinical chemist, is not being able to randomly access the assay. A single-channel analyzer, such as that shown in Fig. 1.2*c*, requires a larger number of samples to be collected for an assay of a single analyte (say glucose in serum). Another analyte (e.g., urea) in the same batch of samples, can be analyzed only after the analyzer has been reprogrammed (by changing manifold components, reagents, etc.). Consequently, multichannel analyzers have been designed with up to 20 parallel channels, each for a different analyte. The aspirated sample is split into parallel channels, which run simultaneously all the time. The unfortunate feature of such a design, besides its complexity, is its inability to perform a single selected test.

Therefore, most recently, a random-access, segmented, continuous-flow analyzer has been introduced (Fig. 1.3). This single-channel system aspirates not only samples (S), but also reagents (R1 and R2). Since reagents are randomly accessible from a rotating reagent tray, any required sequence of assays, single or multiple, can be performed on a sample that has been aspirated into the air-segmented stream, together with appropriate reagents, thus forming a train of tests, each test consisting of a number of segments of sample/diluent/air/buffer/reagent/air (Fig. 1.3*a*). After a 5-min delay, sample and reagents are brought together within a "vanish-zone" section of tubing, where the tube widens allowing the homogeneous mixing of sample with reagents to initiate the formation of the species subsequently to be monitored by a series of fiber-optic flow-through detectors. As an additional precaution to avoid carryover, the air segmentation is augmented by oil "encapsulation," the purpose of which is to minimize fluid slip at the Teflon walls of the channel and to impose additional barriers between segments of sample and reagent solution. This Chem 1 System of Technicon is claimed to have a very high throughput—up to 1800 tests per hour with up to 24 random accessible tests, but the price paid in terms of complexity of instrument design is high. The 5.5-m-long channel takes a sample 14 min to traverse, and with an aspiration rate of five samples per minute, it holds a train of 70 samples at any one time. If a failure occurs, all these samples would be lost. The system is more complex than it appears to be, since an additonal parallel channel is used for electrolytes, a system of valves is used to control sample aspiration and addition of air-segmented buffers, the sample probe has an elaborate system (using Freon) for dispensing the oil, and the construction of the random-access reagent carousel is mechanically demanding. While useful for processing of large series of assays in a clinical laboratory, the instrument and its concept is not suitable for general use.

However, the continuous-flow approach is the most flexible way to handle solutions, and, therefore, it is worthwhile to investigate it further. As was explained previously, the conventional continuous-flow analyzers (Fig. 1.2*c*) use air-segmented streams [1.8] to preserve the identity of individual samples, which are successively aspirated from their own containers and then pumped through the system, where the flowing stream is regularly segmented by air bubbles delivered by a separate pump tube. After the reagent is added, the stream is passed through a reaction coil where the contents of each individual segment are homogeneously mixed, and a sufficient time is given for the chemical reaction to come to or near completion. Consequently, the flowthrough detector records a peak with a flat top (Fig. 1.4). This peak consists of rise and fall curves, which represent transition between two different "steady-state" conditions (*A*

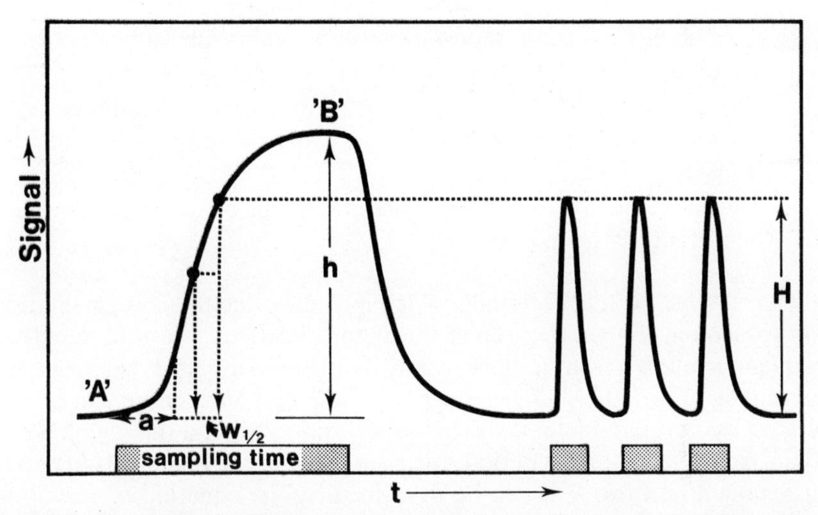

Figure 1.4. The conceptual difference between a steady-state signal (left) and a transitional FIA response curve (right): In an air-segmented continuous-flow system, the transition from baseline ("*A*") to a steady-state level (*B*") requires at least five "half-wash times" ($W_{1/2}$) to elapse, which in turn requires a sample volume of five to be aspirated into the system. The steady-state height (*h*) is then relatively independent of physical and chemical kinetics of the system. In contrast to that by injecting into a FIA system a sample volume corresponding one $W_{1/2}$ value, a transition signal with peak height *H* is recorded, which, however, depends critically on the physical and chemical kinetics of the FIA system. This is why well-controlled dispersion of a nonsegmented FIA system is the key to its successful operation and the understanding of nonsegmented flow systems, which yield much higher sampling frequency than segmented streams do [1]. ("*a*" is the lag phase observed in air-segmented streams, which is absent in FIA systems.)

and *B* in Fig. 1.4). The distance between *A* and *B* is the peak height *h* and is related to the concentration of analyte.

Air segmentation's beneficial effect of preventing carryover has been so obvious that the necessity of introducing air bubbles was never really doubted, although the drawbacks of their presence in the flowing stream are well known:

a. Because of the compressibility of air, the stream tends to pulsate rather than to flow regularly.
b. The streams have to be debubbled before they reach the flow cell or before being repumped.
c. The size of the air bubbles has to be controlled.
d. The pressure drop and flow velocities vary in the presence of air for different tubing materials.

e. Air bubbles in plastic tubes act as electrical insulators, supporting a buildup of static electricity that disturbs potentiometric sensors.

f. The efficiency of dialysis, gas diffusion across a membrane, and solvent extraction is lowered as a result of a decrease in the effective transfer surfaces.

g. The movement of the carrier stream cannot be exactly controlled or instantly stopped or restarted.

That the lack of exact timing, which is mainly due to air segmentation and to sample transport through the pump, leads to a loss of sampling frequency follows from a closer analysis of the AutoAnalyzer response curve (Fig. 1.4, left) [1]. The rise part of the response curve can be described by two parameters: the half-wash time $W_{1/2}$ and the lag phase a. Apart from the lag-phase period, the rising part of the response curve has an exponential form that can be described by the equation $h = h'(1 - e^{-t/\chi})$, where h is the distance from the baseline A measured at time t, h' is the distance between the baseline and the steady state B eventually reached, χ is a constant, and t is the time elapsed from the onset of the rise curve. The fall curve is the reverse of the rise curve. The half-wash (or half-rise) time $W_{1/2}$ equals 0.69χ. To minimize the slight irregularities in the movement of the air-segmented stream, 95% or better achievement of the steady-state plateau is normally required.

To summarize, air-segmented continuous-flow systems, similarly to batch analyzers, emulate manual solution handling, since these systems rely on attaining steady state and homogeneous mixing.

It is the purpose of this book to show how, by changing these two traditional concepts, continuous-flow assays can be simplified and new analytical techniques can be discovered. The resulting simpler instrumentation, being less costly and complex, is becoming standard equipment in many laboratories in both developed and developing countries [440]. Continuous-flow access will aid process control, will allow testing of materials and chemical sensors, and will become a standard training tool even in an undergraduate course of instrumental analysis [59, 139, 253, 299, 333, 446, 727, 1224, 1265].

The availability of a simple and versatile technique will change the perception that the use of an automatic assay is justified only if a large number of samples is to be analyzed. The advantages of automated solution handling, which will replace manual operations like pipetting, dispensing, mixing, and separating, will result in a drastic reduction in the volumes of sample and reagent solutions that are required, with no loss of reproducibility. Thus material and time willl be saved, and less energy will be needed to produce the reagents, to heat the sample, and to obtain the analytical readout. FIA is a truly microchemical technique; therefore, a well-designed FIA instrument does not fill the top of a small laboratory bench and will certainly become even smaller. Furthermore, the ability of FIA to control volumes, mixing patterns, and residence times exactly allows the use of such chemical procedures that are not suitable or reliable when performed manually, because reagents or reaction products are not stable in time. Kinetic assays can be performed by FIA in a novel way, and the advantage of kinetic discrimination [1122] leading to increased selectivity is a new facet yet to be fully exploited.

REFERENCES

1.1 Anonymous, *Biomed. Insight,* **9** (1978) 107.

1.2 G. W. Ewing, *Instrumental Methods of Chemical Analysis,* 4th ed., McGraw-Hill, Kogakusha Ltd., Tokyo, 1985.

1.3 R. L. Pecsok, L. D. Shields, T. Cairns, and I. G. McWilliam, *Modern Methods of Chemical Analysis,* Wiley, New York, 1976.

1.4 G. D. Christian and J. E. O'Reilly, *Instrumental Analysis,* Allyn and Bacon, Boston, 1986.

1.5 T. L. Shirey, *Clin. Biochem.,* **16** (1983) 147.

1.6 A. Zipp and W. E. Hornby, *Talanta,* **31** (1984) 863.

1.7 N. G. Anderson, *Science,* **166** (1969) 317; Am. J. Clin. Pathol., **53** (1970) 778.

1.8 L. T. Skeggs, *Anal. Chem.,* **38** (1966) 31A.

CHAPTER

2

PRINCIPLES

Believe me, my dear colleague, when I say it is so, then it is so.

W. NERNST

2.1 FLOW INJECTION ANALYSIS

Flow injection analysis (FIA) is based on the injection of a liquid sample into a moving, nonsegmented continuous carrier stream of a suitable liquid. The injected sample forms a zone, which is then transported toward a detector that continously records the absorbance, electrode potential, or other physical parameter as it continuously changes due to the passage of the sample material through the flow cell [1, 153].

An example of one of the simplest FIA methods, the spectrophotometric determination of chloride, is shown in Fig. 2.1; this is based on the release of thiocyanate ions from mercury(II) thiocyanate and its subsequent reaction with iron(III) and measurement of the resulting red color. (For details, see Section 6.4.) The samples, with chloride contents in the range of 5–75 ppm chloride, are injected (S) through a 30-μL valve into the carrier solution containing the mixed reagent, pumped at a rate of 0.8 mL/min. The iron(III) thiocyanate is formed on the way to the detector (D) in a mixing coil (0.5 m long, 0.5 mm inside diameter), while the injected sample zone disperses in the carrier stream of reagent. The absorbance A of the carrier stream is monitored continuously at 480 nm in a micro flowthrough cell (10 μL volume) and recorded (Fig. 2.1*b*). To demonstrate the reproducibility of the analytical readout in this experiment, each sample was injected in quadruplicate, so that 28 samples were analyzed at seven different concentrations of chloride. As this took 14 min, the average sampling rate was 120 samples per hour. The fast scan of the 75- and 30-ppm sample (shown on the right in Fig. 2.1*b*) confirms that there was less than 1% of the solution left in the flow cell at the time when the next sample (injected at S_2) would reach it, and that there was no carryover when injecting the samples at 30-s intervals.

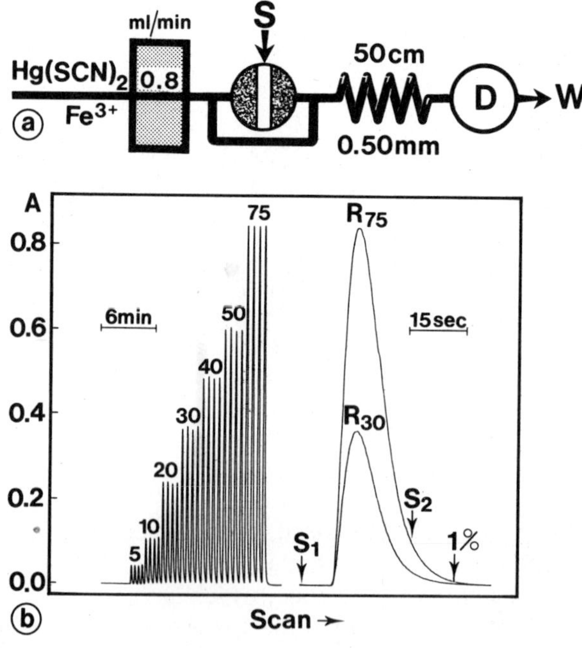

Figure 2.1. (*a*) Flow diagram for the spectrophotometric determination of chloride: *S* is the point of sample injection, *D* is the detector, and *W* is the waste. (*b*) Analog output showing chloride analysis in the range of 5–75 ppm Cl with the system depicted in (*a*). To demonstrate the reproducibility of the measurements, each sample was injected in quadruplicate. The injected volume was 30 μL, sampling rate was approximately 120 samples per hour. The fast scan of the 30-ppm sample (R_{30}) and the 75-ppm sample (R_{75}) on the right show extent of carryover (less than 1%) if samples are injected in a span of 38 s (difference between S_1 and S_2).

This simple experiment [4] proves that air segmentation, homogeneous mixing of sample with reagent solution, and attainment of a steady state, which were thought to be essential prerequisities for performing continuous-flow analysis, can be replaced by a new concept. What remains unchanged, however, is that all samples are sequentially processed in exactly the same way during passage through the analytical channel, or, in other words, *what happens to one sample, happens in exactly the same way to any other sample*. This is why any practical FIA assay is initiated by a calibration run, which is followed by the analytical run, and, after a chosen interval, the controlling calibration is repeated (Fig. 2.2). The diagrams shown in this figure are serial assays of soil extracts, performed at the Analytical Laboratory of CENA, where FIA was first applied routinely on a large scale [440]. Incidentally, these two assays used different

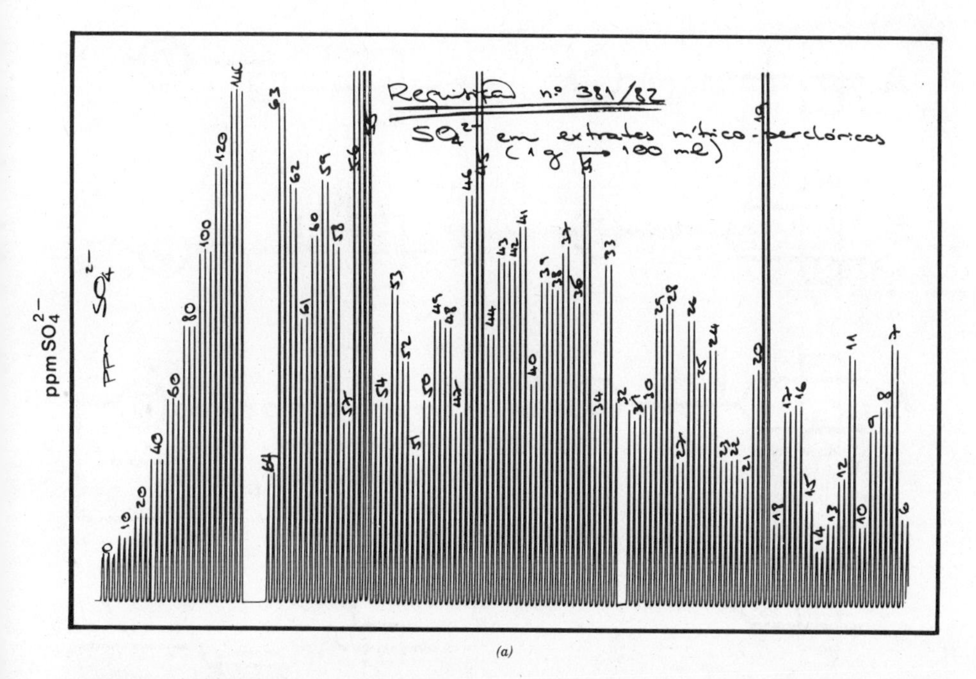

(a)

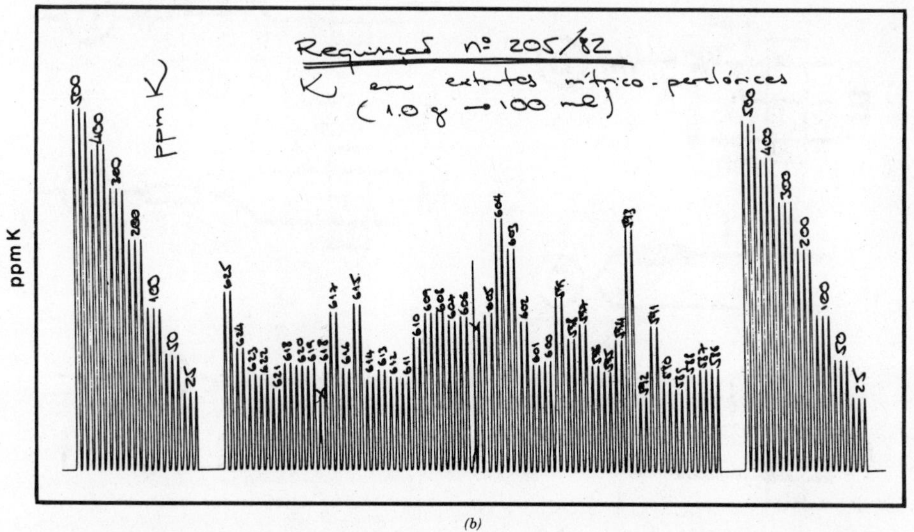

(b)

Figure 2.2. Routine assay of (*a*) sulfate in soil extracts by spectrophotometry [355] and (*b*) of potassium in soil extracts by flame photometry [43] as performed by the Analytical Laboratory of the Centro de Energia Nuclear na Agricultura, Piracicaba, Sao Paulo, Brazil. Note that the large series of routine assays, all performed in duplicate, are bracketed by serial calibration of standards injected in triplicate.

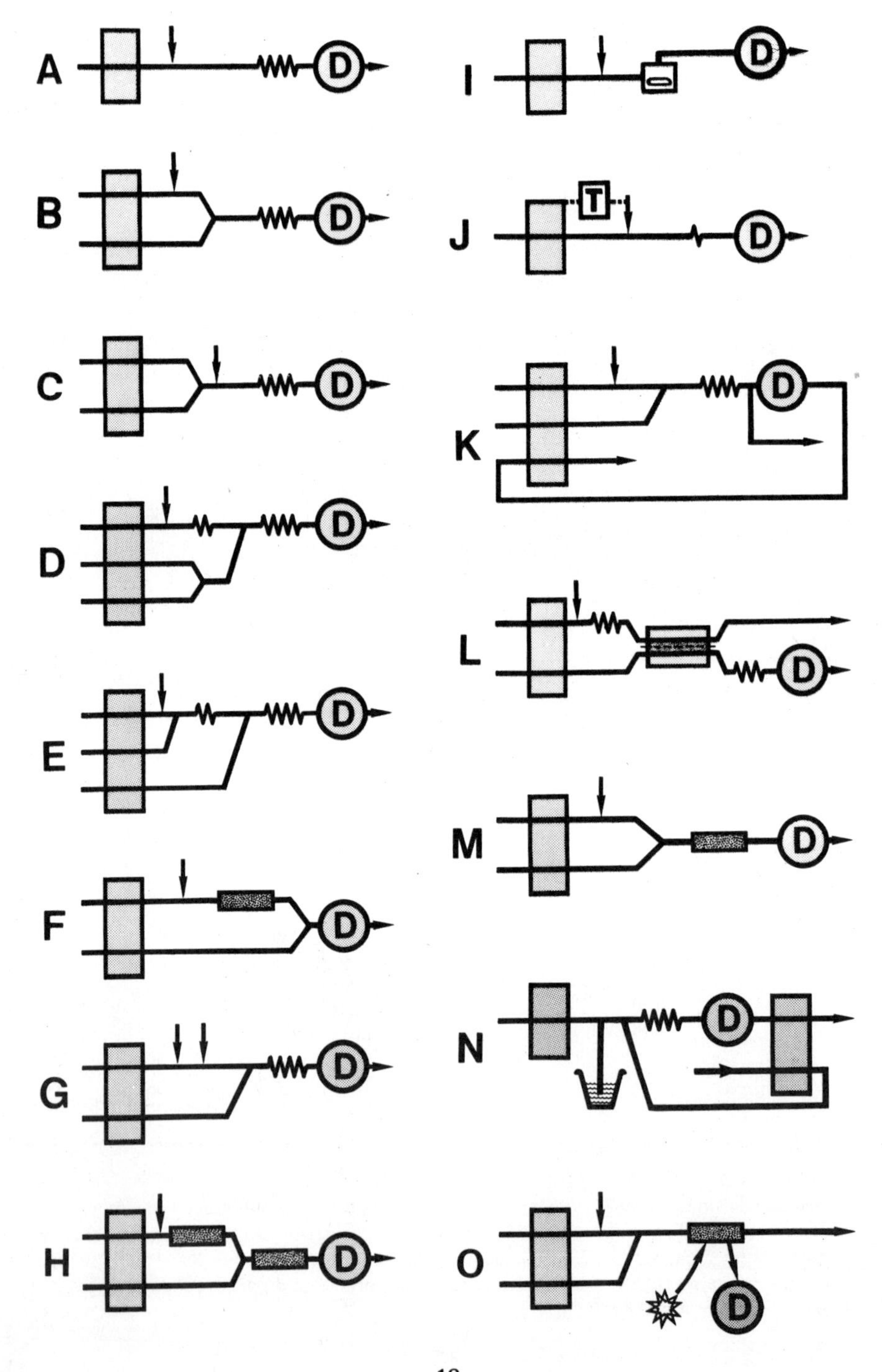
A
B
C
D
E
F
G
H
I
J
K
L
M
N
O
D
T

detectors: nephelometry for sulfate and flame photometry for potassium and different flow systems.

Since its introduction, the versatility of FIA has allowed the method to be adapted to different detectors (Chapter 5) and techniques (Chapter 4) using numerous manifold configurations (Fig. 2.3*A*–2.3*O*). Therefore, optimization and design of the flow channels to achieve maximum sampling frequency, best reagent and sample economies, and proper exploitation of the chemistries involved are possible only through a better understanding of the complex physical and chemical processes taking place during the movement of the fluids through the FIA channel.

In addition to the simple assay described previously, FIA systems may be designed to dilute or to preconcentrate the analyte; to perform separations based on solvent extraction, ion exchange, gas diffusion, or dialysis; to prepare unstable reagents *in situ*; and to dilute the reagents to the concentration suitable for a given assay. Moreover, since FIA is being used increasingly for such diverse tasks as process monitoring and as a new information-gathering tool in an industrial, agricultural, pharmaceutical, research, or clinical laboratory, there is a need to define the design parameters that would allow scaling of the FIA system, so that the injected sample volumes could be selected to be anywhere between a few microliters and several milliliters, depending on the availability of sample material to be analyzed or the cost of reagents to be used. For a rational design of these diverse FIA systems, the common underlying concepts of the flow injection method must be identified.

2.1.1 Principle of FIA and Function of a Flow System

The simplest flow injection analyzer (Fig. 2.4*a*) consists of a pump, which is used to propel the carrier stream through a narrow tube; an injection port, by means of which a well-defined volume of a sample solution is injected into the carrier stream in a reproducible manner; and a microreactor in which the sample zone disperses and reacts with the components of the carrier stream, forming a species that is sensed by a flow-through detector and recorded. A typical recorder output has the form

Figure 2.3. Types of FIA manifolds. *A*, single line; *B*, two line with a single confluence point; *C*, reagent premix into a single line; *D*, two line with a single confluence point and reagent premix; *E* three line with two confluence points; *F*, packed reactor in line; *G*, double injection and single confluence (zone penetration); *H*, sequential reactors (with immobilized enzymes); *I*, single line with mixing chamber; *J*, single line, stopped flow; *K*, solvent extraction; *L*, dialysis, ultrafiltration, or gas diffusion; *M*, two lines, single confluence, incorporating a packed reactor; *N*, hydrodynamic injection (type *I*); *O*, two line, one confluence, optosensing on solid surface.

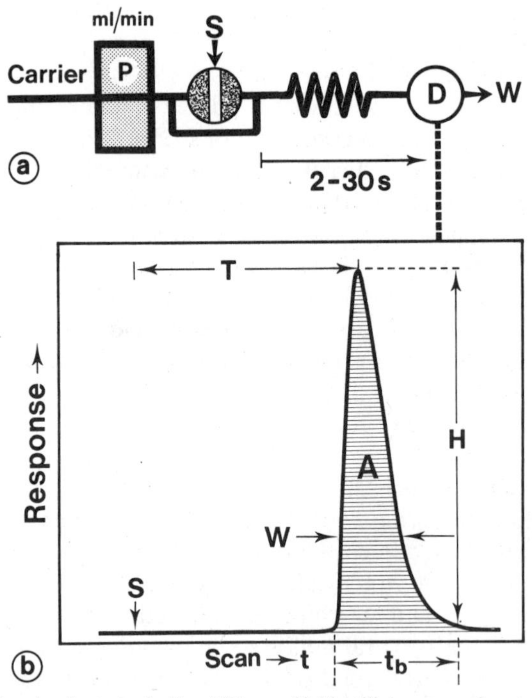

Figure 2.4. (*a*) The simplest single-line FIA manifold utilizing a carrier stream of reagent; *S* is the injection port, *D* is the flow cell, and *W* is the waste. (*b*) The analog output has the form of a peak, the recording starting at *S* (time of injection t_0). *H* is the peak height, *W* is the peak width at a selected level, and *A* is the peak area, *T* is the residence time corresponding to the peak height measurement, and t_b is the peak width at the baseline.

of a peak (Fig. 2.4*b*), the height *H*, width *W*, or area *A* of which is related to the concentration of the analyte. The time span between the sample injection *S* and the peak maximum, which yields the analytical readout as peak height *H*, is the residence time *T* during which the chemical reaction takes place. A well-designed FIA system has an extremely rapid response, because *T* is in the range of 5–20 s. Therefore, a sampling cycle is less than 30 s (roughly $T + t_b$), and thus, typically, two samples can be analyzed per minute. The injected sample volumes may be between 1 and 200 μL (typically 25 μL), which in turn requires no more than 0.5 mL of reagent per sampling cycle. This makes FIA a simple, automated microchemical technique, capable of having a high sampling rate and a minimum sample and reagent consumption.

It follows from the preceding discussion that FIA is based on a combination of three principles: *sample injection, controlled dispersion* of the injected sample zone, and *reproducible timing* of its movement from the injection point toward and into the detector. Thus, in contrast to all other methods of instrumental analysis, the chemical reactions are taking place while the sample material is *dispersing within the reagent*, that is, while the concentration gradient of the sample zone is being formed by the dispersion process. This is why the concept of dispersion, controlled within space and time, is the central issue of FIA. Since the preceding processes are strictly reproducible in all sequentially occurring sampling cycles, what happens to one injected sample happens in exactly the same way to all other subsequently injected samples, as was noted previously.

FIA is a general solution-handling technique, applicable to a variety of tasks ranging from pH or conductivity measurement, to colorimetry, titrations, and enzymatic assays. To design any FIA system properly, one must consider the desired function to be performed [20]. For pH measurement, or in conductimetry, or for simple atomic absorption, when the original sample composition is to be measured, the sample has to be transported through the FIA channel and into the flow cell in an *undiluted* form in a highly reproducible manner. For other types of determinations, such as spectrophotometery, fluorescence, or chemiluminescence, the analyte has to be *converted* to a compound measurable by a given detector. The prerequisites for performing such an assay is that during the transport through the FIA channel the sample zone is mixed with reagents and a sufficient time is allowed for production of a desired compound in a detectable amount. It should be reemphasized, however, that neither homogeneous mixing nor reaching of chemical equilibrium ("steady state") are conditions sine qua non for a flow-injection-based assay to be performed successfully.

Assays based on chemical separations have been performed in more complex FIA systems, using dialysis, gas diffusion, solvent extraction, and ion exchange (cf. Chapter 4) in order to obtain higher selectivity. Additionally, when required, concentrated sample solutions have been diluted extensively so that the resulting signal could be accommodated within the dynamic range of the detector used. Conversely, in trace analysis, the sample material has been preconcentrated in the FIA channel. All these diverse requirements have been met by manipulating dispersion of the sample zone, and, therefore, the dispersion coefficient D (cf. Section 2.2.1) is one of the three parameters allowing the rational design of an FIA system.

2.1.2 FIA Readouts and Peak Dimensions

When recorded, the transient signal observed by a detector during the passage of the dispersed sample zone has the form of a peak, the height

H, width W, or area A of which contains the analytical information (Fig. 2.4*b*).

In the absence of chemical reactions (such as in a simple atomic absorption measurement) when the detector responds linearly and instantly to the injected species, it does not make any difference whether peak height, area, or peak width is being measured, since they all yield useful information (Fig. 2.4*b*), although the concentration of the injected material is related to each of these parameters in a different manner. The same comments apply to assays based on production of a measurable species (such as in spectrophotometry) provided that via the dispersion process an excess of reagent is available throughout the entire length of the dispersed sample zone. (For exceptions see Section 2.4.4.)

Peak height is the most frequently measured peak dimension, since it is easily identified and directly related to the detector response, such as absorbance, potential, or current, and via such lin(C) detectors to concentration of analyte, that is,

$$H = kC$$

where k is a proportionality constant.

In addition to measuring the distance between peak apex and baseline, a vertical readout also can be taken at any section of the ascending or descending part of a peak. Such an additional peak dimension ($H' = k'C$) is exploited in the FIA gradient techniques (cf. Sections 2.4 and 2.5). Further information or peak dimensions can be obtained by taking vertical measurements not only at one detector channel, such as a fixed wavelength, but by scanning an entire spectrum.

Similar to peak height, *peak area* is related directly to the response of a detector, that is,

$$A = kC$$

yet this type of readout suffers from two drawbacks, which are due to its integral character. Peak area A cannot be related to spectra or concentration gradients, and it grossly distorts the readouts of log(C) detectors (such as ion-selective electrodes), since that portion of response which is close to the baseline disproportionally weighs much more than the portions of readout close to the peak apex.

Peak width, being proportional to the logarithm of the concentration, has a wide dynamic range, but it is less precise than the peak height or area measurement. Being horizontal, it cannot be related directly to spectra, but it yields the readout as a time difference (W or Δt) between the rising and falling edges of the peak (cf. Figs. 2.4*b*, 2.16, and 2.22). FIA titrations and peak-width-based FIA applications are gradient techniques that rely on the *horizontal* peak dimension, which can be located at any FIA peak sliced horizontally at one or several selected levels from the baseline (cf. Section 2.4.4).

Thus, a *FIA readout* can be obtained in several ways, of which peak-height measurement has been the most popular so far. The reason for this choice is its simplicity, since peak height obviously is the parameter most convenient to locate. Other readouts have to be identified by means of a time scale allocated to the dispersed sample zone, with the origin at the time of injection (gradient dilution, gradient scanning, or gradient kinetics), or via a selected level of a detector for peak-width measurement.

Time control is essential for reproducibility of any of the previously discussed readout options, and time t is therefore one of the three important parameters in the design of any FIA system.

2.2 DISPERSION OF THE SAMPLE ZONE AND DESIGN OF A FIA SYSTEM

The purpose of this section is to review the most important variables involved in the design of the flow system and to describe these variables by means of the dispersion coefficient, residence time, and dispersion factor. The same parameters will then be used in Sections 2.4 and 2.5 to identify sample dilution and reagent concentration at different readouts as used in various gradient techniques.

2.2.1 Dispersion Coefficient

Let us consider, at this stage, a simple dispersion experiment. A sample solution, contained within the valve cavity prior to injection, is homogeneous and has the original concentration C^0 that, if it could be scanned by a detector, would yield a square signal the height of which would be proportional to the sample concentration (Fig. 2.5, left). When the sample zone is injected, it follows the movement of the carrier stream, forming a dispersed zone whose form depends on the geometry of the channel and the flow velocity. Therefore, the response curve has the shape of a peak reflecting a continuum of concentrations (Fig. 2.5, right), forming a concentration gradient, within which no single element of fluid has the same concentration of sample material as a neighboring one. It is useful, however, to view this continuum of concentrations as being composed of individual elements of fluid, each of them having a certain concentration

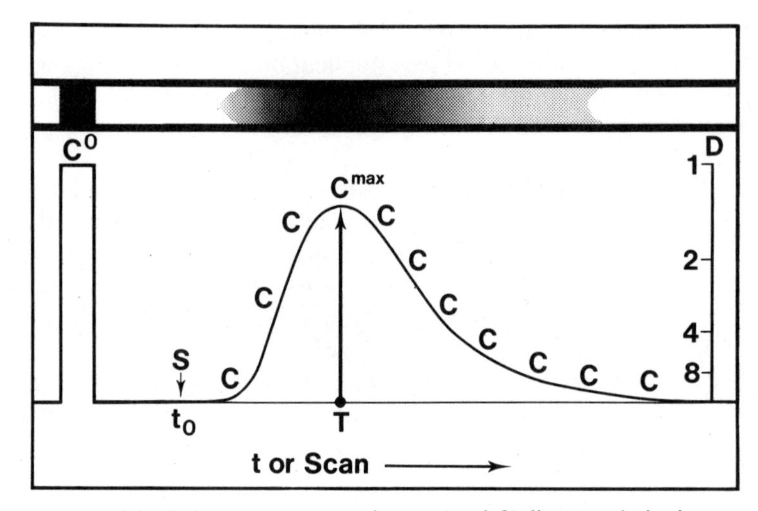

Figure 2.5. An originally homogeneous sample zone (top left) disperses during its movement through a tubular reactor (top center), thus changing from an original square profile (bottom left) of original concentration C^0 to a continuous concentration gradient with maximum concentration C^{max} at the apex of the peak.

of sample material C, since each of these elements is a potential source of a readout (cf. Sections 2.4 and 2.5).

In order to design an FIA system rationally, it is important to know (a) how much the original sample solution is diluted on its way toward the detector and (b) how much time has elapsed between the sample injection and the readout [20, 153]. For this purpose the dispersion coefficient D has been defined as the ratio of concentrations of sample material before and after the dispersion process has taken place in that element of fluid that yields the analytical readout:

$$D = C^0/C \tag{2.1}$$

If the analytical readout is based on maximum-peak-height measurement, the concentration within that (imaginary) element of fluid, which corresponds to the maximum of recorded curve (C^{max}) has to be considered. Thus, by relating C^{max} to the original concentration of injected sample solution C_S^0 (Fig. 2.5)

$$D_S^{max} = C_S^0/C_S^{max} \tag{2.2}$$

sample (and reagent) concentrations may be estimated. For convenience, unless otherwise stated, D_S or D will be used for the sample dispersion coefficient, while D_R will be consistently used for the reagent dispersion coefficient. Also note that the definition of the dispersion coefficient considers only the physical process of dispersion and not the ensuing chemical reactions, since D refers to concentrations of sample material *prior to* and *after* the dispersion process alone has taken place.

Generally speaking, for any readout, when, say, $D = 2$, the sample solution has been diluted 1:1 with the carrier stream. It will be shown later (cf. Section 2.3) how D_R, the dispersion coefficient of the reagent (and reagent concentration), changes with the D value in a single-line and a two-line FIA system. In this manner, D, in conjunction with T, which is the time elapsed between sample injection and the peak maximum (Fig. 2.4), adequately describes the concentration ratios in a FIA system and allows comparison of sample/reagent dilutions and of reaction conditions between conventional, that is, manual (batch), and FIA procedures. For convenience, sample dispersion has been defined [10, 153] as *limited* (D = 1–3), *medium* (D = 3–10), and *large* ($D > 10$), and the FIA systems designed accordingly have been used for a variety of analytical tasks.

The simplest way of measuring the dispersion coefficient is to inject a well-defined volume of a dye solution into a colorless carrier stream and to monitor the absorbance of the dispersed dye zone continuously by a colorimeter. To obtain the D^{max} value, the height (i.e., absorbance) of the recorded peak is measured and compared with the distance between the baseline and the signal obtained when the cell has been filled with the undiluted dye. Provided that the Lambert–Beer law is obeyed, the ratio of respective absorbances yields a D^{max} value that describes the FIA manifold, detector, and method of detection. (An exercise dealing with dispersion measurements is given in Chapter 6, while examples of FIA systems using limited, medium, or large dispersion are described in detail in Chapter 4.)

To conclude, it follows from the foregoing that the FIA peak is a result of two kinetic processes that occur simultaneously: the *physical* process of zone dispersion and the *chemical* processes resulting from reactions between sample and reagent species. The underlying physical process is well reproduced for each individual injection cycle; yet it is not a homogeneous mixing, but a dispersion, the result of which is a concentration gradient of sample within the carrier stream. The dispersion coefficient D is, similarly to the concept of a theoretical plate in chromatography, a fictitious concept, which does not correspond to any actual concentration within the dispersed sample zone. The reason is that a dispersed sample zone is not composed from discrete elements of fluid, but only imagined by a detector at any one time as an apparent discrete section of the concentration gradient accessible to detector sensing. Thus, a spectrophotometric detector integrates over the length and width of the flow cell path, while an ion-selective electrode senses only the adjoining solution layer, integrating the potential over the whole electrode surface. Bearing this limitation in mind, the dispersion coefficient is a very convenient parameter.

It follows from the foregoing that the D value is always related to a certain time, which is the period elapsed from the moment of sample injection, t_0, to the moment, t, when the dispersed element of sample material passes through the observation field of the detector. Thus, for the peak maximum, D^{max} is related to time T, a parameter which is significant in two ways: (1) as a time interval, which allows us to identify and to select that element of fluid which is to be the source of our readout, and (2) as the time available for chemical reactions to proceed and to form a species to be detected. As follows from Fig. 2.4*b*, this time interval always begins at the moment of injection at S, and, for peak-maximum-height measurements, concides with T. For practical reasons, in order to maintain a high sampling frequency, this time is limited to a maximum of 30–40 s during which all chemistries must be executed. This is the principal limitation of FIA, compared to segmented streams where up to 15-min reaction times have been maintained routinely. Fortunately, most analytical applications do not require long incubation periods.

2.2.2 Peak Height and Sample Volume

Let us first consider a one-line FIA system (Fig. 2.4*a*) in which the pumping rate Q is 1.5 mL/min, the tube length L is 20 cm, and the inner diameter of the tube is 0.5 mm. By injecting increasing volumes of dyed solution, a series of curves will be recorded (Fig. 2.6*a*), all starting from the same point of injection S, where the height of the individual peaks will increase until an upper limit "steady state" has been reached. At this final level the recorded absorbance will correspond to the concentration of undiluted dye C^0, and $D = 1$. The rising edge of all curves coincides and has the same shape regardless of the injected volumes, and, therefore,

$$\begin{aligned} C^{max}/C^0 &= 1 - \exp(-kS_v) = 1 - \exp(-0.693n) \\ &= 1 - 2^n = 1/D^{max} \end{aligned} \qquad (2.3)$$

where $n = S_v/S_{1/2}$ and $S_{1/2}$ is the volume of sample solution necessary to reach 50% of the steady-state value, corresponding to $D = 2$. By injecting two $S_{1/2}$ volumes, 75% of C^0 is reached, corresponding to $D = 1.33$; and so on. Therefore, $D = 1$ can never truly be reached; yet, injection of five $S_{1/2}$ volumes results in $D = 1.03$, and injection of seven $S_{1/2}$ volumes

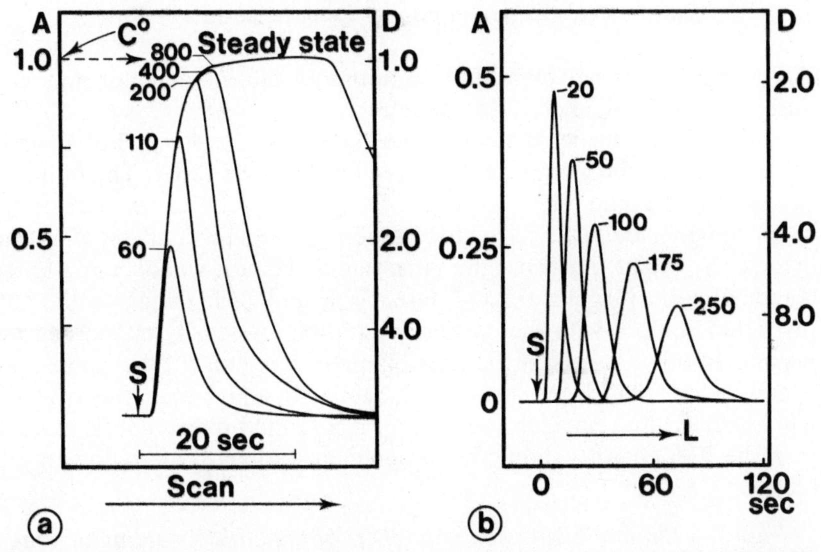

Figure 2.6. (*a*) Response curves as function of injected sample volume. The peak height increases with the volume of the sample injected into the FIA system until a steady-state signal is reached. All curves recorded from the same starting point S, with sample volumes of 60, 110, 200, 400, and 800 μL. Note tht $D = 1$ for steady state, and that the peak width increases with injected volume. (*b*) Dispersion of the injected sample zone in a FIA system as function of the tube length traversed. The sample volume is 60 μL; L is given in centimeters; the tube inside diameter is 0.5 mm; and Q in all experiments is 1.5 mL/min [20].

results in $D = 1.008$, corresponding to 99.2% of C^0 or $C^{\max SS}$. Since the concept of steady state is not used in FIA, the maximum sample requirements will not exceed two $S_{1/2}$ for limited dispersion and less than one $S_{1/2}$ in all other applications. Since the first portion of the rising curve might be considered nearly linear up to approximately 50% C^0 (i.e., $D = 2$), it follows that for FIA readouts with medium and large dispersion coefficients, the peak height is directly proportional to the injected volume. Therefore [153]:

> ***Rule 1.*** *Changing the injected sample volume is a powerful way to change dispersion. An increase in peak height—and in sensitivity of measurement—is achieved by increasing the volume of the injected sample solution. Conversely, dilution of overly concentrated sample material is best achieved by reducing the injected volumes.*

It has been shown that $S_{1/2}$ does not depend on the method of injection (valve injection or impulse injection), but it is a function of the geometry and of the volume of the flow channel [20] (cf. Fig. 2.10, 2.11, and 3.15).

2.2.3 Peak Height, Channel Length, and Flow Rate

The microreactor between the injection port and the detector may have different length, diameter, and geometry.

Let us first consider the coiled tube (Fig. 2.4*a*, and Fig. 2.8*B*), which so far has been the geometry most frequently used in FIA. The influence of coil length L and inner diameter of the tubing d on the dispersion has been studied in detail [20]. The use of tubing of a small diameter will result in lower $S_{1/2}$ values, because the same sample volume will occupy a longer length of tube (θ), since $S_v = \pi(0.5d)^2\theta$, and will, owing to the thus restricted contact with the carrier stream, be less easily mixed and dispersed. In other words, if the tube diameter d is halved, the sample will occupy a fourfold-longer portion of the tube (θ), and, hence, the $S_{1/2}$ value will be four times smaller. Therefore, if a limited dispersion is desired, then the FIA channel should be designed according to the following rule [20]:

> ***Rule 2.*** *Limited dispersion value D is obtained by injecting a sample volume of minimum one* $S_{1/2}$ *into a manifold consisting of the shortest possible piece of a narrow tube connecting the injection port and the detector.*

Even if a medium dispersion FIA system is required, it is economical to use narrow channels. The sample and reagent economy is improved when narrow channels are used, because, for the same linear flow velocity, the pumping rate Q in a tube of diameter d is only one-fourth of that required for a tube of diameter $2d$. On the other hand, practical considerations prevent the use of channels with too narrow a bore or too tightly packed reactors because:

1. The flow resistance will increase, thus preventing the use of a low-pressure drive (peristaltic pump, air drive, piston burette).
2. The system might easily become blocked by solid particles, which will make its use impractical for assay of clinical, agricultural, and similar sample materials unless the samples are meticulously filtered in a preliminary step.
3. The flow cell of the most frequently used detector, the spectrophotometric detector, must have an optical path with an inside diameter of 0.5–1 mm, because, otherwise, sufficient light energy will not pass through, unless a powerful light source is used. The use of narrower manifold tubing than flow cell diameters will cause an undesired nonuniformity in the flow pattern in the detection area.

To summarize, the optimum internal diameter of tubes connecting the injection port and the detector is 0.5 mm, although 0.75-mm inside diameter is useful for the construction of systems with large dispersion, and 0.3-mm inside diameter for systems with limited dispersion.

The mean residence time $\overline{T}$ will depend, for a reaction system made of tubing of uniform internal diameter, on the tube length L, the tube diameter d, and the pumping rate Q:

$$\overline{T} = \pi(d/2)^2 L/Q = L/F = V_r/Q \tag{2.4}$$

where F is the linear flow velocity and V_r is the reactor volume. Thus, when designing systems with medium dispersion where the sample has to be mixed and made to react with the components of the carrier stream [10, 153], in the first place one would tend to increase the tube length L in order to increase $\overline{T}$ and thus T. However, one can expect that dispersion of the sample zone will increase with the distance traveled, and this band broadening will eventually result in loss of sensitivity and lower sampling rate. Thus, if one injects, say, 60 μL of a dye solution into a colorless carrier stream pumped at a rate of 1.5 mm/min through a tube of 0.5 mm inside diameter, one obtains, upon increasing the tube length from 20 to 250 cm, a series of curves, the height of which decreases with the increase of tube length (cf. Fig. 2.6*b*). It has been shown [20, 153, 183] that dispersion in a FIA system caused by the flow in the open narrow tube increases with the square root of the mean residence time $\overline{T}$ (or the distance traveled, L), and that for constant flow velocity the residence time is proportional to the dispersion to the second power:

$$D^{\max} = 2\pi^{3/2} R^2 D_f^{1/2} \overline{T}^{1/2} / S_v. \tag{2.5}$$

where D_f is the axial dispersion coefficient and R is the tube radius. It should be noted that in agreement with Eq. (2.3), Fig. 2.6*a*, and Fig. 3.15, $D^{\max}$ will, for $D^{\max} \geq 2$, decrease linearly with increase of S_v, that is, the peak height H in this region will increase proportionally with S_v.

Thus, although the decrease in peak height and the zone broadening become progressively smaller relative to the distance traveled, the increase in T obtained through an increase of length L is not worthwhile above a certain limit. Instead of increasing L [and thus V_r, Eq. (2.4)], one can decrease the pumping rate Q and keep L as short as is practical. Because of the physical distances between the individual components of the FIA system (injection port, reaction coils, dialyzer, flowthrough cell, etc.), a compromise must be made, and therefore, in practice, the overall length of a well-designed FIA manifold is between 10 and 100 cm of 0.5

mm tubing. Residence times up to 20 s can readily be obtained by selection of the flow rate, and longer residence times may be obtained by means of intermittent pumping, so that for a certain part of the working cycle the forward movement of the carrier stream is stopped [20]:

Rule 3. *The dispersion of the sample zone increases with the square root of the distance traveled through the tubular conduit and decreases with decreasing flow rate. Thus, if dispersion is to be reduced and the residence time is to be increased, the tube dimensions should be minimized and the pumping rate should be decreased. The most effective way to increase the residence time and to avoid further dispersion is to inject the sample into a flowing stream and stop the stream's forward movement, then resume pumping after a sufficient reaction time has elapsed.*

The validity of Rule 3 is illustrated in Fig. 2.7 where 30 μL aliquots of dye solution were repeatedly injected into a single-line FIA system, while the carrier stream was pumped at a rate of 2.0 mL/min through a 100-cm-long (0.5-mm inside diameter) tube. While the first sample (*a*) moved continuously, thus residing in the system for 10 s, the following samples, *b*, *c*, and *d*, reached the detector at times $T = 20$ s, $T = 30$ s, and $T = 40$ s, respectively, because the sample zones, following a delay of 5.5 s, were stopped in the reaction coil for 10, 20, and 30 s, respectively.

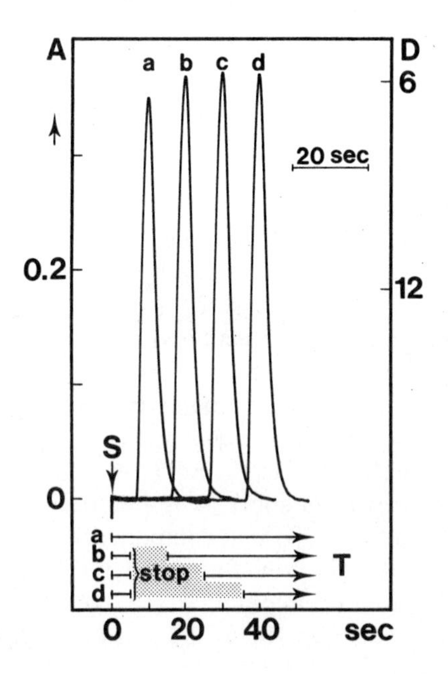

Figure 2.7. Dispersion coefficient *D* as function of different stopped flow periods. As can be seen, *D* is practically independent of the residence time *T* during which the carrier stream is stopped. Curve *a* was recorded with continuous flow; curves *b*, *c*, and *d* were recordered with intermittent flow, where the injection was followed by a delay period of 5.5 s during which the carrier stream transported the sample zone into the coil between injector and detector, after which a stop period of 10, 20, and 30 s followed, during which the zone was held, and then pumping was resumed, thus carrying the dispersed zone through the detector for recording of the peak [153].

It is interesting to note that the first sample (*a*) dispersed slightly more than did samples *b*, *c*, and *d*, which were stopped in the coil. The decrease in the *D* value of the stopped samples is caused by radial diffusion, which affects the laminar flow profile while the zone stands still. (See the discussion of Taylor's effect in Chapter 3.) The principles of stopped-flow technique are outlined in Section 2.4.3 while its practical applications and experimental details are described in Section 4.3.

2.2.4 Peak Height and Channel Geometry

The coiled tube has so far been the most frequent geometric form of the FIA microreactor. However, it is useful to review all channel geometries (Fig. 2.8). These are *straight tube* (*A*), *coiled tube* (*B*), *mixing chamber* (*C*), *single-bead string reactor* (*D*), 3-D or *"knitted" reactor* (*E*), and *imprinted meander* (cf. microconduits Section 4.12) or combinations of these geometries.

The function of these reactors is to *increase the intensity of radial mixing*, by which the parabolic velocity profile in the *axial direction,* formed when the sample zone is injected into a laminar flow of carrier stream, *is reduced*. Thus, (a) the reagent becomes more readily mixed with the sample, and (b) the axial dispersion of the sample zone is reduced.

Relaxation of the laminar profile in the radial direction is best achieved by creating a local turbulence whereby the direction of flow is suddenly changed. In this way the elements of fluid that are lagging because they are close to the walls of the channel are moved into the rapidly advancing

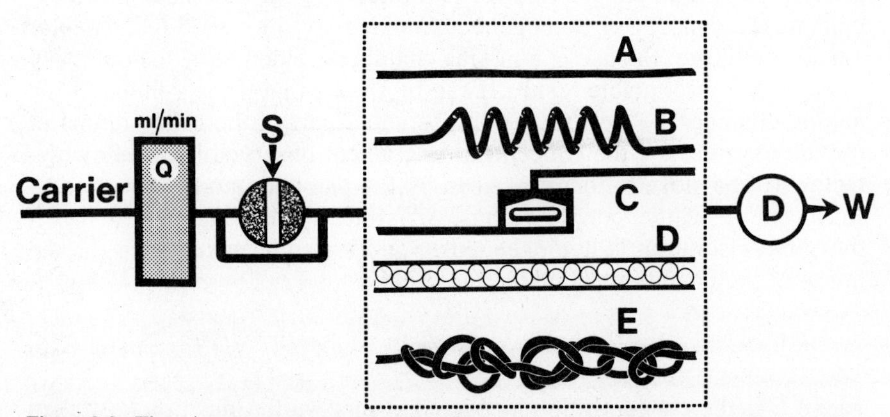

Figure 2.8. The microreactor geometries most frequently used in FIA: *A*, straight open tube; *B*, coiled tube; *C*, mixing chamber; *D*, single-bead string reactor (SBSR); and *E*, knitted reactor. (For an example of the imprinted meandering reactor, see Chapters 3 and 4.)

central streamline, while those elements of fluid that have advanced in an axial direction because they are close to the central streamline are reshuffled closer to the tube wall. The more frequently this process is repeated, the more symmetrical the concentration gradient within the dispersed sample zone will be, and the peak shape will change from an asymmetric to a symmetric (Gaussian) one (cf. Chapter 3).

In a straight tube of uniform diameter (Fig. 2.8*A*), the parabolic profile (Fig. 3.4) formed by laminar flow remains undisturbed up to a flow velocity not normally reached in a typical FIA system, and since the radial diffusion occurring in the time frame of an FIA experiment is not sufficient to offset the axial dispersion initially formed during sample injection, an asymmetrical peak is recorded (Figs. 2.4*a* and 2.10*a*).

A coiled tube (Fig 2.8*B*) is the most frequently used reactor geometry, since it can conveniently accommodate any length of tubing in an experimental setup and also because secondary flow within the coiled tubing promotes mixing in the radial direction (Chapter 3). The result is a more symmetrical, higher, and narrower peak than if an identical straight tube length had been used. The tighter the coiling of the tube is, the more pronounced this effect will be [73]. Also, the longer the tube is, the larger the number of mixing stages can be accommodated and the more symmetric the peak obtained will be. However, it follows from Fig. 2.6*b* that with the Q and R used normally in FIA, a symmetrical peak is obtained only if L is longer than 1 m, and since a typical FIA coil is shorter, a typical FIA peak in a coiled reactor is asymmetrical.

A mixing chamber (Fig. 2.8*C*) with a magnetic stirrer has been used in continuous-flow designs that preceded FIA; it sought to achieve homogeneous mixing of sample and reagent [2.1, 2.2]. At that stage of development the concept of homogeneous mixing still prevailed (cf. Chapter 1) and, therefore, the use of a mixing chamber seemed to be the only way to obtain a reproducible result. It can be shown that if the volume of the mixing chamber V_m constitutes a substantial part of the total volume of the flow system V_r, the concentration gradient observed by the flow detector will be close to that described by the one-tank model (Chapter 3). Furthermore, if the sample volume S_v is much smaller than V_m, and if the sample solution is homogeneously and instantly mixed with the solution in the chamber, then [20, 153, 1245]

$$CV_m/C^0S_v = e^{-t/\bar{T}} \quad \text{or} \quad S_{v/Vm} = \ln 10 \log(D^{max}/D^{max} - 1) \qquad (2.6)$$

where C is the concentration measured at time t after injection has taken place and $\bar{T}$ is the mean residence time. This concentration profile is purely exponential, and for $t = 0$, $C = C^{max} = C^0S_v/V_m$. By expressing

$\overline{T}$ as V_m/Q and by defining $t_{1/2}$ as the time span during which any C decreases by one-half,

$$t_{1/2} = 0.693 V_m/Q \tag{2.7}$$

it follows that it will take $4t_{1/2}$ or $2.8V_m/Q$ to reach the baseline within 6%. The similarity of behavior between a mixing chamber and the injection process is not incidental, since both processes conform for certain flow geometry to the same flow model (Chapter 3).

The influence of a mixing chamber on the dispersion can be seen from Fig. 2.9, where curve A was recorded by injecting a dye solution into a single-line FIA system (Fig. 2.8A, Q = 0.75 mL/min, S_v = 50 μL, L = 50 cm of 0.5-mm inside diameter). Curve B was obtained with at the same conditions as curve A, but with a mixing chamber (V_m = 1.9 mL) incorporated into the FIA system. Peak A is 18 times higher than peak B, the height of the latter being only 2.77% of C^0, the original concentration of the injected dye. The length of the sampling cycle increased six times from 30 seconds to 3 min, making the method unsuitable for rapid serial assays. Therefore, the use of a mixing chamber should be avoided when designing a FIA system for serial assays or whenever the sampling frequency is the highest priority.

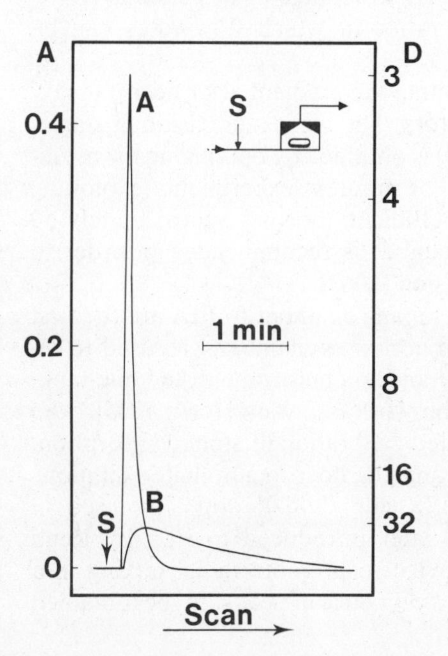

Figure 2.9. Influence of a mixing chamber on the dispersion of the sample zone in a single-line FIA system. Without a mixing chamber (curve A) the peak is narrower and higher, yielding higher sensitivity of measurement and higher sampling frequency than the same FIA system into which a mixing chamber (V_m = 1.9 mL) has been included (curve B). A is the absorbance, and D is the dispersion coefficient.

Rule 4. *Any continuous-flow system that includes a mixing chamber generates larger dispersion and yields a lower measurement sensitivity than a corresponding channel without a mixing chamber. A system with mixing chamber will also have a lower sampling frequency, unless the pumping rates are increased, which, in turn, requires large sample and reagent volumes.*

If the sample solution is too concentrated to be measured directly, the first choice should be a reduction of the injected sample volume (Rule 1), because the other alternative, an increase of V_m or V_r, is less economical.

However, there is the following exception to Rule 4. When applying FIA to monitoring or industrial processes, there are practical limitations in the use of tubular reactors, since the solutions to be monitored are often very viscous and nearly always far too concentrated to be measured without being extensively diluted. Therefore, according to Gisin et al. [1063], the use of a mixing chamber is justified for the following reasons: First, substantial variations of injected volume, hold-up volume of the chamber (V_r), and flow rate do not adversely affect the precision over a large dynamic range of dispersion coefficients (D up to 2000). Next, sample conditioning (density, viscosity, ionic strength, buffer capacity, etc.) is very efficient, because active stirring provides better compensation of matrix effects than the confluence technique commonly used with tubular reactors. Confluence also yields a fairly low range of dispersions (D^{max} up to 50, due to relatively limited ratios in which flow rates can be selected).

To summarize, experiments exploiting the gradient at or near its maximum can be done using tubular reactors. The highest precision of single-zone gradient techniques (Section 2.4) is obtained by optimizing the asymmetry of the concentration profile. For gradient experiments exploiting elements of fluid with dispersion coefficients beyond approximately 50 up to as large as 2000, a mixing chamber is recommended in order to ensure optimal throughput and precision [1063].

Additional examples of the use of mixing chamber to FIA are the first version of FIA titrations [10], where a mixing chamber was used to increase the width of the injected zone and to conform with the one-tank-in series model (see Section 4.10); the work of Tyson [1062, 1245], who used the mixing chamber for automated calibration in atomic absorption; and Stewart [427], who applied a mixing chamber to automated dilution.

A single-bead string reactor (SBSR)(Fig. 2.8*D*), originally used in postcolumn derivatization [2.3] and later introduced to FIA by Reijn et al. [146], is the most effective device to promote radial mixing in a tubular reactor. The SBSR allows symmetrical peaks to be obtained within the time domain and channel length of a typical FIA experimental setup, and it prevents peak broadening so effectively that it allows accommodation of several consecutive sample zones in the same FIA channel without intermxing. A special feature of the SBSR is that the surface to liquid ratio in this reactor is much higher than that of a tubular reactor of 0.5 mm inside diameter. This may be advantageous, if liquid–surface interaction is desirable, as when using immobilized reagents, enzymes, or packed reactors (cf. Section 4.7). If, however, surface phenomena are undesirable (adsorption, or occlusion of sample or reagent material or both), large surface–liquid area is a drawback. Also, small air bubbles and solid particles tend to be trapped in SBSR, which may increase carryover and flow resistance.

A three dimensionally disoriented (3-D reactor), or "knitted" reactor by Engelhardt and Neue [2.4], has been introduced to FIA by Engelhardt and Klinkner [567, 568] (Fig. 2.8*E*). This reactor is most simply made by tightly and irregularly knotting a suitable length of tubing. The chaotic movement of the carrier stream through a spatially disoriented path promotes radial dispersion almost as effectively as does SBSR without its drawbacks, since the ratio of surface to liquid is the same as the one of a straight or coiled tubing and since air cannot be trapped in the smooth channel. The effectiveness of a 3-D reactor compared to straight line and coil is seen from the dispersion experiment shown in Fig. 2.10.

It follows from the foregoing that sudden, irregular changes in the di-

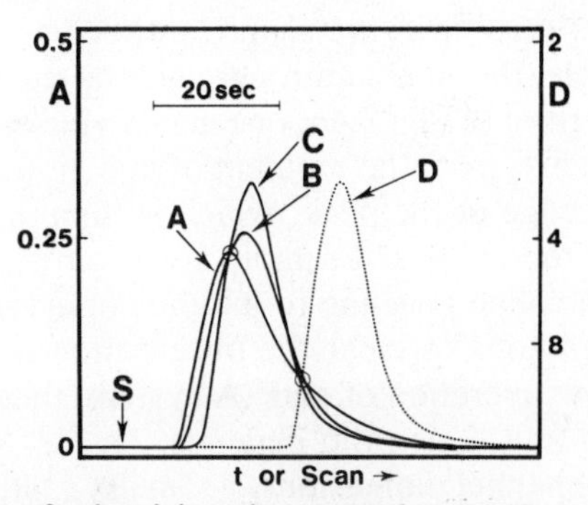

Figure 2.10. Dispersion of a dye, injected as a sample zone (S_v = 25 μL) into: *A*, straight tube; *B*, coiled tube; *C*, "knitted tube"; and *D*, a SBSR reactor. The reactor volumes (V_r = 160 μL) and pumping rates (Q = 0.75 mL/min) were identical in all experiments. The same piece of Microline tubing (L = 80 cm, 0.5 mm inside diameter) was used in experiments *A*, *B*, and *C*. (The injected dye was bromthymol blue, carrier stream 0.1 M borax and wavelength 620 nm, cf. Chapter 6.) The SBSR reactor was made of 0.86 mm inside diameter tube filled with 0.6-mm glass beads. Note that the isodispersion points on the peaks were recorded with microreactors made of identical length and diameter, but different geometry.

rection of flow are desirable features of the FIA channel. Therefore:

> ***Rule 5.*** *In order to reduce axial dispersion, which leads to lower of sampling frequency, the flow channel should be uniform, without wide sections (which behave like poorly mixed chambers), and should be coiled, meandered, packed, or 3-D disoriented. Other types of sudden changes in direction of flow should also be included in the design.*

If two streams merge (Section 2.3.2), one may also wish to increase the intensity of radial mixing so that the reagent will penetrate more readily into the central streamline of the dispersing sample zone as close as possible to the merging point. This can be achieved by designing a mixing point in which the direction of flow is suddenly changed, either by constructing one or several sharp bends or by using a short 3-D or SBSR reactor downstream a short distance from the merging point (cf. Chapter 3.2.3).

2.2.5 Peak Width, Dispersion Factor, and Sampling Frequency

So far we have been concerned with the mixing ratio of sample with the surrounding carrier stream in terms of concentration changes as observed as *vertical* information on the FIA response curve. However, as we learned in the previous section, the peak shape and its width at the baseline are important features, since the concentration changes in the *horizontal* direction, that is, versus time, are significant, being related to the length of the sampling cycle, the sample throughput, and the sample and reagent consumption. Referring to Fig. 2.4*b*, one may divide each sampling period into two time domains: from the moment of injection S (at time t_0) to the rise of the leading edge of the peak, which is the time necessary for the leading element of the dispersed sample zone to reach the detector, and the second domain called baseline to baseline interval, t_b, necessary for the entire dispersed zone to clear the observation field of the detector. For continuous-flow operation of an FIA system, these two time periods are interdependent and their ratio depends, besides on injected sample volume S_v, on the channel dimensions, geometry, and pumping rate. Obviously, the narrower the dispersed zone is, the higher the sampling frequency is and better reagent and sample economies achieved. It follows from the previous section that in order to optimize sample throughput, and to decrease sample and reagent consumption, the geometry of the FIA channel has to be designed in such a way that the ratio of axial to radial dispersion is minimized (Chapter 3).

It was found useful [608] to describe the axial dispersion in a FIA system by the dispersion factor $\beta_{1/2}$, which has been defined as the ratio between the projection of slope of the leading edge of the FIA response curve and the distance between the leading edge and the point of injection t_0, both slope and distance being read at 50% of the steady-state response.

This definition is better understood by means of Fig. 2.11, on which the dispersion factor can be identified as either

$$\beta_{1/2} = S_{1/2}/V_r \quad \text{or} \quad \beta_{1/2} = t_{1/2}/T \tag{2.8}$$

where $S_{1/2}$ and $t_{1/2}$ are the sample volume or time necessary to reach 50% of the steady state [Eq. (2.6)] and V_r or T are volume and residence time characteristics of the FIA channel. In Fig. 2.11 n is the number of $S_{1/2}$ volumes or $t_{1/2}$ pumping times necessary to inject sample solution in order to reach D values indicated on the right-hand ordinate.

The dispersion factor $\beta_{1/2}$ is a dimensionless number, whose magnitude

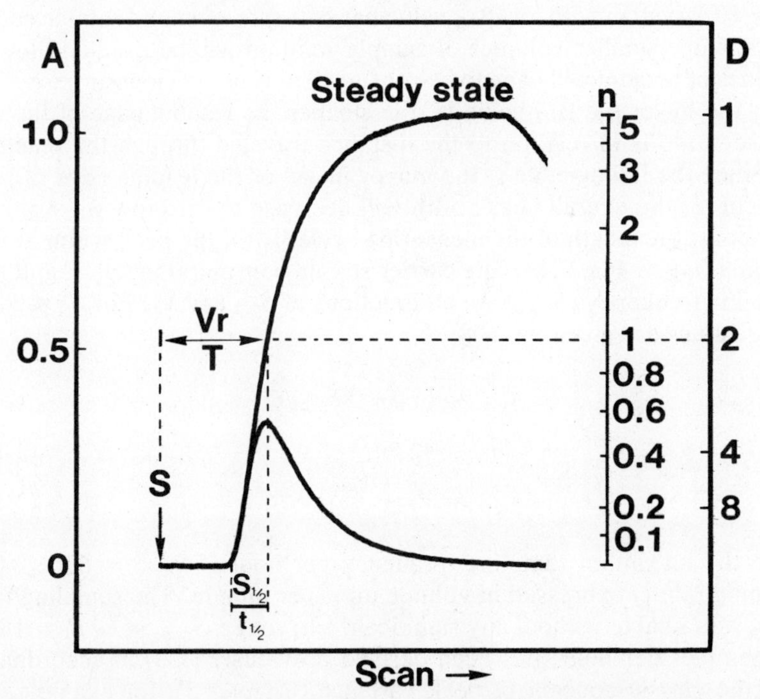

Figure 2.11. The impulse response and the step response obtained in a single-line FIA system by injecting a dye solution of concentration C^0 into an inert colorless carrier stream, the signal being recorded as absorbance A. S is the point of injection and $S_{1/2}(t_{1/2})$ is sample volume (time) necessary to inject sample volume to reach $D^{max} = 2$. V_r is the reactor volume, T is the residence time, and D is the dispersion coefficient. [608].

depends on the channel geometry, length, diameter, and flow rate, and the lower its value is, the more intense the radial transfer rate is and the better the system performs. Straight channels or coiled tubing, or imprinted meandering channels of, say, sinusoidal form, or any channel of repeatable geometry, may be viewed as consisting of a number of geometrically identical sections N with additive values of σ^2, or $(S_{1/2})^2$. Since any deviation of geometry from a circular straight pipe will disturb the laminar pattern of the liquid flowing through the tube, the axial dispersion per reactor volume V_r will decrease from a straight pipe to a coiled pipe to a semicircular imprinted sinusoidal channel and is, as was shown previously, smallest in a tube either packed with single beads (the SBSR reactor of Reijn et al.) or in the knitted or 3-D coiled reactor. This is illustrated with reference to Fig. 2.10, where by extrapolation to $D = 2$ the following $\beta_{1/2}$ values were found: straight tube (A), 0.60; coiled tube (B), 0.56; knitted tube (C), 0.35; and SBSR (D), 0.25. Compared to these values, the mixing chamber (Fig. 2.9) has a $\beta_{1/2}$ value as high as 2.

An FIA system with low $\beta_{1/2}$ value has better reagent and sample economies, since smaller volumes of sample solution will be used and lower volumes of reagent will have to be consumed during each measuring cycle. Also, the lower the $\beta_{1/2}$ value is, the steeper the leading edge of the response curve is in relation to the distance traveled through the reactor, and since the falling edge is the mirror image of the leading edge of any FIA curve, the overall peak width will decrease toward low $\beta_{1/2}$ values. Therefore, the length of the measuring cycle (t_{cyc}), the peak width at the baseline (t_b, cf. Fig. 2.4b), the carrier stream consumption (r_{cyc}), and the sampling frequency (S_{max}) are all functions of $S_{1/2}$ and V_r. For $S_v \ll S_{1/2}$, these values are given by [608]

$$t_{cyc} = (V_r + 4S_{1/2})/Q \tag{2.9}$$

$$t_b = 4S_{1/2}/Q \tag{2.10}$$

$$r_{cyc} = V_r + 4S_{1/2} \tag{2.11}$$

while the maximum sampling frequency per hour is $S_{max} = 60t_{cyc}$, for pumping rate Q expressed in volume units per minute. The sampling frequency attainable without any significant carryover, $S_{max} = 60/t_b = 15Q/S_{1/2}$ samples per hour, has been derived previously [153] in accordance with the classic concept of peak variance (Chapter 3), for example, as applied to chromatography, where sufficient peak resolution is obtained when the distance between adjacent peak crests corresponding to $k = 4$ (Fig. 2.12) is reached. For FIA an approximation that $S_{1/2} = 2\sigma$ has been made [153, 608]. This approximation, although not in strict accordance

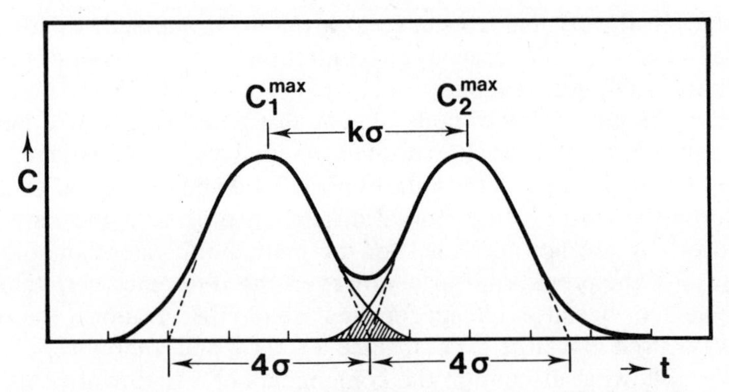

Figure 2.12. Factors influencing the maximum sampling frequency. The hatched area shows the peak overlap, which for $k = 4$ in this figure is 4% of the respective areas.

with the statistical treatment of the dispersion process, since the typical FIA peak is skewed, is useful and practical. The relation between peak statistical moments (i.e., mean residence time and σ^2 variance) and dispersion factor $\beta_{1/2}$ is further discussed in Chapter 3.

> ***Rule 6.*** *To obtain maximum sampling frequency the flow system should be designed to have minimum $S_{1/2}$ and should be operated by injecting the smallest practical sample volume S_v. FIA systems with minimum dispersion factor $\beta_{1/2}$ will require the least volume of reagent solution and will yield maximum sampling frequency in relation to residence time of the zone continuously moving through the channel.*

2.3 DISPERSION OF A SAMPLE ZONE IN A REAGENT STREAM

So far we have considered dispersion of a sample zone in a single-line FIA system (Fig. 2.4*a*), which results in a concentration gradient of the sample material as the one shown in Fig. 2.4*b*. When a chemical reaction is to take place between the sample solution and a reagent contained in the carrier stream, for example, as needed in spectrophotometry or fluorescence, a mixing must take place, the ratio of reactant concentrations depending on the dispersion within the analyzer channel. Similar to the value of the sample dispersion coefficient D, the reagent dispersion coefficient is defined as

$$D^R = C_R^0/C_R \qquad (2.12)$$

where C_R^0 is the original reagent concentration (as pumped into the carrier channel) and C_R is the reagent concentration in the element of fluid that yields the analytical readout.

Since the net concentrations of sample C_S' and reagent C_R' materials in said element of fluid are the result of a mutual dispersion and the ensuing chemical reactions, the concentrations defined above, C_S and C_R (which would be the result of the mutual dispersion process alone), are higher because they are not corrected for the mutual equivalent amounts consumed. For the present purpose, however, this difference in C values can be neglected, because, except for the case of FIA titrations, the reagent is always used in a large excess, since $C_S^0 < C_R^0$ and, therefore, $C_S \ll C_R$. For the following discussion the components of S and R are considered to react in the stoichiometric ratio 1:1.

2.3.1 Single Line

Let us consider a single-line system (Fig. 2.13*a*), where the carrier stream contains a reagent (C_R^0) into which sample (C_S^0) is injected yielding concentrations C_S and C_R, which are obtained by mutual dispersion of sample zone and carrier stream as observed by the detector (D). The time scan of the sample (S) and reagent (R) material as observed within the detector at a time t elapsed from the moment of sample injection t_0 is shown in Fig. 2.13*b*, together with the respective D_S and D_R values. Obviously, when the sample concentration is highest (C_S^{max}), the reagent concentration is at its lowest (C_R^{min}), and, therefore, whenever D_S approaches 1, the reagent concentration approaches zero. Consequently, a species to be measured cannot be formed in the center of the sample zone whenever $D_S = 1$, because in the absence of reagent a double peak is formed [153 (p. 137), 1124, and 1393] and peak height at C_S^{max} will not yield a straightforward calibration curve.

It has been shown [817] that

$$1/D_S + 1/D_R = 1 \quad (2.13)$$

and, therefore,

$$D_R = D_S/(D_S - 1) \quad (2.14)$$

from which it follows that in a one-line system the sample and reagent concentration lines cross at $D = 2$ (where $D_R = D_S$). Furthermore, if at any point within the single channel the reagent concentration should be

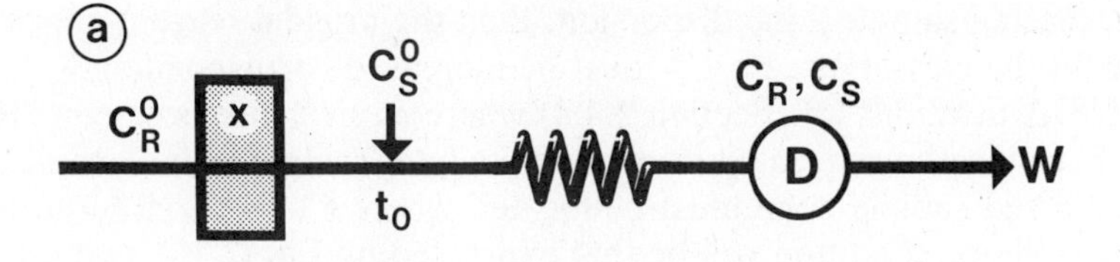

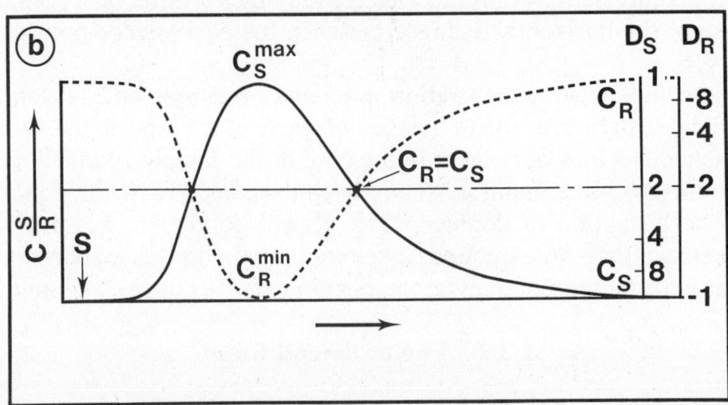

Figure 2.13. Mutual dispersion profiles of sample (S) and reagent (R) in a single-line system. (a) Single-line manifold where sample of original concentration C_S^0 is injected into a carrier stream of original concentration C_R^0 at time t_0. The detector senses the dispersed zone within which the mutual concentrations of sample and reagent are C_R and C_S (if no chemical reaction has taken place). (b) Mutual concentration gradients of C_R and C_S.

equal to the sample concentration, then the following conditions must be fulfilled:

$$C_S = C_S^0/D_S \tag{2.15}$$

or

$$C_S^0 = C_R^0(D_S - 1) \tag{2.16}$$

In other words, if a sufficient excess of reagent is to be maintained throughout the whole sample zone (say at least fivefold stoichiometric excess), and if the original sample and reagent concentrations are equal, then medium dispersion at C^{max} must be obtained by the means outlined in the previous section. If sensitivity of measurement is to be increased by decreasing sample-zone dispersion, then the original reagent concentration in the carrier stream C_R^0 must be increased correspondingly.

For FIA titrations (cf. Section 2.4.4) where equivalence is sought [10, 1124] between the reagent (titrant) and analyte (sample), an element of the dispersed sample zone must be located where $C_S = C_R$. Therefore, the equivalence condition will be that expressed by Eq. (2.16), and if the reagent concentration C_R^0 is kept constant while the concentration of the injected samples is increased, then, for the equivalence to be maintained, D_S at the equivalence point must increase, which will be observed as an increase of the horizontal distance between the equivalence points (Section 2.4.4).

To conclude, the concentration gradients of sample and reagent in a single-line system are mirror images of each other, and if the D value approaches 1, a lack of reagent in the core of the sample zone will occur. This is why two-line manifolds have been designed [3, 10, 20, 153], they avoid the formation of double peaks. It will be shown, however, that double peaks may be exploited in several novel ways as a source of additional information on analyte composition (cf. Sections 2.4.4 and 2.5).

2.3.2 Two or Several Lines

The dispersion of sample and reagent solution in a two-line FIA system differs from a single-line system in two ways:

1. The dispersion coefficient D_S of the sample cannot reach a value of 1, even when a "flat" or "a steady state" is observed.
2. The dispersion of reagent is not influenced by the dispersion of the sample zone, because the reagent is being added to all elements of the dispersing sample zone in the same amount.

This is better understood by referring to Fig. 2.14, showing at the top (*a*) a two-line FIA system, and below (*b*) a concentration profile of the reagent (C_R^0, pumped at a rate of y mL/min), and of the sample (C_S^0) injected into a carrier stream of a diluent (H_2O), which is being pumped at a rate of x mL/min. The respective concentrations C_S and C_R as observed within the detector (D) after passage through the reactor are represented by the lines showing that D_R remains constant throughout the entire scan, while D_S varies due to sample-zone dispersion. If the original concentrations of sample and reagent are equal, $C_R^0 = C_S^0$, then

$$C_R = C_R^0/D_R = C_R^0[y/(x + y)] = C_S^{\max SS} \qquad (2.17)$$

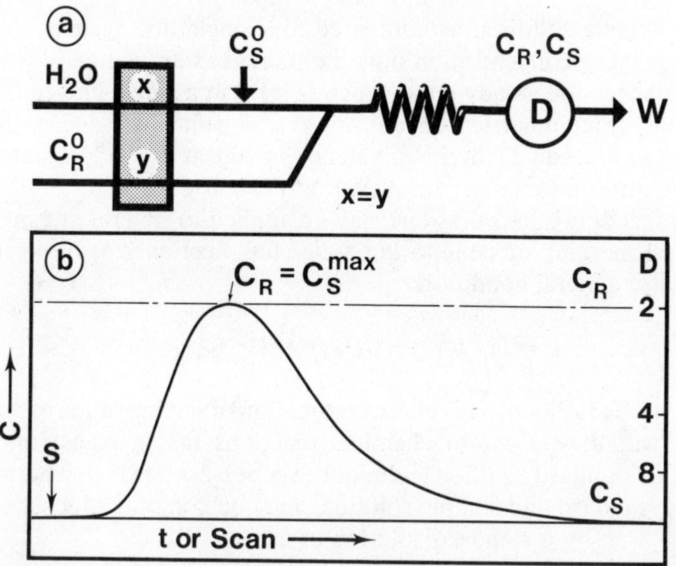

Figure 2.14. Mutual dispersion profiles of sample (S) and reagent (R) in a two-line system. (a A two-line manifold where sample of original concentration C_S^0 is injected into an inert carrier stream that merges downstream with a reagent stream of original concentration C_R^0; the pumping rates of these two streams are equal ($x = y$). The detector senses the dispersed sample zone, which is dispersed first into the inert carrier stream and then into the reagent stream. In the detector the reagent and sample concentrations are C_R and C_S (if no chemical reaction has taken place). (b) While the dispersed sample zone yields a concentration gradient of C_S, the reagent concentration remains constant throughout the scan and $D_R = 2$.

where $C_S^{\max SS}$ is the sample concentration at steady state, obtained at $S_v > 5S_{1/2}$ and

$$C_S^{\max SS} = C_S^0[x/(x + y)] \tag{2.18}$$

or for sample/reagent equivalence at any point within the channel downstream from the confluence point (when $C_R = C_S$) and when $x = y$:

$$C_S^0 D_R = C_R^0 D_S \tag{2.19}$$

Thus, compared to a single-line system it is easier to maintain an excess of reagent throughout the entire sample zone, even when D_S is close to steady state. This is because $D_S^{\max SS}$ cannot be lower than 2, and, therefore, when $x = y$, the original reagent concentration must be only slightly higher than that stoichiometrically corresponding to the concentration of injected sample solution, as compared to a single-line system where the original reagent concentration must be increased dramatically whenever D_S approaches the steady state, since $D_S^{\max SS}$ in a single line equals to 1. If original stoichiometric concentrations and pumping rates in the two-line system are equal, then equivalence is met at $C_S^{\max SS}$, which means that FIA titrations can be performed only when $C_S^0 > C_R^0$.

The preceding considerations may be applied to several line manifolds or several merging or penetrating zones (cf. Section 2.5) by keeping in mind that a general condition

$$1 = (1/D_S) + (1/D_{R1}) + (1/D_{R2}) + \cdots$$

will always be fulfilled. This allows estimation of mixing ratios of reactants as they would be *prior* to chemical reactions taking place, or for the purpose of standard addition technique (Section 2.5.2 [817]), where mixing ratios of standard and sample solutions may be computed by substituting $R1$, $R2$, . . . , by a standard solution of analyte [817].

2.4 FIA GRADIENT TECHNIQUES: SINGLE ZONE

In the FIA applications mentioned in the previous section, the analytical readout has been derived from the peak *maximum height*. Yet the concentration gradient formed from the injected sample zone contains an infinite number of elements of fluid containing different concentration ratios of sample and carrier stream. The gradient techniques described in this, and the subsequent section, exploit the fact that *any element of fluid* along the zone gradient corresponds to a well-defined and reproducible dispersion value, so that the dispersion coefficient of the sample $D_S = C_S^0/C_S$ and dispersion coefficient of reagent $D_R = C_R^0/C_R$ within each of these elements can be related to a well-defined delay time (t_1 to t_{11} in Fig. 2.15), which had elapsed from the moment of injection (t_0). In this way any vertical distance between baseline level and recorded signal level, including the one at the peak apex (i.e., peak maximum $C^{\max}$) represents a potential analytical readout.

In addition to this *vertical* dimension, an FIA peak has also a *horizontal* dimension—the peak width. Again, a number of levels may be identified (Level 1, 2, and 3 in Fig. 2.16) each of which corresponding to two elements of fluid, with identical dispersion coefficients, one of them being situated at the leading edge of the peak, the other one on the trailing edge of the peak. The readout, peak width (W), can be identified by means of a selected level of response—as observed by the detector of choice.

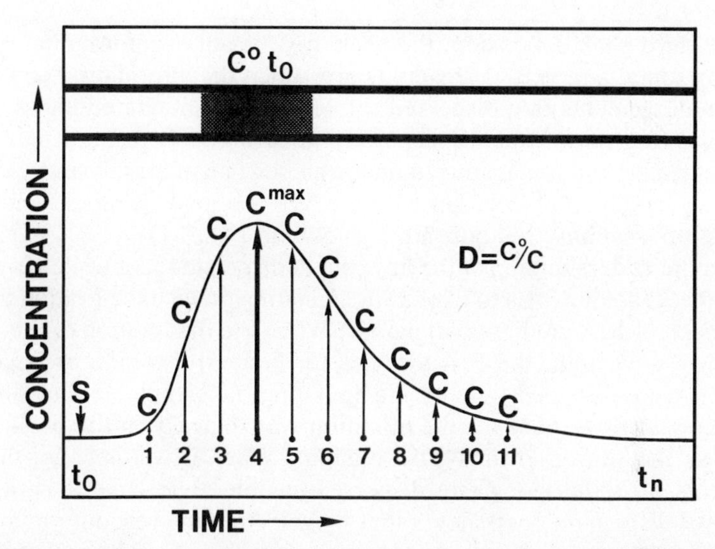

Figure 2.15. In addition to peak height a dispersed sample zone yields an infinite number of vertical readouts t_n, some of them being shown as vertical lines at delay times t_1 to t_{11}. Thus, any dispersed sample zone is a matrix of concentrations versus time, defined by C^0, t_0, and t_n.

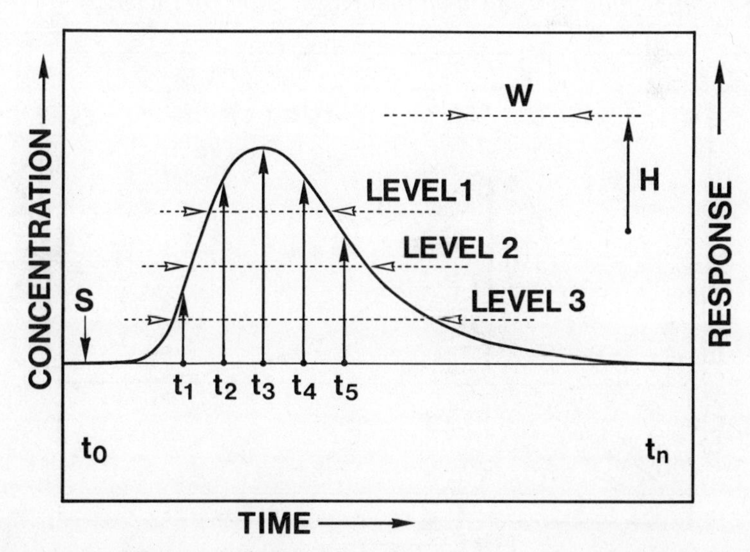

Figure 2.16. In addition to vertical readouts t_1, t_2, t_3, . . . , related to peak height H, a number of horizontal readouts, related to peak width (W), can be obtained by selecting different detector response levels (Level 1, 2, 3, . . .).

The third peak dimension, the peak area, is not as information rich as the horizontal and vertical readouts are, since the information contained within all elements of a dispersed sample zone is integrated into a single readout. One may visualize peak area measurement to be a numerical "homogenization" analogous to that achieved in batch analyses by homogeneous mixing. The relation between various peak readouts and concentration–response relationship is shown in Fig. 2.17.

It is the understanding of the key role of dispersion, controlled in space and time, that allows us to exploit the resulting time/concentration matrix for design of FIA gradient techniques. When the first edition of this monograph was written, the FIA systems for limited, medium, or large dispersion were designed only by manipulating the sample volume and the geometry of flow, because the attention was focused on the peak maximum as the source of analytical readout. When it was recognized that elements of liquid along any dispersed sample zone have such a wide range of dispersion coefficients that by selecting the readout via a delay time, a wide range of options became available, an additional rule was formulated [338]:

Rule 7. *Since any dispersed sample zone is composed of a continuous concentration gradient, a desired degree of dispersion can be conveniently selected by locating the analytical readout within that element of fluid that has a suitable dispersion coefficient.*

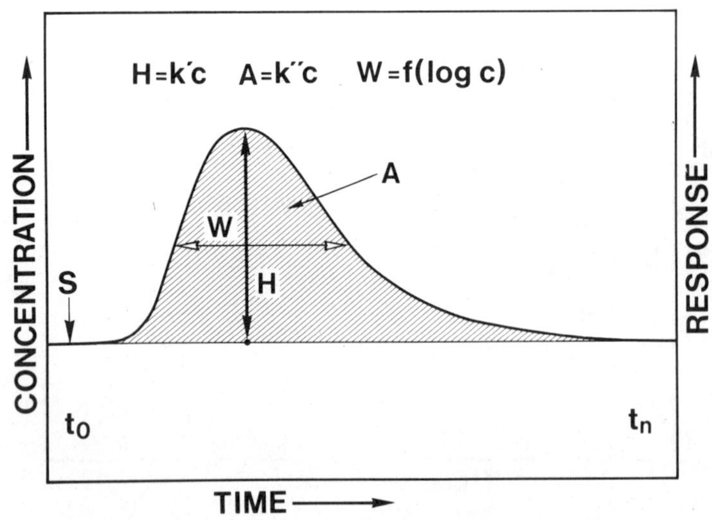

Figure 2.17. Peak height *H*, peak width *W*, and peak area *A* yield useful readouts, which for certain FIA techniques are identical (and therefore redundant), while for other FIA techniques yield different (nonredundant) information, thus adding additional dimensions to the peak readouts.

The concentration gradient can be formed by injecting a single sample zone, as is described in this section, or by injection of two or several zones, as is described in the next section. The resulting readout may be—for both experimental approaches—either a single peak, as exploited in most FIA techniques designed to date, or a multiple peak, which yields additional information on sample composition.

2.4.1 Gradient Dilution and Gradient Scanning

Gradient Dilution

Gradient dilution [275, 338], which was first described and used in 1982 [264, 338], is based on the selection of a suitable element of the dispersed sample zone to become the source of the analytical readout. This is done by selecting a time delay (e.g., 10, 12, 14, 16, or 18 s), elapsed from the point of injection *S*, as illustrated in Fig. 2.18, right. Since the selected elements of the dispersed sample zone have different dispersion coefficients (D_1, D_2, D_3, . . . , Fig. 2.18, left), which yield, for a series of injected standard solutions, a sequence of corresponding calibration curves with decreasing slopes, the sensitivity of measurement can be

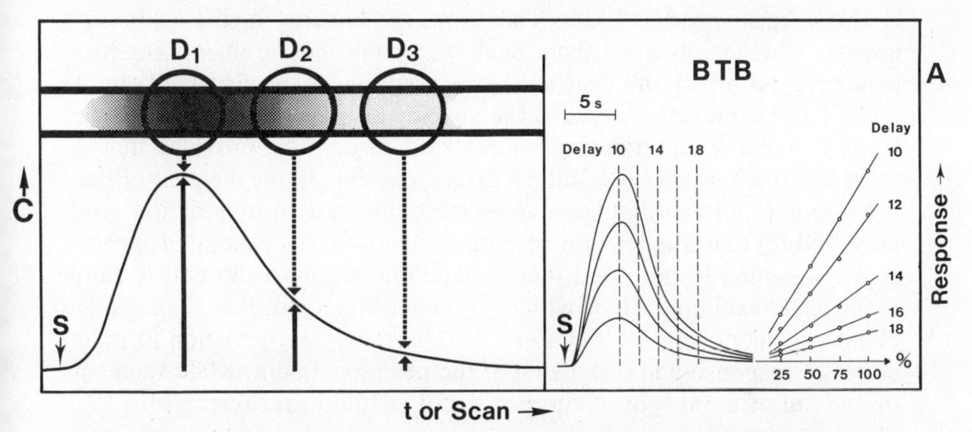

Figure 2.18. Gradient dilution is based on selecting readouts at the tail of the dispersed sample zone (right-hand side of left curve), where elements of fluid can be located within which the sample material has been diluted by the disperson process. The selection may be mechanical, that is, the zone is selected via the delay time, captured by means of a second valve, and then reinjected into a second FIA system. Electronic dilution is achieved by means of a microprocessor, by selecting via the delay time a detector readout from which a calibration curve is constructed (middle curves). The peaks, recorded with four different concentrations of analyte (labeled 25%, 50%, 75%, and 100%) were sliced at different delay times yielding readouts from which calibration curves of different slopes were recorded (right curves).

adjusted. This means that the readouts, which are usually obtained from peak maximum, may, if desired, be replaced by the readouts collected at any other section of the peak, provided that the readings are always taken at exactly the same delay time after injection. This approach is practical if such a wide range of sample concentrations is to be analyzed that some of the samples will exceed the dynamic range of the detector used, and, therefore, the peak maximum cannot be recorded or is distorted. Then a calibration curve obtained with a large delay time, for example, of 14 s (cf. Fig. 2.18), may be used with advantage. Gradient dilution can be performed experimentally in two ways: electronically [264], by programming a microprocessor to collect the readout at a selected time delay, or mechanically [145, 1060], by capturing a selected section of the dispersed sample zone by means of a valve and by reinjecting the sample zone captured into a second FIA system for further dilution and processing.

The nature of the physical process of dispersion on which FIA is based ensures that the calibration curves obtained from data collected at fixed delay times are linear. This applies not only to a simple dilution via a dispersion process, but linear calibration curves will also be obtained if the dispersion is accompanied by a chemical reaction provided that the reaction is linear in nature, because "if a number of independent linear processes are occurring simultaneously in a system, their over-all effect is also a linear process" [20]. Therefore, the linearity of the calibration graphs, whether obtained from peak maximum measurements, or from peak heights corresponding to successive sections of the dispersed sample zone, will be preserved even if the species to be measured is produced by a chemical reaction during sample zone dispersion, provided that (a) there is sufficient excess of all necessary reagents in the element of fluid to be monitored; (b) that there is no inhibition caused by reaction products; and (c) that the detector responds linearly to the generated species. It is interesting to note that these conditions are most difficult to fulfill at the peak maximum, since with high sample concentration, all unwanted situations may occur (cf. Section 2.3). Therefore, the deviation from linearity of response will occur first at the peak maximum, while segments on the tail of the response curve will still yield a linear response.

Gradient Scanning

Gradient scanning is an extension of the gradient dilution in that it uses a dynamic detector, which measures continuously a physical parameter by repeatedly scanning it within a certain range. Originally demonstrated by means of a voltametric detector (current/potential scan [288, 464]), the technique was later extended to spectroscopy [817, 1080] as shown on the emission/wavelength scan in Fig. 2.19.

If the rate of detector scanning is much faster than the movement of the carrier stream, the dispersed zone may be advanced during the scan and still appear to be stationary. If the scanning cannot be performed that rapidly (like, for example, in Fourier transform infrared spectroscopy), of if the chemical reaction occurring is too slow, then selected sections of the zone can be stopped consecutively within the detector (cf. Section 2.4.3). The resulting record has a three-dimensional form, like the three-dimensional emission spectrum depicted in Fig. 2.19*b*, which shows the response obtained when injecting a 40-μL sample solution, containing salts of the three elements Na, Ca, and K, into an atomic emission spectrometer, using a fast scanning monochromator capable of scanning from 350 to 800 nm in 5 ms every 100 ms and starting the data collection 300 ms after sample injection.

The combination of rapid scanning instruments (both optical and electrochemical) with the FIA gradient approach is natural. The tedious manual handling necessary to prepare solutions for batchwise measurement by means of these advanced instruments is not flexible and does not exploit their full potential. Seen from a purely analytical viewpoint, the advantages of gradient scanning should become even more apparent when multicomponent analysis will be performed routinely on unknown samples, using spectrometry. Work has already been done by combining FIA with inductively coupled plasma (ICP) emission spectrometry [188, 206, 518, 565, 1231, 1249, 1254]. An emission spectrum generated by an ICP contains an unavoidable combination of weak and strong lines, which cannot be accommodated within the optimum detector range by conventionally aspirating a single solution. However, if sample is injected into a FIA channel connected to an ICP and the dispersed sample zone is repeatedly scanned while passing into the plasma torch, all lines, regardless of the intensity and analyte concentrations present in the original sample solution, can be evaluated on the basis of a single injection experiment. Combination of the gradient scanning technique and electronic calibration would be a logical extension of this approach.

2.4.2 Gradient Calibration

Gradient calibration [264, 338] is a logical extension of the gradient dilution procedure, since it relies on the strict reproducibility of the same phenomena: physical dispersion of reacting species, accompanied by chemical reactions. Its main goal is to avoid the usual repetitive calibration by means of serially diluted and serially injected solutions, since the information sought is already contained within some of the segments of fluid originating from a sample zone of the most concentrated standard sample material (Fig. 2.20). Thus, by identifying on this curve the time

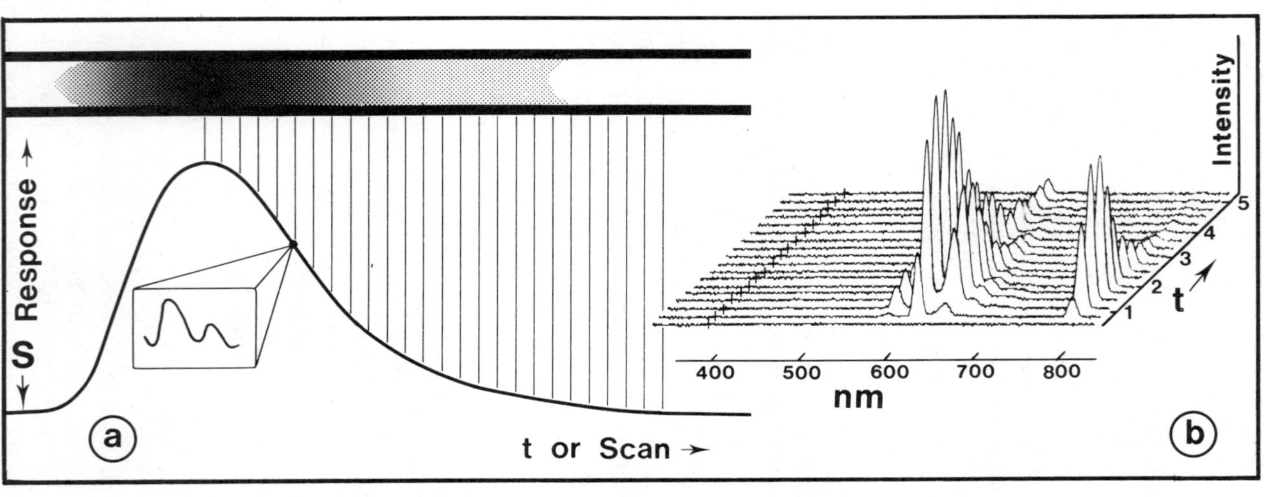

Figure 2.19. Gradient scanning is similarly to electronic gradient dilution based on selection of readouts via delay times during which a detector rapidly scans a range of wavelengths (*a*), thus creating an additional dimension on the time-concentration matrix (*b*), showing a series of successive emission spectra recorded on the ascending and descending part of a dispersed zone, containing Na, K, and Ca injected into an atomic emission spectrometer furnished with a fast scanning monochromator [817].

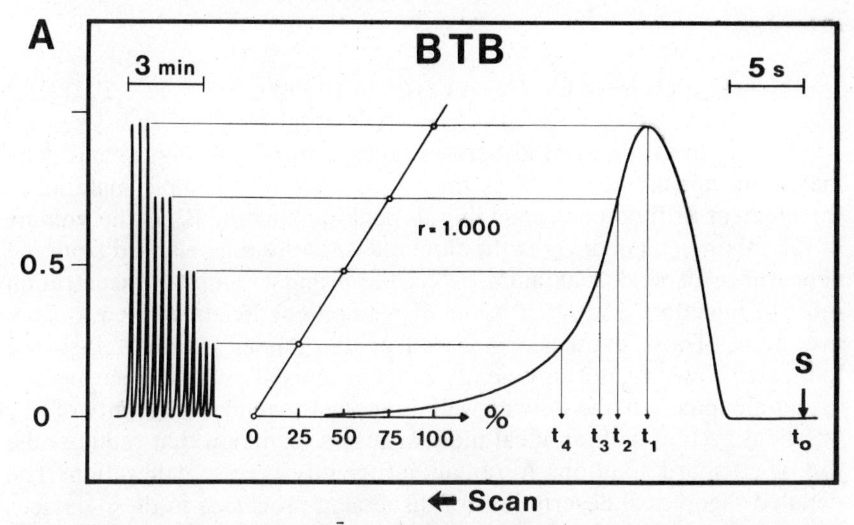

Figure 2.20. The principle of gradient calibration, showing on the left-hand side a conventional calibration run where standards covering 25%, 50%, 75%, and 100% of the range were injected each in triplicate, and on the right-hand side a single injection of the 100% standard, which was recorded at a high chart speed. Via a conventional calibration line (r = 1000), delay times t_1, t_2, t_3, and t_4 are then identified, corresponding to known levels of originally injected standards.

delays associated with those segments of fluid that have concentrations corresponding to the concentrations which would provide the same readouts if the signals were collected from the peak maxima obtained by separate injections of serial dilutions, one can obtain a multipoint calibration curve from a single injection. In other words, the peak maximum value can be supplemented by a series of sequential C values spaced along the gradient and experimentally identified through increasing delay times. Once the formula describing the relationship $C^0 = f(\text{delay time})$ is obtained from a normal calibration evaluation and stored by a computer, all subsequent calibration procedures can be performed by a single injection.

The function f, describing the relationship between concentration C^0 and delay time t, is dependent on the dispersion process taking place in the FIA channel and, therefore, on channel geometry, volume, and flow velocity. In the simplest case, that is, absence of chemical reactions, and when the flow channel conforms with the model of one well-stirred tank (cf. Chapter 3), the concentration–time matrix of the falling edge of the curve is described by:

$$t = (V_m/Q) \ln 10 \log(C^{max}/C) - T \tag{2.20}$$

or

$$t = -(V_m/Q) \ln 10 \log C + (V_m/Q) \ln 10 \log(C^0/D^{max}) - T \quad (2.21)$$

where T is the time elapsed between the point of injection t_0 and peak maximum appearance, C^{max} is the concentration of sample material in the element of fluid corresponding to peak maximum, V_m is the volume of the mixing chamber, Q is the flow rate, t is the time elapsed from the appearance of peak maximum, C^0 is the original sample concentration *prior* to injection, and D^{max} is the dispersion coefficient related to peak maximum. These expressions, based on the model of one well-stirred tank, are, however, not sufficiently exact to be used a priori, since method of sample injection and flow as well as geometrical nonconformity of any practical system with an ideal model cause a deviation that requires the use of standard solutions to obtain sufficiently precise calibration. The detailed theoretical descriptions of dispersion processes in these devices have been developed by Gisin et al. [1063], and by Tyson and co-workers [1022, 1062]. These theories, based on different viewpoints, are most illustrative of the problem and very useful for further development of the gradient calibration and gradient dilution techniques.

If a mixing chamber is not used, as is the case in most FIA systems, it should be noted that the injection process itself results in the formation of a concentration gradient similar to the one stirred tank [183] and that the thus formed sample zone will be transformed gradually into a Gaussian shape as the number of mixing stages increase (cf. Fig. 2.6*b*, and Chapter 3). Calibration by means of known standards is then the only practical way to establish the matrix of $t = f(C^0)$.

2.4.3 Reaction Rate Measurement by Stopped Flow on the Gradient

Except for FIA titrations, the stopped-flow method [20, 264] is, the most frequently used gradient technique. The most practical approach to reaction rate measurement is to halt the reaction mixture within the observation volume of the flowthrough detector by stopping the flow and to measure the change of the signal with time. This is readily accomplished by the FIA stopped-flow technique (Chapter 4.3), which relies on a combination of a suitable dispersion of the sample material within the reagent stream and on the subsequent reaction rate measurement during the stopped-flow period.

The recorded signal contains a portion of the reaction-rate curve, the slope of which is the basis of the calibration curve for a fixed-time kinetic assay. A closer look at the dispersing sample zone (Fig. 2.21, left) shows

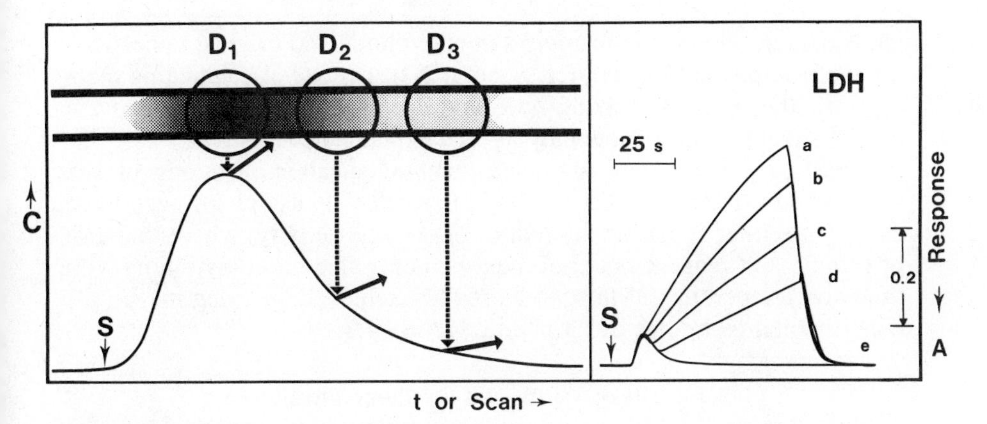

Figure 2.21. The stopped-flow technique as demonstrated on the assay of the catalytic activity of lactate dehydrogenase (LDH). The dispersed sample zone, shown on the left, contains different substrate/enzyme ratios (since $D_1 < D_2 < D_3$), which yield different reaction rate curves. By selecting an appropriate delay time at which the stopped-flow period begins (right), various sections of the dispersed zone can be monitored, yielding reaction rate curves *a–d*.

that the reagent sample ratio (C_R/C_S) is increasing sharply along the trailing edge of the dispersed zone since $D_1 < D_2 < D_3$ (recall the concentration profiles in Figs. 2.13 and 2.14), and, therefore, the slopes of reaction rate curves correspondingly decrease. This is shown on the actual reaction rate curves, recorded during the enzymatic oxidation of pyruvate to lactate, catalyzed by lactate dehydrogenase (LDH), shown on Fig. 2.21, right. These reaction rate curves, with decreasing slope from *a* to *d*, as obtained during the stopped-flow period, have been recorded by spectrophotometric monitoring of the consumption of NADH, obtained by injecting in all experimental runs *a* to *e* the same solution of LDH. Note that the decrease of the recorded slopes is due to an increase of the stop delay time, that is, by selection of which elements of the dispersed zone with decreasing enzyme/substrate ratios that have been trapped in the flow cell. Thus, similarly to the "electronic dilution," optimum sample/reagent concentrations have been chosen for readout optimization. This special feature of the FIA stopped-flow gradient technique—a fine adjustment of the reagent/sample ratio by electronic means—is more convenient than a traditional manual dilution, which is tedious, especially in case of enzymatic assays, where the sample/reagent ratio must be carefully selected to obtain a meaningful result, and when sample and reagent solutions may be scarce. It has been shown that the sample/reagent ratio, that is, substrate/enzyme ratio in our discussion, depends on the geometry and flow rate of the channel, on the type of manifold (single-line or two-line system), and on the injected sample volume and original concentrations of sample and reagent (cf. Section 2.3). Instead of calculating them a priori, the optimum mixing ratio and correspondingly delay time are readily identified experimentally as was explained previously.

There are many clinical and biotechnological applications of the stopped-flow FIA (cf. Chapter 7). This method is useful for enzymatic assays (of either substrate content or of enzyme activity) where the ratio of sample and reagent material must be controlled carefully, along with time and temperature. This can be readily achieved by using the simple experimental technique described in Section 4.3.

2.4.4 Peak Width and Gradient Titrations

Peak Width Measurement

The peak width measurement was originally suggested by Pardue and Fields as an alternative to measurement of peak height [150, 151]. Peak width is related to a time span Δt measured between two elements of fluid, having identical dispersion coefficient, one being situated on the leading edge and the other one on the trailing edge of the dispersed sample zone (Fig. 2.22, top). By injecting standards of different concentrations, different peak widths can be recorded, the slope of the resulting calibration curve depending on the selected level of detector response (Fig. 2.16). Depending on the flow geometry, the calibration curve may conform either with

$$\Delta t = k' \log C^0 + \text{const}' \tag{2.22}$$

if a one-mixed-tank model applies (i.e., $N = 1$), or with

$$\Delta t = k'' f \log C^0 + \text{const}'' \tag{2.23}$$

(if the flow system conforms with a larger number of mixing stages $N > 2$). Owing to this logarithmic dependence, the most frequently emphasized asset of the peak-width-measurement technique is its ability to extend the dynamic range of accommodatable concentrations. However, the price paid is decrease in precision, inherent to this type of response. It is therefore of advantage to combine measurement of peak height and width as shown in Fig. 2.22 (below), where low concentrations of analyte are determined with high precision from peak height, while the measuring range is extended to high concentration of analyte, albeit with lower precision, by peak width measurement, when peak height measurement fails. It is

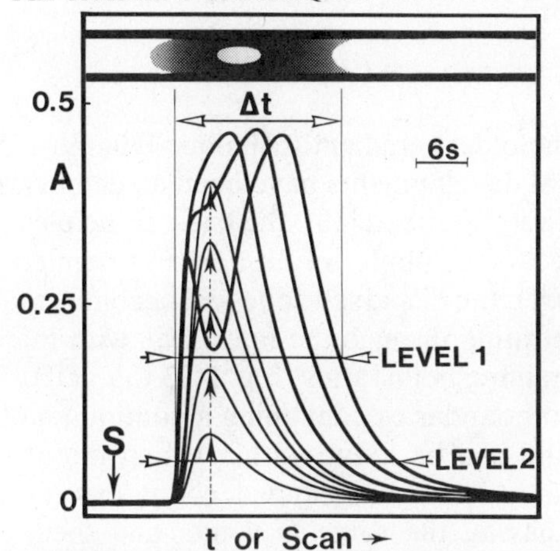

Figure 2.22. Peak width measurement yields as the readout a Δt value that, depending on the selected detector response level, would be more (Level 2) or less sensitive (Level 1). The peak width decreases with decreasing analyte concentration, yet it can be supplemented in this low range by peak height measurement (cf. the vertical arrows). Note that at high analyte concentrations (bold lines), peak height does not yield a meaningful response, while the peak width does. The response curves are spectrophotometric readouts, recorded of the colorimetric determination of iron(II) by orthophenantroline in a two-line FIA system. The injected solutions contained 2, 4, 6, 8, 10, 20, 40, and 80 ppm iron, while the orthophenantroline concentration was equivalent to 12 ppm iron.

interesting to observe that the sensitivity of peak width measurement as well as the detection limit of peak height measurement are limited by the same parameters: baseline stability and detector response, since the detection limit of peak height measurement depends on the minimum vertical detectable signal deviation, while maximum slope of calibration curve is obtained at the peak width measured at a level as close as possible to the baseline. Peak width measurement has recently attracted attention due to its versatility and range extension potential for process control applications. Detailed description of theory and experimental results have recently been given by several authors [1058, 1062, 1063, 1064, 1124, 1127, 1237].

It might appear that peak width merely serves as range extension. Yet, since peak width is related to a time span, which in turn is related through linear flow velocity to volumetric rate, peak width measurements allow flow titrations to be performed in a novel way. Therefore, similarly to classical batch titrations, FIA titrations encompass a domain of determinations, which cannot be performed in any other way, because they are based on *consumption of an equivalent amount of reagent* and, therefore, titrations yield different information than a direct measurement (pH measurement versus titration of a mixture of a weak and strong acid).

FIA Titration

FIA titration is the oldest gradient technique [10, 59, 153, 183, 253, 338]. Although technical developments have been made in various stages since the method was first described [10], the basic principles remain the same: If a sample zone, for example, of an acid (S) is injected into a carrier stream of a base (R), the dispersed zone will become gradually neutralized by the base penetrating through the interfaces with the carrier stream at the leading and trailling boundaries (Fig. 2.23 top, left). Therefore, within each of these two boundaries containing a continuum of acid/base ratios an element of fluid can be found within which the acid is exactly neutralized by the base. These two equivalence points form a pair (A and B Fig. 2.23, left), having the same D value, and their physical distance (measured as Δt for a constant flow rate Q) will increase with increasing concentration of the injected acid, C_S^0, and decrease with increasing concentration of the base, C_R^0, contained in the carrier stream. As in all peak width measurements, the concentration dependence of the time span Δt depends on the concentration profile formed by the dispersion of the zone on the way from the injector to the detector. It has been shown [10, 83, 1062] that for $N = 1$ (Chapter 3), Δt as a function of the concentration

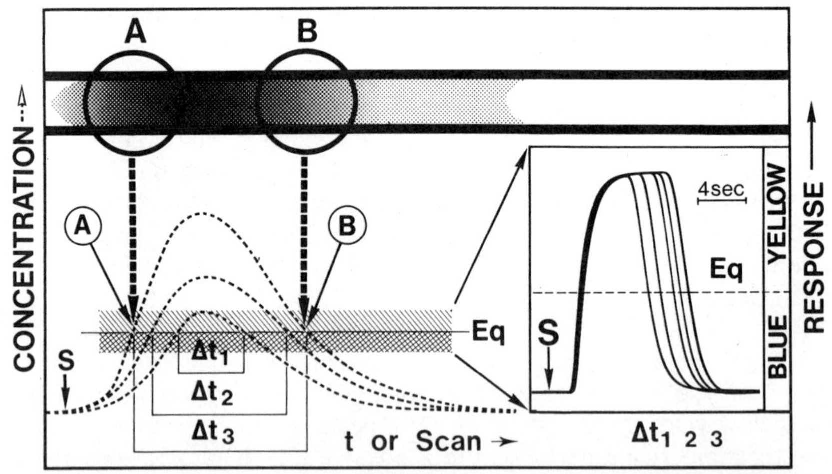

Figure 2.23. Flow injection titrations. The dispersed zone (top left) contains two equivalence points (A and B), the distance between them, Δt, being proportional to the acid concentration in the injected sample zone. By selecting the indicator, the equivalence level (Eq) is set and in this way decreasing values of Δt are obtained for decreasing concentrations of injected acid. Actual titration curves are shown on the right. For details see Ref. 183.

of the sample will be given by

$$\Delta t = (V_m/Q) \ln 10 \log[C_S^0/(C_R^0 n)] + \text{const}_1 \tag{2.24}$$

where n is the molar ratio of reacting components. Originaly [10] the mixing stage was actually a small mixing chamber (equipped with a magnetic stirrer) (cf. Sections 4.9 and 6.7), the volume of which (ca. 1 mL) dominated the flow design. For such an experimental setup

$$\text{const}_1 = (V_m/Q) \ln 10 \log(S_v/V_m) \tag{2.25}$$

where S_v, the sample volume and V_m, the volume of the mixing stage, are easily identifiable physical parameters of the FIA system. Using the dispersion coefficient D, the const_1 can be rewritten [183]

$$\text{const}_1 = (V_m/Q) \ln 10 \log[D/(D - 1)] \tag{2.26}$$

However, further studies have shown that the injection process itself conforms with the "one-mixing-stage" model, and this observation allowed miniaturization and simplification of the FIA system, resulting in elimination of the mixing chamber and implementation of high-speed titrations [183]. Such titrations are performed at medium or even limited dispersion. A detailed treatment of dispersion in a one-tank system for small D values, recently developed by Tyson [1062], requires a modification of const_1 so that for low D values

$$\Delta t = (V_m/Q) \ln 10 \log[C_S^0/(C_R^0 n)] - (V_m/Q) \ln 10 \log(D - 1) \tag{2.27}$$

According to this equation, the detection limit for a single-line system (cf. Fig. 2.13) occurs when the equivalence condition is met

$$C_S^0/(C_R n) = D - 1 \tag{2.28}$$

and when the dispersion coefficients are equal to 2 ($D_S = D_R = 2$), since, then, $\Delta t = 0$. Also $\Delta t = 0$ is reached whenever $C_S/C_R = D - 1$, and, therefore, the detection limit can decreased by further decreasing the dispersion coefficient D. This can be done by increasing the injected sample volume (D increases with S_v, cf. Fig. 2.6*a*), but the price paid is the loss of sampling frequency and of precision. To summarize, use of a mixing chamber in FIA titrations results in a high D value and higher precision, and has some advantages (cf. Section 2.2.4), yet it yields lower sampling frequency and a worse detection limit than titrations performed without mixing chamber at low D values at faster speeds [183].

Technically, FIA titrations can be performed either in a single-line system, as discussed previously, or in a two-line system (cfr. Fig. 2.14 and Section 4.9.). For a two-line system, where titrant R is confluenced to a sample zone dispersed in an inert carrier stream,

$$\Delta t = (V_m/Q) \ln 10 \log[C_S^0/(C_R^0 n) - 1] - (V_m/Q) \ln 10 \log[(D/2) - 1] \quad (2.29)$$

for equal flow rates in both lines and for the mixing chamber volume dominating the overall dispersion process.

A rigorous treatment of the kinetic processes influencing peak width measurement and, therefore, relevant to FIA gradient titrations was developed by Pardue and Jager [1064], who in the third part of a series dealing with FIA peak-width-based measurement used a variable-time kinetic model to predict dependence between Δt and the physicochemical parameters of the system. They subjected their models to exact experimental evaluation, including a predictive kinetic method. It may be expected that work along these lines will continue to be more detailed in future; yet, the majority of FIA peak width and FIA titrations are likely to use an empirical approach, rather than a priori theoretical models, which in spite of their increasing sophistication cannot describe peculiarities of individual experimental setups with a precision sufficient to avoid the necessity of calibration.

So far we have dealt with the concept of FIA titrations, but not with the way in which the equivalence point may be detected. In analogy with traditional batch titrations, the end point can be located by any suitable means (such as an optical or electrochemical detector). If the detector is capable of sensing sample (analyte) reagent (titrant) or product directly, the underlying chemical process is self-indicating [e.g., titration of Cu(II) by EDTA with the aid of a voltammetric detector (cf. Fig. 4.64) or by sensing a change of optical density]. For other types of titrations an addition of a small amount of a suitable indicator to the carrier stream of titrant is necessary (cf. the titration of acid by base using bromthymol blue as indicator and monitoring by a photometric detector, Fig. 2.23, right).

Regardless of which types of titrations will be performed in the FIA mode, the sudden change of slope at the rising and falling edge of the recorded peak will be much more pronounced than the slopes observed in simple peak-width measurement, since all correctly performed titrations are based on an *abrupt* change of detector response at the *end point*. This can be seen readily by comparing the concentration profiles on Fig. 2.23, left, with the corresponding titration curves (right). Points *A* and *B* are the equivalence points where the indicator changes from the blue basic form (as found in the carrier stream of sodium hydroxide) to the yellow acid form when the sample material is in excess. This is why for FIA titrations the requirements on detector stability and linearity of response may be somewhat relaxed compared to measurement of peak width (or peak height).

There is a growing interest in FIA titrations and peak-width measurements for laboratory applications as well as for continuous monitoring of industrial processes, and there is also some unfortunate confusion in the literature concerning the merits of peak-width/peak-height measurement and a discussion of whether the term "FIA titrations" is appropriate [150, 151, 216]. It is correct to say, that seen *macroscopically*, FIA titrations do not conform with the IUPAC definition* of titrations, because the area of the peak corresponds to a larger mass of analyte than is equivalent to the mass of titrant contained within the carrier stream pumped during the interval Δt [If that were the case, Δt would be zero!) Yet, a closer look shows that FIA titrations conform with the IUPAC definition *microscopically*, because when the dispersed sample zone is viewed as a matrix of concentration versus time, then in all FIA titrations a *pair* of elements of fluid can be identified, the distance between said paid of points being Δt, because the substance to be determined (S) has *completely* reacted with standardized solution of a titrant (R) within each *element of said pair* of fluids, while the remaining elements of fluid forming said matrix represent incremental additions of titrant to the analyte.

The similarity of FIA and classical batch titrations is useful to recognize, because such recognition turns our attention to the wealth of chemistries exploited by classical titrations that are now accessible to FIA adaptation. Indeed, all traditional titrations, that is, acid–base, compleximetric, redox, and precipitation, can be performed in the FIA mode. Catalytic titrations and titrations in nonaqueous solutions, including Karl

* From Recommended Nomenclature for Titrimetric Analysis, *Pure and Applied Chemistry* **18** (1969) No. 3: "Titration. The process of determining substance *A* by adding increments of substance *B* (almost always as a standardized solution) with provision for some means of recognizing the point at which all of *A* has reacted, thus allowing the amount of *A* to be found from a known amount of *B* added up to this point, the reacting ratio of *A* and *B* being known from stoichiometry or otherwise." Note that the Committee wisely did not specify whether the incremental addition takes place in a single vessel or series of vessels (i.e., in a single element of fluid or in a series of elements of fluid), continuously or discretely, with kinetic factors involved or at steady state.

Fischer determination of water [870, 1143], benefit from being performed in the protective environment of an FIA channel. In addition to classical titrants, even highly reactive reagents, which are unstable in air or when exposed to light, have been found useful for FIA applications [1112], and will increasingly find applications to FIA titrations, when handled or generated in the protective environment of FIA system.

Rule 8. *All FIA titrations are based on peak-width measurement, but not all peak-width measurements are FIA titrations. The difference is that FIA titrations are based on location of a pair of equivalence points by using indicator or self-indicating chemical reactions, while peak-width measurements rely on a time of flight of a dispersed zone measured at a selected level of detector response.*

2.4.5 Single Zone Yielding a Double Peak

The first examples of recording a double peak when injecting a single zone were described in an early FIA paper dealing with determination of chloride [4]. The injected sample zone was split intentionally into two uneven portions, by splitting the flow channel into two uneven parallel branches, so that the two portions of the split sample zone traveled through different length of tubing and then merged asynchronously in a confluence point prior to passing through the flowthrough detector. Thus, each injection yielded a double peak, which allowed an increase of the dynamic range for the determination of chloride in brackish waters. In the same paper another type of double-peak formation is recorded, that is, when a too large sample volume was injected into a single-line manifold, an artifact, which, owing to incomplete mixing of sample and reagent, adversely affects the simple peak-height measurement.

Since the formation of double peaks recently has become recognized as a source of new information, it is important to discuss the mechanisms by means of which doublet peaks are formed from a single injected zone. The formation of double peaks may be due to physical dispersion processes alone or to a combination of the dispersion process and the chemical processes that form a detectable species.

As a result of the injection process, humped peaks formed by the dispersion process alone can be observed in a single-line FIA system under conditions of laminar flow provided that the reduced time $\tau = D_m t/(0.5d)^2$ is close to 0.04, where D_m is the diffusion coefficient (cm^2/s), d is the diameter of the tube, and t is the time. This observation for FIA was first described by Vanderslice et al. [142], who also recently computed and depicted in an elegant way the microstructure of flow boluses under laminar flow conditions ([1061] and Fig. 3.5). Since generally used FIA systems are designed so that a purely laminar flow is not sustained sufficiently long to achieve τ close to 0.04, this type of humped peak is has been observed very seldom.

Double peaks formed by a combination of the dispersion process and ensuing chemical reactions forming the detectable species have been observed frequently, and the reason for double-peak formation is the lack of reagent in the center of the injected sample zone [153, Fig. 6.6, p. 137]. This inevitably happens in any FIA system whenever the combination of the original sample concentration C_S^0 and that of the reagent C_R^0 and their dispersion coefficients becomes

$$\frac{C_S^0}{D_R} > \frac{C_R^0}{D_S} \tag{2.30}$$

as follows from the equivalence condition for $n = 1$ [cf. Eq. (2.19) and Section 2.3.1, Fig. 2.13 and Eqs. (2.13) and (2.14)]. The result is that the mutual sample/reagent concentration gradient is composed of the following regions: reagent stoichiometrically *equal* to sample (equivalence) and stoichiometric *lack* of reagent to sample (substoichiometry).

In the simplest case the sequence throughout the zone is superstoichiometry, equivalence, substoichiometry, equivalence, and superstoichiometry (Fig. 2.24). Obviously, if the species would react in several

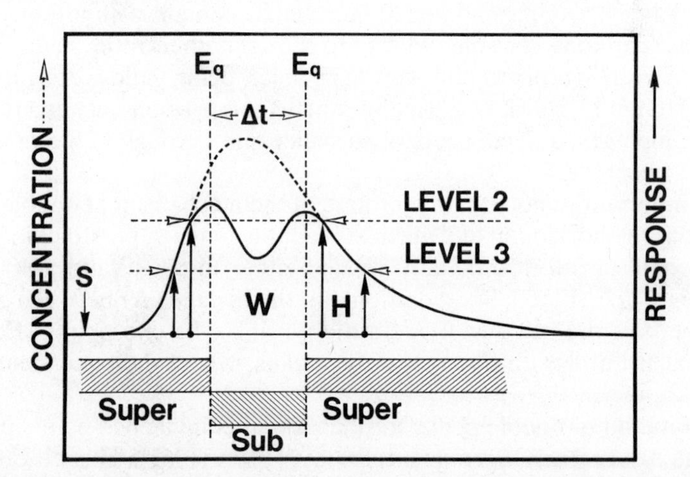

Figure 2.24. A double peak yields three readout regions: reagent excess at the outer edges (super), where the usual vertical and horizontal readouts can be located, two equivalence points (Eq) separated by a time span Δt, and a substoichiometric (sub) region where a wealth of kinetic information awaits further exploitation (cf. Fig. 2.22).

stoichiometric ratios ($n > 1$), the preceding regions would coexist in the dispersing zone in several sequences and the detected zone will show a corresponding number of maxima, minima, and their combinations.

It is interesting to note the difference between double-peak formation in the single- and double-line FIA system. In the single-line manifold, formation of a double peak is more likely to occur, since sample and reagent profiles are *mirror images* of each other and the reagent has to penetrate into the core of the sample zone via the dispersion process through the leading and trailing edge of the sample zone. Thus, the larger the sample volume and the smaller the cross section of the channel are, the more likely the formation of the double peak would be. Whenever a lack of reagent occurs in a single-line system, a trough is formed. In the double-line manifold, all elements of the dispersed sample zone are supplied with equal volumes of reagent (cf. Fig. 2.14), and, therefore, when the equivalent condition is met [Eq. (2.19)], the maximum response is reached and remains constant throughout the substoichiometric region [Eq. (2.29)], yielding a peak with a flat top. Although all elements of the dispersed sample zone are supplied with equal amounts of reagent, a double peak will be observed in a two-line FIA system whenever different chemical species are formed if the sample/reagent ratio changes. An example of such a case is shown on Fig. 2.22 illustrating the spectrophotometric determination of iron(II) by means of orthophenantroline, monitored at 510 nm. The red Fe(II)-(o-phen)$_3$ complex, formed with excess of reagent, yields a smooth peak between 2 and 10 ppm of Fe(II)—thin recording lines. Samples above 20-ppm Fe(II) contain within the core of the dispersing zone sections where the Fe(II)/o-phen ratio leads to formation of Fe(II)-(o-phen) complex 1:1, which being yellow does not absorb at 510 nm [2.5]. Thus, owing to kinetic competition between the 1:3 to 1:1 complexes, a decrease of absorbance and a trough in the recorded peak are observed.

It is important to note that regardless of the mechanism of double-peak formation, the horizontal and vertical readouts in the superstoichiometric regions yield useful information, related either to peak width (W) or to peak height (H) (Fig. 2.24). Additionally, the distance Δt between equivalent points is the basis of FIA titrations, since the end points (Eq) are located at the apexes of the respective peaks, where the equivalent conditions are met.

Exploitation of double peaks for location of equivalence points in FIA titrations has recently been suggested by Tyson [1062]. This elegant approach is useful when the reaction involved is self-indicative, so that the detector yields an abrupt change of signal at the top of a hump.

Another example of an elegant exploitation of the variable sample/

reagent gradient is the FIA mode of performing Job's method of continuous variations. This classical approach for investigation of complex composition has been adapted to FIA by Rios et al. [1224]. The complex sample/reagent gradient techniques, combined with FIA gradient scanning, will allow analyses of mixtures of compounds to be performed much more selectively than by current analytical techniques, since the combination of spectral and kinetic information is very detailed. To conclude, while formation of double peaks due to lack of reagent in the center of the dispersing zone was in the past mostly considered to be an unwanted artifact, it follows from the preceding discussion that double peaks might indeed become a new source of information.

2.5 FIA GRADIENT TECHNIQUES: ZONE MERGING, ZONE PENETRATION, AND MULTIPLE PEAKS

Merging-zone techniques are based on exploitation of the concentration gradients formed when two (or several) zones are injected simultaneously, and then allowed to merge—or penetrate into each other—thus yielding response curves, which either completely or partially overlap each other. The purpose of such an approach is twofold: (1) to save sample and/or reagent solutions and (2) to create a composite zone that is information rich and that may yield a multiple readout.

2.5.1 Sample Zone and Reagent Zone

Improvement of the sample and reagent economy was the goal of the first merging zone design, suggested independently by Bergamin et al. [23] and by Mindegaard [42]. Instead of pumping the reagent stream continuously, a valve furnished with two injection loops injected a reagent zone along with the sample zone at the beginning of each injection cycle. The sample and reagent zones were injected into two separate streams and merged downstream at a confluence point (Fig. 2.25*a*). Thus reagent was used only exactly when needed, and carrier streams made of less expensive buffer solutions (or even water) were pumped between injection cycles. An additional advantage of this approach is the ability to use a carrier solution that also serves as a wash solution (being more acidic or basic, or containing detergents, or complex forming species), which decontaminates the system between injection cycles.

By a proper selection of pumping rates, the sample and reagent zones might merge synchronously, that is, completely overlapping each other (as shown) or asynchronously, so that a section of a reagent zone does

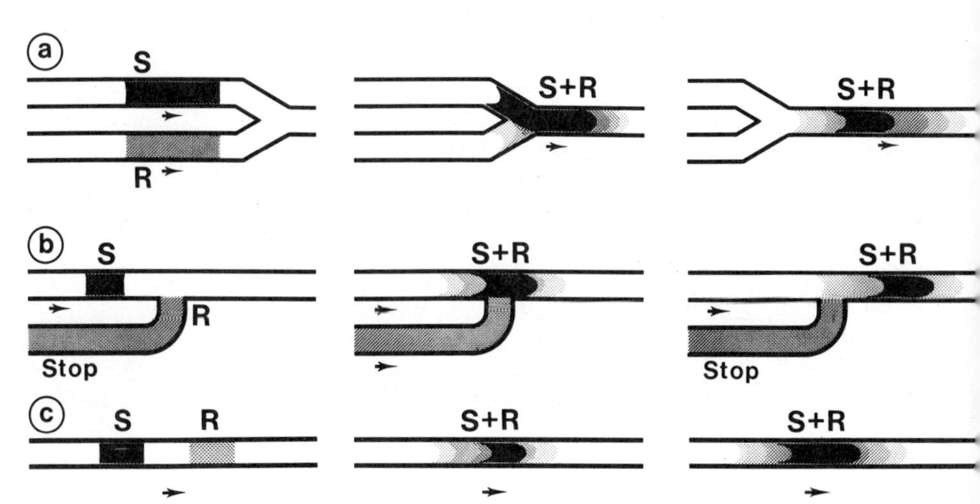

Figure 2.25. The principle of zone merging (*a*); intermittent pumping (*b*); and zone penetration (*c*); showing, from left to right, the sample (*S*) zone and the reagent (*R*) zone separately, during the initial contact, and then merged further downstream.

not come into contact with a sample zone and vice versa. The second approach is more information rich than the first one, since the composite zone contains three pieces of information: reagent blank value, reaction product readout, and sample blank value, accessible as horizontal slices at appropriate delay times. An alternative way to merge zones is to use *intermittent pumping* (Fig. 2.25*b*) [71]. Instead of using a double-loop valve, two pumps and a single-loop valve are used together with an appropriate timing device that allows the stop/go period of the reagent pump to be selected so that sample and reagent zones merge at a confluence point. This approach is more versatile than merging with a double-loop valve, since computer control allows flexibility in the degree of zone overlap and synchronization (cf. Section 4.5).

Zone penetration [378] is the simplest and most precise way to produce a composite zone (Fig. 2.25*c*). It relies on the central streamline in the flow channel moving twice as fast as the medium flow velocity. Thus, when zones of reagent and sample are injected simultaneously close together into the same channel, then the upstream zone will penetrate the downstream zone from behind, and their mutual overlap will increase with the length of travel. By using a double-loop injection valve and a single pump, a very precise degree of merging is achieved, because it is the geometry of the valve and the distance between the injected zones, together with the flow rate of the carrier stream, that determine zone overlap. These parameters are easy to maintain unchanged during prolonged measuring periods rather than the flow velocities in two different channels, generated by two pump tubes (cf. merging zones). Commuting [1060] and six-port valves (Section 5.1) allow complex gradients, composed of up to three zones, to be injected into a single carrier stream. Still another economical way of introducing sample and reagent solutions into a FIA system is the *split-loop injection* technique [848] (Section 4.5.2).

A composite concentration gradient, formed from two zones C_A^0 and C_B^0 by mutual penetration (Fig. 2.26, top), can be exploited for analytical purposes when a range of mixing ratios of the two components A and B yield useful information. This is because the dispersion coefficients D_A and D_B for each zone will have a given value for any given delay time t, and, therefore, the ratio of the D_A/D_B values for these sample zones will remain constant for any given value of t_x. This property of the dispersion

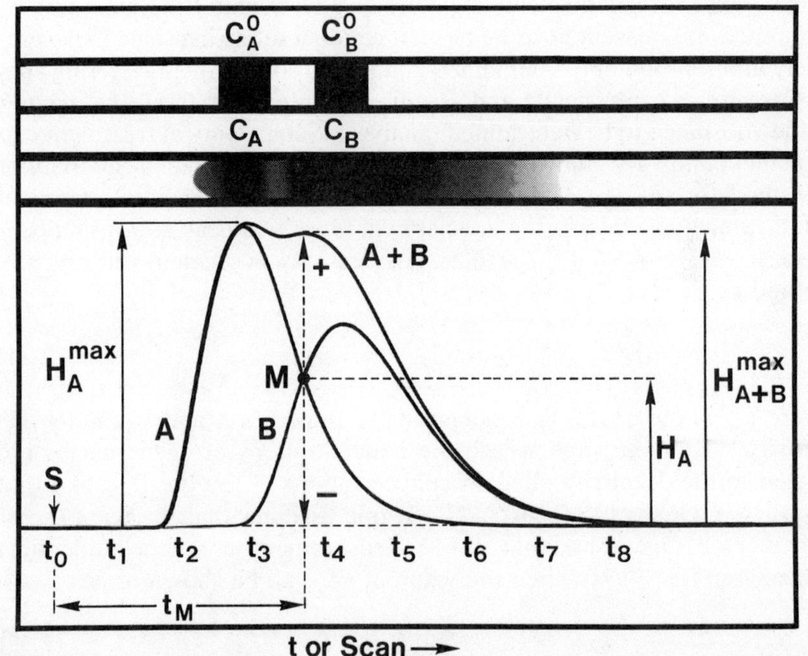

Figure 2.26. Zone penetration as used for interference studies (or for standard addition). Top: separately injected zones C_A^0 and C_B^0 mutually penetrate each other forming a composite zone (C_AC_B), and (below) separate (A, B) and composite (A + B) response curves. The H_{A+B} value measured at delay time t_M on the composite curve as compared to the H_A value of point M reflects the influence of B on the measurement of A (see text for details).

process has so far been exploited for two purposes:

1. To develop a generalized method for quantifying the extent of interference of a foreign species on the assay of an analyte by measuring the selectivity coefficient [378].
2. As a novel approach to performing multicomponent standard addition over a wide, controllable range of standard/analyte concentration ratios [817].

2.5.2 Selectivity Measurement

The *selectivity* of any chemical assay is its ability to provide a result unbiased by the presence of foreign species, and it is therefore crucial to the assay's practical application. Surprisingly, however, very few attempts have been made to describe selectivity of chemical assays in exact terms—except for ion-selective electrodes. To quantify the extent of interference, it is essential to be able to conduct all measurements at precisely and reproducibly maintained conditions so that the interfering species are treated, physically and chemically, in exactly the same manner as the substance to be determined (analyte). Such identical treatment can be achieved in FIA, and it thus makes sense to express the selectivity of a method for species A toward an interfering species B by a numerical value. Since any interfering species will always appear as a pseudoanalyte, it was suggested [378] that the selectivity coefficient for any FIA method be defined by

$$C'_A = C_A + k_{AB}C_B \tag{2.31}$$

where C'_A is the apparent concentration of species A and k_{AB} is the selectivity coefficient, and where the concentrations of each species are related to the originally injected concentrations C^0 by the D value, that is, $D_A = C^0_A/C_A$ and $D_B = C^0_B/C_B$. If one assumes that the response of the detector is linear and that all FIA measurements are made at identical dispersion ($D_A = D_B$), then the value of k_{AB} can be shown to be

$$k_{AB} = (C^0_A/C^0_B)[(H_{A+B}/H_A) - 1] \tag{2.32}$$

where H_A and H_{A+B} are the peak heights recorded by the detector from injection of species A alone and A plus interfering species B, respectively. Thus, the higher the k_{AB} value, the greater will be the interference of the foreign species B in measurement of the primary species A. The interfering effect may be either positive (more species A found than actually present) or negative (depression of signal generated by A in the presence of B), and, therefore, the selectivity coefficient may have either a positive or a negative value.

Zone penetration is an ideal tool for measuring selectivity coefficients, since the method allows readouts to be taken (1) at that vertical slice of the composite zone where the dispersion coefficients are *equal* ($D_A = D_B$ at point M Fig. 2.26, bottom) and (2) at the peak maximum of the pure A component. The concentrations within such a composite zone and position of the point M at a time t_M are readily established by an experiment where zone A is first injected alone and peak A is recorded by a detector of choice. (If chemical reactions are involved, like a reaction with a suitable reagent for colorimetric detection, a suitable manifold and colorimetric detector are used—cf. Section 4.5.2). Next, zone B is injected alone and peak B is recorded. Provided that the same solution of analyte is injected in both runs A and B, and that the detector responds linearly to the injected species, this experiment yields, (a) the isodispersion point M within the time concentration matrix, which is identified via time delay t_M, and (b) the peak height H_A for the response of the pure species A at the time t_M, because it equals the horizontal distance between point M and the baseline (Fig. 2.26).

Furthermore, since the peak maximum for pure A has been found, and since the ratio between H_A and $H_A^{\max}$ is a constant, (because $D_A^{\max}/D_M$ = const), the selectivity constant k_{AB} may be calculated readily from subsequent experimental runs where zone A (the analyte injected at concentration C_A^0) and zone B (the interferrent injected at concentration C_B^0) are injected simultaneously and allowed to penetrate (Fig. 2.26, top). If species B is *not* sensed by the detector, the peak B will *not* show and, hence, $H_A = H_{A+B}$. In the special case when species B will be sensed by detector *exactly* as species A, and when injected at the same concentration ($C_A^0 = C_B^0$), a composite curve $A + B$ as shown in Fig. 2.26, bottom, will be obtained. As seen H_A will increase (+), so that $H_{A+B} = 2H_A$ and k = 2. Generally, from the value of H_{A+B} and the known concentrations of A and B, k_{AB} can be computed from Eq. (2.32). For negative interferences, H_{A+B} will decrease (−) below the value of H_A, and the k_{AB} value will be less than 1.

2.5.3 Standard Addition

The method of using an internal standard, useful whenever the detector response has to be checked continually, can be implemented by the preceding approach, since the peak maximum for pure A may serve as the reference point for each injection cycle. By injecting the same standard solution as zone A in all experiments and injecting samples containing an unknown level of analyte as zone B, the validity of a previously established calibration curve may be checked for each separate sampling cycle, since the ratio of the D values (and, therefore, the ratio of the H values) within the composite concentration gradient remains unaltered through all subsequent measuring cycles. Thus, $H_A^{\max}/H_B^{\max}$ = const and the calibration curve for B is given by $C = (k + H_B^{\max})$. Therefore, $H_B^{\max}$ can be corrected continuously for variations of detector response as reflected in peak height $H_A^{\max}$. In contrast to the preceding selectivity measurement outlined, the degree of zone penetration for the internal standard method should be minimized, since the wider the overlap of zone B with zone A, the higher the value of k, and the calibration curve for $C_B = 0$ will show a higher blank value.

Standard addition is another technique where zone penetration is useful. A sample containing an unknown level of analyte is injected as zone A, while a selected concentration of a standard is injected as zone B. By adjusting a suitable zone overlap a composite zone will be obtained, which may be evaluated by taking the readout either at the equidispersion point (t_M), in which case a formula equivalent to (2.32) can be used for calculations, or it may be taken at any other point along the interface of the two gradients, provided that the ratio of the dispersion coefficients at each selected delay time is known. This is readily accomplished using the same approach as described for selectivity measurement.

Another way of performing standard addition was described [817] in conjunction with multielement analysis in atomic emission spectrometry. The principle is to some extent similar to that outlined for the selectivity measurements, except that the two zones are injected into an inert carrier stream, the sample zone A being chased by such a long zone of standard solution B that $D_B \rightarrow 1$ is reached (Fig. 2.27, top). Thus, if the experimental conditions are selected so that sample zone A is not entirely penetrated by standard zone B, the rising part of the sample peak A is diluted solely with the inert carrier stream (water) and does not contain any standard solution, while the tailing section of zone A is diluted solely with standard solution B. Therefore, if readouts corresponding to sections having identical D values at the front and tailing part of the sample zone are taken (at times t_A and t_{A+B}), the amount of standard added to the sample at t_{A+B} can be deduced, since

$$\frac{1}{D_B} = 1 - \frac{1}{D_A} \qquad (2.33)$$

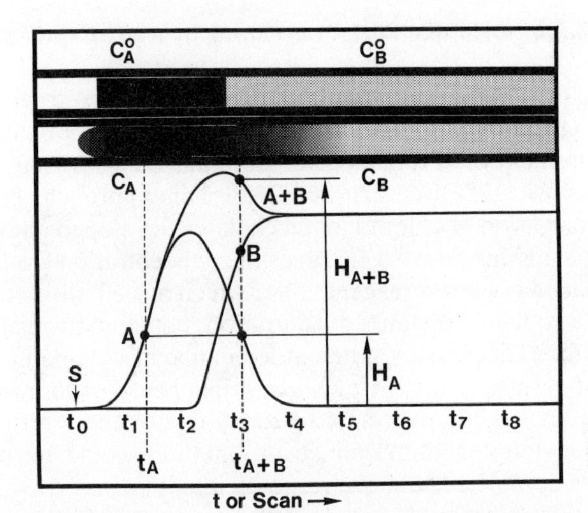

Figure 2.27. Standard addition by zone penetration. Separately injected zones of a sample C_A^0 and a standard C_B^0 solution disperse into C_A and C_B, which mutually penetrate into each other forming a composite zone $A/A + B/B$, yielding a composite peak $A + B$. Readouts at t_A and t_{A+B} yield values for pure A and its mixture with B due to standard addition of B. (See text for details.)

where D_A is the dispersion of the sample at times t_A and t_{A+B}, and D_B is the dispersion of the standard at time t_{A+B} on the tailing part of the peak. Assuming a linear calibration curve for the species determined, the peak heights H_A and H_{A+B} for the responses on the rising part and falling part of the peak can be related to the concentration of analyte in the sample, C_A^0, and standard solution, C_B^0. This leads to

$$C_A^0 = H_{A+B}C_B^0(1 - d)/[d(H_A - H_{A+B})] \tag{2.34}$$

where $d = 1/D_A$. Since the concentration of the standard solution is known and the dispersion coefficient of the section chosen can be measured in the usual manner, the expression $C_B(1 - d)$ will be a known constant, K. Hence

$$C_A^0 = K[H_{A+B}/(H_A - H_{A+B})] \tag{2.35}$$

which indicates that the sample concentration can be determined by standard addition upon a single injection measuring the response H_{A+B} and H_A at times t_A and t_{A+B}, respectively. Since selection of a t value on the composed gradient will provide different sample/standard ratios, the A/B ratio can be "electronically" optimized for obtaining the highest precision.

2.5.4 Multiple Peaks Developed by a pH Gradient

By introducing a pH gradient into a sample zone mixed with a suitable color-forming reagent, a multicomponent readout is obtained, allowing rapid measurement of several species simultaneously present in the same sample zone [87, 135, 162, 220 and 232]. This approach, based on the sequential formation of colored metal chelates developed and generalized by Betteridge and his co-workers, uses the concentration gradient formed at the interface between a reagent zone, which has a built-in pH gradient, and a sample zone of constant concentration, obtained by injecting a large sample volume. This reagent gradient technique has the potential for multicomponent assays, since it yields a multiple peak readout from a single sample zone. It would be interesting to try this method in the zone penetration mode (cf. Fig. 2.27), since the method would become experimentally less complex and more reproducible.

2.6 SCALING OF FIA SYSTEMS

The most frequently used dimensions and volumes of a laboratory FIA system are channel length up to 1 m, channel inside diameter of 0.5 mm, pumping rate up to 2 mL/min, and injected sample volume of 30 μL. For some applications, however, these dimensions have to be increased; such is the case for continuous monitoring in industry or for uses at sea, where the apparatus must be sturdy and the sample may contain solid particles, and, therefore, the use of channels with a larger cross section is advisable, large volumes of sample solutions being readily available fortunately. In other FIA applications sample solution is scarce and reagents may be very expensive and, therefore, the FIA system has to be miniaturized. In miniaturization, the problem is which dimensions and operating parameters can be altered without changing how solutions are handled and data gathered, so that macro- and micro-FIA systems can be rationally compared: the results of a chemical assay in a laboratory FIA system must be able to be transferred to macro- or micro-FIA systems.

2.6.1 Concept of Similarity

We have learned in previous sections that any flow-injection system is built around a flow channel of certain dimensions and geometrical form. It would therefore seem to be an easy matter to miniaturize it by simply scaling down existing manifolds. The difficulty is that such an approach will not yield microchannels, which would behave exactly like macrochannels, because a simple reduction of *all* dimensions does not produce channels that are physically similar. This may be better understood by reviewing briefly the concept of similarity as applied in fluid mechanics [2.6]. The use of similarity for scaling and modeling the behavior of fluids is based on three types of similarities.

The simplest, *geometric similarity*, is similarity of shape, and operates with a scaling factor that is the ratio of any length in one system to the corresponding length in the other system. However, when the channel for FIA is miniaturized, perfect geometrical similarity is impossible to obtain, because the roughness of the walls and other imperfections of the channels cannot be reduced proportionately when the overall dimensions are scaled down. Furthermore, an excessive reduction of the cross-sectional area of the channels is undesirable, because solid particles present in some sample materials (blood, fermentation liquids, etc.) could block a very narrow channel. For the same reason, if one would wish to miniaturize a coil or a packed or 3-D reactor, the concept of geometric similarity could not be applied.

Kinematic similarity is similarity of motion and implies geometric similarity together with similar time intervals. Its scaling factors are velocities and accelerations. This would be a useful tool, if flow-injection systems of geometric similarity were available, because kinematic similarity would allow pumping rates to be adjusted.

Dynamic similarity is a similarity of forces, which could also comprise a scaling factor; this type of similarity includes geometric and kinematic similarities, thus, being a valuable tool for comparison of scaled models (e.g., for aircraft and ships). Outside the field of fluid mechanics, but still involving fluid properties, are *thermal similarity*, which operates through differences in temperatures between model and prototype, and *chemical similarity*, where the fixed ratios of reactants at corresponding points in the flowing streams serve as a scaling factor. Thus, in order for two systems to behave similarly, certain *ratios of like magnitudes* must be fixed. Whatever quantities are chosen, the ratio of their magnitudes (i.e., the scaling factor) is dimensionless. Several scaling factors may be needed to describe a complex system like FIA; but, once identified, they will allow a rational design and comparison of two flow-injection channels regardless of their geometric dissimilarity.

2.6.2 Scaling Factors Applicable to FIA

The FIA response curve is a result of two processes, the *physical dispersion* of the sample material in the carrier stream and the *chemical reaction(s)* forming the species to be measured. Therefore, the *dispersion coefficient* D is a suitable scaling factor, since it describes the ratio of reactants in that element of fluid that is chosen to yield the analytical readout. Because the physical dispersion and the chemical reaction are both kinetic in nature, and because they take place simultaneously, commencing at the same time (the moment of injection t_0), they jointly yield a response curve, the *time* scale of which is the next scaling factor. For those FIA methods that are based on peak-height measurement, the residence time T measured from the point of injection S to the peak maximum (Fig. 2.4*b*), in conjunction with C^{max}, allows (via D^{max}) the known ratios of reactants at corresponding times to be compared even if they are obtained in geometrically dissimilar streams. Thus, D^{max} and T describe *chemical similarity of physically dissimilar* flow-injection systems. More generally, if a flow-injection procedure has been developed, optimized, and tested in a macrochannel, it can be accommodated in a microchannel, provided that both systems yield similar D and t values, regardless of whether we use vertical or horizontal readouts. Going one step farther, if the D and t values are identical for the micro- and macrosystem, then both systems will yield identical readouts, which means that, say, the same spectrophotometric procedure will yield the same peak height (in terms of absorbance) provided that identical concentrations of sample and reagent materials are used in both systems, that the flow cells have identical optical path lengths, and that the wavelength of the radiation to be absorbed is the same. For peak-width measurements, however, peak broadening must also be the same (see following discussion).

While D and t allow comparison of mixing ratios and reaction times, the last scaling factor to be identified must allow comparison of sample, reagent, and time consumption in geometrically dissimilar flow-injection channels, since even though miniaturization of a system results in a decrease of material consumption, it does not necessarily lead to the optimization of sampling frequency, since different channel geometry causes different zone broadening (cf. Fig. 2.10 and Chapter 3). It follows, however, from Section 2.2 that FIA systems with identical *dispersion factors* $\beta_{1/2}$ will yield, regardless of their dimensions and geometrical form, the same sampling frequency and that flow systems with lower dispersion factors will be more economical in terms of material and time consumption.

To summarize, the three scaling factors are the dispersion coefficient D, the residence time t, and the dispersion factor $\beta_{1/2} = S_{1/2}/V_r$. These factors were used for optimization of the design of integrated microconduits [608] (Section 4.12).

While the dispersion factor $\beta_{1/2}$ is determined mainly by the geometry of the flow channel and to a lesser extent by the flow velocity, D may be further manipulated by changing the injected sample volume, while t is effectively manipulated by changing the flow velocity (up to stopped flow). We learned previously that in addition to manipulating these parameters physically, a convenient "fine tuning" of the D and t values can be executed "electronically," by selecting desired sections of the concentration gradients found within the dispersed sample zones.

2.6.3 Scaling and Gradients

It was shown in previous sections how one may choose any element of the dispersed sample zone as a point of readout, if a finer adjustment of the D and t values is required. Therefore, even when two geometrically dissimilar flow-injection systems yield different concentration gradients so that peak heights C_A^{max} and C_B^{max} are quite different (Fig. 2.28), elements of fluid with identical D and t values can be found within them. By injecting identical concentrations C^0 of a dye separately into these two channels, two response curves can be recorded from the same starting point (S). The peak-height measurement for channel A in this example has $D_A^{max} = C^0/C_A^{max} = 2.4$ and $T_A = 7.0$ s, whereas channel B has $D_B^{max} = C^0/C_B^{max} = 3.4$ and $T_B = 12.3$ s. These are the minimum D values obtainable in the given channel geometries, for the given sample volume

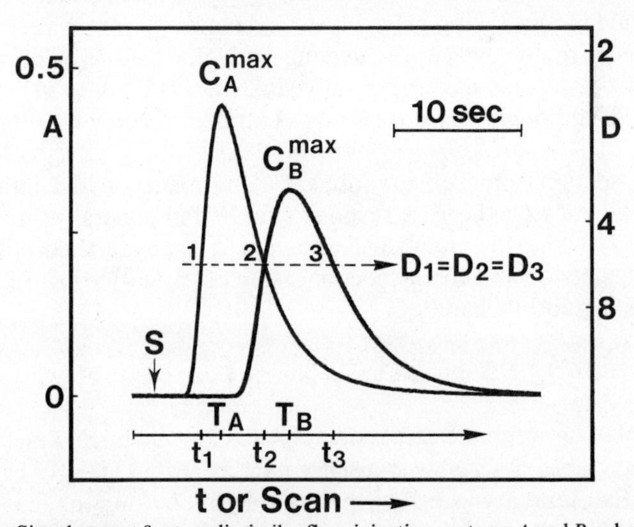

Figure 2.28. Signal output for two dissimilar flow-injection systems A and B, where identical dye concentrations were injected, the responses being recorded from the same starting point (S). The two peak maxima, C_A^{max} and C_B^{max}, are related to the two residence times, T_A and T_B, while times t_1, t_2, and t_3 correspond to fluid elements (1, 2, and 3) with identical dispersion coefficients situated along the gradients of the dispersed sample zones.

injected. Higher D values can be obtained by selecting the readout at times other than those corresponding to the peak maximum. For illustration, isodispersion points 1, 2, and 3 all have identical values ($D_1 = D_2 = D_3 = 5.2$), that is, the same mixing ratio between sample and reagent (contained in the carrier stream). Point 2 also represents identical reaction times ($t_2 = 10.3$ s) for the two systems A and B; thus, the readouts obtained at point 2 will be identical for both A and B channels, regardless of how much these channels may differ geometrically.

2.7 CHEMICAL KINETICS IN A FIA SYSTEM

Most FIA methods are based on the use of chemical reactions, the products of which are measurable by a detector of choice. Indeed, FIA is useful only because it can accommodate such a wide variety of chemistries. Thus, in most cases, a FIA peak is a result of two processes: of the physical dispersion, discussed in previous sections, and of subsequent chemical reactions. These two kinectic processes occur *simultaneously* in any flow system; yet, in FIA their mutual interaction is very complex, since the dispersed zones are not homogeneously mixed, but are composed from concentration gradients formed by *gradual penetration* of reacting species in both axial and radial directions. An exact description of chemical kinetics taking place in FIA system is therefore very difficult, and this is why so few papers dealing with the theory of chemical kinetics in FIA systems have been published [150, 151, 181, 391, 541, 554, 1064, 1065], although this problem is central to further development of FIA.

In this section some observations will be made on the influence of chemical kinetics on the FIA response curve and general principles will be outlined, while the published attempts to analyze theoretically the physical dispersion and chemical conversion of an analyte in a FIA system will be discussed in Chapter 3.

2.7.1 Chemical Conversion at Continuous Flow

The interplay between the dispersion process and chemical kinetics forming the detectable species was observed and reported in early FIA works [2, 4]. At that time it was established that by increasing the length of the mixing coil (L), the peak height will initially increase, and, after reaching a maximum, it will gradually decrease (Fig. 2.29). The importance of reactor length optimization was thus recognized, since too short or too

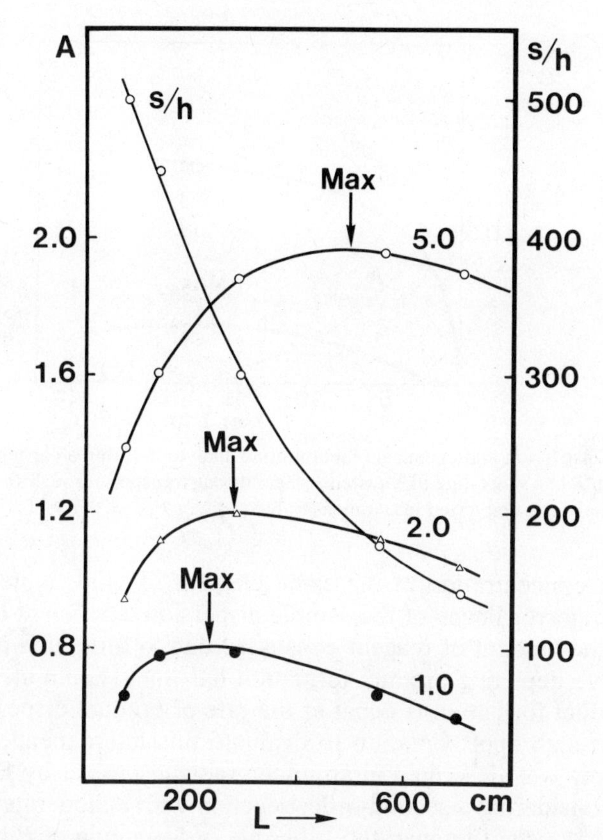

Figure 2.29. The influence of coil length on the sensitivy of chloride determination based on release of thiocyanate from mercurithiocyanate and subsequent formation of Fe(III)thiocyanate (monitored at 480 nm). Single-line manifold, 1, 0, 2.0, and 5 mEq Cl^-/L. Note how the sampling rate decreases with increasing coil length L [4].

long L results in loss of sensitivity, and too long L causes a loss of sampling frequency and wastes reagent.

The relation between sample (A) dispersion and its conversion by reaction with reagent R into a measurable product P

$$A + R \rightarrow P$$

is depicted in Fig. 2.30 as the dependence on the coil length (L), incorporated into a single-line manifold (cf. Fig. 2.13). Curve A depicts the increase of the sample dispersion coefficient in the element of fluid corresponding to the peak maximum, while curve R shows the increase of

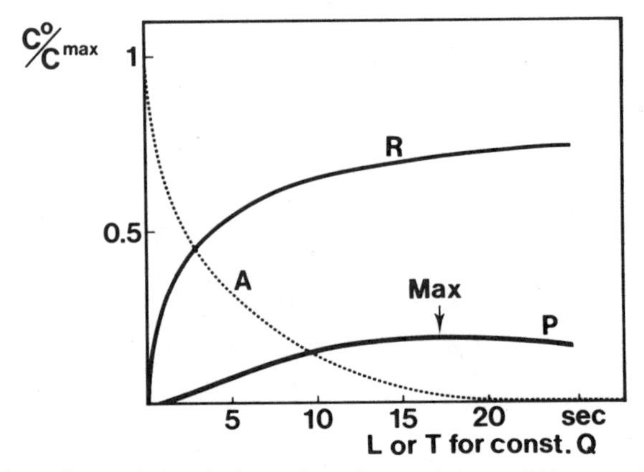

Figure 2.30. Dispersion and chemical reaction of an analyte A within a carrier stream containing reagent R in a single-line FIA system, where during transport through an open tubular reactor the analyte is converted into a measurable product P.

the reagent concentration in the same element of fluid. Note that this curve is the mirror image of the sample dispersion coefficient curve corrected for the amount of reagent consumed due to formation of product P. The curve depicting product formation has a maximum at which the rate of product formation is equal to the rate of product dispersion. The first, very rough approximation to evaluate this interdependence more quantitatively was described in an undergraduate project by Haagensen [153]. By considering a second-order reaction with a moderate excess of reagent at a constant temperature, the rate of formation of P is

$$dC_P \text{ (formed)} = v(C_A, C_R)\, dt = kC_AC_R\, dt \tag{2.36}$$

where k is the reaction rate constant. At time $t = 0$, that is, immediately before injection takes place, the concentration of reagent in the carrier stream is C_R^0 and that of A in the sample solution is C_A^0, while $C_P = C_P^0 = 0$.

After sample injection, the concentrations of A and R at time t, because of the dispersion and reaction taking place, will be given by

$$C_A = \frac{C_A^0}{D(t)} - C_P \tag{2.37}$$

and

$$C_R = \left(1 - \frac{1}{D(t)}\right) C_R^0 - C_P \tag{2.38}$$

where C_P is the concentration of the product at time t and $D(t)$ is the dispersion coefficient at time t. While P is being formed by the chemical reaction between A and R, it also undergoes dispersion simultaneously. This contribution cannot be expressed in the same way as the sample material dispersion coefficient, because at time $t = 0$ there is no product present in the sample zone; an equivalent situation can be simulated as if the product was injected at concentration C_{iP}^0 into a FIA system at time $t = 0$. Then concentration C_{iP} at time t is

$$C_{iP} = \frac{C_{iP}^0}{D(t)} \tag{2.39}$$

which in turn yields, for the rate of dispersion of the injected product,

$$dC_{iP}\,(\text{dispersed}) = -\frac{C_{iP}^0}{D(t)^2}\left(\frac{\partial D(t)}{\partial t}\right) dt \tag{2.40}$$

$$= -\frac{C_{iP}}{D(t)}\left(\frac{\partial D(t)}{\partial t}\right) dt \tag{2.41}$$

where C_{iP} corresponds to the concentration of dispersing product at any given time t. Therefore we may write

$$dC_P\,(\text{dispersed}) = -\frac{C_P}{D(t)}\left(\frac{\partial D(t)}{\partial t}\right) dt \tag{2.42}$$

Hence, for the overall balance of the product at any time

$$dC_P = dC_p\,(\text{formed}) + dC_P\,(\text{dispersed}) \tag{2.43}$$

which in combination with the preceding equations yields for $t = T$ (peak maximum)

$$dC_P = k\left(\frac{C_A^0}{D^{\max}} - C_P\right)\left[\left(\frac{D^{\max} - 1}{D^{\max}}\right) C_R^0 - C_P\right] dT - \frac{C_P}{D^{\max}}\left(\frac{\partial D^{\max}}{\partial T}\right) \mathrm{d}T \tag{2.44}$$

where $D(T) = D^{\max}$. Thus, from $D^{\max}$ (T), k, and the starting concentrations, Eq. (2.42) yields, after numerical integration, the product formation curve $P(t)$, which has a maximum at that residence time where

$dC_P = 0$ (Fig. 2.31). This optimum coil length L will be shorter for faster chemical reactions and longer for the slower reaction rates.

Values of k and $D(t)$ have to be known to compute the product formation curve; these values have to be experimentally determined for a given set of chemical reactions and FIA systems. An example of such an approach is the determination of amino acids with *o*-phthalaldehyde as reagent using fluorescence detection. With this reagent, various amino acids form an intensely blue fluorescing product, which, when measured by a flowthrough fluorimeter incorporated into a single-line system, yields a peak depending on the amount of product formed. These peak heights H, as obtained with increasing coil lengths L, are shown in Fig. 2.31*a*, forming a dependence analogous to the one shown in Fig. 2.29. Since different amino acids react at different rates with *o*-phthalaldehyde, the obtained H–L dependencies have maxima at different L (or T values for constant pumping rate Q). These curves can then be numerically converted by means of Eq. (2.44) into reaction rate curves (Fig. 2.31*b*) using D^{max} values obtained by a dispersion experiment in the same single-line FIA system.

Considerably more refined approaches have been developed. Some of the approaches are based on detailed mathematical models of dispersion processes and the usual kinetic equations for first-order chemical reactions [554, 1065], which are subsequently combined as previously de-

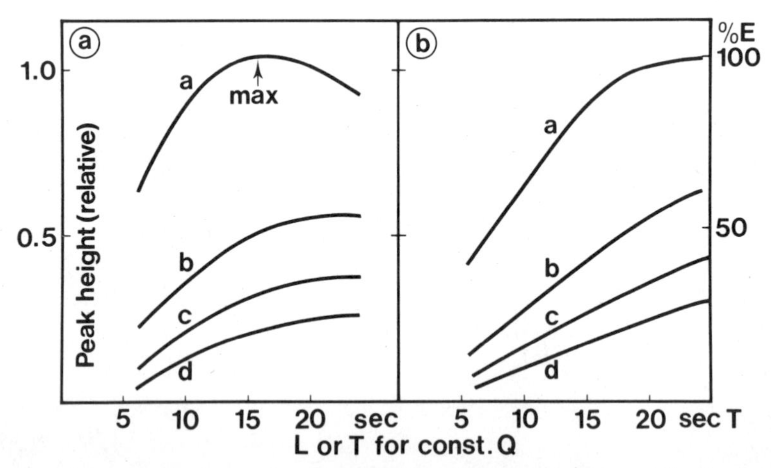

Figure 2.31. (*a*) Dependence of peak height on coil length L measured for the reaction between *o*-phthalaldehyde and four aminoacids: *a*, glycine; *b*, serine; *c*, phenylalanine; *d*, asparagine; (*b*) Reaction rate curves computed from measured D values and from data of (*a*) by means of Eq. (2.44). The increasing slope of the curves from *d* to *a* shows the increase of reaction rate. (%E refers to the degree of conversion reached.)

scribed. These works will be discussed in more detail in Chapter 3; however, since any theoretical consideration developed so far requires a knowledge of physical and chemical constants, which can only be found by experiments, the application of these predictive methods for optimization of real FIA systems is rather limited at present.

2.7.2 Evaluation of Conversion by Stopped-Flow FIA

Since any FIA response curve is a result of physical dispersion and chemical conversion one may write:

$$\text{response} = (C_A^0 K) \times \text{dispersion coefficient} \times \text{conversion factor} \quad (2.45)$$

where K is the proportionality constant between detector response and concentration of the product formed by conversion of the analyte of original concentration C_A^0. As was mentioned previously, the individual peak maxima may be computed from the dispersion and conversion functions of the reactor length, and, for the optimum coil length, it implies that $d(\text{response})/dL = 0$. In practice, however, many approximations must be made in order to make such a computation feasible, and since these approximations accumulate during a search for a manageable mathematical function, the theoretical model becomes progressively more removed from a real FIA system. There is, however, a practical way to circumvent these problems experimentally, that is, by using the stopped-flow method.

Let us consider first a very *fast* chemical reaction, which in a FIA system yields a response curve as the one obtained with the product, if injected as such (Fig. 2.32*a*, curve *A*). This is a response analogous to the familiar dispersion curves discussed in Sections 2.2, 2.3, 2.4, and 2.5, the only difference being that a curve obtained with the injected product *P* truly reflects a response of the detector to the converted analyte in question (rather than to a tracer dye—as discussed previously), and, therefore, the $C_S^0 K$ value becomes incorporated into the peak-height measurement.

Now, if the rate of chemical reaction is *slower* than the rate of sample zone dispersion in a given system, a response curve *B*, as shown in Fig. 2.32*a*, will be recorded, representing a peak *lower* than curve *A*, which would be recorded with the product *P*, if injected as such. (Note that the dispersion curve *A*, for a slowly formed product, could be obtained when preparing the product by mixing the standard and the reagent solutions in a test tube, by letting them react to completion, and by injecting the thus obtained product into the FIA system.) The difference observed between curve *A* and *B* is due to a lower degree of chemical conversion of

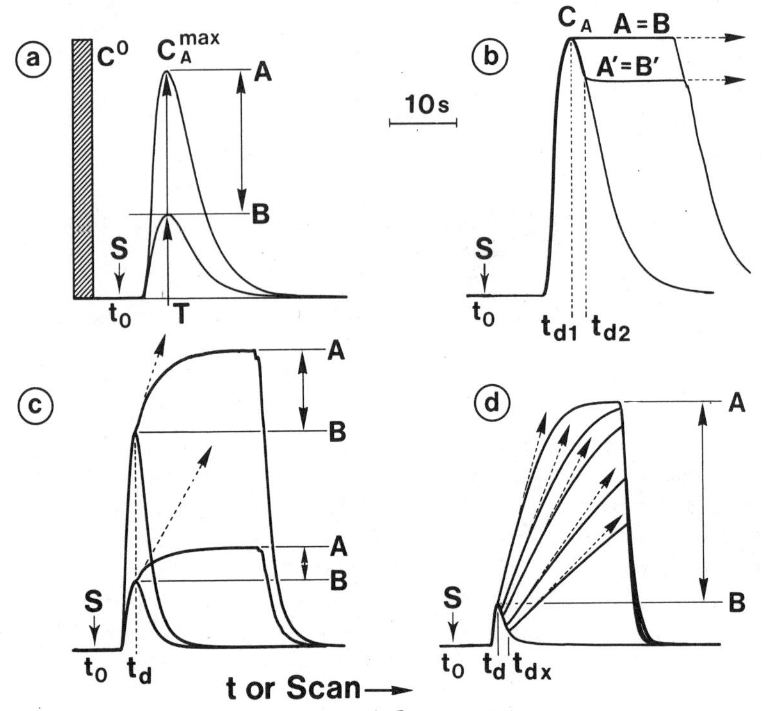

Figure 2.32. Fractional conversion measured by stopped-flow FIA. (*a*) Response curves recorded in continuous-flow mode from the same starting point *S*, where *A* represents a situation where the injected analyte undergoes instantaneous conversion while *B* represents a situation where the analyte at the time of detection is only partially converted into a measurable product *P*. (*b*) Response curves obtained in a continuous-flow mode, and by a stopped flow (horizontal lines), when a measurable product is rapidly formed, so that complete conversion is obtained in all elements of the dispersed sample. Therefore, an additional reaction time, during the stopped-flow period, will not increase the fractional conversion, since $X_A = 1$ and, therefore, $A = B$ and $A' = B'$. (*c*) Peaks obtained by continuous-flow and stopped-flow modes, when a measurable product is formed slowly, so that an incomplete conversion is obtained in all elements of the dispersed sample zone and, therefore, an additional reaction time during the stopped-flow period results in an increase of the fractional conversion and of recorded signal. Hence, $A < B$ at both low and high analyte level. (*d*) A stopped-flow experiment where by increasing stopped-flow delay time t_d to t_x changing substrate/enzyme mixing ratios are obtained, resulting in a decreasing slope of the reaction rate curves (dotted line). While the maximum response level *A* is obtained fastest for the shortest t_d times, longer t_d values yield curves with lower slopes due to different initial mixing ratios. Therefore, fractional conversions must be computed for each of the reaction rate curves separately. *S* is point of injection at t_0, when analyte of original concentration C_A^0 has been injected; C_A^{max} is the concentration of dispersed analyte at peak maximum (corresponding to the concentration of *P* at $X_A = 1$).

analyte into measurable product. This fractional conversion X_A [2.7] of reactant A is defined as the fraction of reactant converted into product:

$$X_A = \frac{N_A^0 - N_A}{N_A^0} \qquad \text{or} \qquad X_A = \frac{B}{A} \tag{2.46}$$

which for a first-order reaction (analyte → product) is

$$-\ln(1 - X_A) = kt \qquad \text{or} \qquad -\ln[1 - (B/A)] = kt \tag{2.47}$$

and for a second-order reaction (analyte + R → product) is

$$-\ln(1 - X_A) = (C_R^0 - C_A)kt - \ln(1 - X_R) \qquad \text{or}$$
$$-\ln[1 - (B/A)] = (C_R^0 - C_A)kt - \ln(1 - X_R) \tag{2.48}$$

and, therefore, the degree of conversion will be higher the higher the reaction rate constant (k) is, and the longer the reaction time t will be.

It would be convenient to assume that a large excess of reagent R originally present (C_R^0) is instantaneously mixed with sample material (A), so that the pseudo-first-order-reaction conditions prevailed and, hence, rate equation (2.47) could be used, but this is seldom the case in a real FIA system. [If it were the case, then the position of the maximum $d(\text{response})/dL = 0$ in Fig. 2.29 would *not* change with the concentration of the injected sample!] Thus, for theoretical prediction of the fractional conversion in a FIA system, a correct reaction rate equation must be selected, and the value of the reaction constant must be known.

However, the stopped-flow FIA technique allows the fractional conversion to be measured experimentally for any FIA system by resolving the contribution of the dispersion process and of chemical kinetics. By stopping the flow (cf. Sections 2.4.3 and 4.3) any element of the dispersed sample zone can be selected and arrested in the flow cell, where the change of response with time [i.e., $d(\text{response})/dt$] can be monitored continuously.

For a *fast* chemical reaction, such as the determination of calcium by *o*-cresolphthalein using colorimetric detection at 580 nm [253], shown in Fig. 2.32*b*, the product formation is so rapid that the response curve is identical with the dispersion curve obtained if product were injected as such. This is readily seen from the zero slope ($dR/dt = 0$) of the reaction rate curve $A = B$ recorded with the element of fluid at the peak top (at C_A^{max}, using delay time t_{d1}). The same stopped-flow experiment repeated at the peak gradient (at C_A, using delay time t_{d2}) again has a zero slope

($A' = B'$), confirming that a complete conversion has been obtained throughout the entire sample zone. Thus, since $X_A = 1$ (at t_{d1}), the degree of conversion and therefore the response cannot be increased by prolonging the resident time T or length L (for continuous pumping), but the response could conceivably be increased by decreasing L, since this would lead to decrease of the D value (cf. Fig. 2.6*b*). Alternatively, an increase of S_V (Fig. 2.6*a*) will increase the response as long as an adequate excess of reagent is maintained.

For *slower* chemical reactions (Fig. 2.32*c*), such as the determination of chromium by 1,5-diphenylcarbazide in acid medium, the stopped-flow experiment conveniently reveals the degree of chemical conversion. By stopping the flow at a delay time t_d, the element of fluid with the highest analyte/reagent ratio is arrested in the flow cell and the initial reaction rate dR/dt (broken line) together with the steady-state level of response A is recorded. Since this level A is the one corresponding to complete conversion of analyte into the product, back extrapolation to t_d shows the difference between B and A and yields the fractional conversion of the analyte. Note that in this experiment the position of the peak maxima for the two injected analyte concentrations (2.5 and 10 ppm Cr) coincide and the fractional conversion for both A/B levels are practically identical ($X_A = 0.72$ and 0.70), indicating that a sufficient reagent excess exists throughout the dispersed zone at both analyte concentrations. Thus any increase of reagent concentration would not increase the degree of conversion, while longer reaction times will. It would be unwise, however, to achieve this by increasing L, since dispersion will also increase, thus partially (or entirely) offsetting the gain in the fractional conversion. However, by stopping the flow, maximum sensitivity of measurement will be most economically achieved.

The diagnostic power of the stopped-flow FIA method is further demonstrated by an example of the estimation of catalytic activity of lactate dehydrogenase (LDH) on the oxidation of pyruvate to lactate, monitored by means of NADH (Fig. 2.32*d*). By selecting increasing delay times t_d to t_{dx} successive elements of the dispersed sample zone with increasing (NADH + pyruvate)/LDH ratios have been arrested in the flow cell (cf. Section 2.4.3) and therefore the thus monitored reaction rate curves (dotted arrows) show decreasing slopes with increasing t_d. All rate curves converge to the same level (A), which is the limit of conversion set either by inhibition of the enzyme by the lactate produced or by the availability of the coenzyme used. If by decreasing the initial concentration of NAD^+ in the carrier stream the reaction rate curves would converge to a lower level than A, then it would prove that it is the coenzyme availability that sets the limit for the reaction rate curve (and product inhibition by the product).

To summarize the stopped-flow FIA technique allows:

a. Exact measurement of the fractional conversion X_A.
b. Optimization of reaction conditions through selection of initial reagent concentrations, of injected sample volume, coil lengths, and pumping rates for continuous-flow FIA by comparing conversion values X_A obtained when changing these parameters.
c. Selection of reactant ratios by selecting delay times t_d and of reaction times by choice of the length of the stopped-flow period.
d. Measurement of reaction rates and investigation of reaction orders.

It follows from the foregoing that the principal limitation of the stopped-flow FIA technique is that it can be used for measurement of kinetic assays only, when the reaction rate is *slower* than the rate of dispersion. A manifold should therefore have a very short channel, which is designed to provide an intense radial mixing, thus ensuring that during mixing of reactants and their subsequent transport into the flow cell, the fractional conversion is minimized. For an exact measurement of k values an additional precaution must be taken, ensuring that the initial concentrations of reactants (C_A^0 and C_R^0) at the initial contact time t_i are known. This is because *every* element of the dispersing sample zone *gradually* moves through the stages of limited, medium, and large dispersion, and, therefore, if a single-line manifold is used, *every* element of A is *gradually mixed*, from the moment of injection (t_0) first with a substoichiometric, then with an equivalent, and, finally, with a superstoichiometric quantity (excess) of reagent, thus making the true initial contact time t_i ill defined (and certainly different from t_0). This does not matter for kinetic assays, since the FIA system is always calibrated by standard solutions and *what happens to one sample happens in the exactly the same way to any other sample*. For exact measurement of reaction rates and investigation of reaction kinetics, however, an instantaneous mixing of reactants at an exactly defined initial time must be ensured. This can be achieved in a specially designed two-line manifold [838] (Chapter 4.3, Fig. 4.15), where the zone of sample A is first allowed to disperse and then merged with a stream of reagent R and instantaneously mixed prior to monitoring of the reaction rate by stopped flow.

2.7.3 Kinetic Assays, Catalytic Assays, and the Kinetic Advantage of FIA

Peak height at continuous flow or fixed-time reaction rate measurement at stopped flow are the most frequently used parameters of the FIA response curve for reaction-rate-based assays.

Use of peak area or peak width at continuous flow has not yet been reported, although it is potentially useful. Most frequent applications of kinetic assays (cf. Chapter 7) are aimed at the enhancement of *sensitivity* of measurement or at the *blank correction* offered by the kinetic discrimination between analyte-nonrelated initial response of sample material and the analyte-related reaction rate response due to formation of the measurable product.

In addition to that also the classical approach to enhancement of sensitivity of determination of traces of species, by letting them act as catalysts to promote formation of a measurable product, has been performed successfully by FIA modes [915, 1009, 1033, 1041, 1121, 1134, 1135, 1159, 1186, 1203, 1216]. These ultrasensitive catalytic assays benefit from the rigorous mixing and timing of all operations performed in a FIA system, and also from the protective environment of the flow channel, where cross contamination and influence of ambient environment is minimized.

Solution handling in the FIA mode is beneficial to all types of assays that rely on controlled kinetic processes, like reaction rates, catalysis, diffusion of ions or gases through membranes, and rates of adsorption and desorption, as well as on response characteristiques of diffusion-controlled detectors. It was Pacey et al. [1073, 1122, 1242] who pointed out that differences in kinetic behavior can be exploited with advantage by FIA. In kinetic discrimination, the differences in the rates of reactions of the reagent with the analyte of interest and the interferents are exploited. In kinetic enhancement, the chemical reactions involved are judiciously driven in the direction appropriate to the analyte of interest. While in batch chemistry the processes are forced to equilibrium, and therefore subtle differences between reaction rates cannot be exploited, small differences in reaction rates with the same reagent in FIA mode will result in different sensitivities of measurement. An example of these kinetic advantages of FIA is the determination of chlorine dioxide in the presence of chlorine with luminol. Chlorine dioxide reacts with luminol extremely quickly, at eleveated pH, while chlorine reacts so much slower that a selectivity factor of over 500 can be achieved. An additional increase in selectivity is obtained by letting the gases diffuse through a microporous Teflon membrane, where the diffusion rate for chlorine dioxide is 3 : 1 times favored over that of chlorine. Thus, in combination with the luminol reaction, 1500 enhancement of selectivity is obtained by kinetic means.

By combining the FIA gradient techniques, dedicated separation techniques, and kinetic discrimination, novel selective methods will be introduced in the future. Some of them will be improved versions of existing methods, while others might be based on rediscovery of old chemical methods, which, although potentially selective, were discarded because they were impractical when handled manually. The work of research teams in the United States [1122] in kinetic discrimination and in the Netherlands [1112] on the use of special reagents is significant in this context.

REFERENCES

2.1 G. Nagy, Zs. Feher, and E. Pungor, *Anal. Chim. Acta,* **78** (1974) 425.

2.2 V. V. S. Esware-Dutt and H. M. Mottola, *Anal. Chem.,* **49** (1977) 776.

2.3 J. F. K. Huber, K. M. Jonker, and H. Poppe, *Anal. Chem.,* **52** (1980) 2.

2.4 H. Engelhardt and U. D. Neue, *Chromatografia,* **15** (1982) 403.

2.5 I. M. Kolthoff, D. S. Leussing, and T. S. Lee, *J. Am. Chem. Soc.,* **72** (1950) 2173.

2.6 B. S. Massey, *Mechanics of Fluids,* 3rd ed. Van Nostrand Reinhold, New York (1975).

2.7 O. Levenspiel, *Chemical Reaction Engineering,* 2nd ed., Chaps. 3 and 10. Wiley, New York (1972).

CHAPTER

3

THEORETICAL ASPECTS OF FIA

Theory guides, experiment decides.

I. M. Kolthoff

Working on a motorcycle, working well, caring, is to become part of a process, to achieve an inner peace of mind. The motorcycle is primarily a mental phenomenon.

R. M. Pirsig, *Zen and the Art of Motorcycle Maintenance*

3.1 THEORY OF DISPERSION RELATED TO FIA

The flow injection analysis (FIA) response curve is a result of two processes, both kinetic in nature: the *physical* process of dispersion of the sample zone within the carrier stream and the *chemical* process of formation of a chemical species. These two processes occur *simultaneously*, and they yield, together with the dynamic characteristics of the detector, the FIA response curve. Simultaneous dispersion and chemical reaction have been studied in flow systems as used in chemical reaction engineering and in chromatography, and, therefore, the theories of these two areas are related to the theory of FIA. This is why most papers about FIA theory have adopted, as a starting point, the classical theory of flow in tubular conduits, with the intention of developing mathematical expressions for peak broadening, mean residence time, and fractional conversion of the analyte to a detectable product.

The physical process of material dispersion is due to the hydrodynamic processes taking place in the flowthrough system and is therefore conveniently investigated by the stimulus response technique, which is based on introduction of a tracer into a flowing stream and on measurement of the dispersion of the tracer as caused by the transport process throughout the system. If the tracer is injected as a zone (stimulus), then the observed response (recorded output signal) reflects the dispersion in the system through the increase of the width of the tracer zone as increased by the combined contributions from convection and diffusion. These two processes occur simultaneously at random and, therefore, the statistical moments of the resulting response curve reflect quantitatively the geometries and forces that influence the moments' magnitude. If the response curve has a Gaussian shape, then its first statistical moment—the mean of the tracer curve—corresponds to the peak maximum. When expressed in time units, the first moment allows, estimation of the *average* time available for chemical conversion, since it constitutes the mean residence time that the tracer material in average has spent in the reactor. The variance of the tracer curve—the second statistical moment—is proportional to the peak width, and for the Gaussian peak, it is the second power of the half-peak width measured at 0.61 peak height (cf. Fig. 3.2). The increase of the second moment caused by transport through the reactor is due to dispersion. The important property of the statistical description of zone broadening is that under certain restrictive conditions it can be modeled mathematically, which allows *a priori* identification and quantification of the forces, geometries, and physical constants contributing to the physical dispersion.

The relation between dispersion and residence time is an important parameter for optimization of all types of flow systems, yet its application varies depending on the purpose in question. Thus, in chemical engineering [3.1] the goal is to obtain maximum conversion of reacting species at optimized economy of materials, energy, and time; therefore, *dispersion is minimized* to achieve maximum conversion at minimum consumption of materials and during the shortest possible cycle of continuous-flow operation. In chromatography the key issue is resolution, which is optimized when the frequency of repetition of the distribution process between the mobile and stationary phases is maximized, while the *band broadening is minimized.* In preparative chromatography this yields maximum purity of the separated fractions, and in analytical chromatography it yields selectivity of the readout at maximum sampling frequency. Accordingly, the postcolumn reactors are also optimized to yield maximum conversion and minimum band broadening when processing the eluate, in order to ensure that the resolution obtained at the column is not degraded by additional dispersion.

It might appear that the FIA flow channel has the same function as a postcolumn reactor, and therefore the flow injection systems could also be optimized solely by means of the predictive models developed for chromatography, simplified for zero retention, or by means of input-response techniques developed for chemical engineering. This is, however, not entirely correct for two reasons.

The first reason is that the mixing in FIA is *nonhomogeneous and directional* (since it yields a concentration gradient in both axial and radial directions), and as a result of this stratification the ensuing chemical reactions take place gradually, while the reagent penetrates the sample gradient during the movement of the dispersing zone through the channel. Therefore, the FIA response curve is not only a result of the processes that occur at the detector location, but also of all the processes that gradually take place upstream in the FIA system at variable reagent concentration. The FIA readout is selected at peak maximum, or at any horizontal or vertical slice. Consequently, the resulting response cannot be described by existing mathematical models, since the response curve reflects only what is viewed by the detector as a result of a nonhomogenized bulk process, while the observed statistical moments yield only what appears as mean (homogenized) values of physical and chemical processes. Random walk models, based on a probabilistic approach, therefore offer more realistic insight into the microstructure of such a stratified flow, although the present modeling is too crude to allow sufficiently precise calculation of the expected response.

The second reason is that flow injection analysis encompasses a much wider range of solution handling tasks than those encountered in postcolumn derivatization or reactor engineering: that is, sample dilution, preconcentration, reaction rate measurement, phase separation, and multicomponent detection with or without chemical derivatization. More significantly, FIA systems are not operated only at continuous flow (as postcolumn reactors are), but also at stopped flow or even at reversed flow. Clearly, evaluation of the FIA response curve obtained at reversed flow or at intermittant flow by its statistical moments is meaningless, while selecting sections of the FIA response curve on the basis of D values to yield limited, medium, or large dispersion of the injected analyte is practical. To conclude, the classical dispersion theory, which uses statistical moments, mixing lengths, and mass transfer rates, and the less traditional concept of dispersion, which describes dilution of reacting species by means of the dispersion coefficient D, should not be viewed as conflicting, but as complimentary.

The wealth of literature, dealing with dispersion phenomena in straight and coiled tubes and packed and mixed reactors [20, 72, 73, 74, 75, 142, 146, 150, 151, 153, 176, 181, 183, 343, 345, 430, 501, 566, 608, 665, 702, 802, 1061, 1063, 1064, 1065, 1226, 1248, 1256, 1273] describes concepts and models, which are useful for gaining insight into what happens during the complex interplay when solutions are mixed and chemical reactions are taking place in an FIA system. The works of Tijssen [73], Berg et al. [74], Neue and Engelhardt [3.2], Betteridge et al. [702], the Dutch group [75, 146, 176, 501], Søeberg [3.3], and Hungerford [3.4] have contributed to a substantial improvement in the design of FIA manifolds by defining the constraints imposed by physics on the continuous-flow design, and by pointing out the areas where improvements can be made. Yet further development of the FIA theory requires further departure from the comparison with postcolumn reactors, by focusing on the specific aspects of FIA, like the mutual dispersion of merging zones, chemical kinetics at varying analyte/reagent ratios, and influence of particle size and solute viscosity on dispersion of suspensions, together with modeling the solute dispersion during repeated stop/flow periods. Such a theory will allow quantification of factors influencing zone broadening in real systems and will reconcile the two definitions of dispersion that are applicable to FIA: *Physics*—mixture of one substance dispersed in another; and "*Statistics*—extent to which values of a variable differ from the mean."

It was Reijn [501] who quoted these definitions from the *Oxford English Dictionary* in connection with the theory of continuous-flow FIA, and who also compared the satisfaction gained from working out his theory with the satisfaction of maintaining a motorcyle. Yet it is the balance that leads to perfection, and therefore only maintaining a bike, or only riding it, cannot be a satisfactory goal.

3.1.1 The Stimulus Response Technique

The transport phenomena in a flowing medium are treated in textbooks on chemical reactor engineering [3.1] and constitute a central part in the theory of analytical separation techniques based on continuous flow, such as chromatography, countercurrent extraction, or multistage distillation. In physical and chemical technology the influence of the shape and size of the reaction vessel and the type of flow on the dispersion process is investigated by the stimulus response technique, which is based on addition of a material, either as a step or as a pulse signal, which can be quantitatively measured at the exit of the vessel. In a vessel with a stabilized flow, two idealized patterns can be visualized: plug flow and a mixed flow, representing two extremes. In reality, neither of these extreme types of flow exists, and the resulting arbitrary flow contains contributions of both these extreme flow types, in different proportions. The type of flow can be described by the residence time distribution curve, which reflects the various times that each element of fluid spent within the flowthrough reactor. This distribution of residence times can be observed at the exit of the vessel for those elements of fluid that entered the vessel simultaneously.

The the so-called F curve (Fig. 3.1) is measured by imposing a step impulse of a dye of concentration C^0 on the fluid stream entering the vessel, with the result that, with a certain delay, an increase of color intensity is observed at the vessel output. A plot of the resulting concentration C as C/C^0 versus time is the F curve, the shape of which charactertizes the type of flow. For the plug type of flow where no mixing occurs (Fig. 3.1*a*, left) the output curve will have the same form as the input curve. The mixed flow, representing the other extreme (Fig. 3.1*a*, center), where instantaneous, homogeneous mixing is taking place within a single stage, will yield a purely exponential concentration gradient, showing a gradual increase of color intensity. In reality, an arbitrary flow will be observed, resulting in an S-shaped curve (Fig. 3.1*b*, right), which will become progressively more symmetrical with an increase in the number of mixing stages through which the sample plug has passed. The slope of the curve is inversely proportional to the axial disperson σ (see Fig. 3.1*b*), and the time span between the input ($t = 0$) and the inflection point on the output of the symmetrical curve is the mean residence time $\overline{T}$.

Alternatively, the signal can be injected as a pulse (ideally infinitely narrow so that its width = 0 while its area = 1), the broadening of which is observed when, with a certain delay, the dyed zone has reached the vessel output. The formalized plot of the observed concentrations C versus time is called a C curve, and its shape characterizes the type of flow (Fig. 3.2*a*). For plug flow the input and output pulses have the same shape; for mixed flow, after an abrupt rise, the color intensity decreases expo-

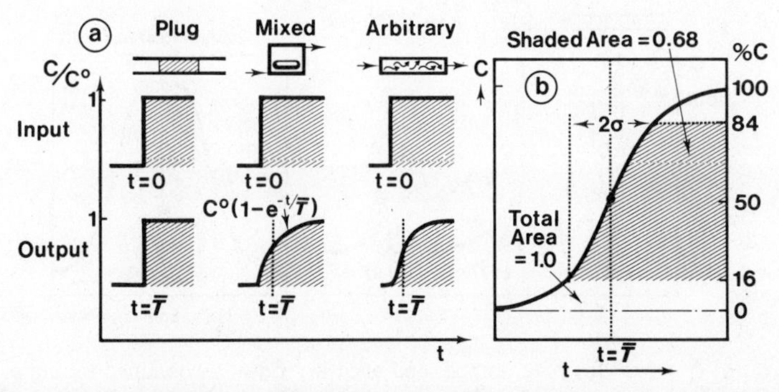

Figure 3.1. F curves for plug, mixed, and arbitrary flow as obtained by imposing step input of dye of concentration C^0 and measuring, as the output, the time record of exit concentration as C/C^0. The two parameters characterizing the curve are $\overline{T}$ (mean residence time that for normalized curves is equal to 1) and σ (axial dispersion).

nentially with time. For arbitrary flow, a peaked curve is obtained, which, with the increase of number of mixing stages through which the sample plug has passed, will become progressively more symmetrical, so that eventually a Gaussian shape is obtained (Fig. 3.2*b*). The standard deviation of this curve equals the axial dispersion σ, and the time span between the input ($t = 0$) and the mean of the output curve is the mean residence time $\overline{T}$.

Thus, the F curve is an integrated C curve:

$$F = \int_0^t C\,dt \tag{3.1}$$

and therefore, for a closed vessel (in which fluid enters and leaves solely by plug flow, that is, with a flat velocity profile), the value of σ and $\overline{T}$ obtained by *either the step or the pulse method will be identical.*

For the investigation of the dispersion in a FIA system, both the step and the pulse methods are *equally* suitable. The F curve is approximated when the dye is introduced for at least $5S_{1/2}$, thus reaching steady state, while the C curve is approximated when the impulse is smaller than $\frac{1}{5}S_{1/2}$, so that its width is negligible compared with the path to be traveled, or its volume S_v is negligible compared to the volume of the FIA reactor V_r. Consequently, the F curve describes the shape of FIA curves for $D^{max} \leq 1.03$ and the C curve those for $D^{max} \geq 7.73$ (see Section 2.2.5). Keeping in mind that the sample plug in a FIA system does not usually pass through a sufficient number of mixing stages so that the FIA response curve is not symmetrical, the concept of axial dispersion σ and the mean residence

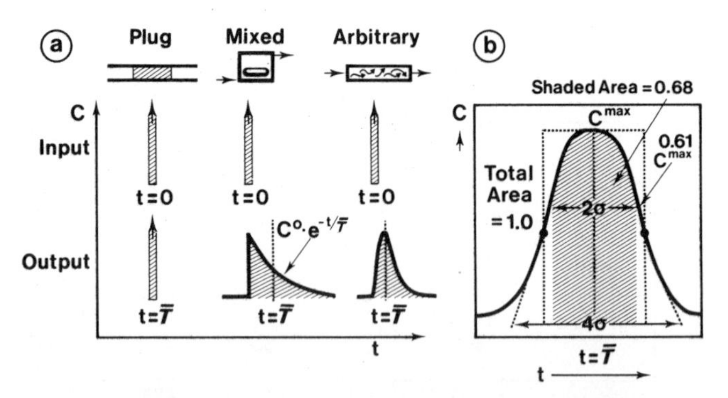

Figure 3.2. (*a*) C curves for plug, mixed, and arbitrary flows as obtained by imposing an idealized instantaneous input pulse of dye and measuring the time record of exit concentration as the output. (*b*) For arbitrary flow, a Gaussian-shaped curve is eventually achieved, which is characterized by the two parameters $\overline{T}$ (mean residence time), for which $C = C^{max}$, and σ (standard deviation equaling axial dispersion).

time $\overline{T}$ can be used to estimate the performance of a FIA system in respect to maximum achievable sample frequency S_{max}, and it also offers a rough guideline for estimation of the expected mean reaction time in a given flow channel design.

The residence time distribution curve (RTD) can be described by its statistical moments, of which the centroid of distribution $\overline{T}$ and spread of distribution σ are the most important numerical values. Thus, for a C curve, the zeroth moment is

$$\text{Area} = (\text{sample zone width}) \cdot C^0_{(to)} \tag{3.2}$$

at the time of injection ($t = 0$), while the first moment is

$$\overline{T} = \frac{\int_0^\infty tC\,dt}{\int_0^\infty C\,dt} \tag{3.3}$$

and the second moment (variance) is

$$\sigma^2 = \frac{\int_0^\infty t^2C\,dt}{\int_0^\infty C\,dt} - \overline{T}^2 \tag{3.4}$$

where the variance is the square of the spread of the distribution and has either the unit time (σ_t^2, in s^2), or volume (σ_v^2, in mL^2) if related to a constant flow rate Q.

In FIA, the sample throughput and the sample dispersion D are related to the standard deviation of the peak, σ. The throughput rate is limited to the maximum value S_{max}, beyond which excessive carryover will make the estimation of peak height unreliable (see Fig. 2.12). Thus

$$S_{max} = \frac{3600}{k\sigma_t} = \frac{60Q}{k\sigma_v} \text{ (samples/h)} \tag{3.5}$$

where k is a factor depending on acceptable carryover. In analogy with chromatographic methods, a distance between adjacent peak crests corresponding to $k = 4$ has been accepted.

The first and second moments can be expressed in either time or volume units. It is important to note that they do not yield useful information on their own. Thus $(\sigma_t)^2$ can be shortened by increasing the flow rate, yet this would increase material consumption. Conversely, decreasing $(\sigma_v)^2$ might decrease sampling frequency. However, the relations between the first and second moments as expressed by the number of mixing stages (N), mixing length (H), radial mass transfer a or a dispersion factor ($\beta_{1/2}$, cf. Section 2.2.5 and Ref. 608) are very useful optimization parameters. It can be shown that this factor for symmetrical response curves is related to standard deviation by

$$\beta_{1/2} = S_{1/2}/V_r = t_{1/2}/T \simeq \sigma(\pi/2)^{1/2} \tag{3.6}$$

It comprises the true width of the injected impulse and the true residence time and, therefore, it yields a realistic estimation of the sampling frequency rather than the classical parameters such as N, H, or σ, which can be obtained only by injecting an infinitely short impulse, but have the advantage of stringent theoretical evaluation.

A useful property of statistical moments is their additivity. Thus, for variances of flow-independent regions it can be shown that

$$\sigma^2_{\text{overall}} = \sigma^2_{\text{inj}} + \sigma^2_{\text{comp 1}} + \sigma^2_{\text{comp 2}} + \sigma^2_{\text{comp 3}} + \sigma^2_{\text{det}} \tag{3.7}$$

while for mean residence times

$$\overline{T}_{\text{overall}} = \overline{T}_{\text{inj}} + \overline{T}_{\text{comp 1}} + \overline{T}_{\text{comp 2}} + \overline{T}_{\text{comp 3}} + \overline{T}_{\text{det}} \tag{3.8}$$

where subscripts inj, comp, and det, refer to injection, component, and detector, respectively. This property of the variancies allows the performance of a FIA system to be predicted from mathemathical models or the contribution of individual system components to be evaluated experimentally, simply by measuring the moments of the response curve in the presence or absence of the investigated component. Alternatively, two identical low-volume flow detectors can be used with the critical component placed between them. Using this so-called two-point measurement, the dispersion and the residence time within the investigated component is obtained from the difference in the moments measured on the first and second response curve.

Additivity of variances [Eq. (3.7)] will in practice not hold if the components are not statistically independent. This means that a perfect radial mixing must occur between the components, or else the dispersion in one component will influence the dispersion in the next component. Furthermore, Eq. (3.7) implies that variances are commutable, which means that the order of components in the manifold may be changed without affecting the overall variance. Again this is not true for flow-dependent regions, as it has recently been demonstrated experimentally [1226]. Thus, dispersion is not predictable for the most commonly used FIA manifolds, which contain relatively geometrically different channel sections (injector, reactor, diffusion unit, confluence points, detector, etc.), within which the flow has not yet stabilized. Even packed reactors normally used in FIA are much too short to accommodate a sufficient number of mixing stages and, therefore, experience and theoretical conclusions from chromatography must be applied with caution.

In analogy with chromatographic systems, Poppe and co-workers [72, 75] have observed that the total peak broadening in FIA is the sum of the contributions from the injection process, the flowthrough reactors and connectors, the holdup volume of the flowthrough detector, and the time constants of associated electronics. These processes can be described by the individual peak variances:

$$\sigma^2_{\text{overall}} = \sigma^2_{\text{injection}} + \sigma^2_{\text{flow}} + \sigma^2_{\text{detector}} \tag{3.7}$$

The influence of the injection process and of the injected sample volume on $\sigma^2_{\text{overall}}$ are discussed later in this chapter. The influence of the detector on $\sigma^2_{\text{overall}}$ is far more complex, since in theory one should consider, first, whether the detector is a cup mixing type or an integration detector. Then the response characteristics of the detector [i.e., c = log(response) or $c = k \cdot$ response], the detector volume, and the speed of its response must be considered (cf. Chapter 5). Provided that the detector and its electronics fulfill all requirements, which means that $\sigma^2_{\text{detector}}$ will be at least five times smaller than the σ^2 due to the injection and the transport, the influence of the detector on $\sigma^2_{\text{overall}}$ may be neglected.

3.1.2 The Tanks-in-Series Model and Nonlaminar Flow

This model is based on the view that the liquid flows through a series of ideally stirred tanks of equal size [3.1] or through a network of parallel ideally stirred tanks. The parameters of this model are N, the number of mixing stages (tanks) through which an element of fluid has passed, and $\bar{T}_i$, the mean residence time of the element of fluid in one mixing stage. For one tank the normalized C curve has the form

$$C = \frac{1}{\bar{T}_i} e^{-t/\bar{T}_i} \quad (N = 1) \tag{3.9}$$

(see mixed flow, Figs. 3.1 and 3.2), while the system with any number of serial mixing stages is described by the statistical G function, written as the C curve,

$$C = \frac{1}{\overline{T}_i}\left(\frac{t}{\overline{T}_i}\right)^{N-1} \frac{1}{(N-1)!} e^{-t/\overline{T}_i} \tag{3.10}$$

For large N the curve approaches a Gaussian shape (Fig. 3.3), whereas for decreasing N the peak becomes increasingly skewed. The mean residence time $\overline{T}$ of the element of the tracer material in the system is determined from individual residence times $\overline{T}_i$ by $N\overline{T}_i$, and coincides with the appearance of the maximum of the ($N + 1$, that is, the next) C curve at the exit. Provided that the C curve has a Gaussian shape (i.e., $N \gg 10$), the variance of the C curve is

$$\sigma^2 = N\overline{T}_i^2 = \frac{\overline{T}^2}{N} \tag{3.11}$$

which normalized to $\overline{T}$ is

$$\sigma^2 = \frac{1}{N} \tag{3.12}$$

An important feature of this model is that it formally covers the transition from mixed flow to plug flow, and, therefore, the C curves (Fig.

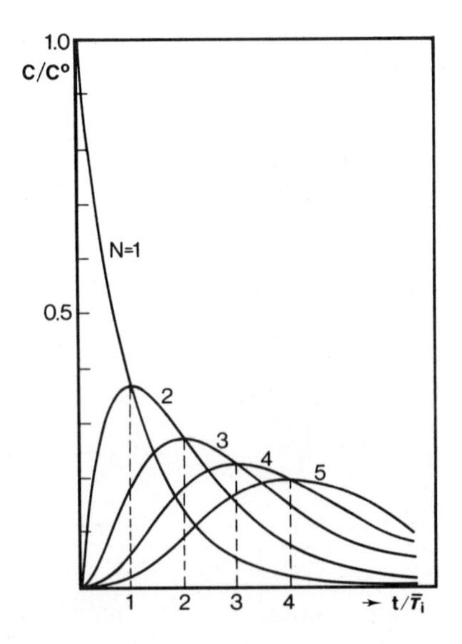

Figure 3.3. C curves obtained for one, two, three, four, and five tanks in series, where $\overline{T}_i$ is the mean residence time in one tank. (From Ref. 3.1 by courtesy of John Wiley & Sons, Inc.)

3.3) resemble the FIA curves (for $S_v < S_{1/2}$), cf. Fig. 2.6*b*. The extreme cases of this model, that is, one mixing stage [Eq. (3.9), curve $N = 1$, Fig. 3.3] and the multistage system ($N \gg 10$) are very useful for describing and designing FIA manifolds. Thus, the one-tank model, first used to create a well-defined exponential gradient for FIA titrations performed at *large* dispersion [10], has been applied to fast FIA titrations performed at *medium and limited* dispersion [183], for exponential dilution, automated calibration [1127], and high-precision extensive dilution [1063].

The tanks-in-series model has been used for theoretical description of fluid behavior in different types of FIA reactors. This concept highlights the influence of radial mixing on the physical and chemical processes, and the importance of the uniform flow geometry, by showing that a larger number of uniform mixing stages connected in series yields symmetrical RTD curves with standard deviations that decrease with increasing N relative to the path traveled. This is described by relating the height of an (imaginary) mixing stage H to the length L (or volume) of the path traveled. This can be done either in units of volume (mL) or time (s):

$$H_v = \frac{V_r}{N} \quad \text{or} \quad H_t = \frac{\overline{T}}{N} \tag{3.13}$$

where V_r is the combined volume of all mixing stages and $\overline{T}$ is the mean residence time. In analogy with the concept of plate height, the *mixing length* H has been introduced, since [Eq. (3.11)]:

$$H = L/N = L(\sigma)_t^2/\overline{T}^2 \tag{3.14}$$

The important feature of this approach is that it allows estimation of the *intensity* of radial mixing from the mixing length H, or more suitably, from the radial mass transfer constant a, which is the reciprocal value of the mean residence time of the individual tank, $\overline{T}_i$. It was Hungerford [3.4] who focused attention on this link between *radial mass transfer rate* and mixing length and thus reemphasized the importance of this parameter on the design of the FIA reactor. The radial mass transfer rate can be written as

$$a = 1/\overline{T}_i = Q/V_i = F/H = N/\overline{T} \tag{3.15}$$

and the higher the a value is, the more efficient the radial mixing is in a given component of the system.

Thus, in formal analogy to chromatographic column, the FIA component (e.g., reactor) with the highest a and N will yield the lowest $\sigma/\overline{T}$

ratio—and hence maximum sampling frequency for given residence time. Note that this conclusion is in harmony with the concept of the dispersion factor, which has been defined as

$$\beta_{1/2} = S_{1/2}/V_r = t_{1/2}/T \simeq \sigma(\pi/2)^{1/2} \quad (3.6)$$

and which when lowest, yields an optimized FIA channel in terms of sample, reagent, and time economies. The current theories of flow in tubular reactors use variance and mixing length H as parameters. They are reviewed in the next sections with the aim of identifying optimum reactor geometry achieveable within design constraints such as flow rate, backpressure, overall dimensions, and material economy.

3.1.3 Laminar Flow and Radial Mixing in FIA

In a straight tube the dispersion of the sample plug is the result of redistribution of material through countless repositionings of the elements of fluid in *axial* and *radial* directions caused by the twin process of convection and diffusion. The resulting concentration of the solute within the carrier stream is the result of the relative contribution of these two effects on the originally homogeneous cylindrical sample plug. The velocity profile—and extent of radial mixing—primarily depends on the type of flow, that is, turbulent or laminar. The laminar flow prevails at Reynolds number (Re) lower than 2100,

$$\text{Re} = 4\rho Q/\pi 2R\nu \quad (3.16)$$

where ρ is the density in grams/mL, R is channel radius, Q is the volumetric flow rate, and ν is the viscosity in poise. A typical FIA system, having an inside diameter of 0.05 cm, and Q between 0.5 and 3 mL/min has a Re value of 20–130. At these conditions the streamlines are symmetric around the center of the channel, where the fluid moves twice as rapidly as the mean flow velocity, while toward the walls the velocity tends toward zero. Therefore, within milliseconds after injection the velocity profiles acquires a parabolic centrosymmetric distribution

$$F_i = 2F(1 - r_i^2/R^2) \quad (3.17)$$

characteristic for Poisseulie flow (F_i is the velocity of the individual streamlines, r_i is an individual streamline's distance from the central streamline, F is the average flow velocity, and R is radius of the tube). Consequently, the sample material is distributed along the parabolic profile as shown in Fig. 3.4. Indeed, if an element of the sample material would remain in the same streamline F_i, that is, at the same distance from the center of the tube, then the axial dispersion would indefinitely increase with the distance traveled, because elements close to the center will move progressively along the tube axis, while the last traces of sample material situated close to the wall would never be swept out of the tube. What makes FIA possible is the existence of forces that promote *radial* dispersion, since they facilitate repositioning of elements of sample material from the original streamline. The forward movement of sample elements is retarded when they move away from the central streamline, and it is accelerated when they move toward it. Consequently, if such a reshuffling occurs at random, the axial dispersion is reduced, and the more intense the radial movement will be, compared to the forward convective motion, the lesser the dispersion of the sample material will be per unit of length traveled.

The radial dispersion can be promoted by introducing turbulent flow or secondary flow, by diffusion alone, or by a combination of these effects.

In turbulent flow the streamlines cross, the movement is chaotic, and mixing in all directions is equally intense and rapid, resulting in fast averaging of the individual velocity profiles and plug flow with minimum dispersion. Exploitation of this breakup of viscous forces by inertial forces, which occur at high Re values, is not practical for FIA, since Eq. (3.16) requires either very high flow rates, which are uneconomical, or use of long very narrow tubes and high linear velocities, which have as a practical constraint a high backpressure [14].

Secondary flow, combined with molecular diffusion promotes the radial transport in a powerful way. Its intensity increases with geometrical disorientation of the flow path. Thus, coiling of the tubes induces a gentle

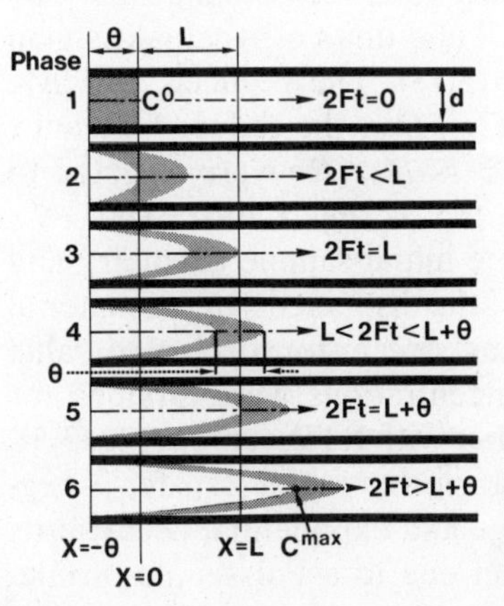

Figure 3.4. Dispersion of a sample plug (width θ) by laminar flow in the absence of diffusion, exhibiting the progressively increasing dispersion due to Poiseuille flow. (According to Ref. 1354, by courtesy of the *Journal of Flow Injection Analysis*, Japan.)

secondary flow, while knitting of the tubes—or meandering of the channel— promotes an intense streaming in radial direction. Packing of tubes with single beads is the most effective way to promote an intensive radial mixing.

Molecular diffusion was identified as the important factor influencing overall design of the flow channel already in the first theoretical paper on FIA [20]. Since then its role in FIA processes has been clarified and mathematically described in a number of detailed papers. Works of Vanderslice et al. [1061], Wada [1065] and Hungerford [3.4] focused attention on molecular diffusion as the sole tool for redistribution of sample elements in radial direction. Their results, observations, and conclusions are reviewed in the next section.

3.1.4 The Diffusion Model and Dispersion in a Straight Tube

The concentration profile formed from the originally homogeneous sample plug is the result of the combined convective and diffusion effects, the relative influence of which can be described by the convective–diffusion equation:

$$D_{\mathrm{m}}\left(\frac{\partial^2 C}{\partial x^2} + \frac{\partial^2 C}{\partial r^2} + \frac{1}{r}\frac{\partial C}{\partial r}\right) = \frac{\partial C}{\partial t} + 2F\left(1 - \frac{r^2}{R^2}\right)\frac{\partial C}{\partial x} \tag{3.18}$$

where D_{m} is the molecular diffusion coefficient, C is the concentration, x is the distance along the tube, r is the radial distance from the tube axis, R is the tube radius, t is the time, and F is the average flow velocity. This equation has been solved using initial boundary conditions shown in Fig. 3.4 by Vanderslice et. al. [142, 1061] for FIA systems, whose aim was to obtain a detailed picture of the microstructure of the dispersing zone. The authors used the method of Bate et al. [3.5] and units of Anathakrishnan et al. [3.6] for numerical solution of Eq. (3.18). The reduced velocity, described by the Peclet number, $\mathrm{Pe} = R(2F)/D_{\mathrm{m}}$, the reduced distance $X = D_{\mathrm{m}}x/R^2(2F)$, and the reduced time $\tau = D_{\mathrm{m}}t/R^2$ were selected to approximate the range of FIA conditions, that is, $\mathrm{Pe} > 1000$, $0.004 < x < 1.0$, and $0.002 < \tau < 0.8$. Using the same initial sample geometry and concentration, the relative concentration of the dispersed sample material was calculated for each point in the moving stream for a selected value of the reduced time τ. The calculated concentrations were grouped for the purpose of illustration and drawn together with RTD curves (Fig. 3.5). For $\tau < 0.01$, when the contribution of diffusion to radial transfer is negligible, a sharp rise at the peak leading edge and exponential decay at its tail, characteristic of convective dispersion due to a Poisseuille profile,

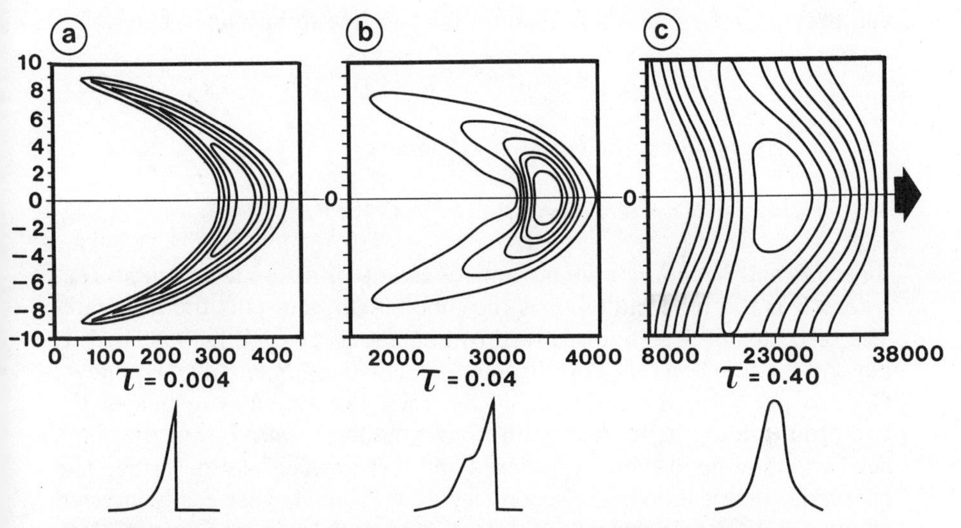

Figure 3.5. Dispersion of a sample plug in a straight tubular channel as affected by molecular diffusion with increasing residence time (top) and corresponding RTD curves (bottom). From left to right $\tau = 0.004$, $\tau = 0.04$, and $\tau = 0.40$. For details see text. (According to Ref. 1061, by courtesy of Elsevier Scientific Publishing Co.)

are observed. For long residence times, $\tau > 0.4$, a nearly Gaussian peak is observed, while for shorter residence times a peculiar double-humped peak is observed. These humped peaks were observed by the same authors in their earlier work [142], where an attempt was made to predict the peak width by the baseline to baseline value Δt_b. This value was, surprisingly, found to be independent of the sample size within $0.002 < x < 0.8$, leading to expressions in real time units:

$$\Delta t_b = \frac{35.4R^2 f}{D_m^{0.36}} \left(\frac{L}{Q}\right)^{0.64} \tag{3.19}$$

and for the peak leading edge arrival time t_a:

$$t_a = 109R^2 D_m^{0.025} (L/Q)^{1.025} f^{-1} \tag{3.20}$$

where f is an accommodation factor. Both equations are useful for determination of the diffusion constants D_m and confirm the previous observations [20] that the dispersion increases with the square of the tube radius and the square root of the traveled length. Vanderslice's results were later questioned by Gómez-Nieto et al. [802], who found by extensive experiments that f *is not a constant*, as the above equations imply, and they proposed another equation for baseline to baseline value:

$$\Delta t_b = 56.7R^{0.293}L^{0.107}Q^{1.057} \tag{3.21}$$

and for dispersion coefficient D at any time t:

$$D = C^0/C = 2.342L^{0.167}Q^{-0.206}R^{0.496} \tag{3.22}$$

for constant injection volume and constant diffusion coefficient D_m, where R is the tube radius, L is the tube length, and Q is the flow rate. The discrepancy was not resolved by ensuing discussion [802, 803]. It is beyond the scope of this work to go into detailed arguments, yet it appears in retrospect that the discrepancy between experimental results of the two groups could be traced to differencies in the *geometries* of the channels, which none of them considered and therefore did not describe. The parameter in question is the secondary flow, which is very easily induced by even a minuscule curvature of the tubing used (cf. next section). Both the U.S. and the Spanish groups used plastic tubing, and one can only speculate that the tubing was coiled for convenience, or was allowed to curl during measurement. In addition, unspecified tube connectors, which might have induced additional radial flow, were undoubtedly used.

These results are, however, useful for future FIA work since:

1. They reveal the difficulty of reconciliation the theory with an experiment.
2. The work of Vanderslice et al. [142] shows that FIA systems can be used for measurement of diffusion coefficients—provided that the flow geometry of the FIA system is kept constant (preferrably a straight rigid inner polished tube) and calibration can be made by means of a solute with known D_m value.

It is well known that Taylor's treatment of axial dispersion is valid only when differences between axial dispersion and molecular diffusion D_m are small due to the choice of experimental conditions (i.e., when R is small, t is long, and F is very low). It follows from Taylor's work that the radial diffusion will start to affect the velocity profiles if

$$t \geq R^2/3.8^2D_m \tag{3.23}$$

or for the time variance when

$$(\sigma_t)^2 = R^2t/24D_m \tag{3.24}$$

Thus for $D_m = 7 \times 10^{-6}$, the minimum time for Taylor's effect to begin to develop is 6.2 s for $R = 0.025$ cm. This minimum time limit is just about met in FIA, since tubes of more than 0.5 mm ID are seldom used, while typical residence times are 20 s or more. Further decrease in channel radius and flow rate will lead to progressively more intensive redistribution of sample material in the radial direction by molecular diffusion and therefore decrease in the flow rate in a straight narrow tube will lead to a decrease in dispersion (cf. Rule 3, Section 2.2.3).

At this juncture one cannot resist pointing out the advantage of the stopped-flow FIA, since in the extreme, when the carrier stream ceases to move, the axial dispersion of the sample zone practically stops (except for a contribution caused by molecular diffusion—which is negligible within the stopped-flow period used in FIA), and therefore the dispersion coefficient D and peak width as expressed by σ_v remains constant—and independent of t.

Returning to the continuous-flow methods, performed at constant flow rate, it is interesting to investigate aspects of the idea of superminiaturization, which is based on substantially increasing the contribution of diffusion to radial mixing by designing the FIA system to operate entirely in the Taylor–Aris domain so that $\mathrm{Pe} \Rightarrow 100$ and $\tau \Rightarrow 1$. It was Tijssen [73] who was first to point out and to compute the consequences of decreasing the radius of the FIA channel from the millimeter to the micrometer range. He, however, did not consider the simultaneous downscaling of L and Q together with R, and this fine notion, conceived by Hirschfeldt and pursued by Hungerford [3.4] exploits in an elegant way the diffusion phenomena in full by using a short, narrow, and straight channel. *A priori*, there are indisputable advantages in miniaturization since:

1. The reactor volume scales with the square of the channel radius.
2. The radial mixing speed scales as the inverse square of the channel radius—if diffusion is the dominant radial mass transfer force.
3. The sampling frequency scales as the reciprocal square root of the channel length.

Therefore, a superminiaturized FIA system will not only be extremely economical in terms of sample and reagent consumption, but it will also allow high sampling frequencies, since diffusion will rapidly reshape the parabolic profile that the sample zone has acquired during the injection. The efficiency of this process can be estimated from the relationship between mixing length and linear velocity F:

$$H_0 = (2D_m/F) + (R^2F/24D_m) \quad (3.25)$$

where the first term describes the contribution of molecular dispersion to axial dispersion—negligible for F values of 1 to 25 cm s^{-1}, and D_m values of 10^{-7} to 10^{-5} cm^2 s^{-1}. Thus the second term predicts that the mixing length will decrease with decrease of tube radius R and flow velocity and with increase of molecular dispersion. The optimum flow rate according to Hungerford [3.4] was found to be

$$F_{\text{mean, opt}} = 4(3)^{1/2} D_m/R \tag{3.26}$$

so that the minimum mixing length is

$$H_{\text{min}} = R/(3)^{1/2} \tag{3.27}$$

while the radial mass transfer rate is

$$a = 1/\overline{T}_{i\ \text{min}} = D_m/R^2 \tag{3.28}$$

which shows that the narrower the channel is, the shorter the mixing length will be. Thus, for example in an FIA system, where the inside diameter is 125 μm, for a solute with $D_m = 1 \times 10^{-2}$ cm^2 s^{-1}, the optimum flow rate $F_{\text{mean, opt}} \cong 0.1$ mm/s and H_{min} is as short as 4×10^{-2} mm, and in theory a Gaussian shape of the response curve will be observed after the zone has been traveling through only 1 mm of tubing! Such a channel diameter will yield a residence time of 5 min, being only 3 cm long, with a pumping rate of about 1 nL s^{-1} (1 nanoliter/second) and will be capable of accommodating a train of samples, within the achievable sampling frequency of up to several hundred samples per hour—depending, naturally, on the D_m value of the sample solutes!

There is, however, the other side of the coin to be considered. From a theoretical viewpoint two drawbacks are evident:

1. Downscaling increases the surface area/reactor volume ratio, since $A_{\text{surf}}/V_r = 2/R$. While this is an advantage when interaction of solute with a stationary surface of the capillary reactor is desirable (e.g., in tubular chromatography), it is a serious drawback for FIA applications, where adsorption of solute from a nanoliter volume of an injected sample will seriously affect the response curve.

2. Since the molecular diffusion is the sole driving force in this design, the differencies in D_m values, which for FIA relevant solutes can be as much as 10^2, will dramatically influence the radial transfer rate [Eqs. (3.25)–(3.28)] and minimum mixing length. Thus the higher-molecular-weight solutes will be radially transferred less readily—and mixed with solutes of lower molecular weight. Whether these differences in mixing

rates will have an adverse effect on the reproducibility of determination by yielding flow-rate-sensitive double-humped peaks, or whether they can be successfully eliminated by flow design and by calibration on the instrument, only experiments will show.

This brings up the question of practicability of superminiaturization. Nanoliter per second flow rates are difficult to realize and a ten- or hundredfold increase of Q to more realistic pumping range of microliters per minute will require a corresponding increase of L to 30–300 cm, with increasing backpressure limitations due to increase of F at low R. Another more serious limitation lies, however, in the maximum permissible sample volume and detector size. The reason is that the injected sample volume must be much smaller than V_r and the detector volume should less than $0.5\ \sigma_V$. This means that p-fold decrease in R, would demand a p^3 decrease in injected sample volume and a p^4 decrease in detector volume. Thus the detector volume is presently the true physical constraint to superminiaturization, since for even a "large" superminiaturized FIA (R = 100 μm, L = 10 cm, and $Q \cong$ 10 μL/min), the detector volume must not exceed 1 μL! The interest in superminiaturization of FIA systems is, however, growing, and present technological barriers will be removed eventually. It is therefore useful to review all the theoretical limitations of superminiaturization as most recently discussed and clearly summarized by van den Linden [1118]. On the experimental side the very first truly superminiaturized FIA system has been described by Olesik, Frensch, and Novotny [1259], who by using components borrowed from capillary chromatography constructed a system where 0.2-μL plug of sample was injected into a 250-μm-ID fused silica capillary 1 m long using a high pressure IR flow cell of 2 μL holdup volume.

To conclude, straight short capillary reactors would exploit in an elegant way the diffusion forces to promote the radial transfer of solutes and mixing of reacting components. However, double-humped peaks will be formed during the dispersion processes under such conditions and these peak forms will be propagated throughout subsequent chemical reactions (cf. Section 3.2). Therefore, unless superminiaturization is the goal, external dynamic forces combined with a suitable reactor geometry should be applied to increase the radial mass transfer rate by promoting flow in the radial direction.

3.1.5 Dispersion in Coiled Tubes and in 3-D Reactors

The geometrical form of the tubing from which the FIA reactor is made has a distinct influence on the solute dispersion, as seen in Fig. 2.10. The reason is that helical coiling, or irregular "knitting" of the tubular conduit, introduces secondary flow perpendicular to the main axial flow, thus increasing the intensity of radial mass transfer. At very low flow velocities the centrifugal forces are weak, and the dispersion is similar to that in a straight tube, while at high flow velocities the dispersion is markedly decreased and this effect is magnified by the sharpness of the tube curvature. As the parabolic flow pattern is deflected toward the outer wall, a recirculating pattern is formed (Fig. 3.6, top). At higher flow velocities the stream is divided into two symmetrical halves (Fig. 3.6, bottom), which through their virtually separate behavior effectively divide the cross-sectional area in half, and at the same time enhance the intense movement of solute in the radial direction. A mathematical treatment of this behavior leading to calculation of a correction factor was first given by Golay [3.23].

Even more effective than helical coiling is the three-dimensional disorientation of the flow path obtained by "knitting" the tubing (Figs. 2.8 and 2.10). Although the flow patterns in this type of reactor are not yet well described, the effectiveness of 3-D disorientation is likely to be caused by formation of the flow pattern shown in Fig. 3.6, which, however, is formed, disrupted, and reestablished frequently along the flow path, thus forcing the streamlines to change direction along with the frequently changing position of the outer wall. The thus formed three-di-

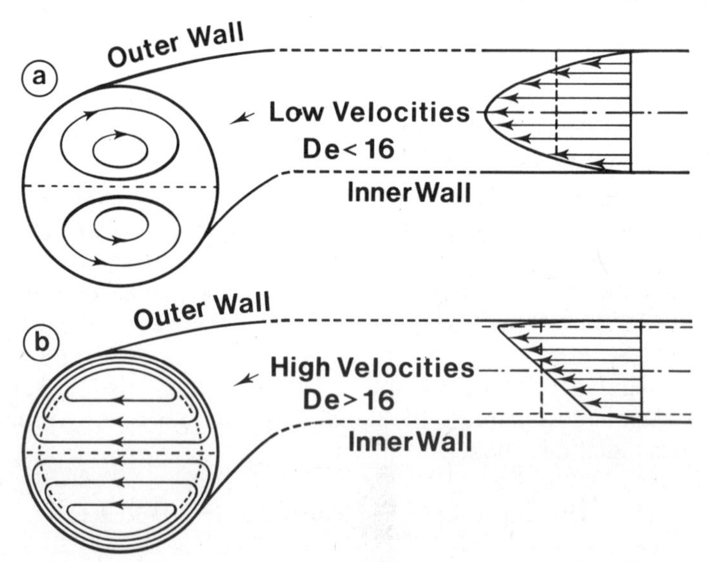

Figure 3.6. Secondary flow patterns and velocity profiles in coiled tube at (*a*) low and (*b*) high flow velocities. (According to Ref. 73, by courtesy Elsevier Scientific Publishing Co.)

mensionally disoriented flow effectively decreases axial dispersion. The energy for radial mass transfer in coiled and 3-D reactors comes from the kinetic energy of the moving liquid, and, therefore, solute dispersion is affected by the same factors, that is, by flow velocity, tube curvature, tube radius, and viscosity.

Coiled tubes were used in FIA at the very early stage [1–10], but coiling was done in analogy with reactor design of segmented flow, for convenience, and in order to keep the reactor geometry well fixed, rather than with the aim of decreasing axial dispersion. It was Tijssen [73] who was first to point out this advantage of tube coiling for FIA, and to predict theoretically the behavior of solutes in coiled FIA reactors. Mass transfer in coiled tubes has, however, been studied by various groups already in the last century [3.7], while within the last 10 years the specific aim of such studies was to design postcolumn reactors. Works of Hofman and Halasz [3.8], and of Austin and Seader [3.9] are examples of investigations of flow patterns in various coil types, while Trivedi and Vasudova [3.10] have shown that coiling may reduce axial dispersion up to 500 times. Most recently Leclerc et al. [1248] have published a theory describing dispersion curves in coiled reactors, while Søeberg [3.3] has developed a mathematical model, based on equations of motion and continuity, which allowed him to compute velocity and concentration profiles and to depict their microstructures by computer graphics as well as to calculate RTD curves.

The efficiency of reshaping of the parabolic profile, which the sample zone has acquired during injection, can be estimated by comparing the mixing length H_0 in a straight tube

$$H_0 = (2\,D_m/F) + (R^2F/24D_m) \tag{3.25}$$

with the mixing length in the presence of secondary flow

$$H = (2D_m/F) + (2\kappa R^2F/D_r) \tag{3.29}$$

by means of the H/H_0 ratio, or since

$$H = L(\sigma_t)^2/\overline{T}^2 \tag{3.30}$$

also from the corrected ratio of the variances $[(\sigma_0)^2/(\sigma)^2] - 1$ of straight and coiled tubes at otherwise identical experimental conditions.

In Eq. (3.29) the κ value is the velocity profile factor and D_r is the radial dispersion coefficient, containing contribution from molecular diffusion and secondary flow and therefore κ and D_r are both functions of flow velocity and reactor geometry. The degree of coiling is described by the aspect ratio $\lambda = R/R_c$, where R_c is the coil radius. Therefore, the Dean number De = Re $\lambda^{1/2}$ is used as a velocity parameter together with the Schmidt number (Sc = ν/D_m, where ν is the kinematic viscosity in cm^2/s). If such a simple model holds, then, when plotting $[(\sigma_0)^2/(\sigma)^2] - 1$ versus De^2Sc, a single curve for all examined values of λ, R, and D_m is to be found (Fig. 3.7). For very small values of De^2Sc, the dispersion is governed only by molecular diffusivity up to the value $De^2Sc = 256$, and there is not much difference in band broadening between the coiled and straight tubes. With increasing flow velocity, however, even very slight tube curvatures yield such reduction in zone dispersion that unless a correct aspect ratio is introduced, the band broadening would be significantly different from the one calculated by the Taylor–Aris theory. At a De value of 16, when secondary flow has developed, the gain in dispersion reduction for coiled tubing is about 4, which is optimum for FIA designs. Beyond this a decrease in the tube radius or aspect ratio, and/or increase of the flow velocity, results in increased pressure drop, and the cost of equipment becomes prohibitive. The dependency of the radial dispersion in coiled tubes on flow velocity has been the subject of numerous theoretical and experimental studies, of which the works of Tijssen [3.11; 3.12], are the most detailed. For a critical summary, the thesis of Lillig [3.13] should be consulted. Søeberg recently developed a mathematical model, which, in contrast to previous attempts, covers the entire range of velocities and coil geometries, including the 3-D disorientation [3.3]. The model is based on three moment and two mass balance

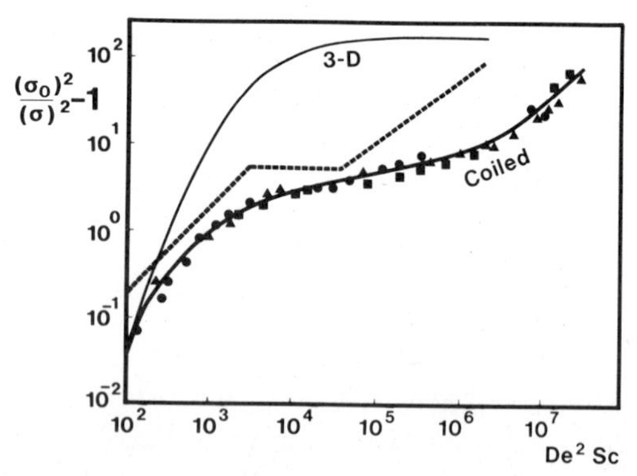

Figure 3.7. Radial dispersion in coiled tubes at low and medium flow velocities. The points represent experimental data [3.13]. The broken line is the model of Tijssen [3.11, 3.12], while solid lines are the model of Søeberg [3.3].

equations:

$$\frac{\partial u}{\partial t} + u\frac{\partial u}{\partial r} + \frac{v\partial u}{r\partial \phi} - \frac{v^2}{r} = \frac{2}{\mathrm{Re}}\left[\frac{\partial}{\partial r}\left(\frac{\partial ur}{r\partial r}\right) - \frac{2\partial v}{r^2\partial \phi} + \frac{\partial^2 u}{r^2\partial \phi^2}\right] - \frac{\partial P}{\partial r} + w^2\frac{R}{R_c}\cos(\phi) \quad (3.31)$$

$$\frac{\partial v}{\partial t} + u\frac{\partial v}{\partial r} + \frac{v\partial v}{r\partial \phi} + \frac{vu}{r} = \frac{2}{\mathrm{Re}}\left[\frac{\partial}{\partial r}\left(\frac{\partial vr}{r\partial r}\right) + \frac{2\partial u}{r^2\partial \phi} + \frac{\partial^2 v}{r^2\partial \phi^2}\right] - \frac{\partial P}{r\partial \phi} - w^2\frac{R}{R_c}\sin(\phi) \quad (3.32)$$

$$\frac{\partial w}{\partial t} + u\frac{\partial w}{\partial r} + \frac{v\partial w}{r\partial \phi} = \frac{2}{\mathrm{Re}}\left[\frac{\partial}{r\partial r}\left(r\frac{\partial w}{\partial r}\right) + \frac{\partial^2 w}{r^2\partial \phi^2}\right] + \frac{F}{4}\left(1 - r\frac{R}{R_c}\cos(\phi)\right) \quad (3.33)$$

$$\frac{\partial ur}{r\partial r} + \frac{\partial v}{r\partial \phi} = 0 \quad (3.34)$$

$$\frac{\partial c}{\partial t} + u\frac{\partial c}{\partial r} + \frac{v\partial c}{r\partial \phi} + w\frac{\partial c}{\partial x} = \frac{2}{\mathrm{ReSc}}\left[\frac{\partial}{r\partial r}\left(r\frac{\partial c}{\partial r}\right) + \frac{\partial^2 c}{r^2\partial \phi^2} + \frac{\partial^2 c}{\partial x^2}\right] \quad (3.35)$$

the three velocity profiles in cylindrical coordinates being in order:

radial: $u = u(r)\cos(\phi)$ (3.36)

tangential: $v = v(r)\sin(\phi)$ (3.37)

axial: $w = w_1(r) + w_2(r)r\cos(\phi) + w_3(r)r^2\cos(\phi) + w_4(r)r^3\cos(3\phi)$ (3.38)

In the concentration profile x is normalized with R and t with R/W_0. The difference between this and all previous models is that it contains harmonic components. The axial velocity in Eq. (3.33) is normalized by W_0:

$$W_0 = 2\int_0^1 wr\,dr = 1 \quad (3.39)$$

Thus, the concentration profile is described by

$$C = [1/(4\mathrm{kt})^{1/2}]\mathrm{e}^{-(x-E(r,\phi)-w_0t)^2/4kt} \quad (3.40)$$

the harmonic components being included in E:

$$E = \mathrm{ReSc}/8[e_1(r) + e_2(r)r\cos(\phi) + e_3(r)r^2\cos(2\phi) + \cdots] \quad (3.41)$$

The solution of the axial profile of velocities w is shown in Fig. 3.8*a* for De = 100, showing translocation of the velocity profiles toward the outer tube wall due to centrifugal forces. The amplitude of the secondary flow was computed as the length of the vector obtained by combination of the tangential and radial velocities (normalized to axial velocity). Thus, the intensity of mass transfer in radial direction can be visualized (Fig. 3.8*b*) from the equivelocity lines that show the strongest radial flow close to the tube wall. The same model was used to compute axial velocity profiles for increasing De values (Fig. 3.9) showing—as observed from the center and the side of the coil—the deformation of the parabolic profile with increasing flow velocity for selected increasing values of De in the range 0–160.

The concentration profiles were obtained with Eqs. (3.40) and (3.41), combined with (3.35):

$$\frac{\partial E}{\partial t} + u\frac{\partial E}{\partial r} + \frac{v\partial E}{r\partial\phi} + w_0 - w = \frac{2}{\mathrm{ReSc}}\left[\frac{\partial}{r\partial r}\left(r\frac{\partial E}{\partial r}\right) + \frac{\partial^2 E}{r^2\partial\phi^2}\right] \quad (3.42)$$

$$k = \int_0^1 \left\{\frac{2}{\mathrm{ReSc}}\left[1 + \left(\frac{\partial E}{\partial r}\right)^2 + \left(\frac{\partial E}{r\partial\phi}\right)^2\right]\right\} dz \quad (3.43)$$

$$k_s = \frac{2}{\mathrm{ReSc}}\left[1 + \left(\frac{\mathrm{ReSc}}{2}\right)^2 \frac{1}{48}\right] \quad (3.44)$$

For zero radial velocity ($u = 0$) the simplest solution for a straight tube is obtained [Eq. (3.44)], showing the familiar Taylor factor 1/48, while k = Pe. Equations (3.40) and (3.41) yield, when the area is normalized to 1, RTD curves, in the simplest case the typical C curves (Fig. 3.10*a*), while, when the factor E [Eq. (3.41)] is included, C curves with isodispersion points are obtained for both mixing cup (Figure 3.10*b*) and area average detector (Fig. 3.10*c*). This is a very important feature of Søeberg's model, since it exhibits agreement with experimental results not previously described, where isodispersion points were obtained when the tube geometry (Fig. 2.10), or the solution viscosity (Fig 3.22), or the diffusion coefficient was varied, while all other experimental parameters were kept unchanged.

Geometrically distorted tubular reactors are designed to promote a sudden change of flow direction, which would increase the radial mass transfer. These so-called 3-D reactors are much more efficient than simple

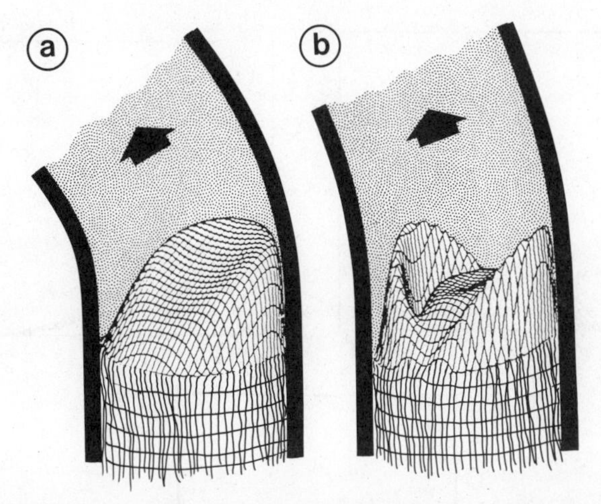

Figure 3.8. Dispersion in coiled tubes according to Søeberg's model [3.13]. (*a*) Equivelocity profiles in axial direction, showing the deformation of the parabolic profile by centrifugal forces. (*b*) Equivelocity profiles in radial direction showing the intense radial mass transfer rate close to the wall (De = 100; for other details see text).

coiled reactors, and, although they are less efficient than packed reactors, they are without doubt more practical to use and easier to make. The milestone in their development was the work of Neue and Engelhardt [3.2] who, by knitting Teflon tubing, produced the the most effective three-dimensionally disoriented reactor. His work was the culmination of a long

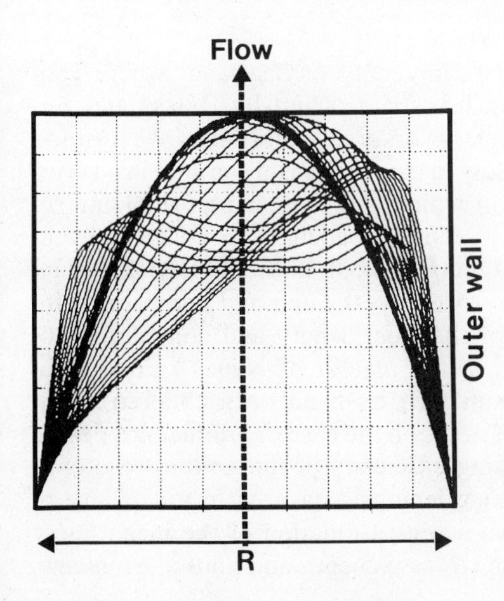

Figure 3.9. Axial velocity profiles in a coiled tube according to Søeberg's model [3.3], computed for increasing De values showing the gradual deformation of the parabolic profile by centrifugal forces. Note the formation of the through in the profile at high De, beginning to appear at De > 16.

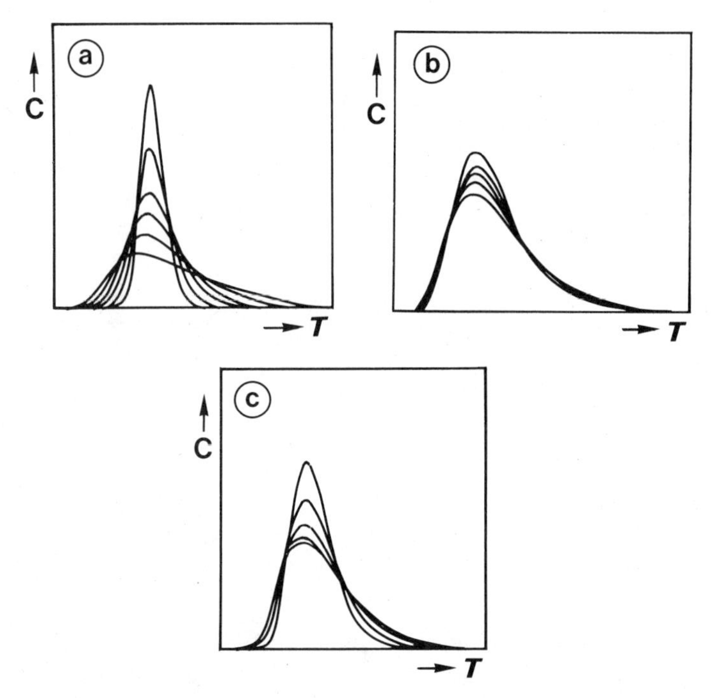

Figure 3.10. Residence time distribution curves in a coiled tube computed with the aid of Søeberg's model [3.3]. (*a*) For $E = 0$ the familiar Taylor's solution is obtained. When the function E is included, then RTD curves showing isodispersion points are obtained for both mixing cup detector (*b*) and area average (across the flow) detector (*c*).

term research of the group at the University of Saarland, where geometrically distorted reactors were investigated first by Halasz and Engelhardt and by a number of their co-workers. A thorough investigation of the geometrically disoriented flow in capillaries is described in a series of papers [3.13], [1248], the aim of which was to design postcolumn reactors for liquid chromatography. Band broadening in coiled (gewendeled), sqeezed (gequetched), beaten (gedrilled), knitted (gestrickte), and stitched (gestickte) capillary tubes was investigated and compared with the dispersion in straight tubes of the same length and comparable diameter. From all these geometries, the reactor of Neue, knitted from Teflon tubing (0.5.mm ID, 1.5 mm OD) by hand on a children's toy, "Strickliesel," is the best tradeoff between the ease of producing a device and its performance. The surprising efficiency of the knitted reactor is seen in Fig. 3.11, where the mixing lengths measured in a capillary of 0.56 mm ID and 10 m long are shown as a function of the mean linear flow velocity F for straight, coiled ($R_c = 8.5$ cm), and knitted geometry.

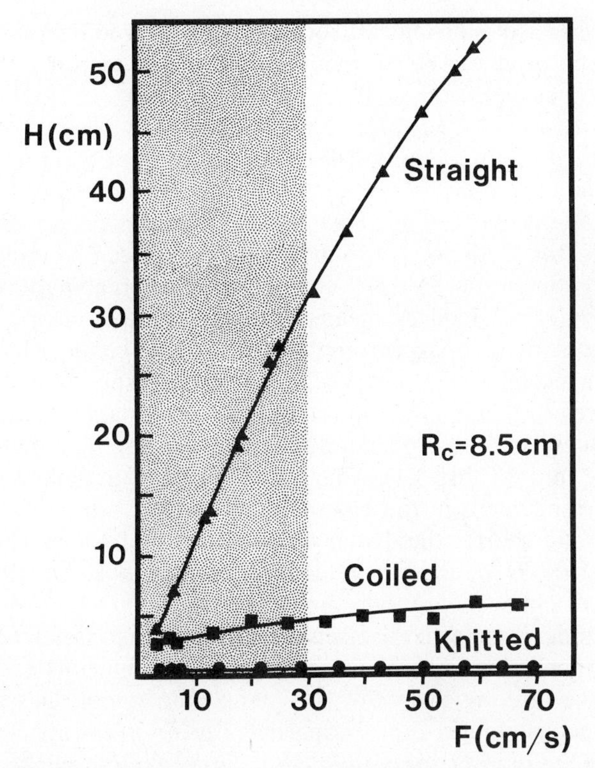

Figure 3.11. Dispersion in straight, coiled, and knitted tubes as function of the linear flow velocity *F* (cm/s). Tube length is 10 m; inside diameter is 0.56 mm. Shaded area covers flow velocities typical for FIA. (According to [3.13], by courtesy of the author.)

Interestingly, the knitted capillary is very efficient over the whole range of flow velocities, the mixing length being on average only 0.47 cm over the full range of flow velocities. A knitted capillary with a smaller diameter (0.36 mm ID) was found to have the mixing height (*H*) only about 3 mm, while the same capillary stitched by a machine had the mixing height about 2.5 mm. Machine knitting allowed production capillaries up to 40 m in length, a feature important to postcolumn derivatization and when slow chemical reactions are to be accommodated in the FIA system. For usual FIA applications, however, much shorter reactors (0.5–1 m) are used, since prolonged residence times are not required. Therefore, knitting may be replaced by simply knotting the tube as closely as practical in order to promote a radial mixing.

The theoretical treatment of dispersion processes in the geometrically disoriented reactors is still in an early stage, yet it appears, that by attenuation of the second harmonic component in Eq. (3.4) by a suitable factor, the effect of sinusoidal disorientation achieved by knitting can be simulated and predicted to be about 25 times more effective than coiling (Fig. 3.7, top curve).

Imprinted geometrically disoriented reactors have been developed for FIA-integrated microconduits [608] with the aim to obtain maximum intensity of radial mixing in a channel, which can be reproduced easily, with a high precision and at a low cost in a solid plate, so that the FIA reactor, injector, detector, and other components can be integrated into one unit (cf. Section 4.12.). It follows from the foregoing that optimum radial mixing is obtained by means of maximum flow disorientation per unit of length traveled. Accordingly, a sinusoidal channel has been designed, with a semicircular cross section, which in spite of its large cross-sectional area (0.8 mm^2), had in the original design a mixing length comparable to a coiled reactor made of tube of a cross section that was much smaller (0.2 mm^2, Table 3.1). This is because the streamlines undergo a drastic rearrangement in the channel between the adjacent bends (Fig. 3.12). Thus the shorter the distance (l) and the more acute the angle (ϕ) is, the lower the H, a, and $\beta_{1/2}$ values will be. The most recently designed microconduit channels are therefore *Z*-formed, rather than sinusoidal.

To conclude, geometric disorientation is the most preferrable form of the FIA channel, since it combines ease of production with excellent radial mixing, while offering a low flow resistance and smooth passage so that unwanted microbubbles or solid particles are not entrapped. Also the volume/surface ratio of these reactors is larger than in packed reactors, which is an important feature for most FIA applications, where interaction between the solutes and the channel surface is not desired.

Table 3.1. Comparison of Dispersion Factors for Five Types of Reactors of Different Geometry

Type of Reactor	Cross-Sectional Reactor Area (mm^2)	V_r (meas.) (μL)	$S_{1/2}$ (meas.) (μL)	Dispersion Factor $\beta_{1/2}$
Microline[a]				
straight	0.2	172.9	70.6	0.41
coiled	0.2	193.7	66.4	0.34
3-D coiled	0.2	184.0	41.5	0.23
SBSR[b]	(0.6)	243.5	69.2	0.28
Meandring μ-conduit unit	0.8	135.6	46.0	0.34
Z-Formed μ-conduit unit	0.8	150.0	46.0	0.29

[a] 75 cm, 0.5 mm ID

[b] Single-bead string reactor (40 cm, 0.86 mm id) filled with 0.5 mm glass beads; the effective cross-sectional area is different from the nominal one.

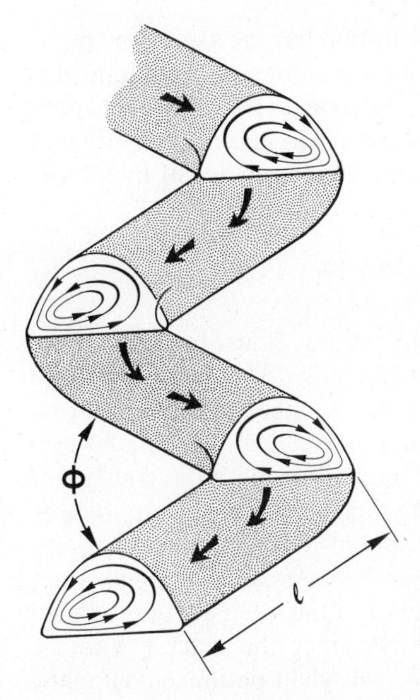

Figure 3.12. Secondary flow pattern in an imprinted Z-formed channel. Note the change in flow pattern as the streamlines turn alternatively from clockwise to anticlockwise at right and left bend. ϕ is the angle of the bend and l is the distance between the bends.

3.1.6 Dispersion in Packed Tubes and in an SBSR Reactor

An effective way how to disrupt the parabolic profile created by the forward movement of the sample zone in laminar flow is to let the zone pass through a reactor packed with an inert material in the form of small spheres. It was van den Berg et al. [74] who were first to suggested this approach for FIA applications and to support their approach with a detailed theory. The real breakthrough tailored to FIA needs came, however, when Reijn et al. [146] in continuity with previous work of Huber [2.3], designed the single bead string reactor (SBSR).

Reactors packed with inert spheres can be visualized as consisting of a series of uniform mixing stages, and can be therefore described by the tanks-in-series model. By rewriting Eq. (3.11) the mixing height H can be obtained from the peak width at the base (W) and from the mean residence time $\overline{T}$, since

$$H = L/N = L/(\overline{T}/\sigma)^2 = \tfrac{1}{16}L(W/\overline{T})^2 \tag{3.45}$$

For $N \geq 30$ the RTD curve will become Gaussian, and $\overline{T}$ will become the peak maximum time T. It is not incidental that these relations remind us of the expressions for the number of theoretical plates, since a reactor packed with inert particles in a FIA configuration has the same flow properties and therefore behaves similarly as a chromatographic column with a nonretained solute passing through it. The similarity is, however, only physical, since the chemical applications are different. In FIA, efficient mixing is sought in the radial direction, and the parameter of interest is the dispersion coefficient D^{max}:

$$1/D^{max} = C^{max}/C^0 = (N/2\pi)^{1/2}(S_v/V_r) \tag{3.46}$$

for small α values ($\alpha = S_v/V_r$) and large N values. Thus, for given α the D^{max}—and the peak height—is a function of N only. Also, the peak width and the sample throughput rate depend on the number of mixing stages, since the axial dispersion in time units (σ_t) is, for $\alpha \leq 0.1$, related to a number of mixing stages by $N^{1/2} = \sigma_t/\overline{T}$. In analogy with the concept of resolution in chromatography, even at the highest sampling rate the distance between adjacent Gaussian peaks should not be smaller than $4\sigma_t$ yielding the minimum permissible time span between adjacent peak maxima to be $\Delta T = N^{1/2}/T$. Again the importance of the intensity of the radial mass transfer rate intensity a is highlighted, since the reactor with the highest N, compared to residence time T, will yield optimized sampling frequency.

The influence of particle diameter (d_p), column diameter, and flow velocity on the dispersion in a packed FIA reactor was described by van den Berg and his colleagues, who, in continuity with their previous work on the use of packed reactors in column chromatography [3.14], concluded that an equation based on Hiby's experimental relationship [3.15] can be used to evaluate the $\sigma_t/\overline{T}$ ratio:

$$\frac{\sigma_t}{\overline{T}} = \left(\frac{(2\gamma D_m/f) + \lambda_1 d_p/[1 + \lambda_2(D_m/fd_p)^{1/2}]}{L}\right)^{1/2} \tag{3.47}$$

where γ is the tortuosity factor, f is the interstitial fluid velocity, and λ_1 and λ_2 are constants characterizing the geometry of the packed bed. This equation, together with the Darcy's equation,

$$\Delta p = \frac{f\eta L}{k_0 d_p^2} \tag{3.48}$$

which was used for calculation of the pressure drop Δp over the reactor (where k_0 is the permeability of the reactor and η is the viscosity of the carrier liquid), allowed the identification of optimum d_p for given combinations of σ_t and $\overline{T}$ for a fixed reactor length L and acceptable pressure drops (Fig. 3.13). Surprisingly, reaction times up to 5 min at sampling rates up to 120 samples/h (for σ_t = 5 s) are predicted for a 25-cm-long reactor, requiring a pressure drop of only 1 atm, well within the working capability of a peristaltic pump.

In an experiment devised to explore the applicability of the packed bed concept in FIA, an even shorter reactor was used (L = 7 cm, R = 4.6 mm), packed with particles of d_p = 40 μm, requiring only Δp = 0.4 atm to maintain a flow rate of 0.55 mL/min. For 30 μL samples, a value of σ_t = 1.8 s was found, corresponding to a sampling rate of 333 samples/h for a residence time of 35 s. Thus in theory a train of samples could be established, where within the packed reactor three samples are found at the same time. An actual experiment, however, did not fulfill these expectations, as, probably due to channeling, the actual sampling rate reached was 120 samples/h.

SBSR is constructed from a tubing (0.75 mm ID) filled with impervious glass beads, which are so large that they nearly fill the tubing (d_p = 0.6 mm). Such a reactor has a void fraction of 0.56 and typically contains 1738 beads per meter of length. Within the FIA working flow velocities, the pressure drop can, according to Reijn [3.17], be described by Eq.

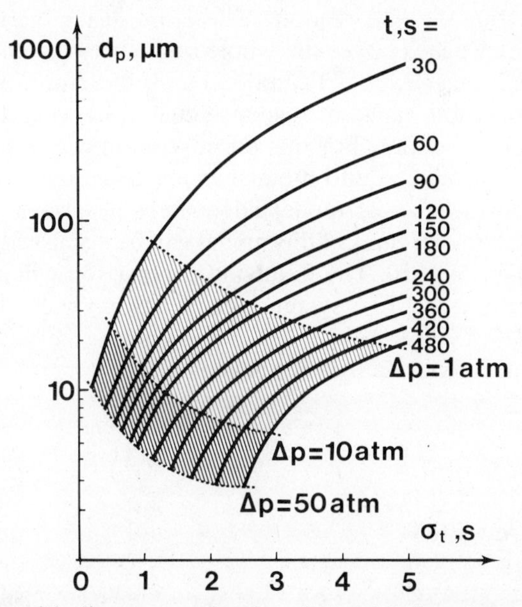

Figure 3.13. The particle diameter as a function of the axial dispersion σ, for various residence times t assuming constant reactor length L = 25 cm. The dashed lines connect points with equal pressure drop. The hatched areas indicate pressures that cannot be sustained by a peristaltic pump. (From Ref. 74 by courtesy of Elsevier Scientific Publishing Co.)

(3.48), requiring a pressure drop approximately 10 times lower than the reactors designed by van den Berg. The most striking feature of the SBSR is that the peak height and volumetric dispersion are virtually independent of the flow rate.

To conclude, packed reactors are the most effective means of promoting radial mass transfer in a flowing stream. They have, however, much lower volume/surface ratio than tubular and 3-D reactors, which is a disadvantage if interactions between solutes and surfaces are undesirable. The smaller the d_p is, the larger N, surface area, and pressure drop will be. In general, short reactors (L up to 25 cm) are practical; beyond that length, the use of packed reactors becomes progressively more difficult.

3.1.7 Influence of Sample Volume

So far, the FIA response curves have been discussed as C curves obtained with an infinitesimal narrow input pulse. While this approach is well suited for description of the relation between axial and radial dispersion for different reactors, it is insufficient for description of a real FIA response curve, since the selection of the injected sample volume S_v is the main tool for optimizing the FIA readout. Therefore, using the terminology of this chapter, FIA peaks cover the whole range between the C curve and the F curve (cf. Figs. 2.6a, 3.1, and 3.2) and, therefore, a useful theory must cover the whole range of response curves between the ideal pulse input and ideal step input. For this purpose an injection parameter $\alpha = S_v/V_r$ must be introduced into the dispersion equations.

Since the tanks-in-series model adequately describes most FIA response curves, two extremes of this model will be discussed, that is, when $N = 1$ and when $N \gg 10$. The following comparison will allow estimate of the behavior of a real FIA system, which has a N value between these two extremes.

For one tank the normalized C curve is

$$C = \frac{1}{\bar{T}} e^{-t/\bar{T}} \qquad (N = 1) \tag{3.9}$$

which can be rewritten

$$C^{\max}/C^0 = \alpha\, e^{-t/\bar{T}} = \alpha\, e^{-\chi} = 1/D^{\max} \tag{3.49$'$}$$

from which the dispersion coefficient $D^{\max}$ can be computed, for a given α value, from a known system constant χ. Thus, if $\alpha = 1$ and $\chi = 0.693$, then $D^{max} = 2.0$, or if $\alpha = 1$ and $\chi = 1$, then $D^{max} = 2.72$ (cf. Fig. 3.3). The system constant for $N = 1$ is simply the ratio of volume of the liquid (Q) that passed through the mixing stage at a given time (t) to the volume of the mixing stage V_m ($\chi = Qt/V_m$). A careful examination of this relation reveals [183] that as the α value increases, the volume of the mixing stage V_m also increases and at values of $\alpha \to 1$, it surpasses the physical volume of the reactor V_r, so that $V_m = 50\%\ S_v + V_r$ (cf. Figs. 3.14 and 3.15). Consequently, the peak height increases at high α values less steeply than predicted by Eq. (2.3) and/or (3.49) and the mixing stage volume V_m is strictly speaking a constant only for a given α value. Equation (3.49) rewritten for peak maximum yields

$$C^{max}/C^0 = 1 - \exp(-0.693 S_v/S_{1/2}) \qquad (2.3)$$

which has been discussed in Chapter 2 and which is shown in Fig. 3.15, curve A.

For a rectangular block input function, transformed during passage through at least 10 mixing stages ($N \gg 10$), Sternberg derived a mathematical description, which was evaluated numerically by Schifreen et al. [3.16], yielding the $1/D$ value as a function of the injected plug width θ, related to dispersion σ_t (Fig. 3.16). More recently Reijn [3.17] developed an exact relationship for a rectangular impulse using N, a, and the injection parameter α:

$$C/C^0 = \tfrac{1}{2}(\mathrm{erf}[(at - N)/(2N)^{1/2}] - \mathrm{erf}[(at - a\xi - N)/(2N)^{1/2} \qquad (3.50)$$

where ξ is the normalized injection parameter ($\alpha N/a$) and erf is the error function. The peak maximum occurs when $t = t_{ave} = (N/a)\ [1 + \alpha/2]$.

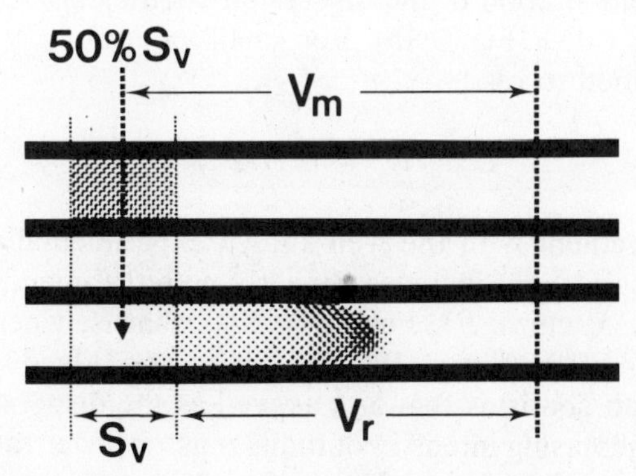

Figure 3.14. Schematic representation of the relationship between the mixing stage (V_m), sample volume (S_v), and reactor volume (V_r) for short tube lengths [183].

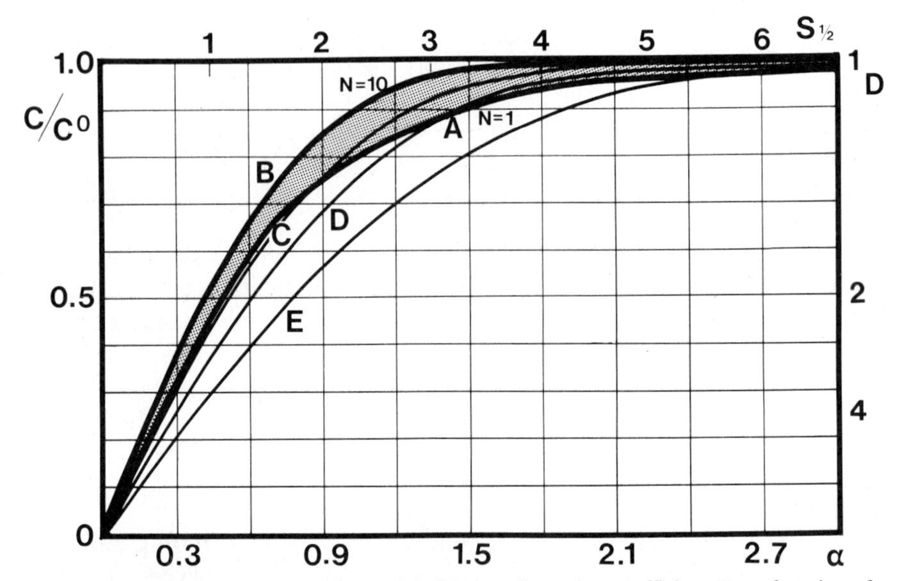

Figure 3.15. Theoretical curves for peak height, or dispersion coefficient D, as function of the injected sample volume. Curve A [Eq. (2.3)] is based on the tanks-in-series model (exp(f) for $N = 1$) when S_v is expressed in $S_{1/2}$ units (top axis). Curve B–E are erf(f) for $N = 10$, $N = 7$, $N = 5$, and $N = 3$, while α ($=S_v/V_r$) is the injection parameter [Eq. (3.51)]. Curve B also conforms with Eq. (3.53). Shaded area covers typical FIA conditions. (Courtesy Carsten Ridder, Chem. Dept. A., Tech. Univ. Denmark.)

Hence:

$$C^{\max}/C^0 = \text{erf}[2\alpha N^{1/2}(2^{1/2})] = 1/D^{\max}, \tag{3.51}$$

which allows computation of the dispersion coefficient $D^{\max}$ from the N value and the α value (Fig. 3.15). For small values of $\alpha N^{1/2}$, Eq. (3.51) is further simplified to read:

$$C^{\max}/C^0 = \alpha(N/2\pi)^{1/2} \tag{3.52}$$

which is in agreement with the well-known experimental result that for $S_v < S_{1/2}$ the peak height increases linearly with the injected sample volume (cf. Fig. 3.15, curve B). For $S_v = S_{1/2}$, that is, when $\alpha = S_{1/2}/V_r = \beta_{1/2}$ and thus $C^{\max}/C^0 = 1/D^{\max} = \frac{1}{2}$, Eq. (3.52) leads to $S_{1/2} = V_r/2(N/2\pi)^{1/2}$, which confirms that $S_{1/2}$ as well as the dispersion factor $\beta_{1/2}$ decrease with increasing intensity of radial mass transfer rate a and therefore with increase of N. It was Hungerford [3.4], who derived from the preceding considerations the dependence between peak height and $S_{1/2}$

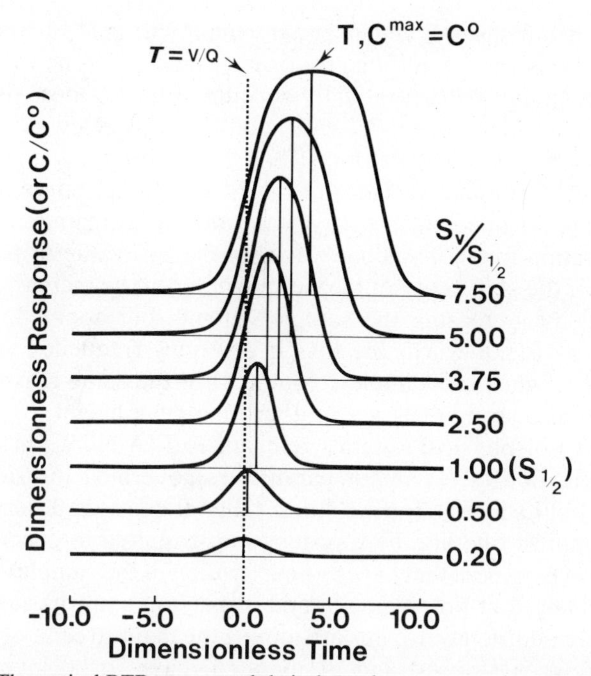

Figure 3.16. Theoretical RTD curves and their dependence on sample volume expressed in $S_{1/2}$ units. [According to R. S. Schifreen, D. A. Hanna, L. D. Bowers, and P. W. Car, *Anal. Chem.*, **49** (1977) 1932, Copyright 1977 American Chemical Society.)

for a Gaussian peak:

$$1/D^{\max} = \operatorname{erf}(\tfrac{1}{2}[\alpha/\beta_{1/2}]) = \operatorname{erf}(\tfrac{1}{2}[S_v/S_{1/2}]) \qquad (3.53)$$

and has shown that $\beta_{1/2} = 1.344N^{-1/2} = 1.344H^{1/2}L^{-1/2}$, which predicts that the *dispersion factor is proportional to the reciprocal square root of the channel length* when the tanks-in-series model applies, and therefore plots of $\beta_{1/2}$ versus $L^{1/2}$ can be used to verify whether the tanks-in-series model applies for a given experimental setup, as well as allowing calculation of N and H without measuring the variance.

To summarize, there is a agreement between the theoretical and experimentally confirmed relationships [cf. Eqs. (2.3) and (3.51) and Fig. 3.15, curves A, B), yet the theory of the injection process is too approximate to allow prediction of the FIA response curve *a priori* with sufficient precision. The reason is that the two discussed extremes ($N = 1$ and $N \gg 10$) do not describe real FIA systems, which typically have $5 < N < 10$. For typical FIA peaks, which are skewed, Eq. (2.3) and a model for $N = 1$ is better suited than the model with $N \gg 10$. Although models for

$2 < N < 10$ might show even closer agreement with real FIA peaks, they will nevertheless not be sufficiently exact, since differencies in injector construction cannot be theoretically accounted for. A more detailed description of the one-tank model with modifications relevant to FIA are found in Tyson [1062] and Pardue [1064].

To conclude, minimal variance of the FIA response curve is achieved by minimizing the injected sample volume and by designing the flow path and the injection mode in such a way that the $S_{1/2}$ value or $\beta_{1/2}$ value is minimized. If the sensitivity of measurement has to be increased, this can be achieved by increasing the sample volume, but above $4S_{1/2}$ further increase of S_v is not worth the loss in sampling frequency and in consumption of reagent and sample solutions. For the same reason, nothing is gained by measuring peak area rather than peak height.

The theory of physical kinetics relevant to FIA allows in its present form identification of the major parameters governing the sample zone dispersion, but its limitations are still serious. It does not describe exactly even the simplest one-line FIA system and is limited to description of a zone that moves at constant speed. Acceleration of the sample zone during injection, changes of flow velocity and dispersion patterns at confluence points, where additional streams are joining the main stream, or dispersion when the flow is stopped and accelerated, have to be investigated in greater detail (cf. Section 3.2).

Experimentally obtained values of D and $D^{\max}$, together with t and T, are at the present state of art the parameters that describe mixing of components and reaction times in a real FIA system with sufficient precision, while α, $\overline{T}$, and $S_{1/2}/V_r$ offer only an estimate of the physical optimization of the system and of the maximum achievable sampling frequency.

The value of the theoretical models reviewed is that they allow the application of the wealth of information and experience from chemical reactor engineering and from analytical chromatography to FIA. The differences between the practical goal of these three disciplines, however, have to be kept in mind.

3.2 CHEMICAL KINETICS IN FIA SYSTEMS AND OTHER THEORETICAL ASPECTS TO BE ADDRESSED

While the models and theories summarized in the previous section describe satisfactorily the physical processes taking place in various reactor types, the problems reviewed here are open ended, cover the ground less systematically, and reveal how much the theory has lagged behind experiments and practical application of FIA. Thus this section contains many more questions than answers. Yet this lack of knowledge is exactly what makes research a long lasting challenge and what gives equal opportunity to newcomers and experienced researchers to make future discoveries. It is therefore hoped that the topics mentioned here will serve as inspiration and as a starting point for further work. The literature concerning these topics is scattered or nonexistent. Exceptions are works dealing with chemical derivatization in chromatography, where some of the design parameters have been dealt with. The monograph of Frei and Lawrence [3.18] is an excellent source of useful references, while yielding a comprehensive view of the development within this discipline so relevant to FIA.

3.2.1 Chemical Kinetics in Straight, Packed, and Coiled Reactors

Only few theoretical studies have been published dealing with this topic. Following the first attempt of Haagensen ([153, p. 25], cf. Section 2.7), Painton and Mottola [541] published a computer simulation based on laminar dispersion and reaction rate equations. Their work illustrates the complexity of the problem and shows the discrepancies between theory and experiment. While their arguments are not followed easily, the random walk model of Betteridge et al. [702] gives a clear description of the interplay between physical and chemical kinetic, both verbally and visually, since the model is also available in colored computer graphics, which unfortunately can be shown here only in a still and black and white version (Figs. 3.17 and 3.18). Although approximate, owing to the low number of elements and steps simulated in its present form (Fig. 3.19), this original approach should prove successful in visualizing the majority of processes relevant to FIA. More exact mathematical models dealing with chemical kinetics in FIA system have been presented by Shih and Carr [905] for a coiled reactor, by Reijn et al. [554] for a packed reactor, and by Wada et al. [1065] for a straight tube reactor.

The computer simulations of chemical kinetics in a straight tube reactor [1065] were based on an equation combining diffusion, convection, and reaction terms. The sample dispersion without chemical reactions gave very similar results to that of Vanderslice [1061], yet the value of that paper is that it expanded the study to computation of FIA response curves for fast and slower chemical reactions. The numerically evaluated equation was similar to that of Vanderslice [1061], however with inclusion of a term for reaction rate. Two model systems were chosen and spectrophotometrically monitored in a FIA system with appropriately con-

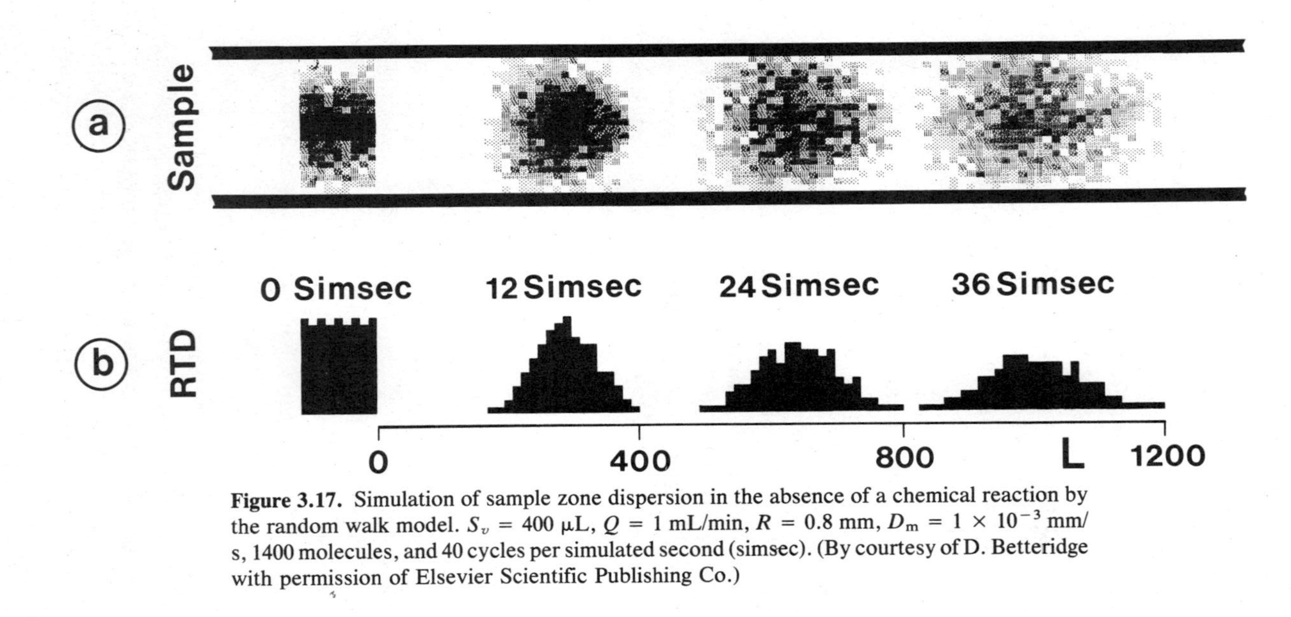

Figure 3.17. Simulation of sample zone dispersion in the absence of a chemical reaction by the random walk model. $S_v = 400\ \mu L$, $Q = 1$ mL/min, $R = 0.8$ mm, $D_m = 1 \times 10^{-3}$ mm/s, 1400 molecules, and 40 cycles per simulated second (simsec). (By courtesy of D. Betteridge with permission of Elsevier Scientific Publishing Co.)

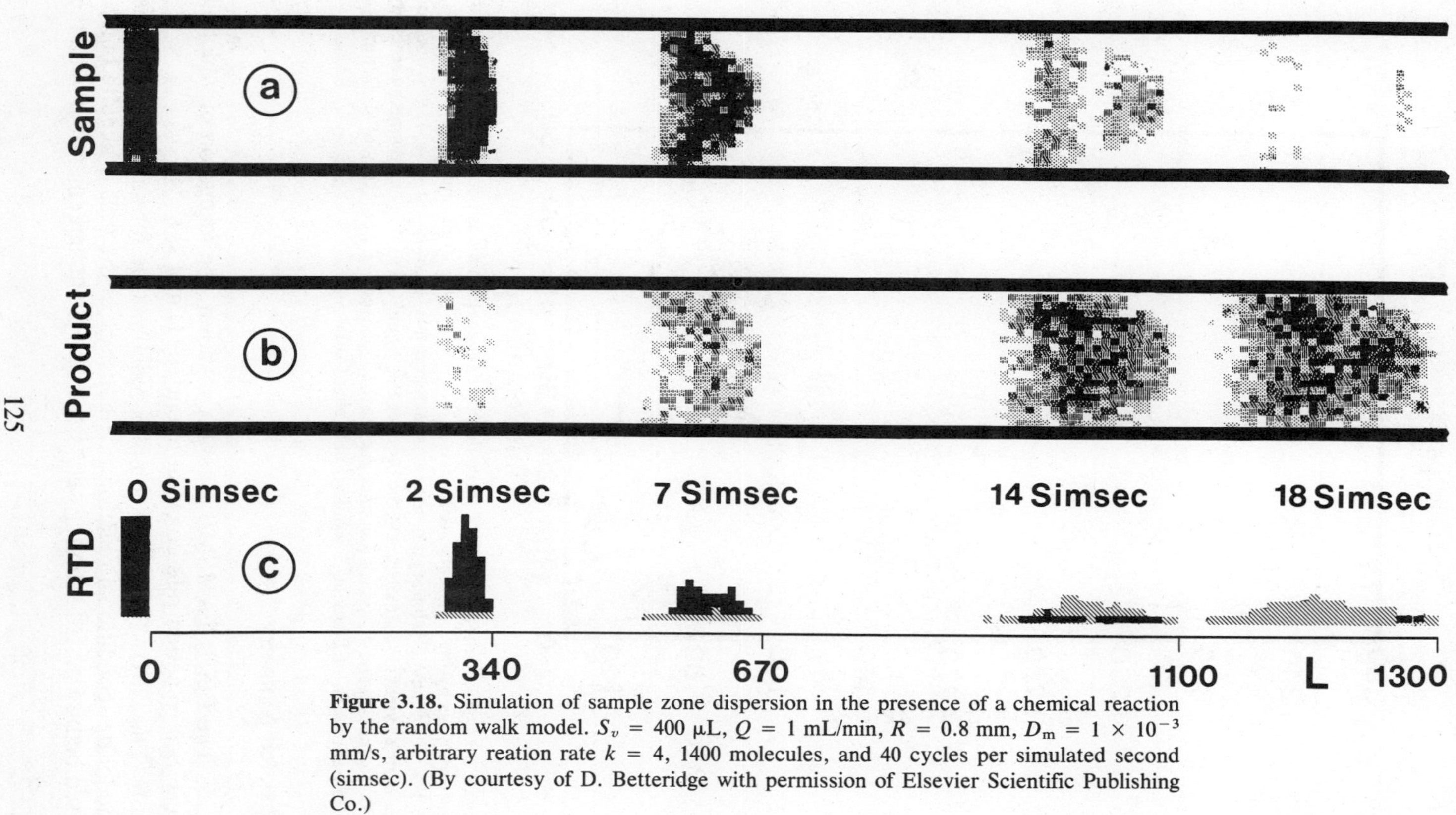

Figure 3.18. Simulation of sample zone dispersion in the presence of a chemical reaction by the random walk model. S_v = 400 μL, Q = 1 mL/min, R = 0.8 mm, $D_m = 1 \times 10^{-3}$ mm/s, arbitrary reation rate k = 4, 1400 molecules, and 40 cycles per simulated second (simsec). (By courtesy of D. Betteridge with permission of Elsevier Scientific Publishing Co.)

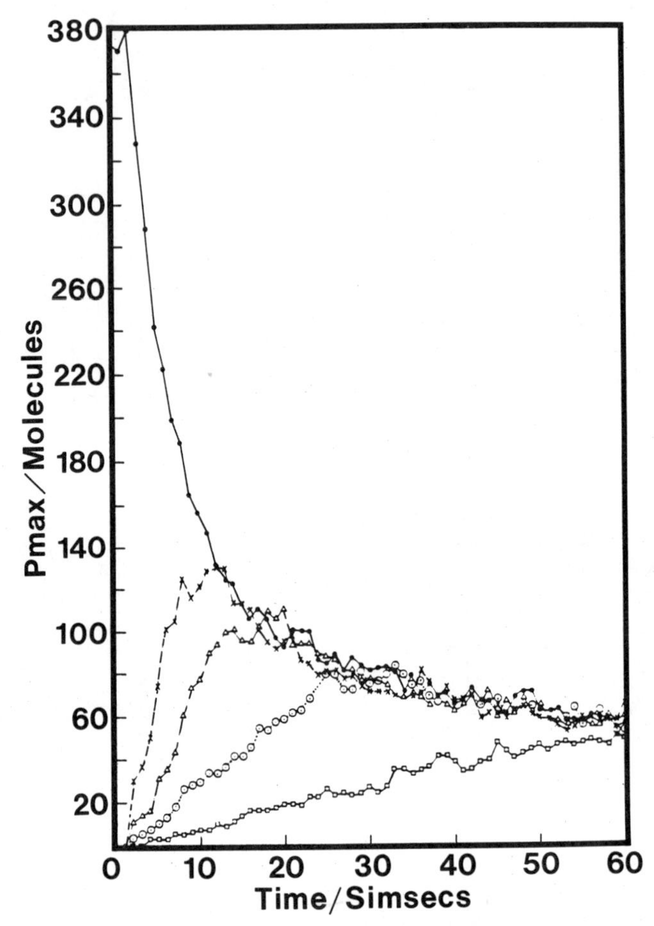

Figure 3.19. Simulation of product formation by chemical reactions occurring at decreasing reaction rates by the random walk model: S_v = 100 μL, Q = 1 mL/min, R = 0.8 mm, D_m = 1 × 10^{-3} mm/s. Arbitrary reation rates increase: k = 0.25, k = 1, k = 4, k = 16. For the highest yield (P_{max}/molecules) the product P is simulated as if injected already reacted; 1500 molecules and 40 cycles per simulated second (simsec). (By courtesy of D. Betteridge with permission of Elsevier Scientific Publishing Co.)

structed geometry:

a. The Cu(II)-TAMSMB complex reaction, the formation of which is very fast (reaction rate constant about 10^8 L mol^{-1} s^{-1}).

b. The Ni(II)-TAMSMB complex reaction, the formation of which is moderate (conditional reaction rate constant in the 0.05*M* acetate buffer used being 6 × 10^3 L mol^{-1} sec^{-1}). An excellent agreement between the experimental and simulated signals were found for both the fast and the slow reactions. The response curves exhibit the humped form characteristic for the straight tubular reactor, the hump being gradually smoothed out with increase of residence time (Fig. 3.20)—as predicted by theory. A very similar pattern is observed for the slow chemical rection, which, however, having a lower conversion factor yields correspondingly reduced response curves in vertical direction. In addition to computing the peak shape concentrational microstructure is also presented in this detailed work (Fig. 3.20, right), again showing patterns characteristic for this type of reactor. Thus, important information is gained: the chemical reaction does *not* appreciably alter the concentration gradients and patterns formed by dispersion in the given reactor geometry.

The chemical kinetics in a packed reactor have been studied in a detail by Reijn, Poppe, and van der Linden [554] with the aid of the tanks-in-series model. They based their theory on three assumptions: that the number of tanks N in the reactor is constant; that the chemical reactions are (pseudo) first order; and that "adequate" mixing is ensured.

When a chemical reaction between analyte A and reagent R results in a detectable product P formed in the presence of a large excess of R, one can write:

$$\frac{d[A]}{dt} = -k'[A] \tag{3.54}$$

where $k' = k[R]$. They derived in an elegant way the RTD curve for the product P and its statistical moments and discussed their magnitude for

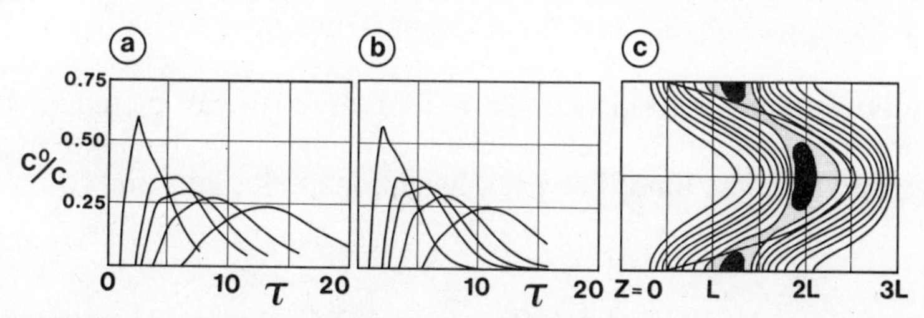

Figure 3.20. Experimental *a* and simulated *b* response curves for formation of a Cu(II)-TAMSB complex in a straight tubular reactor. Values of $(\theta + L)/\theta$ and Q were: (1) 2.0, 0.58; (2) 3.0, 0.29; (3) 3.5, 0.23; (4) 4.3, 0.17; and (5) 6.0, 0.12. Note the presence of the humped peak at $\tau = 2$ for both the simulated and the experimental curves. The simulated concentration profiles c for $\tau = 2$ show the concentration maxima, which, when perpendicularly viewed by the detector, yield a humped peak. (According to Ref. 1065 by permission of Elsevier Scientific Publishing Co.)

two extreme cases. First for a very fast chemical reaction (when $k' \to \infty$):

$$\lim m_t = N/a = \overline{T} \quad \text{while} \quad \sigma_t^2 = N/a^2 \tag{3.55}$$

where we recognize the familiar equations for the first (m_t) and second (σ_t^2) moments for a tracer *if injected as such*. Then for very slow reactions (when $k' \to 0$):

$$\lim m_t = (N + 1)/a \quad \text{while} \quad \sigma_t^2 = (N + 1)/a^2 \tag{3.56}$$

which for large N are identical with the moments for a very fast chemical reaction confirming, that the moments of the RTD curve are *independent* of reaction rate. This important conclusion is in agreement with common sense, since it is obvious that the injected sample zone would disperse in the same way whether it reacts or not. Thus, although the response curve may be in some cases narrower than an RTD curve depending on the rate of formation (or disappearance) of a sensed species, *it is the RTD curve which governs the sampling frequency*, and, therefore, it is of no use to inject sample zones at a frequency higher than the one found from the variance or from the $\beta_{1/2}$ value obtained by a dispersion experiment. For the same reason the term "chemical dispersion" [181] is ill conceived.

It is significant that the preceding conclusions of Reijn et al., obtained for SBSR reactor, are in agreement with the results that Wada et al. [1065] obtained with a straight tube reactor, thus confirming the important conclusion that chemical reactions do not alter the dispersion of the analyte in the reactor. Therefore, the experimental values (D, t, $\overline{T}$, and σ^2), obtained with a nonreacting tracer alone, are well suited for the description of dispersion in any FIA system in the presence or absence of chemical reaction.

It follows that the residence distribution function can be modified by the conversion factor $[1 - \exp(-k't)]$ to obtain the vertical readout for any given residence time. For peak height and delta injection ($S_v \ll V_r$) Reijn et al. [554] found for the SBSR reactor:

$$P_{\max} = \frac{S_v C_A^0 Q N^{1/2}}{V_r(2\pi^{1/2})} [1 - \exp(-k'T)] \tag{3.57}$$

which relates the amount of product in the element of fluid corresponding to the peak maximum, $P_{\max}$, and therefore, *if product P is monitored the peak height will increase with increase of the reaction rate. If, however,*

A is monitored, then the peak height will decrease with increase of the reaction rate (cf. Wada et al.s experiments [1065] described previously).

For a rectangular block input, describing the zones as normally injected in FIA, the parameter $\alpha = S_v/V_r$ must be introduced, which leads to very complex functions which according to the authors are awkward to evaluate and therefore only extreme values of k' were considered:

$$\lim_{k' \to \infty} P_{\max} = C_A^0 \alpha (N/2\pi)^{1/2} \tag{3.58}$$

$$\lim_{k' \to 0} P_{\max} = C_A^0 \alpha (N/2\pi)^{1/2} k't \tag{3.59}$$

which again confirms that for fast reactions the peak height is the same as if the analyte has been injected as if it had already reacted, while for slower reactions an increase of residence time t and of reaction rate will lead to increase of $P_{\max}$.

The degree of conversion was computed using Eq. (3.55) by means of basic relations for a reactor with large N with the result that at constant flow rate the maximum achievable conversion is 71%. Remarkably, Shih and Carr [905] computed, for a FIA reactor of an unspecified geometry, the maximum conversion to be 71.5% using a different, very elegant mathematical approach.

One might wonder why the present theories, although yielding useful information on system constraints (such as the degree of conversion, pressure drop, reactor length, sampling frequency, etc.), are inadequate to yield an exact description of the response curve, so that calibration or evaluation of the system performance (by an "empiric") dispersion experiment could be eliminated. The reason is that boundary conditions used in these theories are so restrictive that theoretical considerations become too remote from the actual experiment. Thus, for values of $\alpha > 0.3$—which are very frequently encountered in FIA—*none* of the three assumptions postulated above are met, since in a single-line FIA system, as discussed by Reijn et al. [554], the reagent penetrates the sample zone *gradually from both ends of the injected zone, while product is formed in the presence of gradually increasing concentration of reagent B*. More refined approaches such as the work of Hungerford [3.4; 838] are the first step toward recognizing the complexity of the kinetic processes taking place in an FIA channel.

To conclude, the theoretical works combining chemical kinetics and dispersion phenomena in FIA reactors outline the limits of the continuous-flow concept and emphasize the importance of radial mass transfer rate [a, Eqs. (3.55) and (3.56)] on optimization of the reactor design. The theoretical limit of 70% conversion is, however, not the ultimate upper limit achievable in FIA system. As discussed in Chapter 2, the stopped-flow approach performed at low D values, using highly concentrated reagent, will lead to conversions above 90%. In theory close to 100% conversion can be reached using reactors packed with solid reagents [566] operated at stopped flow. Similarly, when optimizing the sampling frequency and maximizing the resident time, alternatives to continuous flow should be considered.

3.2.2 Chemical Kinetics in a Stirred Tank

The radial mass transfer in a single stirred tank is the most intense encountered in all conceivable reactor designs, since it is, in addition to molecular diffusion and flow rate, promoted by an external force exerted by the mixer on the liquid in the mixing chamber. This is why Gissin et al. [[1063] recommend the mixing chamber as a high-precision device for extensive dilution of viscous materials. Another feature of the stirred tank is that its holdup volume V_r is a well-defined physical parameter, which together with the injected sample volume S_v defines the system exactly, thus avoiding the need for artificial boundary condition. In FIA context the concept of single stirred tank has found its applications in peak width measurements and in FIA titrations (cf. Chapters 2 and 4) and in the works of Tyson [1062], Gisin et al. [1063] and Pardue and Jager [1064], which all deal with the theory of physical and chemical kinetics. Being useful sources of information, they should be consulted for the detailed mathematical treatment of the following qualitative description, where the tank function is divided into three distinct phases.

As the sample begins to enter the gradient chamber, the species will react with the reagent in the tank until all reactant present in the tank is consumed. In the next phase the concentration of the sample material in the tank will rise until all sample solution has entered the chamber. In the third phase the concentration of the sample material in the tank will decrease as a result of dilution and reaction with the reagent flowing into the tank. Kinetically, the process during the first phase is pseudo-zero-order because the sample zone is flowing into the tank at a fixed rate and $V_r \gg S_v$. The process in the second phase is pseudo-first-order, being a purely physical process of dilution, while the third phase is described by as combined first-order–zero-order [1064]. The net result is that the peak width is a linear function of log C_A^0, with flow rate and sample and reactor volumes as remaining physical variables. While the model has been solved for FIA only for fast chemical reactions, solutions for finite reaction rates and different types of reaction orders are given in a texts on chemical reaction engineering [3.1].

To conclude, the FIA response curve obtained in such a system, which conforms to the tanks-in-series model, reflects the reaction rate at which the detected species is formed—or consumed—and on the physical parameters of the system. The RTD curve, however, in analogy with what applies to tubular and packed reactors, is the function of the physical parameters alone. Thus even if the response curve is narrower than the RTD curves obtained by the tracer experiment, it is of no use to try to reach the thus seemingly achievable higher sampling frequency, since the reactant—or lack of reagent persisting in the chamber—will result in carryover and in loss of reproducibility.

3.2.3 Mixing, Diffusion, and Dialysis Units; Detectors

The confluence of two streams is a typical FIA configuration, which was shown to be feasible in very early FIA works [3–5], thus being crucial to the subsequent wide acceptance of the technique. It was established at that time [20] by a number of dispersion experiments that peak height and peak area decrease in proportion to the flow rate of the added stream, that the peak width in volume units increases, while in time units the peak width remains unchanged, since the increase of the linear flow velocity offsets the volumetric increase of the peak width (Fig. 3.21). The resulting concentration profiles at different flow rates are discussed in Section 2.3.2, but little is known about the rate at which they are established and about the structure of microgradients in mixing tees. The proper construction of mixing units has been of a major concern in postcolumn chromatography, since differences in flow rates, viscosities, and densities of the merging streams cause an extra broadening effect and in extreme cases (such as laminar flow and difference in density), the confluenced streams may run parallel downstream of the mixing point for up to ten centimeters without intermixing. According to Frei [3.18] the most favorable configuration of the confluence is achieved when the three streams meet at the top of an arrow into which the streams enter through the arms of the arrow and exit through its stem, the angles between the arm and the stem being 30° or less. This configuration has therefore been adopted in Chemifolds™ of the FIAstar analyzer (cf. Section 5). The mixing mechanism in the tee is not yet explained, but the random walk model might offer insight into the microstructure of the concentration gradients. It should be mentioned that combination of a mixing tee with a very short (ca 1 cm) SBSR [2.3] is an attractive approach to mixing streams of widely differing viscosities.

Transfer of solutes from a donor stream, which is separated from the acceptor stream by a permeable membrane, was studied by van der Linden for gases [430] and by Bernhardsson, Martins, and Johansson [852].

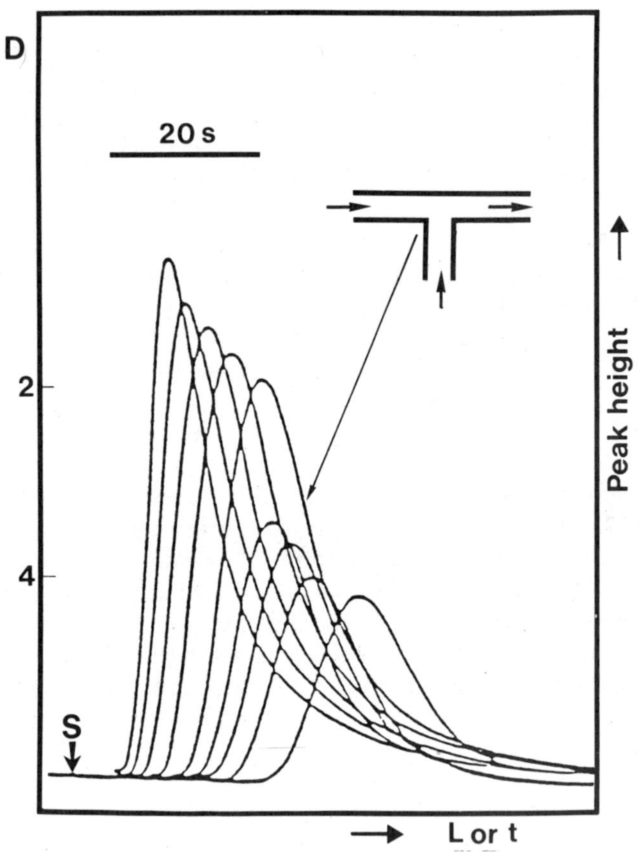

Figure 3.21. Confluence of two streams pumped at equal flow rates and the corresponding RTD curves obtained with a sample zone containing a tracer dye. Peaks were recorded sequentially with a single-line manifold of L = 25, 50, 75, 100, and 125 cm and of 0.5-mm-ID tubing, then a mixing tee was included, by means of which a second colorless stream was added and L was then increased to 150, 175, 200, and 250 cm. Note the abrupt decrease of the peak hight, peak area, and unchanged peak width due to addition of the second stream [20]. (By permission of Elsevier Scientific Publishing Co.).

These parallel-plate devices have identical design for gas diffusion and dialysis, the only difference being the membrane material (Sections 4.6.2 and 4.6.3), and therefore the theory of these two systems has a common feature as far as transport of the analyte from the bulk toward the membrane surface is concerned.

For gases diffusing through a microporous hydrophobic membrane van der Linden [430] used the tanks-in-series model, by assuming that the groove of the separation unit, having a volume V_r, consists of a series of hypothetical tanks N and that the transfer of the species across the membrane is proportional to its concentration and its residence time. The transfer coefficient was not specified, but it comprises a lump contribution of diffusion from the bulk to the membrane, partition between the donor stream and the membrane, diffusion inside the membrane together with the corresponding reverse steps on the acceptor stream side. Two extremes were discussed: when the vapor pressure of the gas is low and its solubility in the donor stream is high. Then the peak height will be directly proportional to the residence time and inversely proportional to the flow rate. At the other extreme, when the gas diffuses rapidly and is readily soluble in the acceptor stream, the peak height will be independent of flow rate. The theory was supported by a number of experiments using gases of different solubilities and permeabilities.

Dialysis was studied in detail by Bernhardsson, Martins, and Johansson [852] with aid of three different transport models: the tanks-in-series model, the plug-flow model, and the laminar-flow model together with Cooney's theory of mass transfer across a membrane. They have shown, in agreement with theory, that the dialysis efficiency decreases with increasing flow rate and depth of the channel. They pointed at the existence and influence of pressure difference between donor and acceptor streams—a factor that is often neglected. To conclude, the efficiency of both gas diffusion and dialysis units is enhanced by having the channels as shallow as practically possible with a geometry promoting radial mass transfer rate.

Use of hollow fibers for construction of FIA gas diffusion, dialysis, and ion-exchange units is one of the most promising areas of future FIA development and miniaturization. The concepts, construction, materials, and devices borrowed from ionic chromatography are excellently suited to FIA needs. It is beyond the scope of this work to speculate on the outcome of the marriage of these two techniques, and it is up to the reader to enjoy speculations along these lines. They will come readily in mind while reading the recent monograph of Tarter [3.19] and especially the inspiring chapter by Dasgupta [3.20].

The variety of detectors used in FIA is wider than that in chromatography (cf. Chapter 5) and therefore criteria valid for chromatographic detectors are not generally applicable for all FIA detectors. However, the experience gained in design of chromatographic flowthrough cells is valuable for design of FIA detectors. For general theory, the work of Poppe [72] should be consulted, since it constitutes the cornerstone of rational design, describing the detector characteristics in terms of the theory of flow by volume standard deviation (σ_v) and by the detection limit. The theory of amperometric detectors has been discussed in detail by Meschi and Johnson [155] using Taylor's model of flow and by Olsson et al. [1103] in conjunction with the development of an amperometric-based glucose electrode. The theoretical response of this electrode was found in excellent agreement with the experimental results, described in the original paper, and with the results obtained recently in our laboratory. The theory of coulometric detectors has been treated by Meshi et al. [156] as a continuation of previous work with tubular amperometric detectors [155] expanded for current/time response curves under FIA conditions. The peak area (and peak height) was found to be a linear function of the concentration of analyte and therefore single-point standard addition was proposed for instrument calibration. Atomic absorption has been discussed by Appleton and Tyson [1022] who concluded that the nebulizer function conforms best with the model of two parallel tanks. A thorough discussion of the aerosol formation and droplet size distribution illustrates the complexity of nebulization process. The work deals extensively with all parameters relevant to efficient operation of nebulizers in a continuous-flow mode and to automated calibration of an atomic absorption (AA) instrument incorporated into FIA systems. It is therefore useful also for the theory and praxis of inductively coupled plasma (ICP) emission spectrometry.

3.2.4 Heterogeneous Samples; Viscosity and Surface Properties

The ability of FIA systems to handle heterogeneous samples has been recognized since turbidimetry and nephelometry was successfully performed. However, while the dispersion theory of homogeneous solutions has been studied frequently, the behavior of heterogeneous samples processed in a FIA system have been discussed only by Janata and Harrow [881, 882, 1116]. In their pioneering work they correctly point out the problems arising when a significant amount of suspension is present in the injected sample. They suggested the ways of how to compensate for this effect. Basically, in the presence of the solid fraction $h (0 < h < 1)$, the effective dispersion of the soluble material is

$$D^{\text{eff}} = (D - h)/(1 - h) = C^0/C^{\text{eff}} \tag{3.60}$$

where C^{eff} is the concentration of the soluble material after dispersion. Thus the same initial concentration of soluble material C^0 yields various resulting concentrations C^{eff} in presence of various amounts of solid fraction h.

For determination of an unknown analyte in whole blood (containing typically 40% hematocrit) this will pose a problem, unless the instrument is calibrated by a standard with the same hematocrit content. The same would apply for other solid suspensions, like soil samples or industrial slurries. According to the authors, the problem could be overcome by adding a tracer (not present in sample material) to the carrier stream. By measuring the tracer concentration simultaneously with the analyte, the h value can be computed and corrected for.

The above considerations reveal that FIA has yet untried capability for determination of suspended matter in liquid samples. Furthermore, hydrodynamic effects on particles of different size and shape might lead to their rough separation or characterization. Much still has to be done both theoretically and experimentally in this area, related to field flow fractionation.

The influence of viscosity on the sample zone dispersion has been subject to several discussions and studies [7, 162, 224, 362, 363] aimed at establishing its effect and at measurement of viscosity by means of FIA. Generally speaking, with increasing viscosity of the injected sample, the zone broadening decreases and therefore the viscosity of standards and unknown samples either has to be carefully matched, or reduced by appropriate dilution in a properly designed manifold. Most recently an interesting observation was made by Petersson (Fig. 3.22) who established the existence of an isoresponse point when investigating influence of varying viscosity and density of injected glucose samples to which different amounts of glycerine were added. Since a similar isodispersion point (Fig. 2.10) has been observed for different reactor geometries, and since the intensity of radial mass transfer rate is the factor underlying these two phenomena, it will be exciting to see whether a theoretical description of these observations will lead to a prediction of how these practically important isopoints can be predicted, obtained, and exploited.

The influence of the properties of the wall materials, such as wettability and roughness, have not yet been considered. Since the surface/volume ratio increases with decreasing tube diameter, these material characteristics, together with the wall absorptive capacity for analytes, are indeed of importance and their future study will be, undoubtedly rewarding and revealing.

There are few investigations of the temperature effect on the dispersion and chemical kinetics in the FIA system. Since an increase in temperature increases molecular diffusion—and thus radial mass transfer—the physical dispersion will decrease with increasing temperature. When the contribution of external forces on the radial mass transfer is minimized, as in straight very narrow tubes, the temperature effect should be greatest, and it should decrease in reactors with progressively distorted geometry. Betteridge et al. have simulated the temperature effect by a random walk

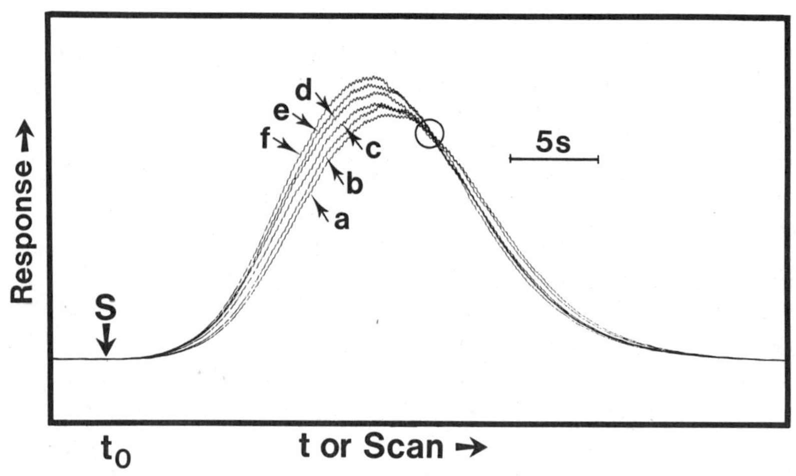

Figure 3.22. Influence of viscosity of the injected sample on the response curves obtained when injecting a glucose sample (200 mg glucose/100 mL) into an FIA system using immobilized enzyme (glucose oxidase) and chemiluminiscence detection. Curves a–f: 0, 10, 20, 30, 40, and 50 v/v% of glycerol. Note the presence of the isoresponse point. (By courtesy of Bo Petersson.)

model [702; 3.21]. Experimental work concerned with the influence of temperature on the dispersion alone and on peak dimensions when fast chemical reactions are involved is under way [3.22], yet further investigations along these lines are much desirable.

REFERENCES

3.1 O. Levenspiel, *Chemical Reaction Engineering*, 2nd. ed. Wiley, New York, 1972.

3.2 U. Neue and H. Engelhardt, *Chromatographia,* **15** (1982) 403.

3.3 H. Søeberg, (unpublished).

3.4 J. Hungerford, Thesis, Univ. of Washington, 1986.

3.5 H. Bate, S. Rowlands, and J. A. Sirs, *J. Appl. Phisiol.* **34** (1973) 866.

3.6 V. Anathakrisnan, W. N. Gill, and A. J. Bardhun, *J. Am. Inst. Chem. Eng.,* **11** (1065) 1063.

3.7 T. Thompson, *Proc. Roy. Soc., Ser. A.* **25** (1876) 5; **26** (1877) 356.

3.8 K. Hofman and I. Halasz, *J. Chromatogr.,* **173** (1979) 211.

3.9 L. R. Austin and J. D. Saeder, *A. I. Ch. E. J.,* **19** (1973) 85.

3.10 R. N. Triverdi and K. Vasudova, *Chem. Eng. Sci.,* **30** (1975) 317.

3.11 R. Tijssen, *Sep. Sci. Technol.,* **13** (1978) 681.

3.12 R. Tijssen, Thesis, Univ. Delft, 1979.

3.13 B. Lillig, *Theorie und Praxis in der Modern Flüssigkeitchromatografie,* Thesis, Univ. des Saarlandes, 1984.

3.14 J. H. M. van den Berg, Thesis, Eindhoven Univ. Technol. 1978.

3.15 J. W. Hiby, in *Proceedings of Symposium on Interaction between Fluids and Particles* (P. A. Rottenburg, Ed.), Institution of Chemical Engineering, London, 1962, p. 312.

3.16 R. S. Schifreen, D. A. Hanna, L. D. Bowers, and P. W. Carr, *Anal. Chem.,* **49** (1977) 1932.

3.17 J. M. Reijn, Thesis, Univ. of Amsterdam, 1982.

3.18 R. W. Frei and J. F. Lawrence, *Chemical Derivatization in Analytical Chemistry,* Plenum Press, New York, 1981, Vols. 1 and 2.

3.19 J. G. Tarter, *Ion Chromatography*, Marcel Dekker, New York, 1987.

3.20 P. K. Dasgupta, "Approaches to Ionic Chromatography," Chapter 6, in Ref. 3.18.

3.21 D. Betteridge, J. T. Sly, A. P. Wade, and J. E. W. Tillman, *Anal. Chem.,* **55** (1983) 1292.

3.22 C. L. M. Stults, A. P. Wade, and S. R. Crouch, *Anal. Chim. Acta,* (to be published).

3.23 M. J. E. Golay, *J. Chromatogr.,* **186** (1979) 341.

CHAPTER

4

TECHNIQUES

Once you have exhausted all possibilities and failed,
there will be one solution, simple and obvious,
highly visible to everyone else.

SNAFU

The methods described in this chapter have been designed to accomplish various analytical tasks, ranging from intelligible sample transport processes in systems where no chemistry is involved, such as those encountered in atomic absorption spectrometry or embodied in procedures for pH measurement, up to chemical separations effected by means of two-phase systems where ions are exchanged, or ions or gases are diffused through membranes, or where analytes on incorporated packed reactors are degraded, converted, or preconcentrated. In all these designs, the concept of controlled dispersion is employed for optimizing the manifold design (Chapter 2). The analytical readout is taken either at the peak maximum or at other parts of the concentration gradient. The techniques described herein are, for the sake of clarity and conceding to the majority of the available FIA literature, divided into sections where the dispersion employed is manipulated with the aim of taking the readout either at the peak maximum or along the gradient. Since this book does not attempt to be comprehensive, although striving to cover the technical concepts of flow injection analysis (FIA), only selected examples of the various techniques are described. For specific applications and/or techniques the reader is advised to consult Chapter 7. The individual components of the FIA systems are described in greater detail in Chapter 5, and hints on operational details are recounted in Chapter 6 together with exercises. Information as to the use of FIA in continuous monitoring and on-line control is presented in Chapter 8 along with future trends in developments of the technique.

4.1 SINGLE-LINE FIA MANIFOLDS

The simplest FIA system consists of one tube through which the carrier stream moves toward the flowthrough detector. Depending on the injected volume S_v, tube length L, flow geometry, and element of fluid along the gradient selected for the analytical readout, *limited*, *medium*, and *large* values of the dispersion coefficient D might be achieved (Chapter 2). Restricting ourselves first to procedures where the source of information is identified at the peak maximum, examples of FIA with limited and medium dispersion are initially illustrated. The large dispersion—obtained, for instance, by means of a mixing chamber—which generally is used to accomplish dilution in order to accommodate the analytical signal within the detector range, may be employed for FIA titrations and is dealt with in Section 4.9.

The *limited* dispersion (D = 1–2) is used when the original composition of a sample solution is to be assayed. In fact, in this case the FIA system serves merely as a means of rigorous and precise transport of the sample material to the flow cell in undiluted form or as a means of reproducible sample introduction. As an example of this approach may serve the determination of metal ions by flame atomic absorption spectrometry (fAAS), the manifold for which is depicted in Fig. 4.1 [428, 583]. In this case the carrier stream (5×10^{-4} M sulfuric acid) is propelled by a pump P, the distance between the injection valve and the nebulizer of the fAAS instrument being reduced to the shortest possible length (20 cm). Figure 4.2*a* shows a typical calibration run for Zn in the range 0.10–2.0 ppm of Zn with each sample being injected in triplicate. The resulting linear calibration graph has a regression coefficient of 0.999, the limit of quantification being 0.026 ppm of Zn. Besides yielding a high reproducibility of measurement, as seen from the recorded peak maxima, the benefits of combining FIA with fAAS is demonstrated by referring to Fig. 4.2*b*, in

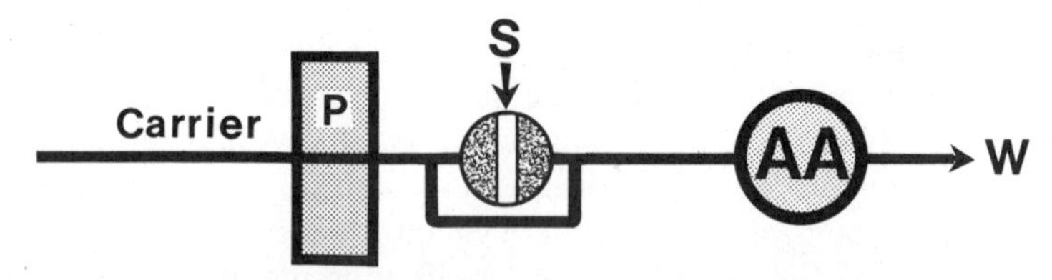

Figure 4.1. Single-line FIA manifold for the determination of metal ions by flame atomic absorption spectrometry (AA). The sample (S) is injected into a carrier stream of diluted acid (5×10^{-4} M sulfuric acid), propelled forward by pump P, and transported to the nebulizer of the AA system, the distance between the injection valve and the AA instrument being reduced as much as possible (length 20 cm) in order to secure limited dispersion of the injected sample.

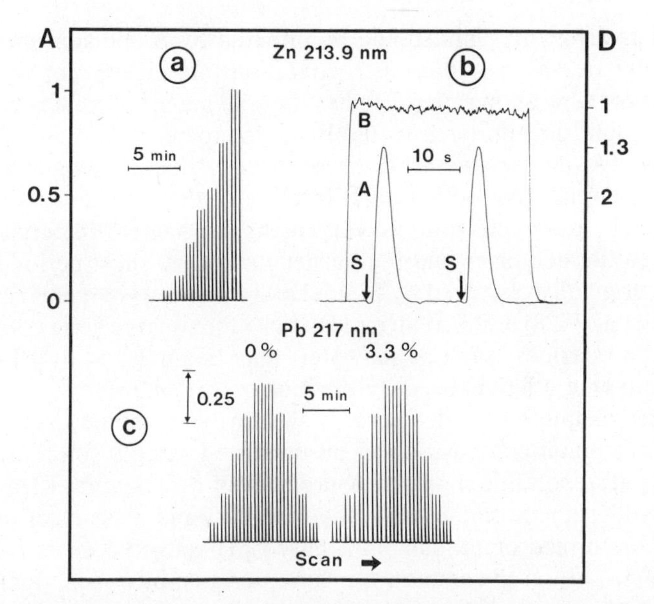

Figure 4.2. Recordings obtained with the single-line FIA flame atomic absorption spectrometry system of Fig. 4.1 using a flow rate of the carrier stream of 4.9 mL/min and an injected sample volume of 150 μL. (*a*) Calibration run for zinc as obtained by the injection of 0.10, 0.20, 0.50, 0.75, 1.0, 1.5, and 2.0 ppm of zinc standards. (*b*) Recorder response for the 1.5 ppm zinc standard as obtained by (*A*) injection via the FIA system and (*B*) continuous aspiration in the conventional mode. For the sake of comparison, the aspiration rate in (*B*) was increased to 4.9 mL/min, corresponding to the propulsion rate used in (*A*), where *S* is the point of injection. *D* represents the dispersion coefficient value, which in (*B*) is equal to 1. (*c*) Calibration run for a series of lead standards (2, 5, 10, 25, and 20 ppm), recorded without (0%) and with (3.3%) sodium chloride added to the standards to simulate, in the latter instance, a matrix of seawater.

which an fAAS signal obtained by continuously aspirating a standard (1.5 ppm of Zn) in the conventional way is compared with the FIA–fAAS readout. Although the FIA peaks only amount to 80–90% of the steady-state signal reached by continuous aspiration, one may perform several sample injections during that time period necessary to obtain a reliable steady-state readout. Hence, it is possible by FIA during a comparable time period to perform several individual injections of sample material, thereby increasing not only the reproducibility of measurement (precision) but also the accuracy.

An important feature in order to obtain optimum performance in terms of sensitivity and precision is that the rate of flow of the carrier stream pumped into the nebulizer is always greater than the natural aspiration rate of the nebulizer, that is, it is necessary to maintain a forced flow into the AAS instrument. This should be effected by propulsion means that are as pulse free as possible. Such conditions are clearly not achieved if the nebulizer aspiration rate is used as the sole means of transport. Therefore, although this approach would be an inexpensive way of combining AAS and FIA, its use should be discouraged. Another advantage is, that since the sampling period is only a fraction of the measuring cycle, that is, the wash to sample ratio is very high, the sample material is only exposed to the nebulizer and the burner for a very short period of time. Thus, as originally observed by Mindel and Karlberg [184] and confirmed by Olsen et al. [428] and Brown and Růžička [583], large series of samples of complex matrices, such as seawater, may be analyzed by FIA–fAAS with no adverse effect caused by the matrix (Fig. 4.2*c*).

Another example of a design of a system with limited dispersion, in which the original composition of an undiluted sample solution is to be assayed, is the potentiometric measurement of pH. Several FIA systems have been designed, employing either capillary pH glass electrodes [20, 71] or miniaturized or tubular PVC-based pH-sensitive membrane electrodes [608, 778]. It is common for these designs, however, that the distance between the injection valve and the detector is minimized. Thus, in Fig. 4.3*a* the connecting tube between the valve *S* and the capillary pH glass electrode (Radiometer G 299 A) was 10-cm, 0.5-mm-ID tubing; furthermore, the sample volume used was relatively large (30 μL). The carrier stream, consisting of 0.4 M NaCl–1×10^{-4} M NaH_2PO_4–1×10^{-4} M Na_2PO_4, was chosen so that its composition fulfills three requirements: (1) It should have sufficient conductivity so that resistance between the glass electrode and the reference electrode, situated downstream, is low (Section 5.4). For this purpose sodium chloride is included. (2) Its pH value should be approximately in the middle of the extreme values to be measured, so that both positive and negative peaks can be obtained and the baseline can thus be in the center of the recorded graph, thereby providing a continuous check on the electrode. (3) Its buffering capacity should be lower (preferably at least 10 times lower) than the buffering capacity of the injected samples, and therefore the buffer used is rather diluted.

In the FIA system, the glass electrode exhibits a nearly theoretical response; that is, a straight-line calibration line is obtained provided that the system is operated at such a sampling rate that the signal is allowed to reach the baseline after each injection, and that at peak maximum the electrode will sense the sample solution of original composition (i.e., not substantially diluted by mixing with the carrier stream). In cases where the sample itself has a very low buffering capacity (such as seawater or soil samples), it is advantageous to increase the sample volume (to, say,

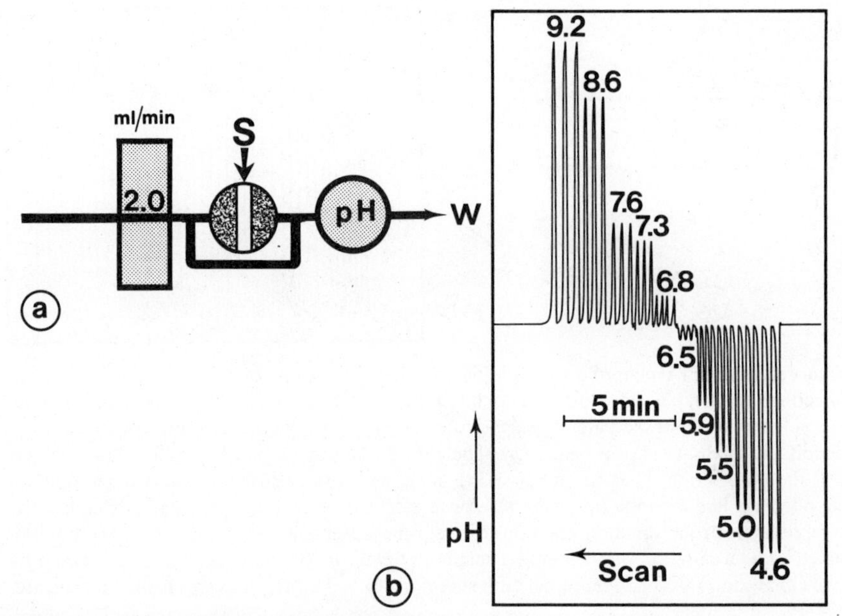

Figure 4.3. (*a*) FIA system with limited dispersion used for rapid pH measurements, comprising a flowthrough capillary glass electrode and a conventional calomel reference electrode. The carrier stream is 0.2 m*M* phosphate buffer in 0.14 *M* NaCl (pH 6.64), each sample (30 μL) being injected in triplicate (*b*).

150–300 μL [778]) so that $D = 1$ will be approached and the effect of the carrier stream buffer will be minimized. In clinical measurements, on the other hand, the relatively high buffering capacity of blood directly allows the use of the experimental setup shown in Fig. 4.3*a*. The greatest advantage of pH measurements performed in this way is the good reproducibility of measurement and its speed (± 0.002 pH unit at 240 samples/h), which yields the readout within 5 s after sample injection (Fig. 4.3*b*). This is the result of well-defined flow geometry, which provides exact timing, and of reproducible sample injection, both of which allow the time-consuming approach of waiting for a steady-state signal to be abandoned.

Limited dispersion has also been used for the potentiometric determination of the calcium activity in serum or the so-called ionized fraction of calcium [21]. The calcium electrode was placed in a specially designed flowthrough cell (Fig. 4.4), which also accommodated the reference electrode, and by incorporating a pH glass capillary flowthrough electrode into the system, it was possible to measure simultaneously the pH and pCa of the injected sample at a rate of approximately 100 samples/h with high reproducibility. Inherent for the latter is that the design of the FIA

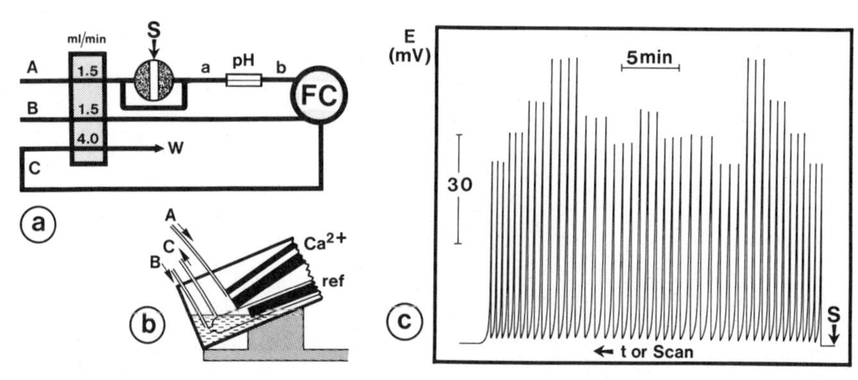

Figure 4.4. (*a*) FIA manifold for simultaneous determinations of pH and pCa. *S*, point of injection (30 μL); pH, flowthrough capillary glass electrode; *FC*, flowthrough cell containing a PVC-membrane-based calcium-selective electrode and the common reference electrode, details of which are shown below (*b*). The carrier solution (*A* and *B*) is TRIS buffer of pH 7.4, the connecting tubes (*a* and *b*) being made as short as possible. The carrier solution supplied via line *B* to the tip of the reference electrode is included in order to stabilize the reference electrode junction potential for sera measurements. In (*c*) is shown the potentiometric determination of the ionized calcium content in six serum samples, bracketed by two calibration runs of aqueous calcium standards (0.5–5 m*M*), all assays made in triplicate.

system secures an effective wash to sample ratio, so that the electrodes are exposed to the individual serum samples for very limited periods of time. Other methods that can benefit when incorporated into a FIA system with limited dispersion are typically electrochemical procedures based on detection via conductometry, amperometry, or voltammetry. This is equally true for enzyme electrodes, in which a membrane containing one or several immobilized enzymes is placed in front of the active surface of an electrochemical detector. The analyte is transported by diffusion into the membrane and there degraded enzymatically forming a product that subsequently can be sensed electrochemically. A condition for obtaining a linear relationship between analyte and signal is, however, that pseudo-first-order reaction conditions are fulfilled, that is, the concentration of converted analyte reaching the detector surface must be much smaller than the Michaelis–Menten constant. Since the degree of conversion might be adjusted by the time the analyte is exposed to the enzyme layer, the amount of converted analyte, and hence the dynamic measuring range, can be regulated by adjusting the flow rate of the FIA system. Additionally, the time of exposure might possibly be exploited for kinetic discrimination, advantage being taken of differences in diffusion rates of analyte and interfering species within the membrane layer of the electrode.

Systems with *medium* dispersion are the most interesting from an analytical point of view. They encompass a large number of procedures, in which one or several reagents are mixed with the sample solution in order to form colored, fluorescent, electroactive, or otherwise marked product, which can be sensed by a flowthrough detector (Tables 7.1–7.3). In this type of determination not only sufficient mixing must take place, but also sufficient time should elapse before the sample zone reaches the detector to ensure that the chemical reactions involved are allowed enough time to produce an adequate amount of detectable product. Again, because of the high reproducibility of the mixing and timing in the unsegmented stream, there is no need to reach chemical equilibrium in order to obtain valid analytical results. Here, however, is the principal limitation of the flow-injection method, because if one chooses, say, 30 s as the maximum acceptable residence time of a sample in the system, then slow chemical reactions may have reached only a fraction of their equilibrium, and fractional conversion will be low within the residence time chosen. This may not give high enough sensitivity for the particular analytical purpose. Surprisingly, very few spectrophotometric methods have so far been encountered where an insufficient change of absorbance, caused by a low conversion, has proved to be an obstacle. This is probably because most analytical methods used in practice are based on fast reactions. Also the samples used in many manual methods have often become very diluted prior to actual measurements, since it is more convenient in classical techniques to pipet and handle large volumes of liquids. In FIA one can, if necessary, choose precisely that combination of minimum dispersion and maximum residence time that will exactly suit a particular chemistry.

Many simple colorimetric procedures requiring medium dispersion can be accommodated in single-line FIA systems by injecting the sample directly into a carrier stream of reagent (Table 7.1). Assay of chloride has already been mentioned (Chapter 2.1, cf. Fig. 2.1). Another example of this approach is the determination of nitrite based on the very fast reaction with sulfanilamide, forming a diazo compound, which is subsequently coupled to *N*-(1-naphthyl)ethylenediamine yielding a strongly colored pink dye, the absorbance of which can be measured at 540 nm [71]. A single-line manifold (Fig. 4.5) allows this determination to be performed at a rate of up to 240 samples/h with a sensitivity limit of 0.25 ppm N–NO_2^- (for high-sensitivity measurements, see Fig. 4.9).

As illustrations of potentiometric determinations based on medium dispersion may serve the assay of nitrate [12, 13, 68], potassium [8, 13] and sodium [8], in which applications the mixing of sample and carrier stream is not aimed at promoting a chemical reaction, but to conditioning of the individual sample solutions before they are exposed to the electrode in order to measure them under optimal operational conditions. Thus, de-

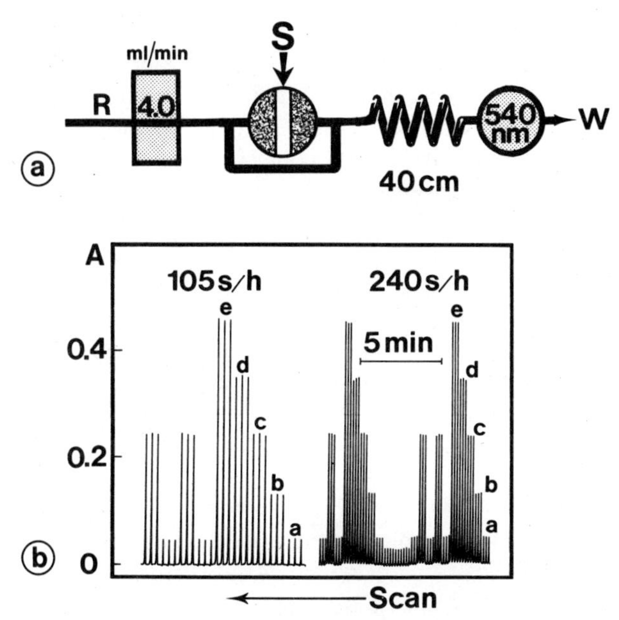

Figure 4.5. (*a*) Single-line manifold for the determination of nitrite. The sample (30 μL) is injected directly into the carrier stream of color-forming reagent (*R*), consisting of sulfanilamide and *N*-(1-naphthyl)ethylenediamine. After having passed the mixing coil, the resulting dye is measured spectrophotometrically at 540 nm. (*b*) Calibration curves for nitrite at two different sampling rates. Standard nitrite solutions containing (*a*) 0.25, (*b*) 0.50, (*c*) 1.00, (*d*) 1.50, and (*e*) 2.00 ppm $N{-}NO_2$ are injected first and then followed by a number of standards of alternating concentrations. Note that even at the higher sampling rate, no carryover is encountered.

pending on their origin, nitrate samples may exhibit greatly variable pH values and different buffer capacities and therefore it is advantageous to adjust the pH of the samples prior to measurement, that is, the carrier electrolyte solution should be a buffer of sufficient capacity. Another example of employing potentiometric detection is the determination of urea [60], which procedure depends on the enzymatic degradation of urea by dissolved urease according to the reaction:

$$Co(NH_2)_2 + H_2O + H_3O^+ \xrightarrow{\text{urease}} 2NH_4^+ + HCO_3^-$$

that is, in a carrier buffer of constant buffering capacity (β), a linear relationship between the ensuring recording change in pH (ΔpH), as monitored by an incorporated glass electrode (Fig. 4.6), and the urea content in the injected sample (ΔC_{urea}) could be obtained, the three entities being interrelated by $\beta = \Delta C_{urea}/\Delta pH$ (cf. also Section 4.3).

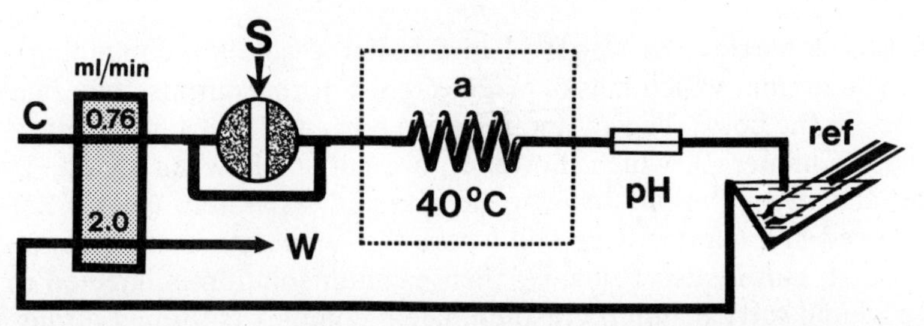

Figure 4.6. Flow-injection manifold employed for the enzymatic determination of urea by potentiometric measurement of pH. The sample *S* (30 μL) is injected into a carrier stream of 1.0 m*M* TRIS buffer in 0.14 *M* NaCl (pH 7.70), containing dissolved urease, and then passed to the reaction coil *a*, placed in a themostated water jacket, where the enzymatic degradation of the injected urea takes place. The sample zone is then led to a capillary glass flowthrough electrode (pH) and finally via the reservoir, housing the reference electrode (REF), to waste, *W*.

The greatest asset of the single-line manifold is its simplicity, which permits the use of simple means of propelling the carrier stream. Therefore, even a constant head or a gas pressurized reservoir [102, 253], which both yield a pulse-free stream allowing excellent reproducibility of measurement, may be employed. However, for all other FIA systems, two or more well-controlled channels are required and therefore a peristaltic pump is the best choice when designing a flexible FIA system.

4.2 TWO- AND MULTILINE FIA MANIFOLDS

More often than not, an analytical scheme requires the application of two or several reagent solutions, which are either *premixed* shortly before use or *added sequentially* because the reagent or the reaction products are incompatible. Initially, it was doubted [167] that FIA would be capable of accommodating procedures that require addition of further reagents downstream, because it was thought that the mixing at the confluence points would not be reproducible. Fortunately, it soon became apparent that the addition of several reagents in sequence was feasible [3], and the theory of dispersion [20] confirmed that to perform analyses successfully, the sample and reagent solutions do not have to, and indeed should not, be mixed homogeneously.

Premixing of reagents immediately prior to sample injection is a very useful technique in cases where a reagent mixture might deteriorate during storage. An example of such an approach is the determination of calcium by means of *o*-cresolphthalein complexone, based on the formation of a purple calcium complex, which absorbs light strongly at 580 nm. The reaction is carried out at pH 11 in a buffered solution containing 8-hydroxyquinoline, which masks magnesium. For this purpose two reagents are used: the color reagent and the base reagent (for details, see Ref. 21; cf. also Chapter 7), which, however, are not stable when mixed. Therefore, the two solutions have to be pumped separately (Fig. 4.7*a*) and combined in a short mixing coil where the temperature of the combined carrier stream is also adjusted; then calcium solution is injected (S_v = 30 μL) and the calcium *o*-cresolphthalein complex is formed during pas-

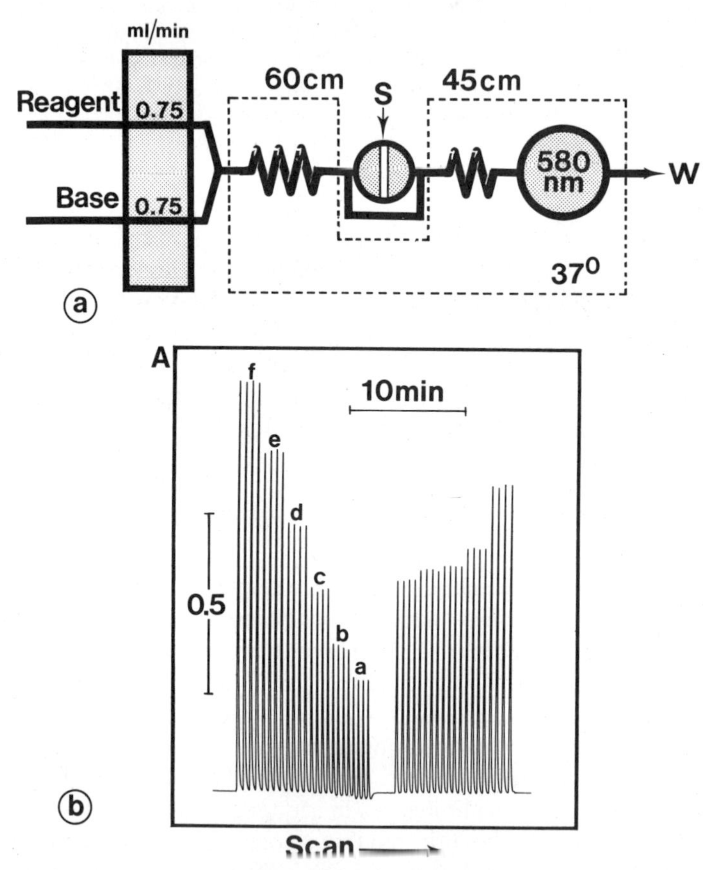

Figure 4.7. (*a*) Two-line manifold for the spectrophotometric determination of calcium. Color reagent (*o*-cresolphthaleine complexone) and base are mixed and preheated in the first coil before the sample (30 μL) is injected and allowed to react in the second coil, the system being thermostated at 37°C. (*b*) Spectrophotometric determination of the total calcium content in serum at a sampling rate of approximately 110 samples/h. From left to right are shown a series of aqueous calcium standards containing (*f*) 25.0, (*e*) 20.0, (*d*) 15.0, (*c*) 10.0, (*b*) 5.0, and (*a*) 1.0 ppm, followed by three human serum samples, a serum pool (DSKK-K77), and the same serum pool sample with standard addition (25 μL of a 500-ppm Ca standard plus 2.50 mL of serum pool). Each standard or sample was injected in quadruplicate.

sage through the second mixing coil and subsequently measured. This approach yields the readout within 15 s and allows a sampling frequency of 110 samples/h (Fig. 4.7*b*) for determinations of calcium in the range 1–25 ppm. Higher sensitivity of measurement can undoubtedly be obtained by using the approach described below (and in Section 6.6), and lower sensitivity might be achieved by reducing the injected sample volume. Note, that in the design of this manifold, Rule 3 (Section 2.2.3) has been followed: the reaction coil is kept short (45 cm) and the pumping rate is low.

Determination of phosphate (Sections 6.5 and 6.6) and sulfur dioxide (Section 4.3) are additional examples of spectrophotometric measurements where reagents are premixed in the FIA system. Many assays will benefit from this simple approach, which eliminates the problem of instability of premixed reagents. Furthermore, it allows the use of a reagent in *statu nascendi* (Section 4.7.3), which through its higher reactivity will speed up formation of a species to be measured. Thus, when adopting a manual procedure to a FIA system, one should always ask why the reagents are added separately in the original procedure: Is it because their stock solutions deteriorate when mixed, or because the first reaction has to take place before the next reagent may be added? Only in the latter case do reagents have to be added sequentially, and this should be done only with some reluctance, because addition of many reagents to the carrier stream requires a complicated manifold.

Sequential use of only two reagents is a simple matter because a manifold for this purpose involves only one confluence point at which a second reagent is added to the sample zone when carried past by the stream of the first reagent. An example of a more complicated procedure, where three reagents are used sequentially, is the determinations of urea, based on the following reaction sequence:

$$Co(NH_2)_2 + H_2O \xrightarrow{\text{urease}} NH_4^+ + NH_3 + HCO_3^- \quad (pH = 7)$$

$$NH_3 + OCl^- \longrightarrow NH_2Cl + OH^- \quad (pH = 11)$$

$$2OCl^- + NH_2Cl + C_6H_5O^- \longrightarrow OC_6H_4NCl + 2Cl^- + OH^- + H_2O$$

$$OH^- + OC_6H_4NCl^- + C_6H_5O^- \longrightarrow \underset{\text{Indophenol Blue}}{OC_6H_4NC_6H_4O^-} + H_2O + Cl^-$$

Here, urea has first to be converted to ammonia with the aid of the enzyme urease, a reaction that has to be performed at pH 7, as the activity of urease decreases significantly at lower and higher pH values. Then follows formation of chloramine in strongly alkaline solution and its further reaction with phenol, resulting in the final product to be measured. As indophenol blue is a redox as well as an acid–base indicator (p*K* 8.1), the pH of the carrier stream must not be lower than 10, and an excess of hypochlorite must be present. Of these reactions the enzymatic degradation of urea and the formation of chloramine is fast, whereas coupling of chloramine with phenol is slow. However, the rate of the latter reaction is increased in the presence of methanol or ethanol [4.1; 5]. This observation was first utilized in a FIA method for determining total nitrogen after Kjeldahlization [5], and it is also the basis for determination of ammonia in water, a method that is accommodated in a two-line FIA manifold. However, for the determination of urea (e.g., in serum), a three-line manifold has to be designed (Fig. 4.8*a*) where the reagents are added sequentially and the reaction product is measured at 620 nm. The analytical readout (Fig. 4.8*b*, left) is available after 14 s, and the signal returns within 1% to the baseline within 35 s, allowing a sampling rate of 110 samples/h. Within that short time, enough ammonia and indophenol blue are produced to allow determination of urea in the clinically interesting range of 2–20 m*M*.

Two-line and multiline manifolds are, of course, now commonplace for FIA methods. In fact, most of the procedures described in the FIA literature (Chapter 7) utilize this approach. Thus, in Ref. 52 is described a turbidimetric procedure for the determination of ammonia in low concentrations with the use of Nessler's reagent, while Ref. 253 recounts the spectrophotometric determination of chromium(VI). Besides being based on one-phase equilibria, multiline manifolds may also involve gas diffusion, solvent extraction, and liquid–liquid phase reactions in packed reactors (see the following sections). It should be emphasized, however, that a FIA system should always be kept as simple as possible, and that a well-designed chemical analysis will often require only the use of a two-line manifold.

Another example is the determination of glucose in human serum with ensuing chemiluminescence detection [330], where glucose is enzymatically degraded by glucose oxidase, giving rise to formation of hydrogen peroxide, which then is reacted with luminol in the presence of hexacyanoferrate(III) to give an emission of light that is measured by a photomultiplyer tube. In this context it should be mentioned, that the use of solution chemiluminescence (CL) in quantitative analysis is very attractive, primarily because of its potentially high sensitivity and wide dynamic

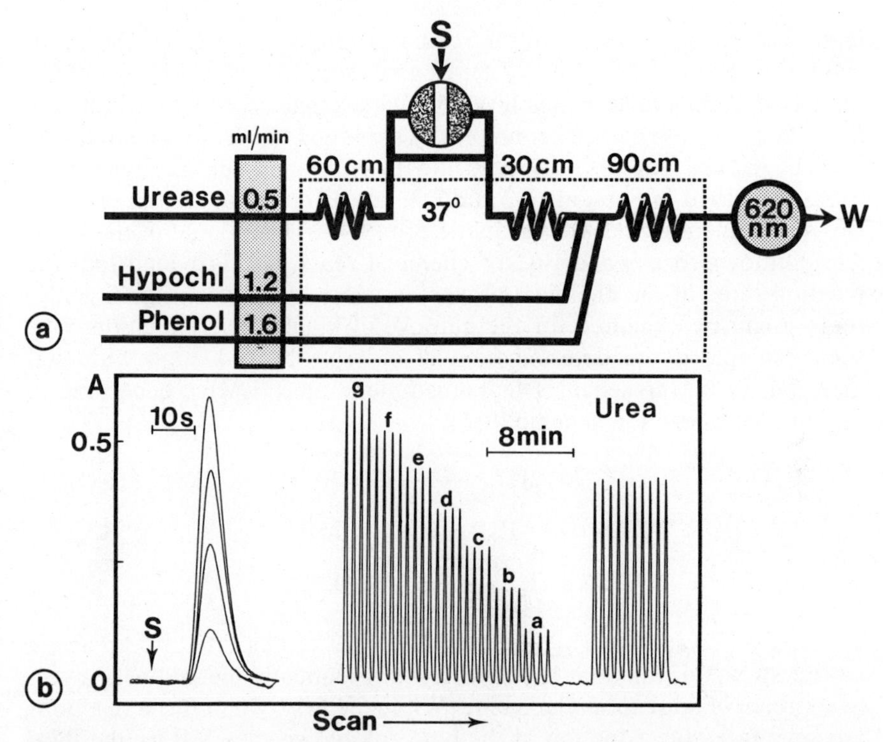

Figure 4.8. (*a*) Manifold for determination of urea in serum where ammonia from the enzymatic conversion of urea is oxidized to chloramine and coupled with phenol. The resulting indophenol blue is measured in a flowthrough spectrophotometer at 620 nm. Sample volume is 30 μL. (*b*) The left side of the recorder output shows a series of peaks registered from the same point (*S*), which confirms that, regardless of urea concentration, the whole measuring cycle is completed within 20 s after sample injection. On the right side, the record shows a series of urea standards (*a*–*g*) in the range 2.0–20.0 m*M* urea, and 10 repetitive injections of a human serum sample containing a normal urea level.

range, but also because the required instrumentation is fairly simple. The inherent lack of selectivity of CL procedures can readily be compensated for by coupling the chemiluminescent reactions to processes that selectively generate species that are prerequisites for the ensuing light emission. Furthermore, CL has an added advantage to most optical procedures: as light is only produced and measured when sample is present, there is generally no problems with blanking. Yet, as CL reactions usually generate transient emissions, it is necessary to execute all measurements at precisely defined and reproducibly maintained conditions so that all samples are treated, physically and chemically, in exactly the same manner. For these reasons CL lends itself to FIA, where all experimental parameters can be so rigidly controlled and reproducibly maintained. This is, in fact, amply reflected in the revitalization of this old detection procedure in recent years (Table 7.1). Contributing to this is, of course, the increased commercial availability of a wide range of enzymes, but particularly that these can be economically exploited in FIA by immobilizing them on packed reactors integrated into the FIA manifolds (Section 4.7.2).

As discussed in Sections 2.2 and 2.3, any FIA readout is the result of two simultaneously occurring kinetic processes, that is, that of the physical dilution process and that of chemical reactions. Provided that the reaction rates of the chemical processes taking place vary significantly, these might be exploited for the purpose of kinetic discrimination. One such example is the determination of chlorate by reduction with titanium(III) in the presence of leukomethylene blue (LMB), according to the following reaction scheme [986]:

$$2ClO_3^- + 10Ti^{3+} + 12H^+ \longrightarrow 10Ti^{4+} + Cl_2 + 6H_2O$$

$$Cl_2 + LMB \longrightarrow MB$$

$$MB + Ti^{3+} \longrightarrow LMB + Ti^{4+}$$

The assay is performed by injecting a sample of chlorate into an acidic carrier stream of titanium(III), which is subsequently merged with a second stream of leukomethylene blue. While the first two of these reactions are very fast, the reduction of the blue colored species MB by the third reaction is slow, that is, the chlorate concentration might readily be quantified via the absorbance due to the color of the MB species generated by reaction 2.

4.2.1 High-Sensitivity FIA

If a FIA procedure is required to reach the same level of sensitivity as a batch procedure, two obstacles have to be overcome: the short reaction time, which is due to the short residence time and may result in a relatively low yield of reaction product, and an excessive dispersion of the sample zone, which results in an unwanted dilution of the species to be measured. Leaving aside at this stage the problem of too short a residence time, it might be helpful to consider an approach by means of which the dispersion can be minimized, yet still remain sufficient for supplying the *middle* of the sample zone with an adequate amount of reagent. Obviously, lack of reagent in the center of the sample zone will result in the absence of product to be sensed (cf. Chapters 2 and 3). Thus, instead of a sharp smooth peak, one will obtain either a "double peak"—which may be well pronounced or may merely appear as a noisy peak signal—or a nonlinear calibration curve at the upper range where the concentration of analyte in the standard solutions is high (see Section 6.9). Although an increase of the reagent concentration in the carrier stream may forsake the formation of humped peaks and widen the linear calibration range, this approach is not the most economical in terms of reagent consumption.

Injecting the sample into a carrier stream of pure solvent and adding the reagent by means of a confluence tee downstream is, however, an effective way to ensure the rapid and adequate mixing of sample with the reagent. Using such a flow arrangement, a high sensitivity of measurement can be obtained by injecting large sample volumes (cf. Rule 1). For example, the spectrophotometric determination of nitrite, mentioned above, which has a sensitivity limit of 0.25 ppm in a single-line system (Fig. 4.5), will become more sensitive if the sample volume is increased. A manifold for this purpose (Fig. 4.9*a*) consists of two lines, the top one carrying water into which the sample is injected and a lower one carrying the reagent. To ensure rapid mixing of the two streams, the confluence point must be properly constructed (Section 3.2.3), by forcing the streams into each other at a 30° angle. Alternatively, a small reactor packed with single beads [146], or a three dimensionally coiled tube [567, 608] may be used. By increasing the sample volume from 30 to 160 μL and the sensitivity of the recorder five times, a 22-fold increase of sensitivity was obtained (Figs. 4.9*b* and 4.9*c*), thus reaching a detection limit of 0.002 ppm nitrite [153]. This large increase in sensitivity is in agreement with the principles expressed previously (Chapter 2) and is generally applicable. A more detailed description of this approach in connection with a practical exercise is given in Section 6.6, where it is employed for the high-sensitivity determination of phosphate based on the molybdenum blue method.

In this context it should be pointed out that the injection of sample into a solvent stream, which subsequently is merged with the reagent stream, is a method well suited for tackling yet another problem, which might arise when a sample exhibits a certain matrix effect. Thus, when the sample composition differs markedly in physical and chemical respects from the reagent solution, the mixing in a single-line system may not be effective, and variable concentration gradients may persist even until the sample zone reaches the flow cell. This may result in an irregular baseline and an overall irreproducibility of measurements. Therefore, for determinations of chloride in, say, solutions of 5 *M* nitric acid, the sample should be injected into nitric acid of similar concentration, using the manifold type shown in Fig. 4.9, rather than directly into the reagent stream as done in the single-line system (Fig. 2.1). Similarly, viscous liquids should be injected into a stream of similar viscosity. Yet, it must be emphasized that often a single-line manifold is adequate. Hence in clinical

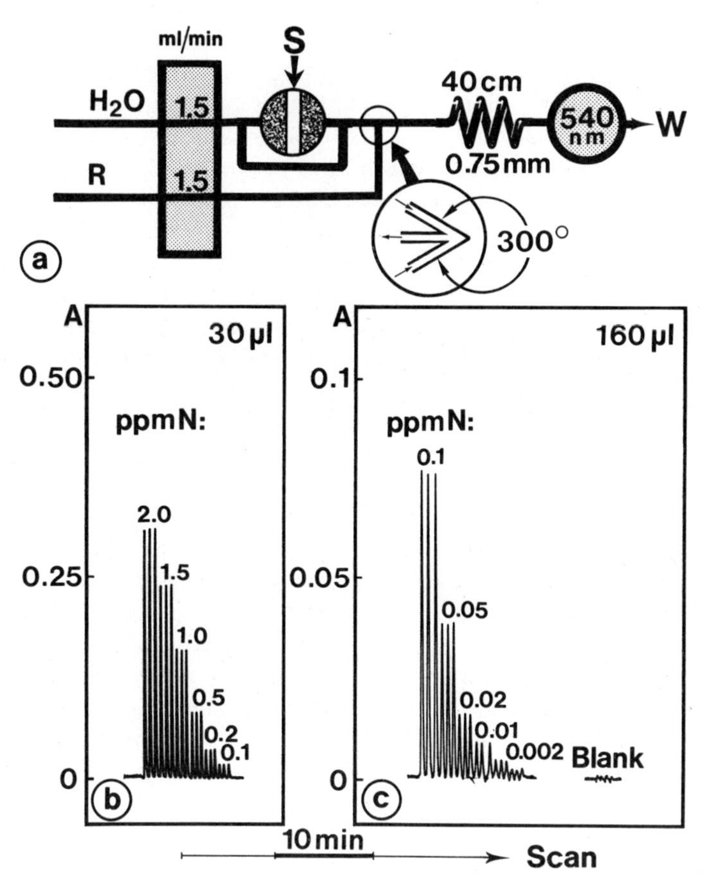

Figure 4.9. (*a*) Manifold for high-sensitivity measurement as used for the determination of nitrite. The aqueous sample containing nitrite is injected into a carrier stream of water to which the color-forming reagent *R* is added in confluence. (The radial mixing within the sample zone being improved by confluencing the streams in the manner shown on the inset.) After passage through a mixing coil, the resulting color is monitored at 540 nm. (*b*) Determination of nitrite in the range 0.1–0.2 ppm N–NO_2 by injecting 30 µL of sample solution in the manifold depicted in (*a*). (*c*) Determination of nitrite in the range 0.002–0.1 ppm N–NO_2 by injecting 160 µL of sample solution in the manifold shown in (*a*). Note the combined effects of increased sample volume and the gain of spectrophotometer amplifier (see the *A* scales) on the sensitivity of measurement.

applications, for instance, the difference in viscosities between serum and reagent solutions does not usually cause any problems, because if a sufficiently small volume of serum is injected (10–20 µL), the sample material becomes rapidly dispersed, and any difference in viscosities between the carrier stream and the sample zone soon becomes negligible. Besides, serum standard materials rather than aqueous standards are generally used for calibration.

Returning to the task of achieving higher sensitivity, the natural question to consider is: How much can be gained by prolonging the reaction time? The simplest way to investigate this parameter, which is different for each chemical reaction, is to stop the flow when the sample is either in the reaction coil or the flow cell (cf. Chapter 2). Thus in the examination of the nitrite method (manifold, Fig. 4.9*a*), the pump was stopped 9 s after the sample of nitrite (containing 1-ppm $N{-}NO_2^-$) had been injected into the carrier stream, and was kept in the coil for 30, 60, and up to 180 s, after which the pump was restarted. In this manner, a series of equally high peaks were recorded (Fig. 4.10), which confirms that the color-forming reaction has already reached equilibrium within 30 s. Then, even a shorter range of residence times was investigated by stopping the sample zone in the flow cell as soon as 15 s after sample injection (Fig. 4.10*b*, curve Stop), which confirms that the reaction had already reached the maximum yield while the sample zone was on its way toward the detector, that is, within $T = 15$ s. These simple experiments were easily performed by following the recorder track at higher paper speed and by switching the pump off and on. By such procedures the yield of chemical reactions can rapidly be evaluated and the reagent composition and manifold design

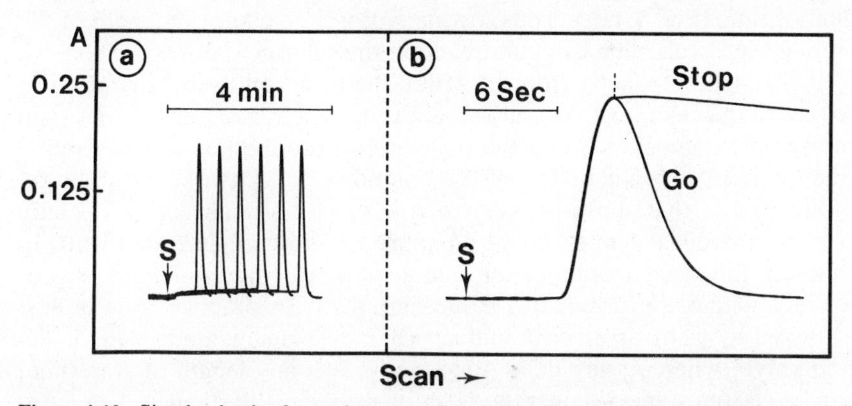

Figure 4.10. Simple check of completeness of reaction between sample material and color-forming reagent, as demonstrated on the reaction between nitrite [1 ppm in (*a*) and 1.5 ppm in (*b*)] and *N*-(1-naphthyl)ethylenediamine using manifold Fig. 4.9*a*. To test the effect of longer residence times, the sample zone was stopped 9 s after injection in the *coil* for periods of 30–180 s, whereupon the pump was reactivated, the peaks being recorded repeatedly from the same start (*S*), (*a*). For shorter residence times (*b*) the signal was recorded continuously (Go) as well as when the sample zone 15 s after injection was stopped within the *flow cell* (Stop). The sample volume in all experiments was identical (30 μL).

optimized. Since its introduction [20] the FIA stopped-flow approach has found extensive use in such optimization, yet its most important exploitation has been its application to quantitative assay, which is discussed in the next section.

4.3 STOPPED-FLOW FIA MEASUREMENT

The stopped-flow approach is, apart from FIA titrations (Section 4.9), the most thoroughly tested gradient technique (Chapter 2.4). It is based on a combination of stopped-flow measurement and of gradient dilution. There are two different purposes for operating a FIA system in the stopped-flow mode: (1) to increase the sensitivity of measurement by increasing the residence time T—and thus the conversion of the measured species—and (2) to record a reaction rate which then serves as the basis for the analytical result. The stopped-flow technique is based on one of the most important observations in FIA (summarized in Rule 3 in Section 2.2.3) postulating that to increase the residence time, one should keep the reaction coil short and decrease the velocity of the carrier stream. In the extreme, when the carrier stream ceases to move, the dispersion of the sample zone will stop (except for a negligible contribution from molecular diffusion) and the dispersion coefficient D will become independent of time (Fig. 4.11*b*). Thus, by an appropriate choice of length of the stop–go intervals, one can gain reaction time during the stop period.

If the sample zone is stopped within the flow cell itself, it is possible to record the change of, say, absorbance, caused by the reaction between the sample components and the reagent constituting the carrier stream. The obvious prerequisite for such a reaction rate measurement to be reproducible is that the movement of the carrier stream can be exactly controlled from the operational pumping rate to complete standstill. In this way the same section of the sample zone can always be held reproducibly within the flow cell for measurement. In practice, this is best achieved by using an electronic timer (Fig. 4.11*a*) activated by the injection valve, whereby any delay time, as well as any length of stop time, can be chosen so that it suits the reaction rate of the particular chemistry. Referring to Fig. 4.11*b* it can be seen that if a sample is injected and stopped within the flow cell, one of two situations may arise: either the signal will remain *constant* indicating that the reaction already has proceeded to equilibrium, or the signal will *increase* revealing that the chemical reaction is still in progress. If the reactant concentrations in the latter case are selected to fulfil the requirements of a zero- or pseudo-zero-order reaction (dC/dt = const) the increase in signal will be a straight line, the

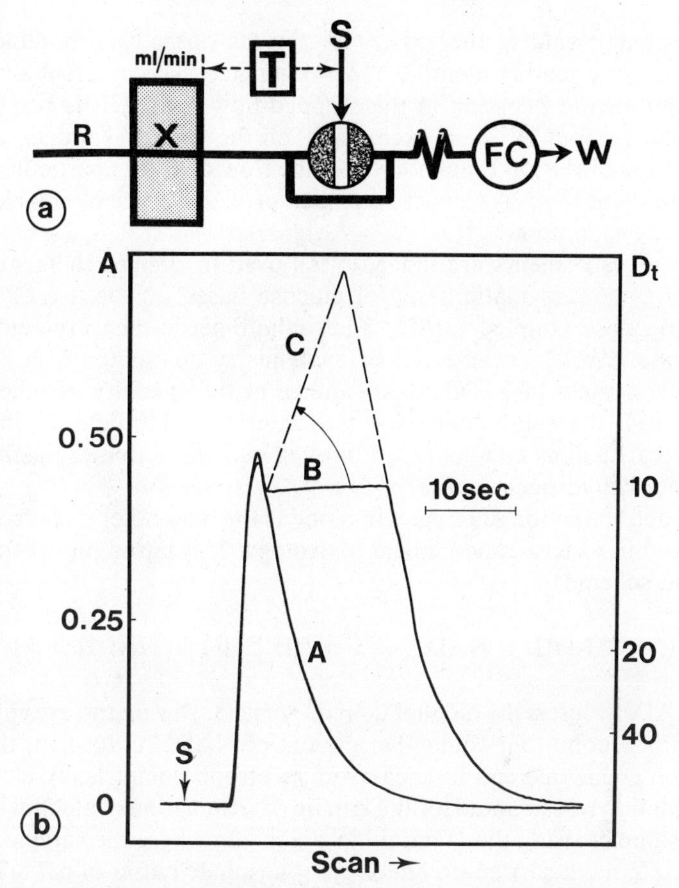

Figure 4.11. (*a*) Simple stopped-flow FIA manifold. When the sample (*S*) is injected, the electronic timer (*T*) is activated by a microswitch on the injection valve. The time from injection to stopping of the pumping (delay time) and the length of the stop time can both be preset on the timer. (*b*) The principle of the stopped-flow FIA method as demonstrated by injecting a dyed sample zone into a colorless carrier stream and recording the absorbance by means of a colorimetric flowthrough cell. All curves recorded from the same point (*S*) by injecting 26 μL of the same dye solution: (*A*) continuous pumping; (*B*) 9 s pumping, 14 s stop, and continuous pumping again; (*C*) the dashed line indicates the curve that would have been registered if a zero- (or pseudo-zero) order chemical reaction had taken place within the flow cell during the 14 s stop interval.

slope of which will reflect the reaction rate of the analyte and hence its concentration. Reaction rate measurements, where the rate of formation (or consumption) of a certain species is measured on a larger number of data points, improves the reproducibility of the assay and secures its reliability. These benefits are obtained because interfering phenomena such as blank values, the existence of a lag phase, and nonlinear rate curves may be readily identified and eliminated. The inherent advantage of the automatic blanking in the stopped-flow approach is because the analytical readout is based exclusively on the chemical processes monitored during the stop time interval in the flow cell. It is particularly beneficial in clinical assays, where samples of widely variable blank values often are encountered.

Rate measurements are indeed often used in clinical chemistry, and, therefore, the enzymatic assay of glucose based on the use of glucose dehydrogenase coupled to the spectrophotometric measurement of the coenzyme NADH became the first chemistry automated in a stopped-flow FIA system [46]. Taking advantage of the specifity of other dehydrogenases, the same coenzyme was later used for determining other substrates, such as ethanol [226], or even enzyme activities, such as that of lactate dehydrogenase [264].

The determination of ethanol in blood is the parameter most frequently measured in forensic and clinical toxicology. It is based on the following reaction scheme:

$$CH_3CH_2OH + NAD^+ \xrightarrow{ADH} CH_3CHO + NADH + H^+$$

where ADH represents alcohol dehydrogenase. During the enzymatic degration of alcohol, an equimolar amount of NADH is formed, the color of which coenzyme can be measured spectrophotometrically at 340 nm. This reaction was adapted for the kinetic determination of alcohol in whole blood samples using the manifold in Fig. 4.12*a* where the sample (30 μL) is injected into a carrier stream of pyrophosphate buffer of pH 8.7, which is then merged with another stream of carrier solution to which alcohol dehydrogenase and coenzyme have been added. After passing through a short mixing coil, the sample zone is directed into the flow cell of the spectrophotometer and stopped. The increase in absorbance during the fixed stop interval is monitored (Fig. 4.12*b*). Bracketed by a series of six standards, the readouts for 10 blood samples (all run in duplicate) are shown, the analytical signal being the small absorbance change atop each peak that was monitored by a computer. Because the absolute height of each peak is due partly to the reaction already having taken place during the transport toward the flow cell, and partly to varying background signal owing to the matrix of the sample solution, any blank value, whatever its origin, is effectively eliminated. Therefore, the only pretreatment necessary for this assay was a 100 times predilution (by buffer solution) so that the sample concentration could be adjusted to the enzyme activity

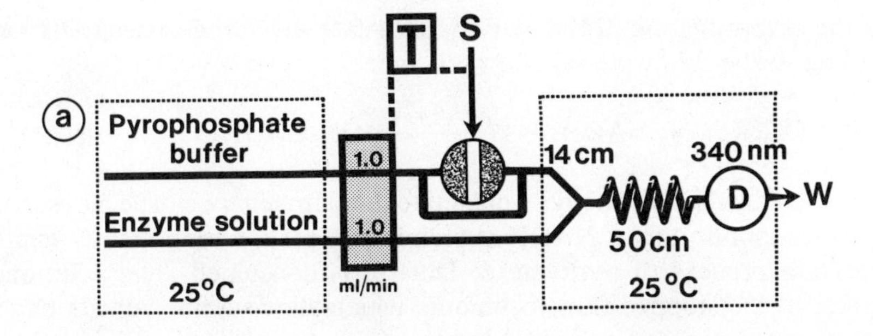

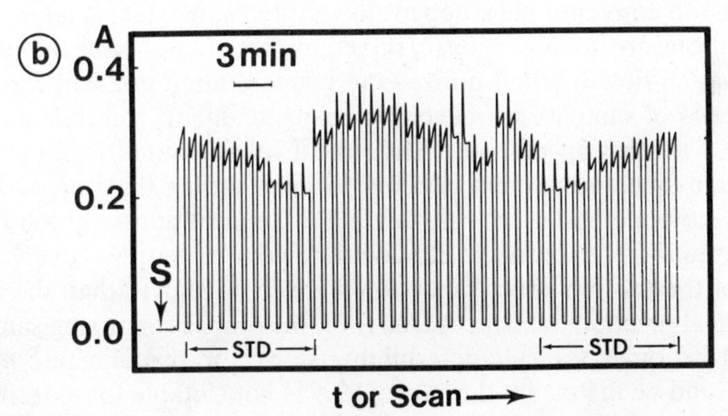

Figure 4.12. (*a*) Flow-injection manifold for the enzymatic stopped-flow determination of alcohol. Samples (30 μL) are injected into a carrier stream of pyrophosphate buffer of pH 8.7. The second channel delivers the same solution to which alcohol dehydrogenase has been added. The broken lines indicate those parts of the manifold that are thermostated. The coil between the confluence point of the reagent streams and the detector allows for a suitable dispersion of the sample zone prior to the stop sequence. (*b*) Recorder output for the stopped-flow determination of alcohol in blood. To the left is shown a series of six aqueous standards (STD) (0–32.0 mg/L) followed by 10 blood samples and a repetition of the six standards, all analyzed in duplicate: stop time, 14 s; delay time, 13 s; total measuring cycle, 40–45 s.

used (90 U/mL). Since the enzyme solution is only needed during that short period during which the sample zone passes the merging point, a drastic reduction in reagent consumption may therefore be achieved either by means of the merging zones approach, or by pumping the buffer–enzyme solution intermittently (Section 4.5).

The potential of the FIA stopped-flow technique may, of course, equally well be exploited by reversing the roles of substrate and enzyme. Thus in Chapter 2.4 is shown a figure (Fig. 2.32) depicting the readout for the determination of the activity of lactate dehydrogenase (LDH) according to the following reaction scheme:

$$CH_3COCOO^- + NADH + H^+ \xrightarrow{LDH} CH_3CHOHCOO^- + NAD^+$$

according to which the activity of lactate dehydrogenase can be measured, again via monitoring of NADH, which in this case, during the enzymatic oxidation process of pyruvate to lactate, is consumed. The additional feature of the stopped-flow technique, which allows a fine adjustment of the reagent-to-sample ratio and hence of the dispersion coefficient D by choosing an appropriate delay time, is discussed in detail in Section 2.4.

In contrast to conventional stopped-flow systems, in which sample and reagent solutions are homogeneously mixed by force in a special chamber, the FIA stopped-flow method utilizes the concentration gradient formed where the ratio of sample and reagent is changing due to controlled axial dispersion as a function of time. Also, the FIA apparatus is simple to construct, and since no reagent is being pumped during the stop period, the stopped-flow method is less demanding on reagent consumption than is the continuous-flow approach. It should be pointed out, however, that if the rate of the sample zone dispersion were to be slower than the rate of the chemical reaction to be measured (i.e., the process of mixing sample and reagent becomes the rate-determining step), the reaction rate measurement would be distorted; therefore, FIA is not suitable for extremely fast reaction rate measurements (see Section 2.6).

In Section 4.1 the potentiometric determination of urea via enzymatic degradation by urease according to the reaction:

$$Co(NH_2)_2 + H_2O + H_3O^+ \xrightarrow{urease} 2\ NH_4^+ + HCO_3^-$$

was mentioned. Although the potentiometric detection appears attractive because of its wide dynamic range, it also has drawbacks. One of these, inherent to all potentiometric detectors, is that the ion to be sensed has to reach the ion-sensitive surface of the electrode, and therefore the speed of the electrode response depends on the intensity of mixing of the measured solution. In contrast to this, spectrophotometric measurements yield instant response to the changes within the bulk of the solution through which the beam passes. Besides, when assaying urea in complex matrices such as serum, the inevitable problem of matrix background signal arises. Therefore, a stopped-flow procedure, employing the manifold shown in Fig. 4.13a, was devised by means of which urea in serum could be determined via enzymatic degradation, the ensuing change in acidity being monitored optically by means of an acido-basic indicator.

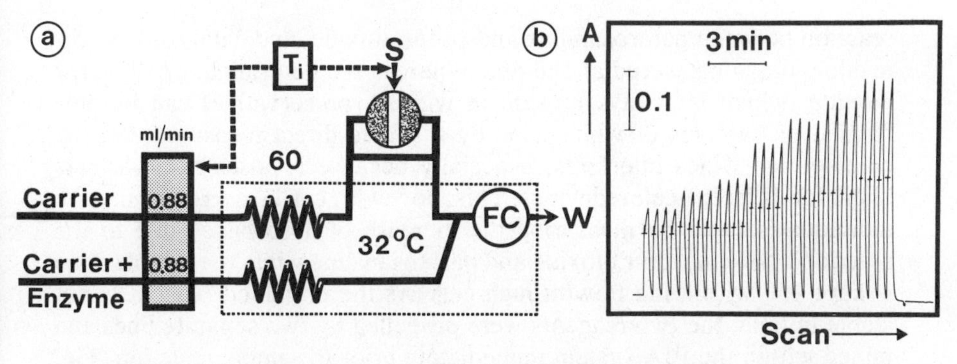

Figure 4.13. (*a*) Manifold for the stopped-flow determination of urea. Samples (30 μL) are injected (*S*) into a carrier stream containing 1.0 m*M* TRIS buffer and 0.5 m*M* phosphate buffer in 0.14 *M* NaCl and embodying 2% mixed indicator solution. The second channel delivers the same solution to which dissolved urease has been added. Both carrier streams are adjusted to pH 7.00 at 32°C. The length between the point of injection (*S*) and the confluence point is 8 cm; that between the confluence point and the flow cell (FC) is 10 cm, allowing a residence time of 6.7 s, and a contact time between sample and enzyme before measurement of 3.5 s, the absorbance of the mixed indicator solution being measured at 580 nm. (*b*) Calibration run for stopped-flow determination of urea, all samples prepared to contain a fixed level of hydrogen carbonate (20 mmol/L) in order to simulate biological assay. Samples containing 0, 1, 2, 4, 6, 8, and 10 mmol urea/L were each injected in quadruplicate. Delay time was 7.6 s; stop time (Δt) was 15 s. Total measuring cycle (including wash period) was 36 s, allowing a sampling rate of 100 samples/h. The start of the stop period can be readily observed on the original recording, but is emphasized by the horizontal lines in this photographic reproduction.

In this work, which is detailed in Ref. 79, it was shown that the rate of the absorbance change during the stop period could be expressed as $dA/dt = (SV_m/\beta K_m)(q/k)$, where S is the substrate concentrations, V_m is the maximum velocity of the catalyzed reaction, β is the buffering capacity of the carrier solution containing the acido-basic indicator, K_m is the Michaelis–Menten constant, and q and k are indicator and stoichiometric constants. Thus, provided that the buffering capcity is maintained at a constant level, the change in pH and hence in absorbance may be directly related to the concentration of urea in the sample. Since the operational range of an acido-basic indicator is restricted to approximately ± 0.5 pH unit within the pK value, a mixed indicator system was prepared, the individual indicators selected so that their color change was identical and the mixture made according to their molar absorptivities, thereby ensuring a linear relationship between absorption and change in acidity over a wide range. A calibration run for the urea assay is shown in Fig. 4.13*b*.

Another interesting example of the FIA stopped-flow procedure is the determination of sulfur dioxide in wine [71], based on the well-known West–Gaeke [4.2] method, in which a purple compound formed by the reaction between pararosaniline and sulfur dioxide, and catalyzed by formaldehyde, is measured at 580 nm. Whereas the determination of sulfur dioxide (which is always present in wine as preservative) can be performed on samples of white wine by a simple direct measurement, the color of red wines interferes, especially because it poses as a variable blank for different-colored wines. This, however, can be corrected in each individual sample by measuring the increase of absorbance due to the reaction between sulfur dioxide and pararosanaline while the sample zone is kept still within the flowthrough cell. As the premixed reagent is not stable in time, the two reagents were propelled by two separate lines and mixed within the FIA system immediately prior to sample injection (Fig. 4.14*a*). With this experimental arrangement sulfur dioxide could be de-

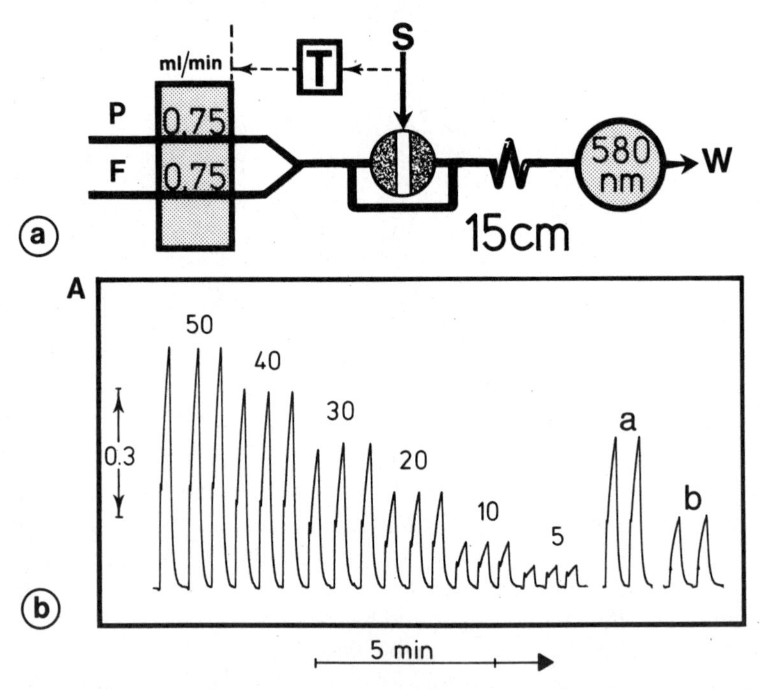

Figure 4.14. (*a*) Manifold for stopped-flow determination of SO_2 in wine. The sample (10 μL) is injected into a carrier solution of pararosaniline to which is added a solution of formaldehyde, which catalyzes the reaction. After mixing in the 15 cm coil, the sample zone is stopped in the flowthrough cell for measurement. (*b*) Calibration record for sulfur dioxide with the system in (*a*). The concentrations are given in ppm SO_2. The delay time was selected so that the samples were stopped shortly after the peak maximum had been passed. As the stop time was identical in all cases (15 s), the analytical result corresponds to the peak increase during the stop interval. To the right are shown recordings for the determination of the free sulfur dioxide contents in two wines (*a*, Touraine Blanc, 29 ppm; *b*, Gumpoldskirchner, 1977, 18 ppm).

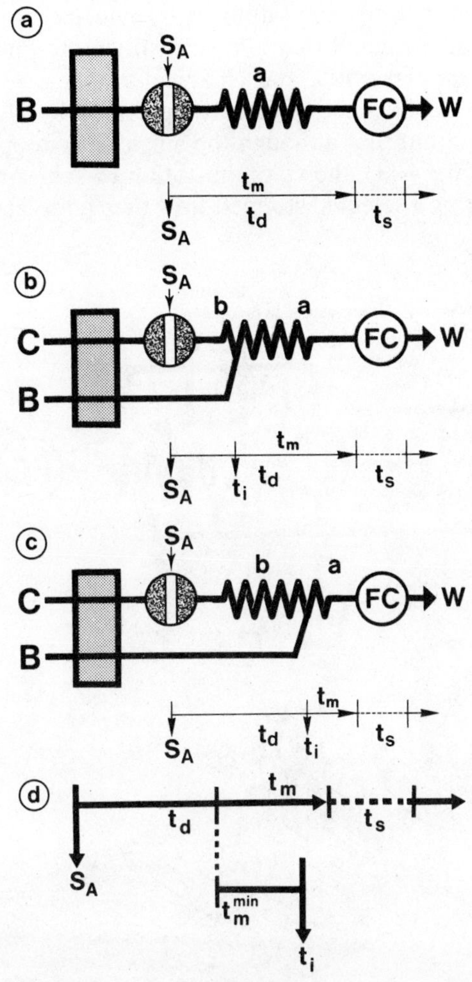

Figure 4.15. Three types of flow channel designs for stopped-flow injection analysis allowing options for selecting initial reaction time (t_i) between injected sample (S_A) and reagent (B), mixing time (t_m), delay time (t_d), and stop time (t_s) of the sample within the flow cell (FC). While the system shown in (a) suffices for assays based on reaction rate measurements (cf. Figs. 4.12–4.14), physicochemical rate studies aimed at determining reaction rate constants require manifolds such as those depicted for slower (b) and faster (c) reactions, the delay time required for sufficient radial mixing within the element of study (t_m^{min}) being a crucial parameter (d) C represents an inert carrier stream.

termined in wines at a rate of 105 samples/h with the analytical readout available 23 s after sample injection (Fig. 4.13*b*).

Inherently, the FIA stopped-flow procedure should be an ideal vehicle to determine reaction rates and rate laws, provided that an experimental approach could be designed that allows resolving the individual contributions of physical dispersion and chemical kinetics. A comprehensive treatment of this problem was recently described by Hungerford et al. [838], who pointed out that although the single-line stopped-flow system (Fig. 4.15*a*; cf. Fig. 4.11) allows optimization of solution conditions for measurement during a selected stopped-flow time interval (t_s) by choosing

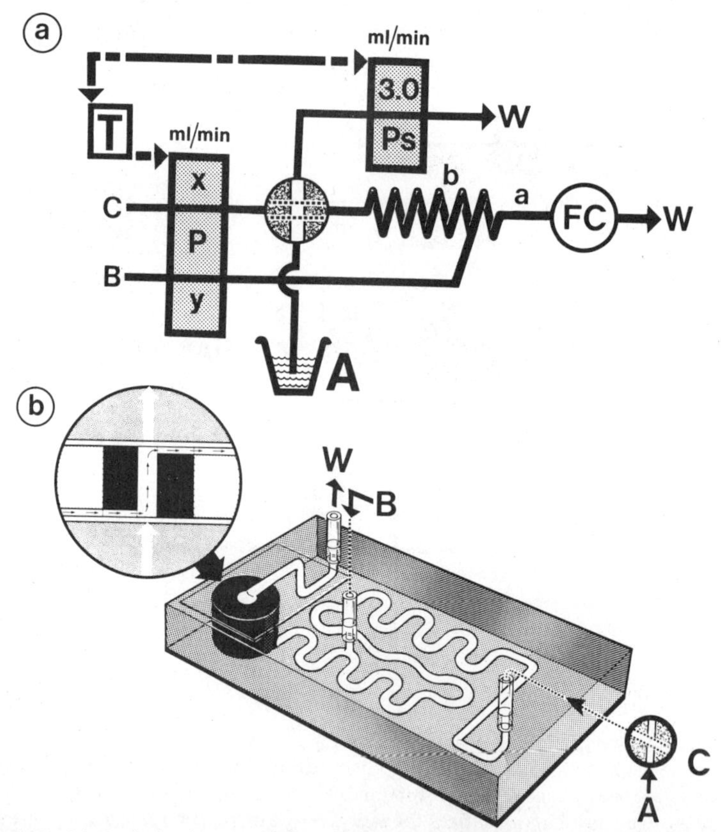

Figure 4.16. (*a*) Optimized stopped-flow FIA system according to the design depicted in Fig. 4.15*c* and (*b*) a detail of the microconduit actually used with integrated optical flow cell and incorporating coils *a* and *b*. The cross section of the flow cell is shown on the inset, light being carried to and from the flow cell via optical fibers. *P* and *Ps* are the manifold and sample pumps, respectively, the operation of which is controlled by the electronic timer *T*. *A*, sample solution; *B* reagent; and *C*, inert carrier stream (buffer).

delay times (t_d) giving the best ratio of sample to reagent (C_A/C_B), and that this approach to kinetic analysis is more convenient than traditional methods because the very precise timing needed in rate-based analytical methods is easily accomplished, it is, nevertheless, unsatisfactory for physiochemical reaction rate measurements because the initial reaction time t_i, initial reagent concentration $C_{A,i}$, although well repeatable, are not known. This is because mixing of the sample A and reagent B by mutual penetration begins at the moment of injection and proceeds continuously, reaching consecutive sections of the sample zone as it disperses within the reagent stream. Thus, all elements of fluid have different, unknown values of t_i, $C_{B,i}$ and $C_{A,i}$. However, when sample A is injected into an

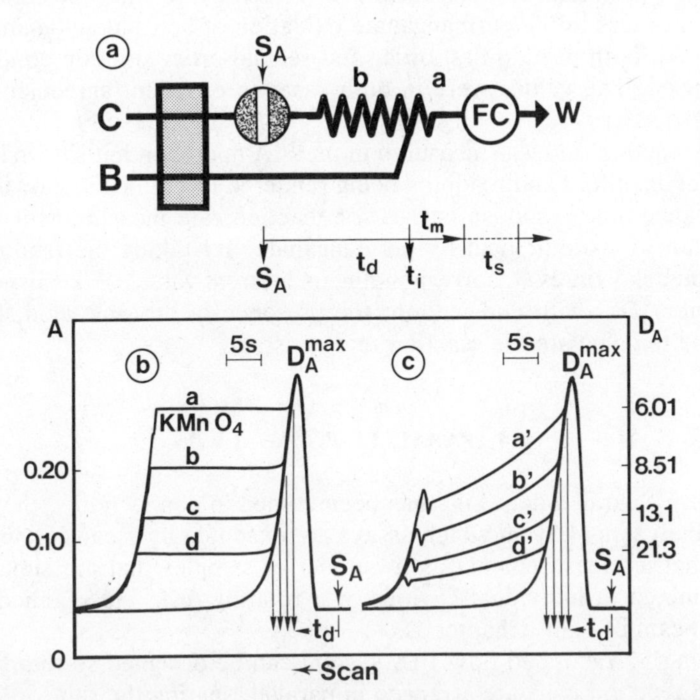

Figure 4.17. Determination of the reaction rate constant for the oxidation of crotonic acid by potassium permanganate. (*a*) Manifold used (cf. Figs. 4.15*c* and 4.16). (*b*, *c*) Absorbance-time response curves actually recorded. The values of the dispersion coefficient D_A were obtained by dispersion experiments. All curves in each set of experiments were recorded consecutively from the same starting point (S_A), with an increasing delay time (t_d = 7, 8, 9, and 10 s, *a–d* and *a'–d'*) with the stopped-flow period t_s = 20 s (additionally, in each series a single run without stop is included). (*b*) $KMnO_4$ (C_A^0 = 8.54 × 10^{-4} *M*) in phosphate buffer in absence of crotonic acid. (*c*) $KMnO_4$ (C_A^0 = 8.54 × 10^{-4} *M*) in phosphate buffer, crotonic acid ($C_{B,i}$ = 2.10 × 10^{-4} *M*) in phosphate buffer, stream *B*. (From Ref 838 by permission of the American Chemical Society).

inert carrier or buffer stream C, and this stream is merged with reagent B in a two-line FIA system (Fig. 4.15*b,c*), each element of the dispersed sample is mixed with the same volume of reagent, and hence the initial reagent concentration $C_{B,i}$ will be well defined via the pumping rates of the C and B lines. Depending on the reaction monitored, the mixing time of sample and reagent (t_m) is a crucial parameter, since this time will determine to what extent the original concentration of sample A (and of reagent concentration B) is consumed by reaction in coil a before the sample zone reaches the detector. Thus, if $t_m \ll t_{1/2}$ of the reaction, the manifold depicted in Fig. 4.15*b* will suffice; yet, for faster reactions, coil a has to be shortened (Fig. 4.15 *c*). By appropriate manipulation of the reaction conditions, it was feasible to measure the rate constants and reaction orders in the permanganate oxidation of benzaldehyde and crotonic acid, both pseudo-first-order and second-order solution conditions being used. The values were in both cases in excellent agreement with literature values.

The reaction rate was measured in the FIA microconduit shown in Fig. 4.16, the manifold and readouts being rendered in Fig. 4.17, showing the absorbance-time response curves for reaction rate measurement of the oxidation of crotonic acid by permanganate. By taking the readouts at different delay times t_d, corresponding to different values of the dispersion coefficient D_A, with and without the presence of crotonic acid, it was possible to calculate the reaction rate constant.

4.4 PARALLEL FIA ANALYSIS

The term "parallel analysis" has been coined in connection with centrifugal analyzers [4.3], in which assays are executed by means of spectrophotometric measurements on a number of samples that are simultaneously mixed in individual cavities of a rotating disk and scanned by a single beam of light (Chapter 1).

Similarly, a stopped-flow FIA system can be designed so that two or several reaction coils are arranged in parallel, sharing the same injection port and flowthrough detector. In this way appropriately dispersed sample zones may be stored in coils, with the aim of increasing reaction time beyond the limit of 30 s, which, from the viewpoint of the lowest sensible sampling frequency (say 75 samples/h), cannot be executed in a single-channel FIA system.

When a parallel FIA system is designed, two conditions must be fulfilled. First, as the channels can be interchanged only when the carrier stream does not move, and because the sample zone always has to be stopped at an exact distance from the injection port, the flow maneuverability and the exactness of timing must be as precise as that achieved for an ordinary stopped-flow method. Second, the individual channels must have identical geometries; otherwise, the dispersion will vary from one channel to the next, and samples of identical composition will not yield identical peaks.

A two-channel system, constructed according to these guidelines (Fig. 4.18*a*), is made of two identical coils (0.5 mm ID Teflon tube of strictly uniform diameter), 85 cm long, into which the carrier stream is alternately directed by means of a two-way valve *V*. The injected sample volume is 25 μL. This system was tested by means of a bromothymol blue dye solution. Using first continuous pumping, the dispersion in the two channels was adjusted by shortening the coil that yielded the lower peak, until identical signals were obtained regardless of which of the two channels was used. After this minor adjustment, which was necessary because the connecting blocks and valve connectors were slightly different in each channel, a perfectly identical dispersion pattern was obtained in both channels (Fig. 4.18*c*). Next, a stopped-flow approach was tried, the purpose of which was to store the sample zone in one of the channels for a fixed period of time. Using channel 1, the stream was pumped continuously (Fig. 4.18*d*, curve *a*), and in the following experiments the zone was stopped after a prefixed delay time t_1, and stored within the coil for increasing periods of time (curves *b–e*), so that the total residence time for the last curve was 50 s. It is interesting to note that this experiment confirms that dispersion decreases, by radial molecular diffusion, when the sample zone does not move. Furthermore, in agreement with Taylor [4.4], this effect is observed after storage in a 0.5 mm ID tube for as short a period as 6.2 s [20], which is indeed seen from curves a and b (see also Chapters 2 and 3 and Fig. 2.7).

After the geometrical identity of both channels and the reproducibility of the stopped-flow manipulation have been established in this way, the conditions of an assay were simulated by alternately storing sample zones (of dye solution) in the channels with the aim of demonstrating which residence time may actually be obtained if a certain sampling frequency of, say, 120 samples/h should be maintained. For this purpose, however, the timing sequence of the electronic timer and the sample injector had to be altered.

When we examine the time distribution of the recorded signal (Fig. 4.18*c*), it is apparent that the peak width at its base in this system is approximately twice as long as the time interval between the sample injection and the moment when the signal starts to rise. In a simple stopped-flow system this time distribution does not present a problem because a new cycle is initiated only when the next sample is injected. The timer for a two-channel parallel stop analysis must, however, run in a fixed go–stop–go cycle, because while one sample zone is being injected (or flushed out) the other is stored, and if the length of the storage period is not always the same, the analyzer will not yield reproducible results–unless, of

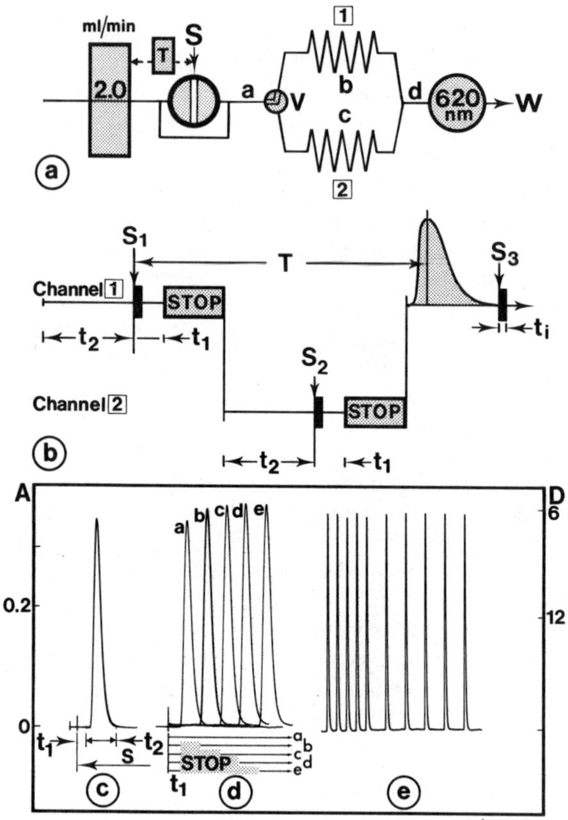

Figure 4.18. (*a*) Two-channel parallel stopped-flow FIA system designed to allow a high total sampling frequency. The injected sample *S* (25 μL) is by means of valve *V* directed alternately to coils 1 and 2 (each 85 cm long, 0.5 mm ID), exhibiting identical dispersion patterns [see (*c*)]. (*b*) Timing sequence for the two-channel system. Sample 1 is injected into the carrier stream over a fixed period of time t_i, and when the sample after time t_1 is within coil 1, the pump is stopped for a prefixed period. During the stop period, valve *V* is turned and sample 2 is loaded into the injection port. The pump is reactivated and after time t_2, sample 2 is injected and directed into coil 2, whereupon the pump is stopped again. The valve is again turned, sample 3 is loaded into the injection port, and after reactivation of the pump, sample 1 is flushed out and detected, the period lasting t_2 seconds, whereupon the timer triggers the injection of sample 3 into line 1; and so on. (*c*) Dispersion patterns of the two identical channels. (The figure actually shows two recording traces, one on top of the other.) (*d*) Peaks obtained by injecting identical samples (of a dye solution) into channel 1 and—after a delay of t_1 = 5.5 s—stopping for increasing periods of time: *a*, 0 s (continuous pumping); *b*, 10 s; *c*, 20 s; *d*, 30 s; and *e*, 40 s. (*e*) Signal output obtained by two-channel parallel stopped-flow operation of the system depicted in (*a*), using a BIFOK FIA 05 apparatus (i.e., the peaks are alternately from channel 1 and channel 2). The total sampling frequency is 120 samples/h (left) and 90 samples/h (right) [153].

course, the storage period is longer than that corresponding to *c* of Fig. 4.18*d*. In other words, stop–go intervals must repeatedly follow each other, and the sample injection (i.e., the turning of the sample valve) should not activate the function of the timer, but must be programmed to take place at a suitable moment within the pumping interval.

This brings us back to the time distribution of the recorded signal within which the two time intervals t_1 and t_2 (Fig. 4.18*c*) must be chosen. From the moment the valve is turned and the sample introduced into the carrier stream (S_1, Fig. 4.18*b* channel 1) until the pump is stopped (channel 1), the time span must be no longer than t_1; otherwise, the sample zone—or at least part of it—would pass the confluence point, line *d* (Fig. 4.18*a*) and even into the flow cell. Now, if t_1 were *longer* or at least equal to t_2, the following sample could be injected right at the beginning of the next cycle, because then the sample zone previously stored in the now-opened channel (2) would be entirely flushed out past the flow cell before the stream is stopped following the t_1 period. This is, however, *not* the case; therefore, the sample injection must be postponed by an interval t_2 from the moment when the next pumping cycle is initiated (see channel 2, Fig. 4.18*b*), so that the preceding sample is flushed out before the next sample is injected (S_2), and then following t_1 the stream is stopped again. Only hereafter, when pumping is resumed, will S_1 emerge from the storage system (channel 1), its total residence time T being equal to $2\ (t_1\ +\ \text{stop})\ +\ t_2$, and be measured. Thus for $t_1 = 5.5$ s and $t_2 = 14.5$ s (a t_1/t_2 ratio deliberately chosen very low to prevent any carryover whatsoever), and for a stop period of 10 s, the residence time will be 45.5 s at a sampling frequency $S_f = 120$ samples/h [where $S_f = 3600/(t_1 + \text{stop} + t_2)$], and for a stop interval of 20 s, the residence time T would be 65.5 s and the sampling frequency 90 samples/h (Fig. 4.18*e*).

Although the foregoing description inevitably implies that the parallel stopped-flow FIA system is complicated in design, the actual analysis is simple to perform, as all that is needed is to fill the injection valve with the sample solution *S* and then turn the directional valve *V* at any time during each stop interval. As far as hardware is concerned, the electronic timer has to be self-activating, so that it continues in a stop–go mode; and the injection valve has to be operated by a motor that can be activated at a given moment and turned back after a well-defined prefixed time interval (e.g., as here, $t_i = 3.0$ s), during which the sample is swept by the carrier stream into the system. These features are incorporated into the BIFOK FIA system (Chapter 5), and the reproducibility tests and dispersion experiments shown in Fig. 4.18 were performed by means of the BIFOK FIA 05 apparatus. The reproducibility of operation of the FIA system in a parallel two-channel mode confirms the total absence of carryover at a sampling frequency of 120 samples/h for a residence time of $T = 45.5$ s (Fig. 4.18*e*). Alternatively, use of a packed reactor of the single-bead-string type (Section 3.1.6) for obtaining comparable residence time and sampling frequency might be considered, as its design and operation is simpler than those of a two-channel stopped-flow FIA, although other problems are encountered [1057].

The residence time can be further prolonged to hours or possibly days in a *multichannel* parallel flow-injection analyzer, where several sample zones are stored in parallel for a desired period (Fig. 4.19). The procedure consists of a load cycle, during which all samples are loaded into a rotating drum; a storage period, during which the reaction takes place; and a measuring cycle, in which samples are measured when flushed from the storage cavities and through the flow cell. As the loading requires no more than 7 s per sample and the measuring period is about twice as long, an *effective* sampling frequency of 120 samples/h can readily be maintained, because a 9 s stop interval is more than sufficient to load the sample and to turn the drum.

The requirements for its mechanical design are strict, as all parallel cavities must exhibit exactly identical flow geometries. Because it was not found practicable to make a larger number of identical coils from a long piece of tubing, the only remaining choice was to bore a number of exact holes of 1.3 mm ID in a Teflon cylinder. This, in turn, limits the volume of liquid that may be stored to approximately 80 μL, which has the consequence that only *part* of the dispersed sample zone might be accommodated in each individual channel, and therefore the front or the tail or both these sections of the sample zone have to be chopped off during the loading operation and flushed from the system during the next loading cycle. By choosing various delay times, various sections of the central part of the dispersed zone can be preserved for storage (Fig. 4.20*a*), and it was found that the maximum amount of sample material preserved for storage and optimum reproducibility of zone retainment is found when part of both the tail and front sections are chopped off (see peaks *d–f*, Fig. 4.20*b*), rather than if only the front or only the tail section is removed. When the optimum delay time had thus been found (for a given pumping rate, sample volume, and flow geometry, here 3.4 s–peak *e*), a final dispersion experiment was performed, using a drum with eight parallel channels, into which seven identical colored zones were loaded (Fig. 4.20*c*), the first part of the record showing the amount of sample material (dye) that was discarded from each sample zone. The drum was then removed from the FIA

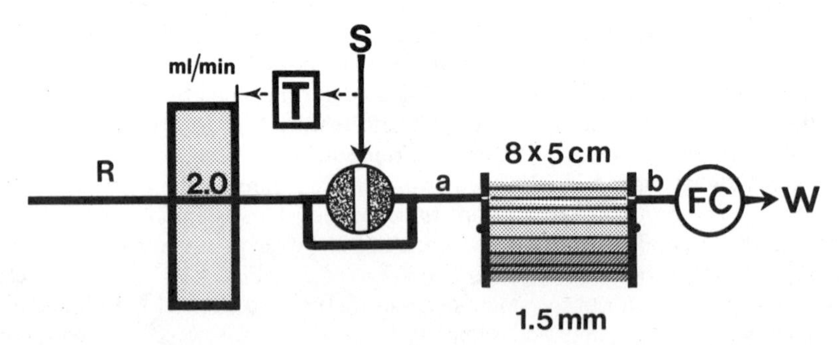

Figure 4.19. Parallel FIA analyzer with eight 80-μL storage tubes. The connecting lines (*a* = 20 cm, *b* = 30 cm) are of 0.5 mm ID. The system was tested by injecting 30 μL aliquots of dye solution, which were measured at 620 nm. *T* is the timer; *W* is the waste line [153].

system, stored for 6 h, reinstalled, and the dye flushed out; this time seven identical peaks were produced, corresponding to the stored portions of the sample zones. (One of the cavities, in turn, had to be used for flushing, loading, and reinstalling the flow in the system after the storage period.)

When the procedure described above was first published [153], the following reservations were made as to the concept of parallel FIA: (a) the method has been tried only by dispersion experiments with a dye and not with any chemistry; (b) it is much more complicated than all other FIA approaches, which is against the spirit of the overall simplicity of FIA designs; and (c) too much room has been allocated to the desription of the parallel method, which has not yet been applied to any practical assay. Therefore, many may prefer an air-segmented continuous-flow system as a practical alternative for $T \geq 1$–10 min. Yet the description of

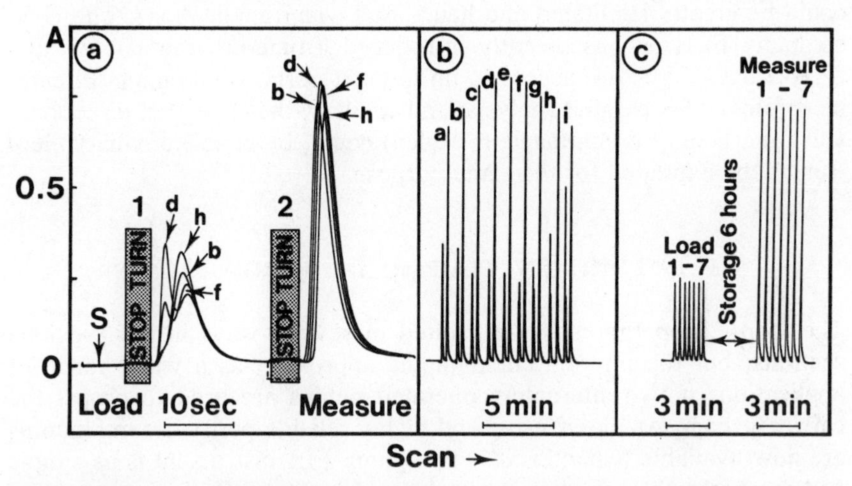

Figure 4.20. (*a*) Storage of a section of a dispersed zone of dye in one of the tubular cavities of the parallel FIA analyzer (Fig. 4.19). The zone was first loaded (Stop/Turn 1) so that when the pump was restarted after a 4 s stop period, the outer sections of the zone were discarded (d, h, b, f, respectively). During the next stop period (Stop/Turn 2) the drum was turned back so that the stored portion of the zone was flushed out and measured (b, d, f, h, respectively). By changing the delay time, and thus the position of the dye zone in the drum, either the tail (b), the front (h), or both tail and front portions (d, f) of the dispersed zone were chopped off. The same experiment was repeated in (*b*), but at a lower paper speed, for a number of delay times (a = 2.6, b = 2.8, c = 3.0, d = 3.2, e = 3.4, f = 3.6, g = 3.8, h = 4.0, and i = 4.2 s), while a fixed stop period of 4 s for the load/turn cycle was preserved. The delay of 3.4 s (curve e) allows optimum repeatability of material storage and was therefore used for sample loading when running a 6 h storage experiment (*c*), where seven samples were loaded *in parallel* in seven tubular cavities (load) and measured after the storage period.

the parallel FIA system is nevertheless included, as it demonstrates the versatility of sample zone manipulations, the feasibility of extremely long storage (under anaerobic conditions), and may thus inspire thoughts, approaches, and instrumental designs beyond the authors' reach.

It is rewarding to observe that the parallel FIA system indeed has inspired thoughts and been brought to practical applications (cf. also Sections 4.7.1 and 4.8). Thus Rocks et al. [1070], faced with determining the activity of acid phosphatase in human serum—for which assay no rapid methods exist—devised a system in which an automated distribution valve could direct samples into one of four holding coils, where they were arrested for a predetermined time. Using an incubation time of 5 min, they achieved a sampling frequency of 48 samples/h with their system, the results obtained being in excellent agreement with those attained by conventional procedures. With the advent of commercially available multiselector valves, it is to be expected that these types of analysis schemes could be greatly facilitated and hence find wider applications. The U.S. company FIAtron has recently introduced a time-programmable valve (FIA-Valve 2500) with access to up to eight ports, which might directly be employed for parallel analysis, and similarly the eight-port directional valve marketed by Pharmacia (Sweden) could, by simple auxiliary electronics, be exploited for the same purpose.

4.5 INTERMITTENT PUMPING AND MERGING ZONES

The stop–go operation was exploited in several ways in the previous sections, but the intermittent pumping approach has a wider range of applications if two alternately operated pumps are used to propel the carrier stream. As inexpensive and highly reliable peristaltic minipumps are now available (Chapter 5), a two-pump FIA instrument is as simple and mechanically reliable as a single-pump system. (In fact, most commercially available FIA systems are provided with two individually operational pumps.) There are many combinations in which two pumps can be incorporated into a manifold, especially if one considers different timing of their stop–go intervals, which may overlap or coincide. Restricting ourselves here to considering systems where the pumps are used exclusively for propagation of carrier stream(s) (systems where one of the pumps is used for sample introduction, for example, by hydrodynamic or split-loop injection are described in Section 5.1), there are several feasible designs based on various combinations, of which the two possibilities described below and in the next section have proven most useful.

The simplest application of a two-pump FIA system is to increase the sampling frequency by increasing the washout speed from the coils and from the flow cell. This is done by using one pump to propel the carrier stream of reagent and a second pump to flush the system with the wash solution (e.g., water). The reagent pump (Fig. 4.21*a*, pump I) propels the slowly moving carrier stream of reagent only until the element of fluid corresponding to the point of readout, here the top of the peak, has been recorded. Immediately after, pump I is stopped, and pump II, delivering the wash solution, is started, generating a higher flow rate than pump I, thus allowing the baseline to be more rapidly reached than if only one

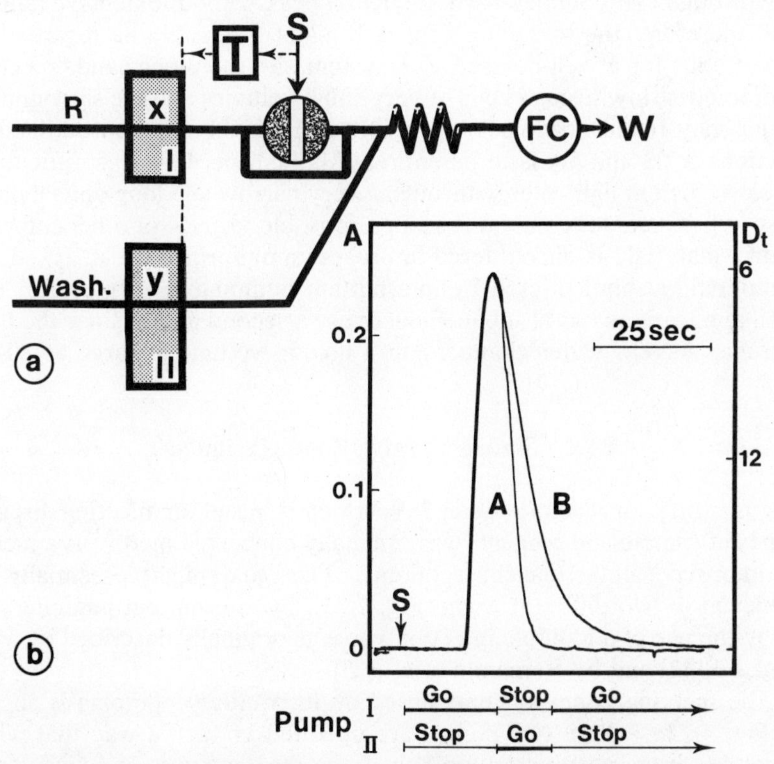

Figure 4.21. (*a*) FIA system operated with intermittent pumping. The sample S is injected into the reagent stream propelled by pump I. After the peak maximum has been recorded by the detector, pump I is stopped and pump II is activated by timer T to wash the sample out of the system [pumping rate $y > x$; see (*b*)]. After a preset time, pump I is restarted and pump II is stopped, thus permitting a new sample injection. (*b*) Recorder tracing obtained by using the system in (*a*) and injecting a tag dye solution (20 μL) into an inert carrier stream; $x = 0.9$ mL/min and $y = 4.0$ mL/min. Comparing the output obtained by this technique (curve A) with that registered by continuous operation of pump I alone (curve B), it is seen that the intermittent pumping in this case allows the sampling rate to be almost doubled.

pump is being used in a single-pump arrangement (see Fig. 4.21*b*, curves *A* and *B*). Because the first pump is started (and the second one stopped) only when the next sample is being injected, this approach allows an increase in sampling frequency and saves reagent solution. A review article, demonstrating applications of the intermittent pumping procedure for various practical analytical tasks, has recently been published by Krug et al. [1060].

As a matter of principle, one should, when designing a FIA system, always try to choose such components which—by forming a uniform flow path, free of undue dead volumes, or mixing chambers, including large flowthrough cell volumes—would yield a peak without extensive tailing, and, therefore, the foregoing approach might ostensibly be regarded as superfluous for a well-devised FIA system. On the other hand, spectrophotometric flowthrough cells of very small volumes and the surrounding optics may readily become very costly if the ideal situation outlined in Sections 3.2.3 and 5.4 is to be entirely fulfilled, because it is difficult to direct sufficient light energy through a very narrow and long optical path. Also, when samples containing suspensions, blood cells, or other complex matrix material, as encountered in process monitoring, are analyzed, an additional washout offered by intermittent pumping is very useful. For such applications it is also beneficial to use a second pump, since the flow cell may have a wider channel and a holdup volume as large as 20–30 μL.

4.5.1 Simple Merging Zones Techniques

The merging approach (Section 2.5), which is based on injecting discrete zones of sample and reagent, was originally conceived merely as a means to improve sample/reagent economy. This goal might essentially be achieved in two different ways (Fig. 2.25): by intermittent pumping [71] or by the use of a multiple injection valve as originally described by Mindegaard [42] and by Bergamin et al. [23].

The merging zones systems based on intermittent pumping is shown in Fig. 4.22*a*, where two pumps are operated in such a way that when pump I is in *go* position, pump II is in the *stop* position, and vice versa. Thus the sample zone is first transported from the injection port by means of pump I; then when a chosen distance from the merging point is reached, pump II is started, which continues to bring the carrier stream forward while the reagent is being added (cf. Fig. 2.25*b*). After the sample zone has passed the merging point, pump I is reactivated while pump II is stopped again. This approach allows the length of the reagent zone to be regulated simply by choosing different go and stop periods by means of

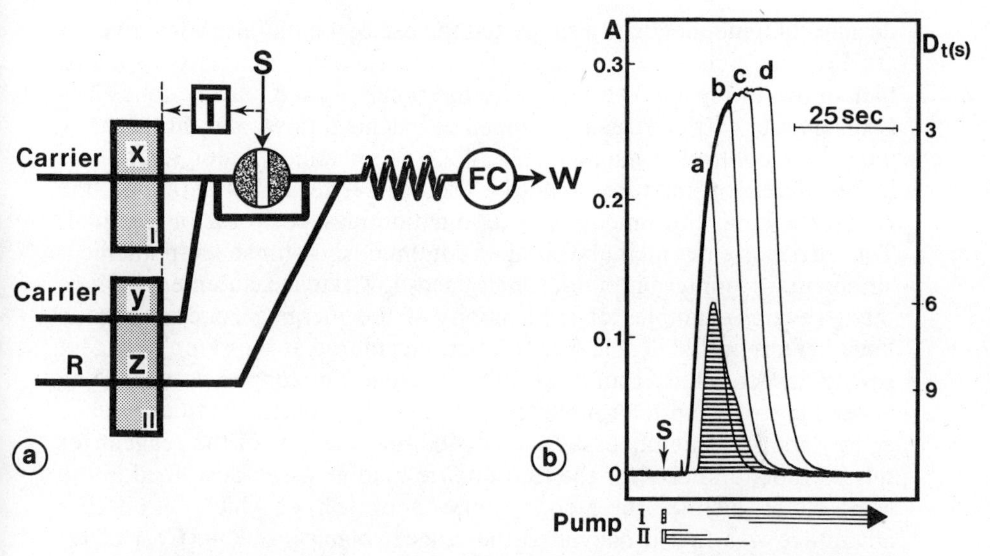

Figure 4.22. (*a*) FIA manifold for merging zones system based on intermittent pumping (cf. Fig. 2.25*b*), operated so that when pump I is in the Go position, pump II is in Stop, and vice versa. After sample injection and a preset delay time, the timer (*T*) stops pump I and activates pump II. Reagent is added at *z* mL/min, while the sample zone is carried forward at *y* mL/min by pump II. After a preset time, pump II is stopped and pump I is reactivated. (*b*) Recorder trace obtained with the system in (*a*) at pumping rates $x = 2.0$ mL/min and $y = z = 1.0$ mL/min. The shaded peak was recorded by injecting 10 μL of a tag dye solution into a colorless stream, pumped by all three lines, *x*, *y*, and *z*. Next, colorless solution was injected as well as pumped by lines *x* and *y*, while dye solution was pumped through line *z*. Using the same delay time in all experiments (0.3 s), but activating pump II for increasingly longer periods (*a*, 5 s; *b*, 10 s; *c*, 15 s; and *d*, 20 s), the "reagent" curves were recorded. It is seen that when the pumping period for pump II increases, the output signal is broadened, and at the same time the signal approaches equilibrium when the input approaches a step function, that is, $D_R = 1$. Obviously, pumping for 10 s (curve *b*), corresponding to 166 μL of reagent per sample, is more than sufficient to cover the shaded area representing the sample zone.

the timer (*T*), and makes it possible to create different concentration gradients on the interface between the sample zone, reagent solution, and carrier stream (Fig. 4.22*b*). Variations on this theme are numerous. Thus, by choosing different lengths of reagent zone, and by letting it overlap in different ways over the sample zone, one may, for example, obtain, on a single injected reagent zone, an individual blank for the reagent alone and for the sample zone alone, as well as the peak height resulting from the chemical reaction between the components of the sample and reagent solutions [42].

In their original publications Bergamin and co-workers [23] and Mindegaard [42] independently suggested the use of a multiinjection valve for the FIA merging zone approach. The purpose of using a valve, such as that shown in Figs. 5.3 and 5.5, is to inject sample and reagent zones into two separate carrier streams pumped at balanced flow rates so that they meet in a controlled manner (Fig. 4.23). As distilled water (or diluted buffer–detergent mixture) might be used as carrier in both streams, the reagent volume consumed per determination may be 30 μL or less [46]. The carrier streams might be pumped continuously—for single-point measurements—or intermittently, for stopped-flow rate measurements. Since then, several examples of applications of the merging zones approach have been reported (Table 7.1), primarily centered at the sample/reagent saving aspect. Thus, Lim et al. [80], in using the stopped-flow merging zones approach for determining albumin by a homogeneous fluorescence energy transfer immunoassay, concluded that the cost of the reagent for this method, and also for the fluorimetric binding assay described in the same paper, justified the merging zones approach, of which yet another advantage was that it alleviated the reagent blank problem. Cost of reagents was ostensibly also the main impetus for Worsfold et al. [835, 935, 1078] in their work on immunoassays, their aim being to provide cheap, rapid, and flexible analytical facilities that could be used even in small clinical laboratories, while Pasquine and Oliveira [532] took advantage of the effective mixing pattern of sample and reagent in the merging zones configuration in order to obtain reliable results in their experiments, which

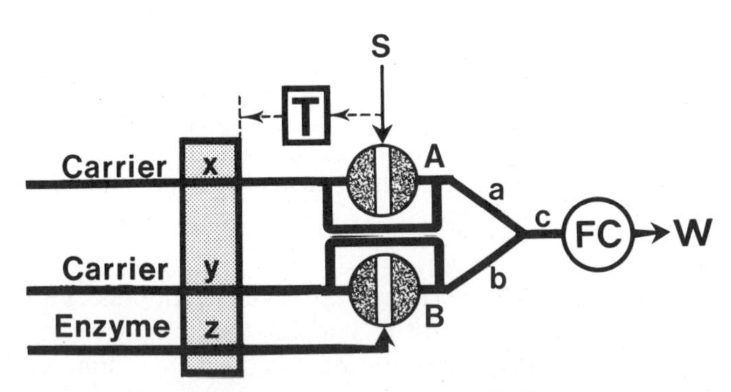

Figure 4.23. FIA manifold for the synchronous merging of two injected zones in a symmetrical system (cf. Fig. 2.25*a*) as used for determination of glucose, where the carrier stream is pumped at equal rates in lines x and y (1 mL/min) through tubes of equal length ($a = b =$ 10 cm, ID = 0.5 mm). The merged zones are then transported through line c (30 cm) to the flow cell (FC). The sample is injected at S through port A, while the reagent (enzyme) is injected through port B, to which the reagent is transported through line z. The volume of injected reagent (enzyme) was 26 μL; the sample was 26 or 10 μL, depending on the required sensitivity. T denotes the electronic timer.

involved very delicate detection by enthalpimetry. In a recent paper [1160], Petersson et al. presented an elegant on-line conversion technique in which an injected sample of sulfide was merged with a zone of cadmium(II) ions acting as a precipitating tag material thereby allowing subsequent quantification of the sulfide content by flame atomic absorption spectrophotometry (for details see Section 4.7.1). However, in the following section it is demonstrated that the simple merging zones approach, and variants of it, carries even more potential, that is, when the technique is used not merely as a convenient way of combining sample and reagent, but advantage additionally is taken of the concentration gradients formed concurrently.

4.5.2 Gradient Techniques with Injection of Two or Several Zones

These techniques are based on exploitation of the concentration gradients formed when two (or several) zones are injected simultaneously—either synchronously or asynchronously—and then allowed to penetrate each other, thereby yielding response curves that either completely or partially overlap (Section 2.5). Since the dispersion coefficient D for each sample zone will have a given value for any given delay time t, the *ratio* of the D values for the sample zones will remain constant for any value of t. This approach has been employed (1) to develop a generalized method for quantifying the extent of interference of foreign species on a given chemical assay through introduction of the term selectivity coefficient [378]; (2) as a novel standardization method to achieve multicomponent standard addition over a wide, controllable range of standard/analyte concentration ratios [817]; and (3) for devising a new, economical way of introducing sample and reagent solutions into a FIA system via the split-loop injection technique [848]. Referring to Section 2.5 where the principles of these techniques were outlined, examples of each of these methods will be briefly described in the following.

The method of *selectivity measurement* is based on the fact that in any chemical procedure for a given species A any interference caused by the presence of an extrinsic species B will always, at least as observed from the detector's point of view, appear as (positive or negative) "pseudo A." Therefore, the extent of this interference can be quantified by the selectivity coefficient, k_{AB}, defined for any FIA method by the equation:

$$C_A' = C_A + k_{AB}C_B \tag{4.1}$$

where C_A' is the apparent concentration of species A, and C_A and C_B are the actual concentrations of A and B, related to the originally injected concentrations C^0 by their D values, that is, $D_A = C_A^0/C_A$ and $D_B = C_B^0/C_B$. If the FIA manifold used for determining the k_{AB} value is devised so that all measurements can be executed in that or those elements of fluids along the concentration gradients where the dispersion coefficients have identical values ($D_A = D_B$), it was shown previously [378] that the value of k_{AB} simply can be expressed by

$$k_{AB} = (C_A^0/C_B^0)[(H_{A+B}/H_A) - 1] \qquad (4.2)$$

where H_A and H_{A+B} are the peak heights recorded by the detector from injection of species A alone and A plus interfering species B, respectively. Thus, the higher the k_{AB} value, the greater will be the interference of the foreign species B in measurement of the primary species A. The interfering effect may be either positive (more species A found than actually present) or negative (depression of signal generated by A in the presence of B), and, therefore, the selectivity coefficient may either have a positive or negative value.

Several designs of the FIA system for determining selectivity coefficients by the gradient method are conceivable [378], the only restriction being that the element of quantification meets the condition of $D_A = D_B$. This may be accomplished by injecting the two species separately by means of a double valve, employing a merging zones configuration where the two zones meet either synchronously or asynchronously. An example of the latter approach is presented in Fig. 4.24, the system being used for studying the interference of nickel on the colorimetric determination of manganese by the formaldoxime method as monitored at 455 nm. Thus, in this setup, advantage is taken of the fact that within that region where the two zones overlap there is inevitably a segment of fluid, characterized by a certain delay time t_M, where the dispersion for both sample zones are equal. This delay time t_M is readily identified by injecting identical sample concentrations into each loop of the valve in two separate experiments, and by recording the resulting peaks from the same starting point. Thus, curves A and B in Fig. 4.24*b* were recorded by injecting 6 ppm Mn alone by sample loop A and B, respectively, from the same starting point (S). The composite curve ($A + B$) was obtained by simultaneous injection of 6 ppm Mn from both sample loops. As the sum of a set of linear processes will yield a linear response, the signal recorded at any composite curve ($A + B$) at delay time t_M (point M where the dispersions for A and B are identical) will be, for a fixed level of A, a function of the concentration of B injected into loop B. Therefore, by injecting the primary species (Mn in this case) by valve A and the interfering species (Ni) by valve B, the composite signal at delay time t_M will,

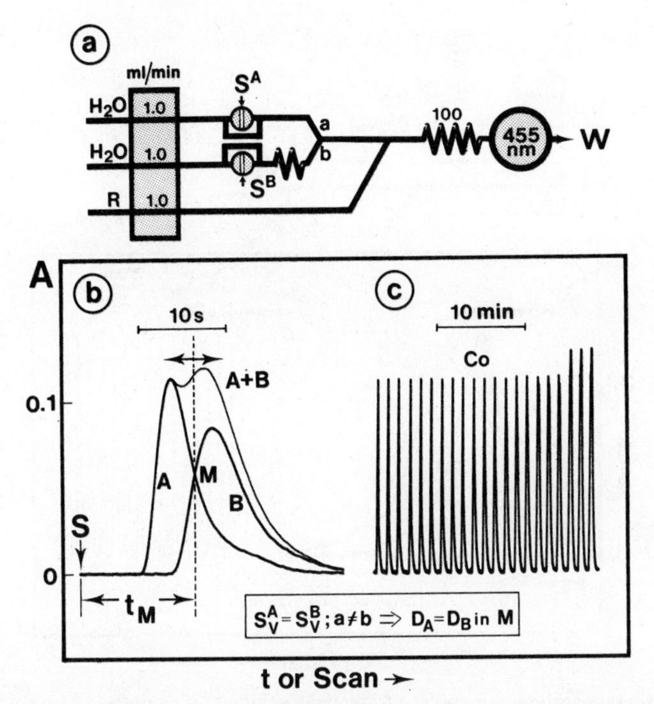

Figure 4.24. Asynchronous merging of samples S^A and S^B injected into two separate carrier streams (water). (*a*) Manifold with $a \neq b$ (a = 13 cm; b = 48 cm). (*b*) *A* and *B* are the recorder outputs obtained by injecting samples of identical volume (S_v) and concentration of manganese(II) (6 ppm) alternately into valves *A* and *B*; the composite curve *A* + *B* is obtained by injecting the samples simultaneously. Quantitative evaluation is done at delay time t_M corresponding to M where $D_A = D_B$. (*c*) Interference of Co(II) on manganese, where C_{Mn}^A = 6 ppm and C_{Co}^B = 0, 5, 10, 25, 50, 75, and 100 ppm; all samples injected in triplicate.

according to Eq. (4.2) for any level of concentrations of the two species, allow the computation of the selectivity coefficient, that is, k_{MnNi} for this particular procedure. Once determined, the value of t_M may be programmed into a microcomputer attached to the FIA system used so that all subsequent readouts can be done at this delay time, thus facilitating the evaluation.

However, for both synchronous or asynchronous merging, the precision is critically dependent on maintaining not only the ratio of the two pumping rates, which is simple to achieve with a peristaltic pump, but also their absolute values, which is much more difficult. Thus, a more practical flow system should allow the two species to be injected separately without the need of keeping the pumping rates in separate lines absolutely constant. Such a system is presented in Fig. 4.25*a*, which

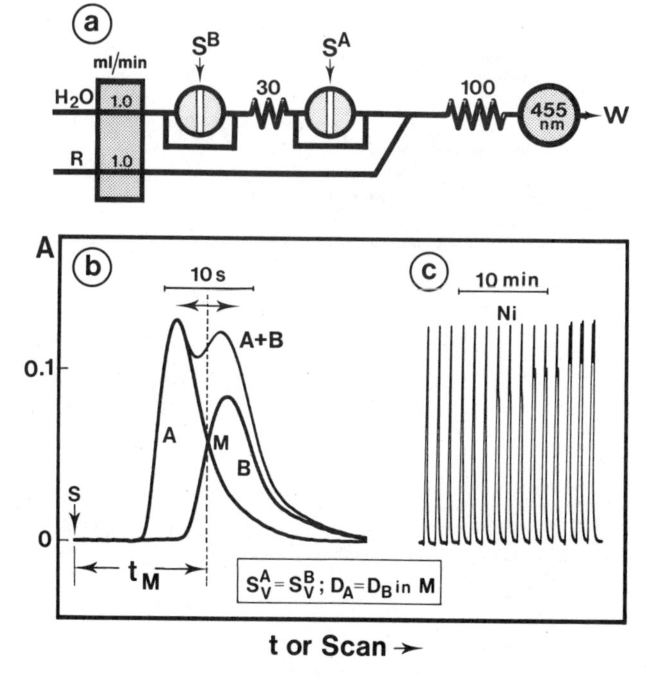

Figure 4.25. Asynchronous merging of identical volumes of samples S^A and S^B injected simultaneously into a one-line carrier system (water) by two separate valves (A and B). (a) Manifold with a 30 cm coil between A and B that delays sample B relative to sample A before both samples are merged with the reagent stream (R). (b) Injection of manganese(II) (6 ppm) alternately by valves A and B (curves A and B), and then simultaneously (composite curve $A + B$). At point M, corresponding to delay time t_M, the dispersion coefficient for both curves is identical ($D_A = D_B$). (c) Interference measurements for nickel, where $C_{Mn}^A = 6$ ppm and $C_{Ni}^B = 0, 2, 5, 7$, and 10 ppm, with each sample injected in triplicate.

shows a single-line manifold furnished with two valves (A and B) spaced by a short length of tubing resulting in asynchronous merging of the two sample zones, that is, during the transport through the system one sample zone will chase the other, the two zones gradually penetrating into each other. Using it for the same chemical procedure as that detailed above, the system was, as before, first calibrated by injecting a Mn standard solution separately by valves A and B (curves A and B) in order to localize the delay time t_M at which $D_A = D_B$ (point M). Then the composite curve, obtained by simultaneous injection of the same standard by both injection valves, was recorded ($A + B$), confirming that the signal at delay time t_M was twice that for the composite curve as for the curves obtained for the two individual injections. Thus, by injecting for the actual experiments of determining k_{MnNi}, the primary species Mn by valve A and the interfering species Ni by valve B, the composite signal at delay time t_M could for any level of concentrations of the two species directly be used for computation of the selectivity coefficient. Note, that use of the delay coil placed between the valves disperses zone B (Ni) to an extent such that when the peak maximum of zone A (Mn) is recorded, the contribution from zone B is zero. Consequently, the peak maximum height of the manganese peak itself will not be affected by increasing concentrations of nickel, yet the shape of the composite curve will become increasingly distorted, eventually leading to the formation of double peaks (Figs. 4.25*b* and 4.25*c*). While a manual execution of this study would demand an extensive dilution series to be prepared and compared, the FIA approach requires only a limited number of solutions. (Provided that the k_{AB} value is constant, it actually only requires a single solution of the primary ion and a single solution of the interfering ion.) Hence, it is not only very rapid, but it costs less in chemicals.

A natural extension of the zone penetration approach is to use it as a vehicle for the *standard addition* technique. If species A is regarded as the unknown sample, known additions of the same analyte in concentrations bracketing the unknown level present in A can be injected as "pseudo species B" (Section 2.5.3). This will allow corrections to be made for (moderate) interfering contributions of unknown foreign species. The readout may then either be taken at the point of equal dispersion (corresponding to delay time t_M in Figs. 4.24 and 4.25), in which case a formula equivalent to Eq. (4.2) can be used for calculations, or it may be taken at any other point along the interface of the two gradients, provided that the ratio of the dispersion coefficients at each selected delay time is known. This is readily accomplished using the same approach as described above [4.5].

Recently, a novel way of performing standard addition in conjunction with multielement analysis in atomic emission spectrometry was described [817; cf. also Fig. 2.27]. The principle is to some extent similar to that outlined for the selectivity measurements, except that the two zones—using the flow system illustrated in Fig. 4.26—are here inserted into an inert carrier stream, the injected sample zone being chased by an "infinitely" long zone of standard solution (Fig. 4.27*a*, top). Thus, if the sample (A) zone is large enough to prevent the penetration of the standard (B) zone, the gradient section on the rising part of the sample peak is diluted exclusively with the inert carrier stream (water, W) and is free of standard solution, while the tailing section is diluted entirely with standard. Therefore, if readouts corresponding to sections having identical D values at the front and tailing part of the sample zone are taken (at times t_A and t_{A+B}, Fig. 4.27*a*), the amount of standard added to the sample at

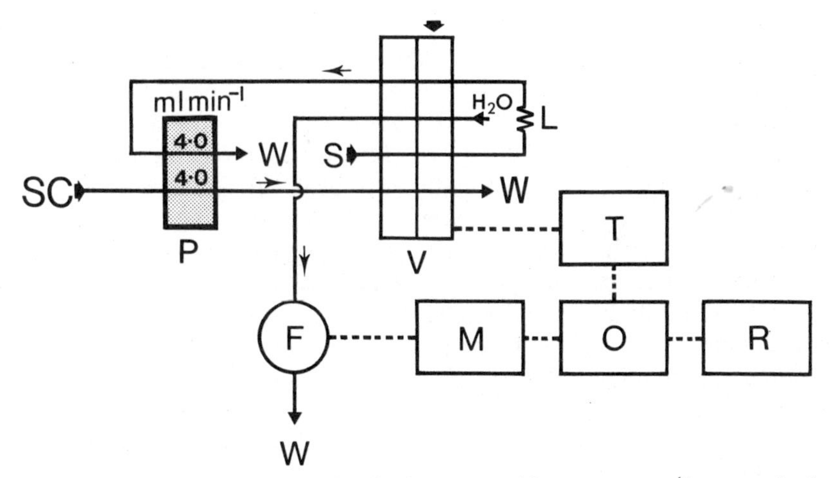

Figure 4.26. Flow system and setup for simultaneous multicomponent gradient scanning by flame photometric FIA embodying standard addition. The sample (*S*) is initially aspirated into loop *L*, which upon turning of the valve (*V*) is propelled forward by pump *P* through the FIA system and toward the detector (*F*), the sample being sandwiched between the inert carrier solution (water) and an "infinitely" long zone of standard carrier solution (SC): *F*, flame nebulizer–burner; *T*, timer; *M*, scanning monochromator; *O*, storage oscilloscope; and *R*, *X*-*Y* recorder.

t_{A+B} can be deduced, since it will hold (Section 2.5) that

$$1/D_A = 1 - 1/D_B \tag{4.3}$$

where D_A is the dispersion of the sample at times t_A (in the water) *and* t_{A+B} (in the standard), and D_B is the dispersion of the standard at time t_{A+B} (within the sample solution) on the tailing part of the sample peak. Assuming a linear calibration curve for the species determined, the peak heights H_A and H_{A+B} for the responses on the *composite* signal output, corresponding to times t_A and t_{A+B} (Fig. 4.27*b*), can therefore be related to the concentration of analyte in the sample, C_A^0, and standard solution, C_B^0. This leads eventually to the equation [817]:

$$C_A^0 = H_{A+B}C_B^0(1 - d)/[d(H_A - H_{A+B})] \tag{4.4}$$

where $d = 1/D_B$. Since the concentration of the standard solution is known and the dispersion coefficient of the section chosen can be measured in the usual manner, the expression $C_B^0(1 - d)/d$ will be a known constant, K. Hence:

$$C_A^0 = K[H_{A+B}/(H_A - H_{A+B})] \tag{4.5}$$

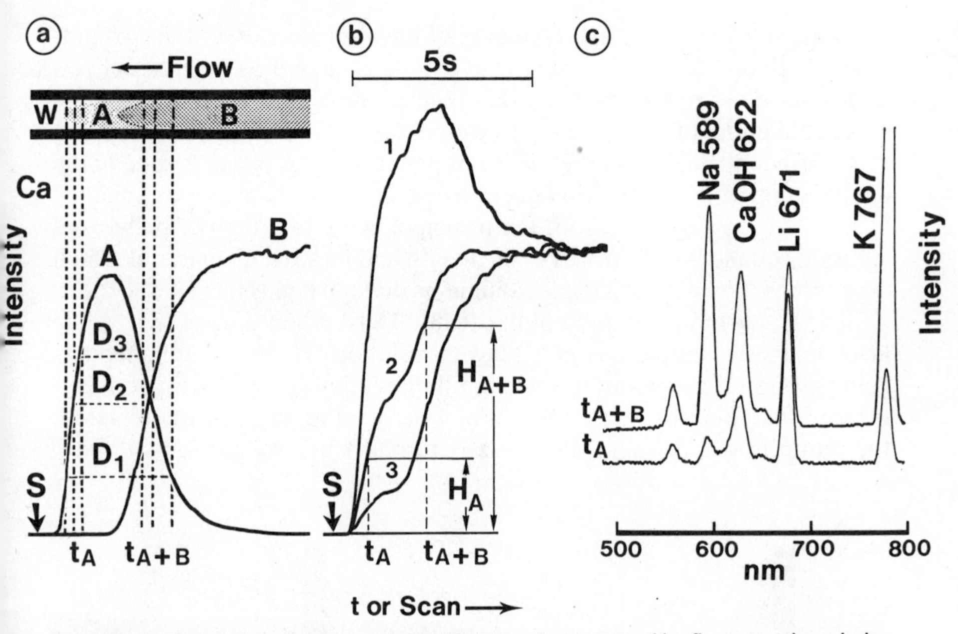

Figure 4.27. Gradient scanning standard addition as demonstrated by flame atomic emission spectrometry. (*a*) Time–intensity recordings at 622 nm of an 80 mg/L standard Ca solution injected as a 100 μL sample (*S*) with water (*W*) as carrier to show the dispersion of the sample zone (*A*). The broken lines indicate the times and dispersions when identical *D* values can be obtained on the rising and falling parts of the FIA curve. *B* is a recording of the same standard used as carrier and distilled water injected as sample. Note that the standard solution is not dispersed into the rising part of the sample zone. At the top is a schematic diagram of the conditions in the flow channel during the actual assay, the injected sample being sandwiched between the aqueous water carrier solution (*W*) and the standard carrier solution (SC) immediately prior to nebulization, t_A, t_{A+B}, and D_1, D_2, and D_3 representing the scanning times and dispersions coefficients, respectively, for points of identical *D* values. (*b*) Time–intensity recordings at 622 nm of 100 μL soil extract samples with 80 mg/L Ca solution as standard carrier. Curves 1, 2, and 3 are composite curves ($A + B$) of soils with high, medium, and low calcium concentrations. Times t_A and t_{A+B} for curve 2 are the points where scans were taken at identical dispersions for the sample zone, H_A and H_{A+B} being the corresponding peak height responses obtained. (*c*) Wavelength–intensity scans at t_A and t_{A+B} of sample 2 in (*b*) using a standard carrier solution containing Ca (80 mg/L), Na, K, and Li. Note the identical heights of the Li 671 nm internal reference line that acts as a check on the equal sample dispersions of the two scanning points.

which indicates that the sample concentration can be determined by standard addition upon a single injection measuring the response H_A and H_{A+B} at times t_A and t_{A+B}, respectively. As different sections on the gradient will provide different sample/standard ratios (Fig. 4.27*a*), the ratio could be optimized for precision by selecting the most suitable section. In the atomic emission experiment reported above (Fig. 4.27*c*) a simple method was employed to check the elements of equal dispersion at the two gradient sections scanned, namely, the use of Li added as an internal reference to the samples, implying that the heights of the recorded Li signal must be identical for each chosen pair of times t_A and t_{A+B}.

A further variation on the theme of penetrating or chasing zone is the introduction of the *split-loop injection* approach, which may be explained by referring to Fig. 4.28, top, the principle being that the normally used external sample loop of a valve, the length and internal diameter of which determines the injected sample volume, is divided (split) into two sections (*a* and *b*) sharing a common outlet [848]. Thus, when the valve is in the load position, sample (*S*) and reagent solution (*R*) are simultaneously being aspirated by means of a pump, filling sections *a* and *b*, respectively, the surplus going to waste (*W*). When the loading has been completed, the pump is stopped and the valve is turned whereby sections *a* and *b*

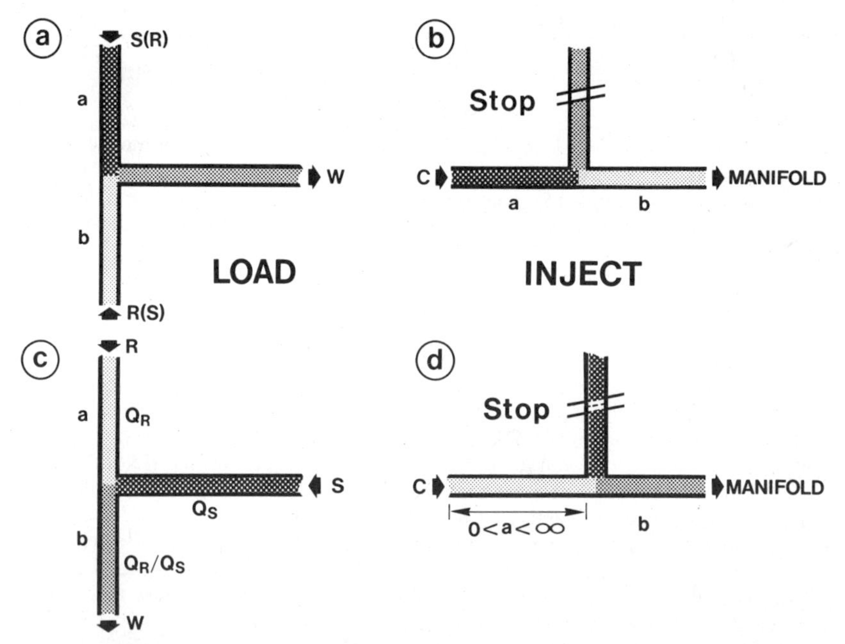

Figure 4.28. The FIA split-loop injection technique, the concept of which is that the (external) loop of an injection valve is split into two sections (*a* and *b*), which in the LOAD position are filled with either sample (*S*) and reagent (*R*), respectively (*a*), or with a mixture of both (*b*). In the first variant, the dimensions of sections *a* and *b* determine the volumes, which, upon turning of the valve to position INJECT (90° turn of the figure), are propelled into the FIA system by carrier solution (*C*); in the latter variant the injected volume is determined by the size of section *b* while the premixed ratio of sample and reagent is a function of the two pumping rates Q_R and Q_S.

become part of the carrier stream line leading to the manifold, while the wasteline is blocked. Both zones are thus injected and swept down the manifold by the carrier stream (*C*), zone *a* (sample) chasing zone *b* (reagent). The dimensions of the lines and the microreactors downstream determine the mutual dispersions of the two zones and hence their overlap. As an alternative, volume *a* may be filled with reagent and volume *b* with sample, thus leading to the eventual chasing of the sample zone by the reagent zone. (This approach is actually employed in the experimental setup shown in Fig. 4.36, Section 4.6.2.)

A further modification of the split-loop injection technique in depicted in the lower part of Fig. 4.28. Here, sample and reagent are simultaneously being aspirated into tube section *b*, the ratio between sample and reagent solutions being determined by the pumping rates Q_R and Q_S. Upon turning of the injecttion valve, this premixed solution of the two constituents is chased by additional reagent solution, the amount of which is governed by the dimensions of tube section *a*. Other variations are possible with the split-loop approach, and this technique is also well suited for standard addition procedures, similar to those mentioned above.

4.6 FIA DETERMINATIONS BASED ON SEPARATION PROCESSES

Many analytical procedures involve a separation step, the purpose of which is to increase the selectivity of determination by separating the analyte from an interfering matrix. In the manual batch mode such separations are always time consuming, laborious, and difficult to perform in the microscale. It is not practical to automate them by a batchtwise mode, and a continuous-flow approach is therefore the only way to automate precipitation, solvent extraction, dialysis, or distillation [4.6]. In the air-segmented systems, however, the presence of air bubbles has a disturbing effect. Thus, when water and organic phases are equilibrated during solvent extraction, the presence of the third phase (air) decreases the contact areas between the two liquid components. Similarly, in dialysis or gas diffusion, the contact area between the dialysis/gas permeable membrane and the donor liquid is diminished in the presence of air bubbles. Additionally, when performing gas diffusion, the gaseous substance emanating from the solution to be measured might be contaminated by gases in the ambient atmosphere (notably by carbon dioxide) unless the air used for segmentation is carefully purified. Thus the advent of the FIA technique has allowed significant advances in the automation of solvent extraction [19, 20], dialysis [6], and gas diffusion [57].

4.6.1 Solvent Extraction

FIA procedures based on solvent extraction require the development of the capability to mix and separate reproducibly two immiscible liquids. It was Karlberg et al. [19] and Bergamin et al. [22] who independently designed and tested a method that proved to be very efficient and successful in a variety of areas (Table 7.3), notably in pharmaceutical analysis. The principle of the FIA solvent extraction procedure is illustrated in Fig. 4.29: The aqueous sample, a component of which is to be extracted, is injected into an aqueous carrier stream to which at point (*a*) the organic phase is continuously added. After passing through the extraction coil (*b*), the organic phase is separated from the aqueous phase at point (*c*) and further led through the flow cell for continuous measurement. The key to successful performance of this method, which has a standard deviation of less than 1%, is the extreme regularity with which small droplets of organic phase are dispersed within the carrier stream and separated for spectrophotometric measurement free of any trace of aqueous phase. Detailed descriptions of the segmenting and separating devices, suggested by Karlberg et al. (cf. Fig. 4.30), have been published [19, 116, 699, 1079].

Based on this phase equilibration various solvent extraction techniques have been suggested. Thus, Kina et al. [31], taking advantage of detection by fluorimetry, designed a system without phase separation. For spectrophotometric detection, Sahleström and Karlberg [1079] recently described an unsegmented extraction system where the aqueous phase, into which the sample is injected, is led into an extraction module together with an organic recipient stream, the two streams being separated by a membrane across which transfer of the analyte takes place. Thus, the

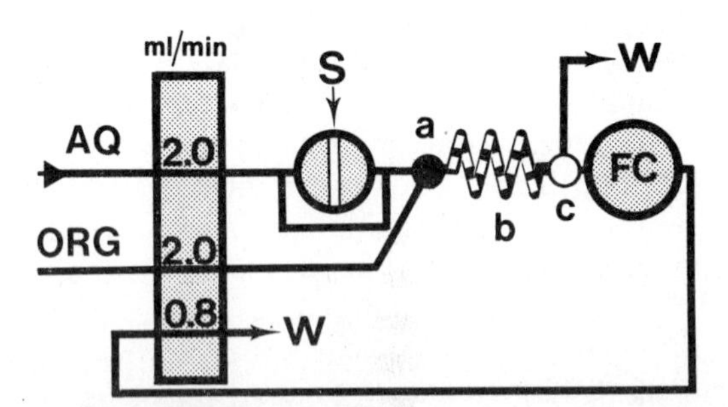

Figure 4.29. FIA manifold for the determination of trace metals by solvent extraction with dithizone in carbon tetrachloride. The aqueous sample *S* is injected into the aqueous carrier stream *AQ*, to which at (*a*) the organic phase ORG is continuously added and after passing through the extraction coil (*b*), made of Teflon tubing, the organic phase is again separated at point (*c*) and carried on through the flow cell for spectrophotometric measurement.

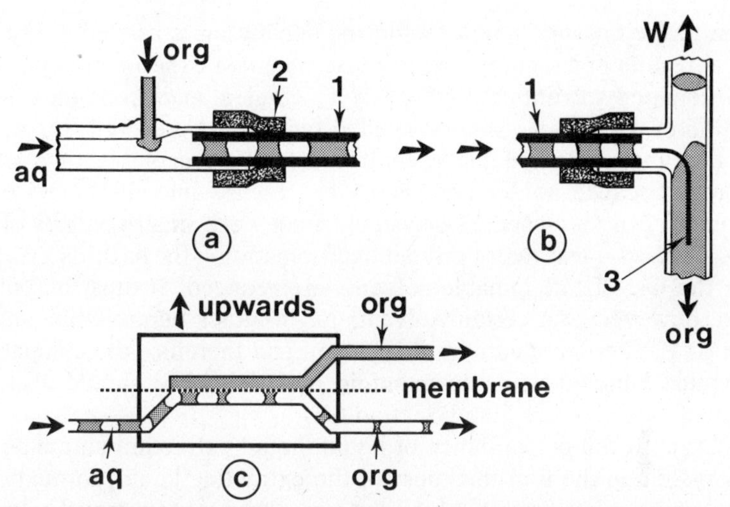

Figure 4.30. Segmentor and separator units used for solvent extraction. In the separator (*a*) the distance between the end of tube (1) and the orifice of the capillary, through which the organic solvent is introduced, determines the size of the aqueous and organic segments. Thus, by adjusting this distance, the extraction can be optimized. (*b*) Simple *T* separator into which a Teflon thread is inserted to facilitate phase separation. (*c*) Membrane phase separator, furnished with a microphorous hydrophobic membrane (such as Teflon), allowing passage of the organic phase with better than 95% efficiency.

segmentation as well as the separation step is avoided because the unsegmented recipient stream is led directly into the detector flow cell. The price paid, however, is a low extraction efficiency due to the restricted area contact. In a most recent publication [1272], the same authors proposed a self-contained liquid–liquid extraction module integrated into a FIA microconduit system, which has very short start-up times, requires low maintenance, and has reduced reagent consumption because of its small internal volume. Also, Audunsson [1329] has presented an ingenious extraction system where aqueous/aqueous extraction is executed via a liquid membrane installed in a dialysis module. Containing a hydrophobic membrane pretreated with a solvent support, this separator permits the transfer of analyte from an aqueous donor stream to an aqueous acceptor phase. In the extreme, if the acceptor stream is arrested while sample solution is continuously pumped through the donor line, a considerable preconcentration in addition to sample cleanup might be obtained prior to analyte detection. A detailed theoretical discussion as to the effects of sample volume, support matrix, types of immobilized solvents, donor flow rate, and partition coefficients of analytes between donor phase and membrane phase on enrichment factor and separation is rendered, the theoretical results being compared to those obtained experimentally.

Automated solvent extraction is very efficient and economical (only a fraction of the organic solvent needed for manual procedures is used by the FIA method), and more environmentally acceptable than manual methods because no solvent vapors can escape into the laboratory atmosphere from a closed FIA system. Since very small volumes of solutions are used—less than 1 mL per determination—the hazards associated with the use of inflammable solvents are reduced. It must be borne in mind, however, that certain solvents might attack pump tubing and Perspex or PVC components of the system, and therefore the compatibility of liquids handled with the manifold materials used should always be checked (see Chapter 5 and Section 6.1).

Crucial to the performance of liquid–liquid extraction automated in a FIA system is the film thickness of the extracting liquid, formed on the walls of the tube through which the two-phase system travels. In order to investigate this topic more closely, Karlberg [1117] developed a photographic technique by which it was possible to measure and study the film thickness left behind a solvent segment as a function of the velocity and the nature of the organic phase [688]. An almost linear relationship was found to exist between the film thickness and the flow velocity. For pentanol in a pentanol/water system, the film thickness was estimated to be 0.03 mm at a linear flow velocity of 5 cm s^{-1} (0.7 mm ID PTFE tubing). For chloroform the film thickness formed under similar conditions was about $\frac{1}{8}$th of the thickness of the pentanol layer. Since the theory of film formation in liquid–liquid segmented systems has not previously been developed, equations describing the film thickness behind one plug in a gas–liquid segmented system were used for a qualitative discussion. Such equations have the form [4.7, 4.8] $d_{film} = \text{const} \times R\,(F\eta/\gamma)^k$, where d_{film} is the film thickness (cm), R is the tube inner diameter (cm), F is the flow rate (cm s^{-1}), η is the viscosity (poise), γ is the surface tension (dyn cm^{-1}) of the aqueous/organic system, and k is an empirical constant ($\frac{1}{2}$ and $\frac{2}{3}$ were used).

Study of this equation implies that the quotient η/γ could be useful in estimating and comparing film-forming properties for different solvents. In the extraction coil the analyte can leave the aqueous segment and enter the film as illustrated in Fig. 4.31. At time "1," the aqueous segment having an ellipsoidal shape is approaching a region (dotted, vertical lines) where the analyte concentration can be followed. When the aqueous segment passes across this region (time "2"), the analyte is extracted into the organic film and C_{film} increases rapidly. When the aqueous segment leaves the observed region, C_{film} starts to decrease because of the dif-

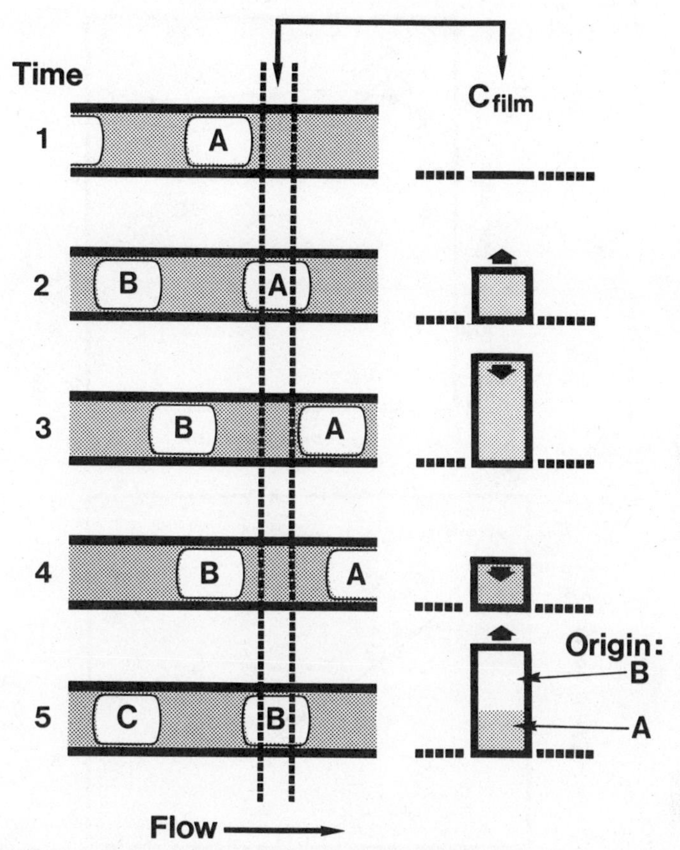

Figure 4.31. Stages during analyte extraction from an aqueous segment, *A*, into the organic film. The analyte concentration in the film, C_{film}, is illustrated for the tube area within the dotted zone. If C_{film} has not returned to a "zero" value when the next water segment, *B*, arrives (time "4"), axial dispersion occurs. (According to Ref. 1117 by permission of Elsevier Scientific Publishing Co.)

fusion of analyte into the organic segment (time "3"). If C_{film} has not reached its original "zero" value by the time the next aqueous segment arrives (times "4"), the net result will be that analyte material originating from the first aqueous segment appears not only in the next organic segment but also in the second, pursuing organic segment. This leads to undesired dispersion and the analyte concentration profile will exhibit tailing.

Therefore, the dispersion process within the extraction coil should, according to the arguments raised in discussing Fig. 4.31, be less significant when the film is thinner. Experimental results reveal that this is

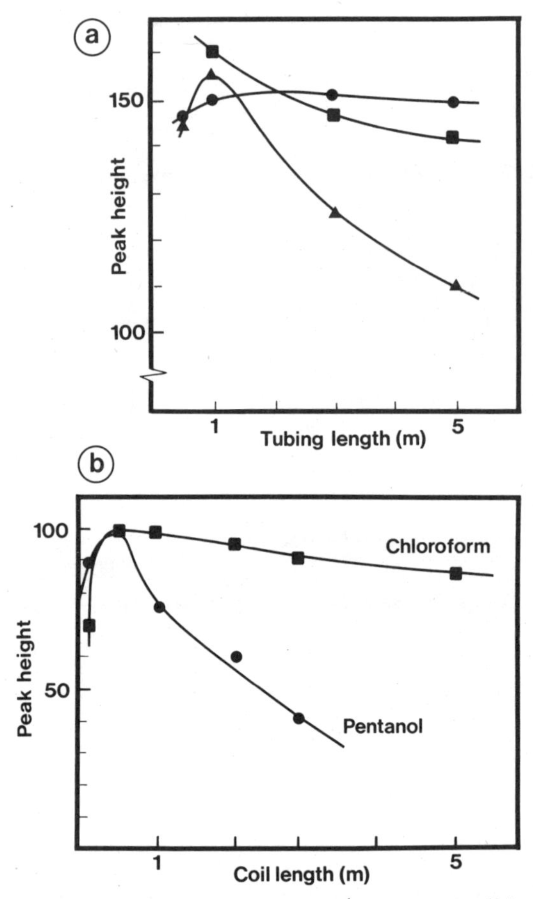

Figure 4.32. (*a*) Flow injection extraction of 1×10^{-4} M aqueous caffeine into chloroform. The influence of the flow velocity on the dispersion for linear flow rates of: (●) 2.7; (■) 5.4; and (▲) 17.8 cm/s. Conditions: 0.7 mm ID PTFE tubing, segment length 6 mm [1117]. (*b*) Influence of the organic phase on the dispersion. Extraction of caffeine into chloroform and of bromocresol green into pentanol. PTFE extraction coils were used in both cases. The total flow of chloroform/water was 1.3 mL/min., while that of pentanol/water was 2.0 mL/min. Since chloroform forms a much thinner film than pentanol on the wall of the tube, a lower dispersion of the sample zone is observed. (According to Ref. 688 by permission of Elsevier Scientific Publishing Co.)

true. Thus, in Fig. 4.32*a* is shown the extraction of caffeine from the aqueous phase into chloroform as a function of different flow velocities. As the flow rate increases, the film thickness is increased, that is, the analyte material becomes more dispersed and lower peak heights are recorded. The same pattern is observed when the thickness of the film is increased as a function of the characteristics of the organic phase. Since solvents distinguished by low viscosity and high interfacial tension lead to small values of d_{film}, extraction, for instance, by chloroform as compared to pentanol results in much smaller dispersion (Fig. 4.32*b*), because chloroform, as stated above, forms a film that has a thickness that is only $\frac{1}{8}$th that of pentanol. Furthermore, in order to reduce dispersion, and hence increase the rate of exchange between the stationary layer and the bulk of the segments, that is, increasing the interfacial area between the film and the segment of the nonfilm-forming phase, the tubing diameter of the extraction coil should be made small, thus giving a short diffusion distance between the stationary layer and the bulk of the segment, and as stressed by Nord [688, 699], short segments resulting in better mixing should be used. A special situation arises if the extraction system is to be used for preconcentration (Section 4.7.1). In this case the flow rate of the stream carrying the analyte necessarily has to be high. This leads to a high flow velocity in the extraction coil, and, therefore, it is particularly important to choose a solvent that, having very low viscosity and high interfacial tension, gives a thin film. A good example of such a solvent is Freon 113 [527].

Thus, being able to manipulate the manifold parameters, flow velocity, inner tube diameter, segment length, and coil length in a reproducible manner, it might eventually be possible to reach conditions at which the *extraction rate* is controlled by the extraction kinetics only. This would mean that FIA extractions could be used as a new tool for the study of extraction kinetics.

As the dimensions of the film layer according to the foregoing is crucial to the extraction rate and thus to the dispersion of the sample in the extraction system, it is obvious that the coil material will exert a definite influence. Tubing material such as PTFE and polyethylene, which allow the organic phase to form a thin film, will therefore improve the conditions for efficient extraction. When, however, the analyte zone appears only in the organic phase, and it is to be transported over a certain distance with minimum dispersion, segmentation of this stream with water is recommended in a coil of glass or steel. The aqueous phase then forms a film on the tube walls, thereby encapsulating the organic segments so that no intersegment mass transfer occurs.

Finally, it should be pointed out that a very critical part of a practical extraction system, with respect to overall dispersion or dilution of the sample zone, is the phase separator [1372]. Thus, Karlberg [116] found an almost linear relationship between the peak height and the fraction of organic phase transported to the flow cell, which observation emphasizes the importance of the incorporation of an efficient phase separator. Separators of the membrane type seem to be preferred for flow-injection extraction systems. Although it may be dangerous to give general guidelines for controlling (i.e., minimizing) the sample zone dispersion in an FIA extraction system at this stage, the two most important aspects are indeed the nature and thickness of the film formed in the extraction coil and the separator efficiency.

4.6.2 Gas Diffusion

Gas diffusion from a donor stream—in which the analyte chemically is converted to a volatile species—into an acceptor stream, where the two streams run in parallel separated by a suitable gas-permeable membrane, is a highly selective technique particularly well suited for adaptation into FIA, because in nonsegmented streams the diffusion unit can be made extremely small and the flow rates may be considerably reduced. Although in the first FIA gas-diffusion method, developed by Baadenhuijsen and Seuren-Jacobs [57] for the determination of carbon dioxide in plasma, a nonporous dimethyl silicone rubber membrane was used, hydrophobic microporous membranes such as Teflon or isotactic polypropylene have proven to be more versatile diffusion barriers, since they can be used for a greater variety of gases.

As an example of a FIA procedure incorporating gas diffusion Fig. 4.33*a* shows the manifold used for assay of total carbonate via determination of carbon dioxide. The aqueous sample containing carbonate/hydrogen carbonate is injected into an acid carrier stream, which then is directed into the gas-diffusion unit (Fig. 4.33*c*; cf. also Section 5.3). During the passage of the sample zone, the carbon dioxide formed in the donor stream diffuses across the membrane and into the recipient stream, which contains a second reagent so that a chemical reaction can occur and the reaction product can be measured by a suitable detector. Using in this case an acceptor stream containing a single, appropriately selected acid–base indicator in the basic form, so that its color will change when the acceptor stream becomes acidic as a result of the acid gas transfer through the membrane, a photometric detection procedure was employed, the wavelength being selected to monitor the acidic form of the indicator. As seen from Fig. 4.33*b*, a linear response was obtained in the concentration region assayed with a sampling frequency of 90 samples/h, the analytical readout being available 15 s after sample injection. If one wishes to obtain a linear change of absorbance over a wider range of linear increase of concentration of analyte, a mixture of appropriately selected acid–base indicators have to be used and the acceptor stream has to have a certain starting pH, while its buffering capacity has to be kept constant [60, 79].

By incorporating gas diffusion into a FIA system, physically separating

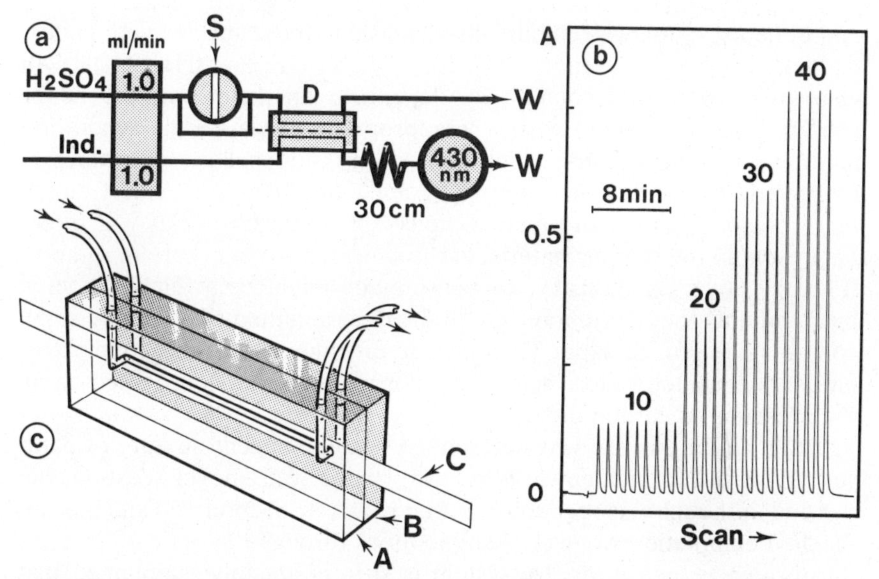

Figure 4.33. (*a*) Manifold for the determination of carbon dioxide, carbonates, or bicarbonates. Sample solution (30 μL) is injected into a carrier stream of 0.01 *M* sulfuric acid. In the diffusion unit *D* (see *c*) the liberated carbon dioxide diffuses through a Teflon membrane into the acceptor stream, consisting of 0.2 m*M* sodium carbonate and 0.2 m*M* sodium bicarbonate (pH adjusted to 8.85) containing an acido-basic indicator (Ind) the color of which, after passage through the short mixing coil, is measured by a spectrophotometer. (*b*) Readouts obtained by injecting samples of sodium bicarbonate in the concentration range 10–40 m*M*. (*c*) Gas-diffusion unit, consisting of two parts, *A* and *B*, each furnished with a shallow groove, separated by a hydrophobic Teflon membrane *C*.

the sample (donor) stream from the recipient stream, several selectivity factors can be exploited. Besides the fact that only few species are sufficiently volatile at room temperature, the most obvious is that by using hydrophobic microporous membranes, aqueous solutions do not wet the membrane surface. Therefore, only gaseous species will pass through the membrane while ionic interferents will be effectively excluded. Furthermore, any matrix effect due to color of the sample solution is eliminated. Another advantage is that the dynamic range for a method easily might be changed by changing the composition of the recipient stream. As to protolytic gases, these might be selectively detected by exploiting appropriate pH values of the donor stream. Thus, by using an *alkaline* stream, Karlberg and Twengström determined ammonia with high reproducibility down to 50 ppb [451]. Garn and Nystrup [4.9] assayed trimethylamine (TMA) in fish extracts achieving a detection limit of 0.03 mg TMA-N/100 mL at a sampling frequency of 92 samples/h. Interference from ammonia and dimethylamine was eliminated by masking these species with glutardialdehyde. In *acid* streams CO_2, SO_2, HCN, HF, and CH_3COOH will emanate in detectable quantities, depending on the pH of the donor stream. Therefore, by choosing an appropriate pH, a certain separation may be obtained, because of differences in the respective pK values of hydrocarbonic acid, generating CO_2, and other stronger acids forming more "acidic" gases. In practice, however, all these species will rarely be present in the sample material at the same time. Yet in certain samples from the beverage industry, carbon dioxide might be available in such high amounts that it will interfere in the determination of relatively small contents of sulfur dioxide. Then instead of using an acid–base indicator, one may consider a more selective color reaction that is not affected by the presence of carbon dioxide.

Thus, sulfur dioxide was determined using a reagent mixture of pararosaniline and formaldehyde [451, 816] by the well known West–Gaeke method; although this reaction is rather slow (cf. Section 4.3) and has not reached completion when the sample flows through the optical detector, the residence time in a FIA system is so reproducibly maintained that this problem can be circumvented, samples and standards being treated absolutely identically in the flow system. An excellent method for measuring sulfur dioxide is via decolorization of malachite green contained within the recipient stream [1230]. Trace quantities of cyanide might similarly be determined using a selective color-forming reaction (see below). In this context it should be mentioned that by pumping the acceptor stream n times slower than the donor stream, the concentration of gas in the acceptor stream will be approximately n times higher than if both streams are pumped at the same flow rate, that is, the gas-diffusion unit may serve as a means of increasing the concentration of very dilute sample solutions, the extent of the preconcentration being exploited to its fullest potential if this approach is carried to its extreme, that is, if the acceptor stream is stopped while a large zone of sample solution contained in the donor stream is passed through the gas-diffusion unit. An example utilizing this avenue is described below.

Another less obvious advantage of gas diffusion is the difference between gases in terms of their permeability through the membrane. Although gas diffusion is based on physical properties such as vapor pressure [430], the membrane used adds some selectivity to the system. As an example, although the vapor pressure of chlorine gas is three times that of chlorine dioxide, the selectivity between them, based on a Teflon membrane only, is 3:1 in favor of chlorine dioxide [947]. Combining this selectivity with that for kinetic discrimination, where advantage is taken of the fact that under FIA conditions species reacting at different rates do so reproducibly but with different apparent sensitivities, very impressive results may be obtained. Thus, Pacey et al. by using detection by chemiluminescence—which detection technique in itself proved to yield a selectivity factor of >500 in favor of chlorine dioxide—achieved an overall selectivity factor of >1500.*

An elegant application of gas diffusion in FIA was recently described by Zhu and Fang [4.10] in connection with a method for determination of total cyanide in waste waters. In their system, the manifold for which is shown in Fig. 4.34*a*, the gas-diffusion unit is made to encompass an integral part of the injection system, that is, the acceptor solution constitutes the injection loop of the valve. Thus, when the chemically pretreated sample solution is propelled through the donor line, hydrogen cyanide generated in the sample solution will pass through the Teflon membrane of the gas-diffusion unit and is absorbed by the arrested acceptor solution. Since the sample solution may be pumped for any preset period of time, the gas-diffusion unit in this case not only serves effectively to separate the analyte from the matrix, but additionally as a means for preconcentration. An interesting aspect of this work is that the authors for the determination of the cyanide exploited kinetic discrimination. Using as color-forming reagent for the spectrophotometric detection isonicotine acid-3-methyl-1-phenyl-5-pyrazolone, which is a widely used reagent for batch assays of cyanide, Zhu and Fang utilized the fact that this reagent has been observed to form a red colored intermediate product, which appears immediately after the mixing of the reagent with sample and fades rapidly away with the ensuing development of the final blue product. (Evolution of this color is slow and requires incubation periods of over 40 min, even at elevated temperatures.) Thus, taking advantage of the reproducible timing in the FIA system, the authors were able to exploit the unstable red intermediate product to develop a fast analytical method, yielding of sampling rate of more than 40 samples/h (Fig. 4.34*b*).

A novel and very ingenious application of gas diffusion has been proposed by Hwang and Dasgupta [1146, 1173], who wanted to devise a system by which, prior to detection of a species—previously formed in an incorporated immobilized enzyme reactor—they could increase the pH of their carrier stream, yet without undue dilution by introduction of

* Speculating on this, Pacey et al. [1073] added this comment: "This is but one example of how FIA can improve the selectivity of a given analytical determination. Not only can existing methods be improved but new methods that could not have been developed without FIA can be utilized. This raises an interesting question: How many of the older chemical methods were discarded because of non-selective behaviour or difficult handling under batch methodology? A careful inspection of this chemistry is in order. It is possible that under FIA conditions these reagents may have another useful life."

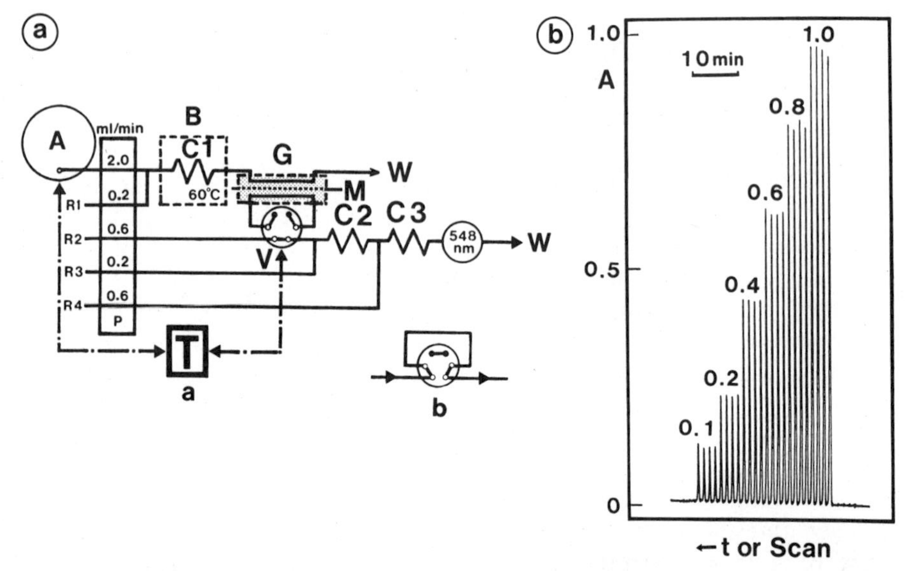

Figure 4.34. (*a*) Automated FIA system for total cyanide determination: *A*, autosampler; *P*, peristaltic pump; *B*, thermostated water bath; *G*, gas-diffusion unit incorporating a microporous Teflon membrane; *T*, timer; *V*, injection valve, *C*1, *C*2, and *C*3, mixing coils; *R*1, *o*-phenanthroline acid solution; *R*2, 0.025 *M* sodium hydroxide; *R*3, chloramine T, buffer solution; *R*4, isonicotinic acid–pyrazolone solution; and *W*, waste. (*a*) Valve in sampling-preconcentrating position; (*b*) valve in injection position. (*b*) Recording tracings obtained by injecting a series of cyanide standard solutions (0–1.00 μg/L) using the FIA system shown in (*a*), each standard being injected in quadruplicate (From Ref. 4.10 by permission of Elsevier Scientific Publishing Co.)

an extra line into the manifold used. This problem was solved by incorporating a so-called "passive" membrane reactor into the system, consisting of a length of perfluorosulfonate cation exchange membrane tubing of small diameter immersed in a solution of concentrated ammonia in a closed flask. During operation, ammonia permeated through the membrane into the carrier stream, raising its pH, the membrane matrix effectively hindering the transport of the negatively charged fluorophore to be detected.

Taking advantage of the selectivity factors of gas diffusion, this approach might, however, be extended even further. Thus, Pacey et al. [1018, 1073] in developing a dual-phase gas-diffusion system demonstrated how this system could be used to generate hydrides for their detection by atomic absorption spectrometry, exploiting the process of gas diffusion occurring from liquid to gas phase in order to utilize gas-phase reactions for detection. Hydride generation procedures suffer from interference in both the liquid and gas phases. While the interferences in the gas phase involve the atomizer and essentially have been eliminated, the interferences in the liquid phase still pose a problem. These interferences fall into two categories. The transition metals precipitate during hydride generation and adsorb the hydrides on their surface, thereby decreasing the signal. The second type of interference is the competition between hydride-forming metals. However, both these types of interferences are functions of the *residence time* of the hydride in solution. Although attempts have been made to use FIA to automate the process in the hope that the kinetics would be in their favor, Pacey et al. very ingeniously realized that a way around these difficulties was to incorporate gas diffusion into the hydride system.

The use of the dual-phase gas-diffusion cell allows the hydride to pass immediately through the membrane into a hydrogen acceptor stream. The effect of this process is a decrease of the contact time between the hydride and any transition metal precipitate. The end result is that the interferences observed for this system versus other flow methods generally are significantly diminished. Since the reduced contact time also favorably influences the interference due to hydride competition, the observed reductions are therefore really a combination of the separation techniques obtained in the FIA system and that of kinetic discrimination.

Inherently, the gas-diffusion unit can be miniaturized, and, therefore, it is suited to be integrated into FIA microconduits (Ref. 608; Chapter 2.6, and Section 4.12), where added advantage can be gained by combining the gas diffusion with the detection facilities offered by the miniaturized optosensors. A system exploiting this combination, used for the determination of ammonia, is depicted in Fig. 4.35*a*, along with a close-up of the gas-diffusion cell [848]. Using the split-loop injection technique (Section 4.5.2) for placing a zone of ammonium chloride and a zone of sodium hydroxide into the manifold, these two zones are, after being dispersed and partly mixed in coil M, led into the flow cell. Here is a miniaturized gas-diffusion unit furnished with a porous hydrophobic membrane (m) (Fig. 4.35*b*). The ammonia emanating from the donor stream (D) will thus change the pH of the acceptor stream (A), which containing an acido-basic indicator will result in a color change of this stream. Placed on top of the transparent cover of the flow cell is a multistrand optical fiber, bifurcated at the remote end to accept the incident light from an external light source and to accommodate a conventional spectrophotometer to monitor the light reflected by the hydrophobic membrane. The membrane is nontransparent white, and serves as an ideal reflecting opaque background at the proximity of which the chemical reaction takes place. Since the indicator solution is transparent, the incident light will traverse the acceptor stream twice (the distance d being 0.13 mm). In order to ensure

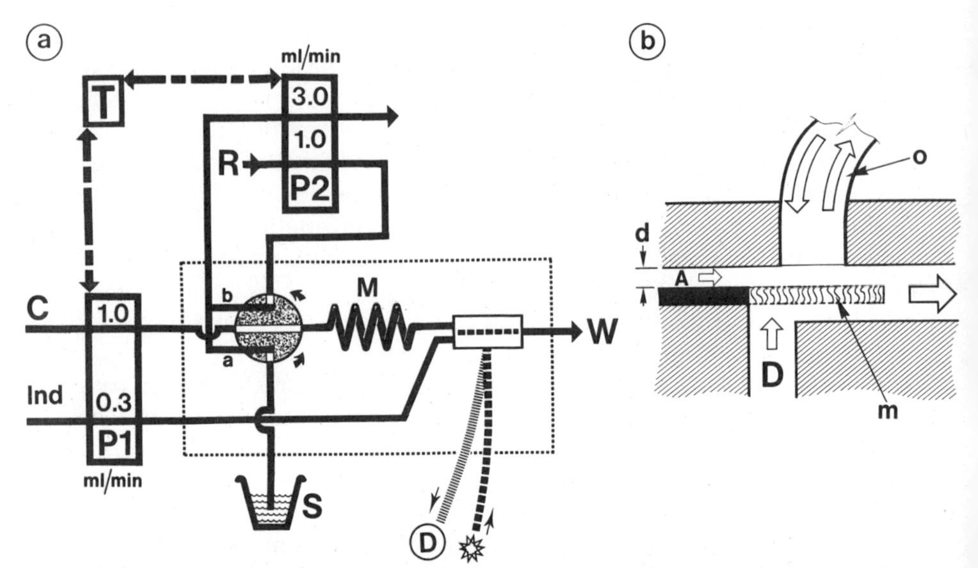

Figure 4.35. (*a*) Manifold for assay of ammonia and urea, comprising two peristaltic pumps (*P*1 and *P*2), a timer (*T*), reagent (*R*), carrier (*C*), and indicator (Ind) channels, mixing coil (*M*, 70 μL), sample channel (*S*), and a valve furnished with a split-loop configuration (cf. Fig. 4.28) comprising sample (*a*) and reagent (*b*) sections. The optosensor (*b*) is connected to a light source and spectrophotometer (*D*) by means of optical fibers. *W* is the channel leading to waste. The boxed area includes the components within the microconduit. (*b*) Details of the optosensor with a gas diffusion membrane (*m*) separating a donor stream (*D*) and an indicator-containing acceptor stream (*A*), the membrane being situated in that section of the channel that is monitored by the optical fiber (*o*).

an effective measuring frequency, the small portion of the acceptor stream (ca. 1 μL) in which the change of pH and of indicator color occur is renewed periodically aftter each measuring cycle. Furthermore, the system might either be operated in a continuous flow or in the stopped-flow mode, the advantage of the latter approach being that a suitable section of the mutually dispersed zones can be selected and contained within the flow cell, whereby the diffusion of gas across the membrane is enhanced. In this manner, the sensitivity of measurement can be increased, the length of the stop period governing the degree of transfer of the ammonia available within the entrapped segment of sample zone.

In Fig. 4.36 is shown the readouts obtained in this system when injecting a series of ammonium chloride standards where, in order to increase the sensitivity, the sample zone was stopped in the flow cell (in this case for 16 s). To the right in the same figure is shown the recorder outputs for the determination of urea, enzymatically degraded to ammonia by means of urease. In the latter case, the chasing zone (*b*) consisted of

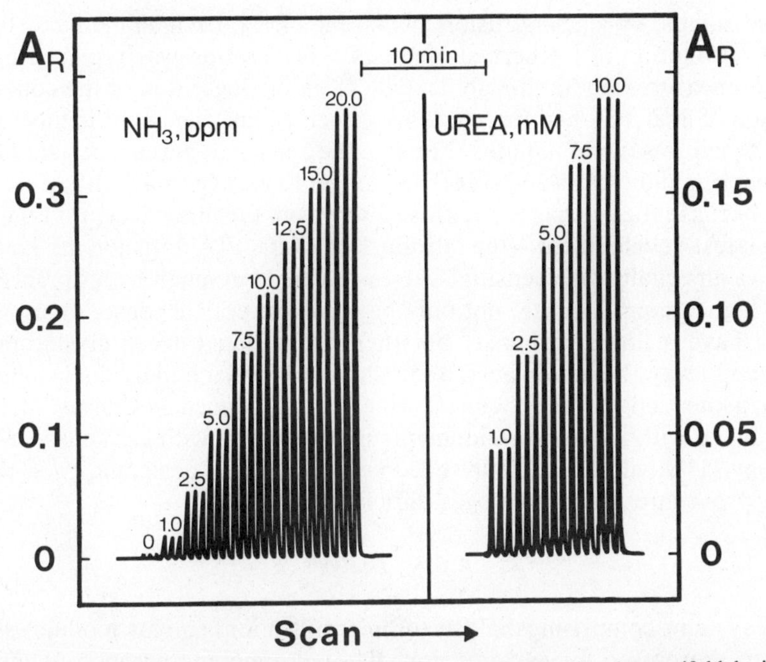

Figure 4.36. Recorder output for the assay of ammonia and urea using the manifold depicted in Fig. 4.35*a*. To the left is shown the response for a series of ammonium chloride standards in the concentration range 0–20 ppm NH_3. At the right is shown the response for five urea standards, enzymatically degraded to ammonia by urease, in the range of 1–10 m*M*. All samples being injected in triplicate.

a TRIS buffer of pH 8.3 containing dissolved urease (alternatively, the enzymatic reaction could be performed by the incorporation of a miniaturized packed reactor containing immobilized urease—cf. Section 4.7.2). The operational pH value was chosen as a compromise between the optimal pH for the enzymatic degradation procedure and a sufficient degree of liberation of ammonia.

To conclude this section, it is intriguing to observe how developments in instrumental analysis have led to an increase of sensitivity and a dramatic decrease of analysis time required by procedures based on isothermal distillation. The method of Conway, developed more than 35 years ago [4.11], still in use in clinical and pharmaceutical laboratories, requires many hours to perform an assay of a volatile species. The samples are kept in small, enclosed chambers containing the donor and the acceptor liquid, respectively, and after the diffusion process has reached equilibrium, the acceptor liquid is titrated. Gas-sensing probes, which operate on the principles of ion-selective electrodes, separated from the liquid sample by a gas-diffusion membrane [4.12], became the next stage in the evolution of isothermal methods. The titration was replaced by a direct measurement (although dependent on the logarithm of the concentration of analyte—and therefore less precise), and the analysis time was shortened to several minutes (because the sensor requires a certain time to reach equilibrium—and somewhat longer to recover to reach zero reading, because the gas has to rediffuse away from the inner acceptor liquid). The latest development—the introduction of the FIA diffusion method—allows an equally high sensitivity to be reached but much better precision and higher sampling rate, not only because the optical sensors normally used have a linear response, but the contaminated acceptor stream is pumped away, being replaced by fresh liquid for each individual sample. Still, a knowledge of the chemical principles introduced by Conway [4.11], combined with further developments in connection with gas sensor technology [11], will undoubtedly become useful when designing even new FIA procedures comprising gas diffusion.

4.6.3 Dialysis

FIA systems comprising dialysis (or microfiltration) employ modules similar to those used for gas diffusion, the hydrophobic microporous membrane being replaced by a hydrophilic dialysis membrane (such as cuprophan, cellulose aceate, or cellulose nitrate). Basically, the incorporation of a dialysis unit into a FIA manifold is to serve one (or several) of the following objectives: (1) separation of an analyte species from unwarranted matrix constituents, (2) as an exact and reproducible means of dilution, or (3) microfiltration performed continuously by transferring species from one stream (the donor) to another stream (the acceptor).

Typically, dialysis is utilized when analyte constituents in complex matrices such as serum [6, 9, 160, 262, 1200] or blood [686, 719] are of interest, permitting diffusable low-molecular-weight species to be separated from macromolecular components like proteins. Although the objective always is to obtain increased selectivity, the reasons depend on the procedure used. Thus, for instance, when using detection by chemiluminescence, the emphasis is primarily placed on eliminating species that might cause quenching [686, 719], while in systems employing immobilized enzyme columns, the purpose of dialysis is to prevent physical blocking of the enzyme reactors and to exclude chemical adsorption phenomena on the active sites of the reactors. Interestingly, van Staden and Basson [160], in developing their FIA procedure for turbidimetric determination of urinary inorganic sulfates, exploited the dilution capability of the dialysis unit [6], which obviated a manual dilution procedure thereby facilitating a sampling rate of 120 samples/h.

An ingenious and elegant application of dialysis is for dispensing very accurately small amounts of reagents to a carrier stream. Using this approach, which more correctly may be termed microfiltration since the addition of reagent is done by means of an applied pressure, Nieman et al. [660, 1011] were able to administer minute amounts of enzyme solutions to an injected plug of analyte, whereby hydrogen peroxide formed by the enzymatic degradation process could be detected by chemiluminescence following coupling to luminol. As the enzymatic reaction and the coupling process occur at very different pH values, advantage was taken of the fact that diffusion of a buffered enzyme solution through the membrane allows creation of a stable pH gradient in the acceptor part of the dialysis unit, which in this case also served as the flow cell. Thus, near the membrane the pH can be maintained at a value that is optimal for the enzymatic reaction, while in the bulk of the analyte solution where the chemiluminescent reaction takes place the milieu is strongly basic, which promotes the light-emitting process. Hwang and Dasgupta [1173] later employed the same scheme, again for addition of enzyme solutions to a sample/reagent stream, yet in their procedure for the ultimate purpose of detection by fluorometry, thereby achieving extreme reagent conservation.

Recently, an inventive use of dialysis in FIA was described by Macheras and Koupparis in connection with studies on drug-protein binding [1268]. Terming their technique "dynamic dialysis," they constructed a system where the acceptor part of the dialysis unit constituted the sample loop of the injection valve (Fig. 4.37). Thus by pumping their protein–drug solution for a selected period of time through the donor part of the dialysis module, the amount of diffusable drug in the acceptor stream could be assayed by subsequent injection into the analytical manifold. Independently, and almost simultaneously, Meyerhoff et al. [1317, 1318] described a similar approach for combining dialysis and preconcentration of analyte in an acceptor stream for subsequent assay, their aim being to prevent unwanted interference in potentiometric measurements. Using this configuration, the selectivity of the anion-responsive liquid–membrane electrodes used as detection devices was greatly enhanced, facilitating the determination of species such as nitrite, NO_x (converted chemically in the donor stream of the dialyzer to nitrite) and salicylic acid.

Within the last few years FIA has found increased application in process monitoring, such as in biotechnology. A common feature of these procedures is that they require (a) representative samples to be drawn from a process solution and (b) the samples have to be purified from

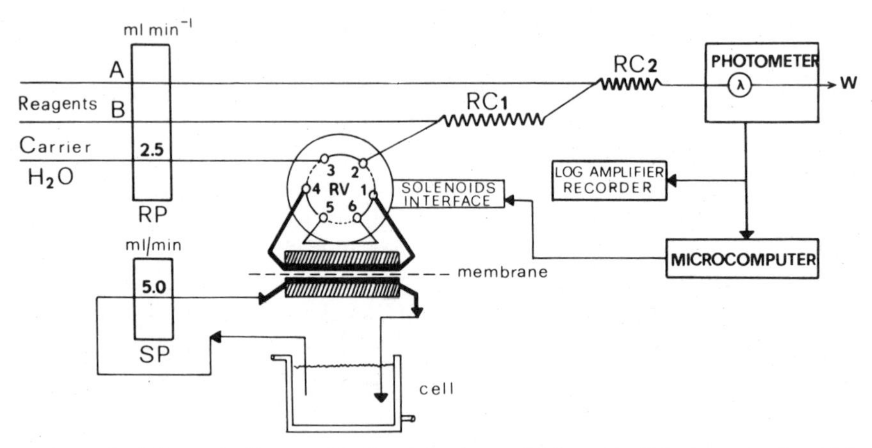

Figure 4.37. Schematic diagram of a flow-injection serial dynamic dialysis system in which the acceptor part of a dialysis unit constitutes the sample loop of a rotary injection valve (*RV*). *RP*, reagent pump; *SP*, sample pump; *RC*1 and *RC*2, reaction coils; and *W*, waste. (From Ref. 1268 by permission of Elsevier Scientific Publishing Co.)

particulates, which otherwise would clog flow channels, columns, and other manifold components. With this aim a Swedish group has in a series of papers investigated the application of dialysis for continuous sampling/on-line use [743, 758, 771, 852, 867]. Also, in order to optimize the design of the dialysis modules, a thorough theoretical study of the parameters affecting the mass transfer across a dialysis membrane was conducted. Using three different theoretical models, numerical solutions for a laminar flow regime, as that encountered in FIA, were obtained by the finite-difference approximation method.

4.7 FIA SYSTEMS WITH PRETREATMENT OF SAMPLE IN PACKED REACTORS

One of the great advantages of the FIA technology is the ease with which additional components can be added to the system to achieve a particular analytical objective. This is most notably exemplified by the incorporation of small packed reactors, which subject an injected sample to appropriate on-line pretreatment in order to facilitate the detection of analyte. Whatever the function, a packed reactor in a FIA system is therefore designed always to perform the same chemistry on each individual sample. This is in contrast to chromatography, where the purpose of the column is to separate each individual sample into several components. Besides, since the FIA apparatus is a low-pressure system, most often operated by simple peristaltic pump(s), the use of solid particles in the form of columns is generally restricted to fairly short units.

Since the first FIA experiment exploiting the incorporation of a packed reactor was published [71] there has been much use of this approach (Table 7.1) for a variety of purposes. Thus, ion exchangers have been employed as column materials in order to preconcentrate an analyte, or to remove matrix components that might interfere, or to convert a sample constituent into a detectable species; immobilized enzymes have been utilized for selective degradation of substrates; and reactors containing oxidants or reductants have been devised to generate reagents *in statu nascendi*, advantage being taken of the protective environment offered by the FIA system to form and apply reagents that are impractical to handle under normal analytical conditions. The function of chemically inactive packed reactors, used in FIA systems in order to promote radial dispersion at the expense of axial dispersion, is discussed in Section 3.1.6.

4.7.1 On-line Preconcentration and Conversion Techniques

The principle of the FIA *preconcentration* technique is illustrated in Fig. 4.38, which demonstrates how reduced dispersion of the injected sample (i.e., $D < 1$) might be achieved by introducing a relatively large volume of sample solution, the analyte content of which is retained on an incorporated miniaturized packed reactor within the FIA channel, from which column the analyte subsequently is released and passed to a detector. The method was originally proposed to enhance the sensitivity of measurement of cationic trace elements in very diluted aqueous samples using

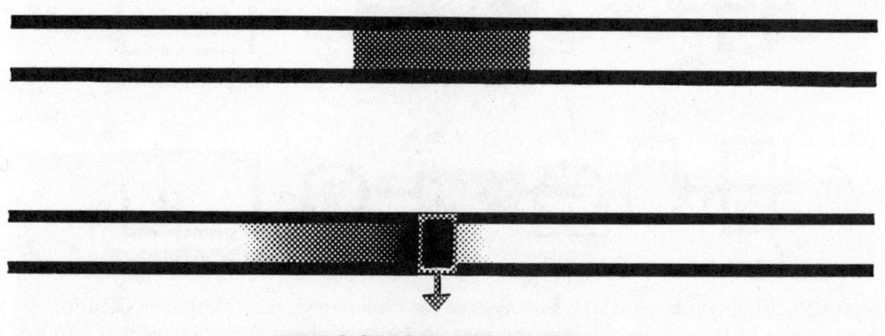

Figure 4.38. Principle of the FIA preconcentration method. A large sample volume (top) is injected and after appropriate dispersion let pass through a microcolumn (shown as boxed area). Thus, locally, concentration of analyte is increased and may be sensed there directly (optosensing) or eluted into a smaller volume (cf. Fig. 4.39).

detection by flame atomic absorption spectrophotometry (fAAS) [428]. Each sampling cycle consists of two separate operations (Fig. 4.39): preconcentration and elution. The simplest, one-channel FIA system has therefore two valves. First, a large sample volume (e.g., 5 mL) is injected by the upstream valve (*S*) into the carrier stream and propelled, within typically 1 min, through a cation-exchanger column (typically 2 cm long, 1.5 mm ID, filled by Chelex-100). Then a small volume of eluting reagent (such as 50 μL of 1 *M* HNO_3) is injected by the second valve (*E*) liberating the 100 times preconcentrated analyte into the detector. The main advantage of using on-line preconcentration is that all samples and standards are subjected to exactly the same treatment from the moment of injection to the moment of detection and that the same ion-exchanger column is used for all samples and standards. As all time events, the geometry of flow and the potentially ensuing chemical reactions are strictly controlled and reproducibly maintained in the FIA system, it is not an absolute prerequisite that quantitative adsorption of analyte species from the injected sample zone is achieved, yet the subsequent elution must be quantitative, otherwise carryover from one sample to the next will occur.

More advanced designs of the on-line preconcentration technique use sophisticated ion-exchange materials [576, 683, 684, 1217], time-based injection rather than valve injection [576, 712, 1228], and, most importantly, a countercurrent operation of the microcolumn [684, 712], an operation mode that prevents any matrix material from entering the flow-through detector even during the preconcentration period. Thus, in Fig.

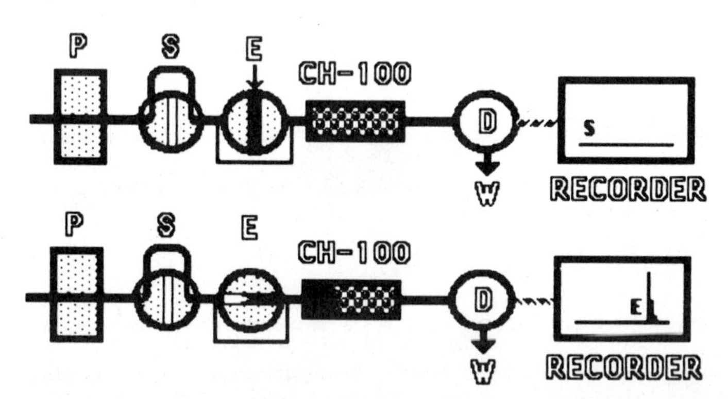

Figure 4.39. Single-line two-valve FIA system for on-line preconcentration as used for the determination of trace amounts of metal ions. A large sample volume is injected into the carrier stream by means of valve *S* and propelled into a small column containing a cation exchanger (Chelex-100), which effectively retains the metal ions. Hence no signal is recorded by the detector. Next, a small volume of eluent (*E*, nitric acid) is injected by the second valve, the metal ions on the column being eluted into a small zone that upon transport to the detector gives rise to a signal before the sample is led to waste, *W*.

4.40*a* is shown the manifold for a fully automated, single-valve two-pump system used for assay of metal ions by fAAS, the components within the boxed area of the manifold being those contained in a microconduit (Fig. 4.40*b*). This system is equipped with an electronic timer capable of sequencing the two incorporated pumps, P_1 and P_2, in alternate stop–go modes, each preconcentration–elution cycle being initiated by the turn of the injection valve. During the preconcentration cycle pump P_1 is going while the sample is being injected by turning the valve (S). The injected

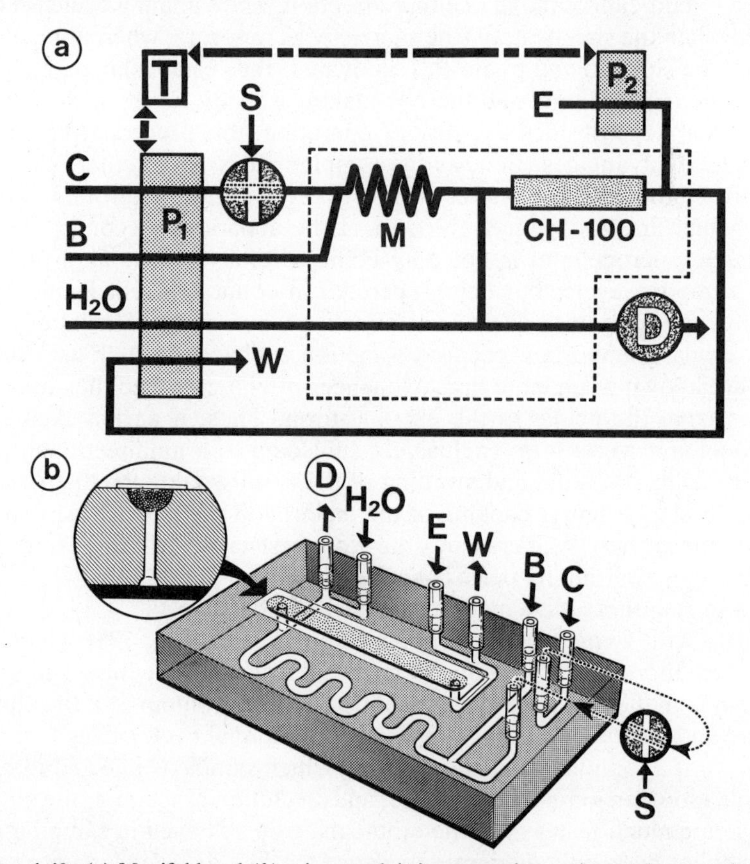

Figure 4.40. (*a*) Manifold and (*b*) microconduit incorporating an ion-exchange column as used for preconcentration of metal ions. The sample, injected into the carrier stream C, is mixed with buffer B to adjust pH before entering the column filled with Chelex-100 (volume 150 μL). Metal ions are retained on the column and are subsequently eluted countercurrently by a small zone of eluent (E, acid) pumped by pump P_2. As the two pumps are operated sequentially, the eluted sample is carried to detector D (atomic absorption spectrophotometer). The manifold components within the boxed area are those contained within the microconduit.

zone is mixed with an ammonium acetate buffer solution (*B*) in coil *M* in order to fix the pH, and then passed through the microcolumn (CH-100) containing the ion exchanger. During this operational step a flow of water is continuously fed to the nebulizer of the fAAS instrument. In the next sequence, pump P_1 is stopped and pump P_2 started, permitting the eluting agent (*E*, acid) to move through the column in the opposite direction and thereby transport the eluted metal ions into the fAA spectrophotometer. It should be noted that when pump P_1 is stopped, the liquids in the thus closed circuit cannot move in either direction. The sampling cycle is completed when the signal ouput (the sample peak) appears, whereupon pump P_2 may be stopped and pump P_1 reactivated, thus establishing a high pH inside the microcolumn and thereby making it ready for the next preconcentration step. Besides the ease of operation, this procedual mode has two added advantages: first, as the sample matrix never enters the fAA spectrophotometer, it obviates the necessity for any background correction, which in turn reduces the cost of the apparatus; second, the ion-exchange microcolumn is not only being carefully regenerated prior to each sampling cycle, but being operated in countercurrent fashion any possible blocking of the column is effectively prevented. The latter point is particularly important because some ion exchangers (such as Chelex-100) are known to undergo drastic changes of volume when they are converted from the acidic to the alkaline form. Thus, if an ion-exchanger microcolumn were to be exclusively subjected to a unidirectional flow, the alternate shrinking and swelling of the resin would eventually cause a progressively tighter packing of the material in the downstream end of the column, thereby increasing the flow resistance of the system and eventually impairing its performance.

Various authors have modified the design shown in Fig. 4.39, devising systems with two preconcentration columns [576, 683, 773] which are operated alternately, advantage being taken of the fact that, while one column is in the preconcentration mode the other column is in the elution mode, and vice versa. However, as the operational cycle of each column consists of a sampling sequence (during which sample is injected into the sample loop), an ion-exchange state, and an elution procedure, the elution step being much faster than the other, the gain achieved in sampling frequency has been only moderate, because the rate-determining operations are the sampling and ion-exchange modes. Exploiting the fact that the exchange column might be considered as a special kind of sample loop in itself, Malamas et al. [576] succeded in improving the sampling frequency by employing time-based aspiration of the sample, that is, the sample solution is pumped through the columns at a certain flow rate for a definite period of time. Fang et al. [684] amended this design by further adding pumping of the eluant directly through the column for a given period of time. Although allowing in their setup for the two columns to be loaded with samples concurrently and in parallel, the two columns cannot, of course, be eluted simultaneously, because the two peaks have to be time-resolved. This was accomplished by using the manifold system depicted in Fig. 4.41, using a sequential scheme whereby one column is eluted while the flow in the other is stopped, and after completion of the first elution, the process is reversed. This design calls for two pumps working intermittently under the control of a dual-channel timer that precisely controls the different stages of the entire process (Fig. 4.41*d*). With this system, the rate of analysis can be almost doubled, but, more importantly, a higher yield of preconcentration can be obtained. Thus, using this manifold scheme up to a hundred times preconcentration of metals ions [Pb(II), Cu(II), Zn(II) and Cd(II)] was achieved, the response being unaffected by matrix constituents.

This is demonstrated in Fig. 4.42, which shows the signal outputs for a series of aqueous Pb(II) standards in the range 25–100 μg Pb/L (Fig. 4.42*A*), and those for a 100 μg Pb/L sample solution prepared in a matrix simulating sea water (Fig. 4.42*B*). By comparing the peak heights, it is seen that the readouts for the 100 μg Pb/L "seawater" sample are, within the experimental error, the same as those obtained for the 100 μg Pb/L aqueous standard [the twin peaks for each concentration level refer to the outputs for reactors *A* and *B* (Fig. 4.41), respectively]. For comparison the response of the same instrument, operated in the conventional mode (by aspiration without FIA preconcentration), is shown in Fig. 4.42*C*. From the figure it is apparent that the direct mode of aspiration requires almost two orders of magnitude higher concentration of sample material in order to reach the same level of signal. The FIA–fAAS system with on-line preconcentration might thus challenge the position of the graphite-furnace technique, because it yields comparable sensitivity for much lower cost by using simpler apparatus and separation mode. Besides, the preconcentration method is an ideal vehicle for increasing the sensitivity of multielement measurements by inductively coupled plasma emission spectrometry [712]. In discussing the operation of their system, Fang et al. [684] pointed to an interesting aspect of the dual-column scheme, that is, instead of incorporating practically identical columns into the system in order merely to increase the sample throughput, for routine applications where samples generally are analyzed in duplicate one could exploit columns of different capacities, thereby extending and checking the dynamic measuring range.

Besides preconcentration, the ion-exchanger columns have found applications for speciation studies (Section 4.8) or for removal of interfer-

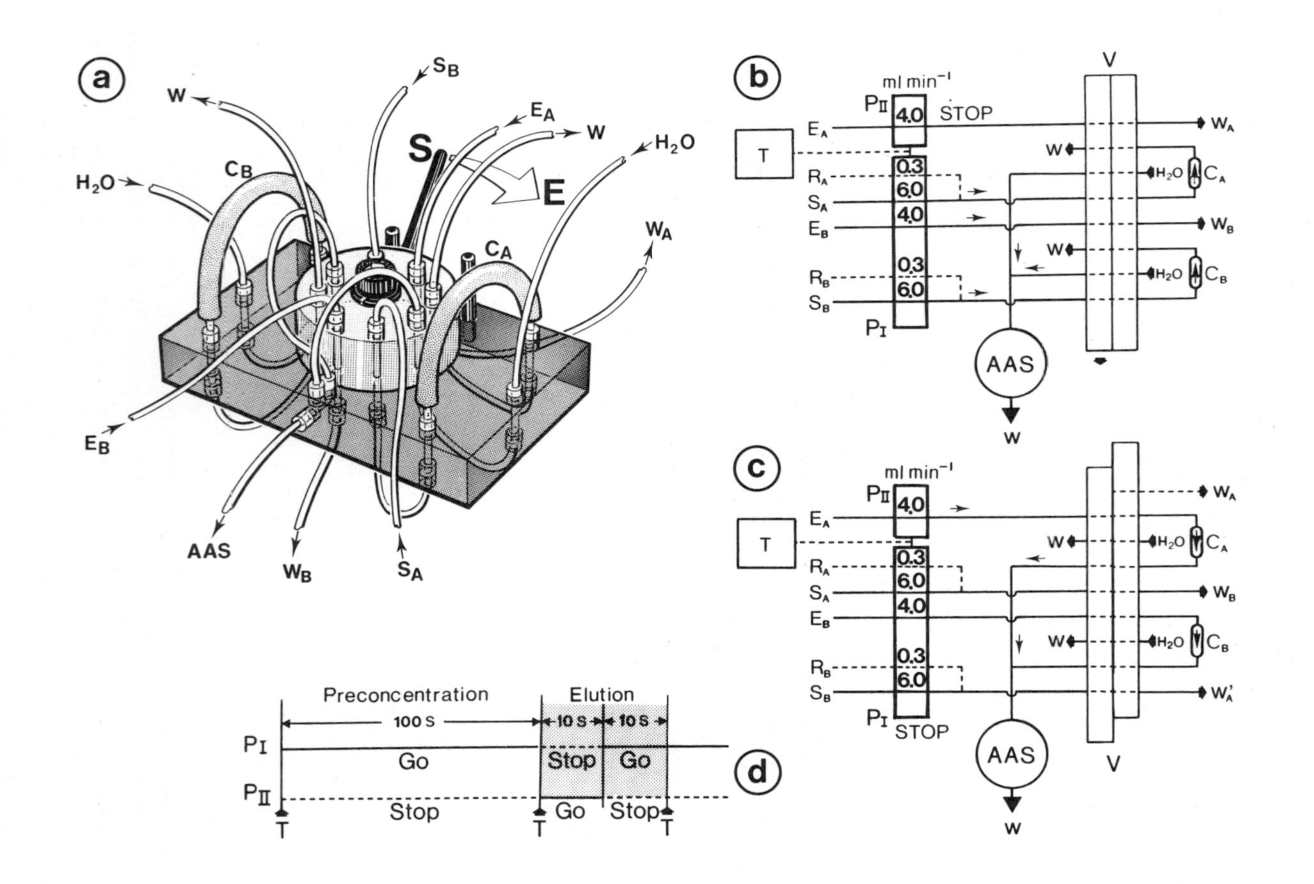

a
W
S_B
E_A
W
H_2O
H_2O
C_B
S
E
C_A
W_A
E_B
AAS
W_B
S_A
b
ml min⁻¹
P_II
4.0
STOP
E_A
T
R_A
S_A
E_B
R_B
S_B
0.3
6.0
4.0
0.3
6.0
P_I
V
W_A
W
H_2O
C_A
W_B
W
H_2O
C_B
AAS
W
c
ml min⁻¹
P_II
4.0
E_A
T
R_A
S_A
E_B
R_B
S_B
0.3
6.0
4.0
0.3
6.0
P_I
STOP
W_A
W
H_2O
C_A
W_B
W
H_2O
C_B
W'_A
AAS
W
V
Preconcentration
Elution
100 s
10 s
10 s
P_I
Go
Stop
Go
P_II
Stop
Go
Stop
T
T
T
d

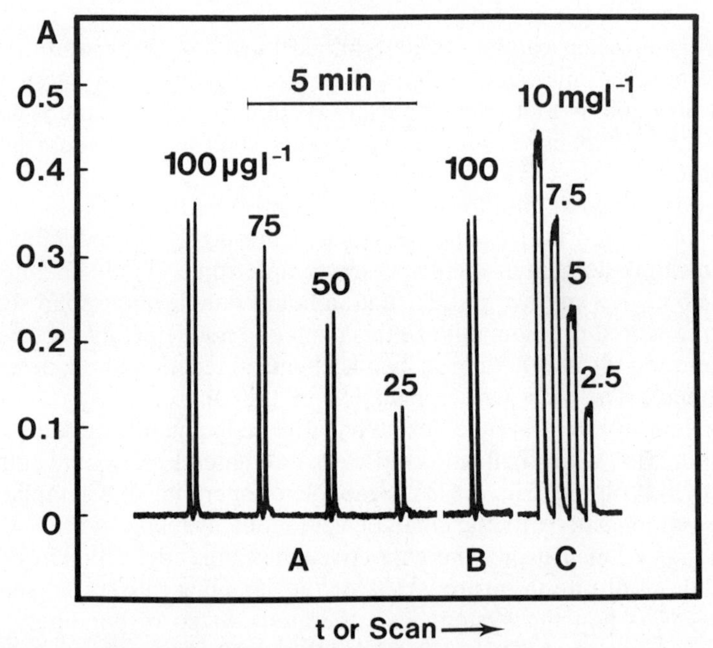

Figure 4.42. Demonstration of the efficiency of the incorporation of an ion-exchange preconcentration column into a FIA system for the determination of lead by atomic abşorption spectrometry. In (*A*) is shown the recordings obtained for a series of aqueous lead standards in the range 25–100 µg/L, using the manifold depicted in Fig. 4.41 comprising a dual-column system, that is, the twin peaks for each concentration refer to the outputs for reactors *A* and *B* respectively. (*B*) is the output for a 100 µg/L lead solution prepared in a matrix simulating seawater. For comparison the response of the same instrument, operated under conventional experimental conditions of direct continuous aspiration, but without the inclusion of the on-line preconcentration column, is shown in (*C*).

←

Figure 4.41. Dual-column on-line ion-exchange preconcentration system with flame atomic absorption detection. (*a*) Drawing of the eight-channel valve actually used, incorporated into a FIA microconduit unit comprising the two preconcentration columns C_A and C_B, each containing a cation-exchange resin. The two positions of the rotor of the valve refer to sampling (*S*) and elution (*E*); other symbols and detailed explanation of the operation are given in the following. (*b*) Manifold for the system in the sampling and preconcentration mode. During this sequence, both columns *A* and *B* (C_A, C_B) are loaded by time-based aspiration of sample material via sample streams S_A and S_B propelled forward by pump *P*I. (R_A and R_B are optional streams in case the samples require pretreatment such as adjustment of pH.) (*c*) Manifold for the system in the elution mode. During this sequence the valve (*V*) is turned [because of the circular arrangement of the valve channels (cf. (*a*)], W'_A represents the same channel as W_A. First, column *A* is eluted by eluent stream E_A, and then column *B* by eluent stream E_B, according to the time-sequencing program for valve operation shown in (*d*). The points marked *T* indicate turn of valve. Note that the elution of both columns *A* and *B* takes place in position (*c*) of the valve, yet successively as pumps *P*I and *P*II are sequenced stop–go and go–stop, respectively. *T* is the timer; *W* is waste.

ences. Thus when calcium is determined by AAS, incorporation of an anion-exchange microcolumn downline from the injection valve allows removal of anions that might interfere (sufate, phosphate, etc.), and the need to add expensive suppressing agents such as lanthanum [497] is eliminated.

An intriguing exploitation of packed reactors in FIA is the *conversion* technique, which utilizes the fact that in the highly reproducible time-concentration domain of the dispersed sample zone, kinetic chemical exchange between any two phases (e.g., a liquid sample and a solid surface) can be executed reproducibly. In this manner a nondetectable species can be converted through a heterogeneous chemical reaction into a detectable component. The very first application of this approach was the determination of nitrate [71] by colorimetry after reducing nitrate to nitrite on a column filled with small zinc chips. An automated FIA system equipped with such a column (Fig. 4.43*a*) is capable of operating at a sampling rate of 180 samples/h, with the analytical readout available within 15 s of sample injection. Despite the short residence time, the efficiency of the reduction of nitrate to nitrite is 5% for the sampling rate of 180 samples/h and 45% when the system was operated—at correspondingly lower pumping rates (A = R = 1.2 mL/min)—at 90 samples/h; thus by decreasing the sampling frequency by a factor of 2 the sensitivity of measurement is increased almost 10 times. Furthermore, despite its small dimensions, the lifetime of the column in terms of days, and in terms of numbers of analyses performed (Fig. 4.43*b*), is prolonged compared to conventional "steady-state" operations because the exposure time to the sample solution is shorter and the wash periods are longer.

The FIA conversion technique has proven particularly attractive for the analysis of anions, either their sum or individual anions, many of which form few useful colored species, so that there is a dearth of direct spectrophotometric methods for their determination. In Fig. 4.44*a* is shown such a FIA system incorporating an anion exchanger resin on the OH − form. The anion-containing sample is injected into a slightly alkaline carrier stream and passed through the column, where an equivalent amount of hydroxide ions are released from the resin causing an increase of the pH of the carrier stream. Subsequent merging with a slightly acid stream containing an acido-basic indicator changes pH and therefore results in a change of the indicator color, which is then monitored by the detector. Thus, less than 0.1 ppm of anions may be measured within 15 s, the sensitivity of determination being adjustable by changing the concentration of base and acid in the balanced carrier and indicator streams. Faizullah and Townshend [1071] used a slightly different approach employing an anion exchange resin in its thiocyanate form whereby the injected

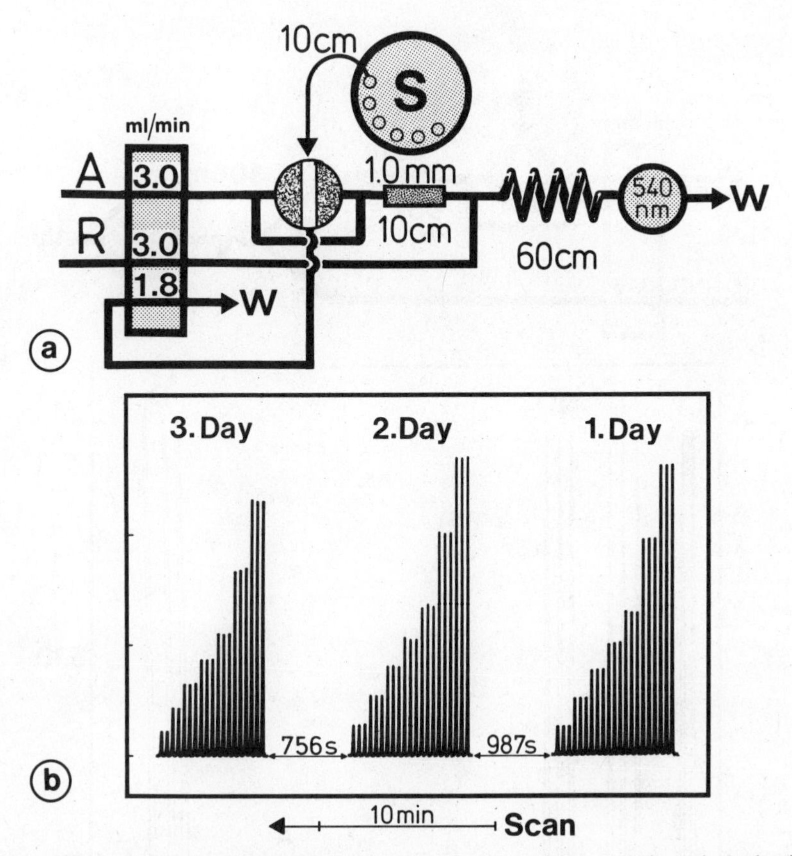

Figure 4.43. (*a*) Automated FIA manifold used for the determination of nitrate by reduction to nitrite with spectrophotometric detection (cf. Fig. 4.5). The sample (30 μL), aspirated from the sample changer *S*, is injected into a 0.25 *M* acetate solution of pH 6.0 (*A*) and then carried to the reduction column, which is filled with small zinc chips. The nitrite produced is then mixed with the color-forming reagent solution *R*. (*b*) Long-term stability test of the nitrate reduction column used in the system shown in (*a*). A series of samples (1.0–10 ppm N–NO_3), each injected in triplicate, were placed in the sample changer and the system was operated continuously for 6 h every day over a period of 3 days (sampling rate 180 samples/h). Not until the third day did the column show signs of gradually reduced efficiency.

anions displaced equivalent amounts of thiocyanate, which subsequently were merged with a stream of Fe(III) leading to formation of the intensely red colored complex. In the same publication the authors presented a similar system based on the phosphate form of the resin with determination of displaced phosphate as molybdenum blue. Mixtures of anions (e.g., chloride and nitrate) were determined by splitting the injected sample, so that part of it was passed through a silver reductor column, where

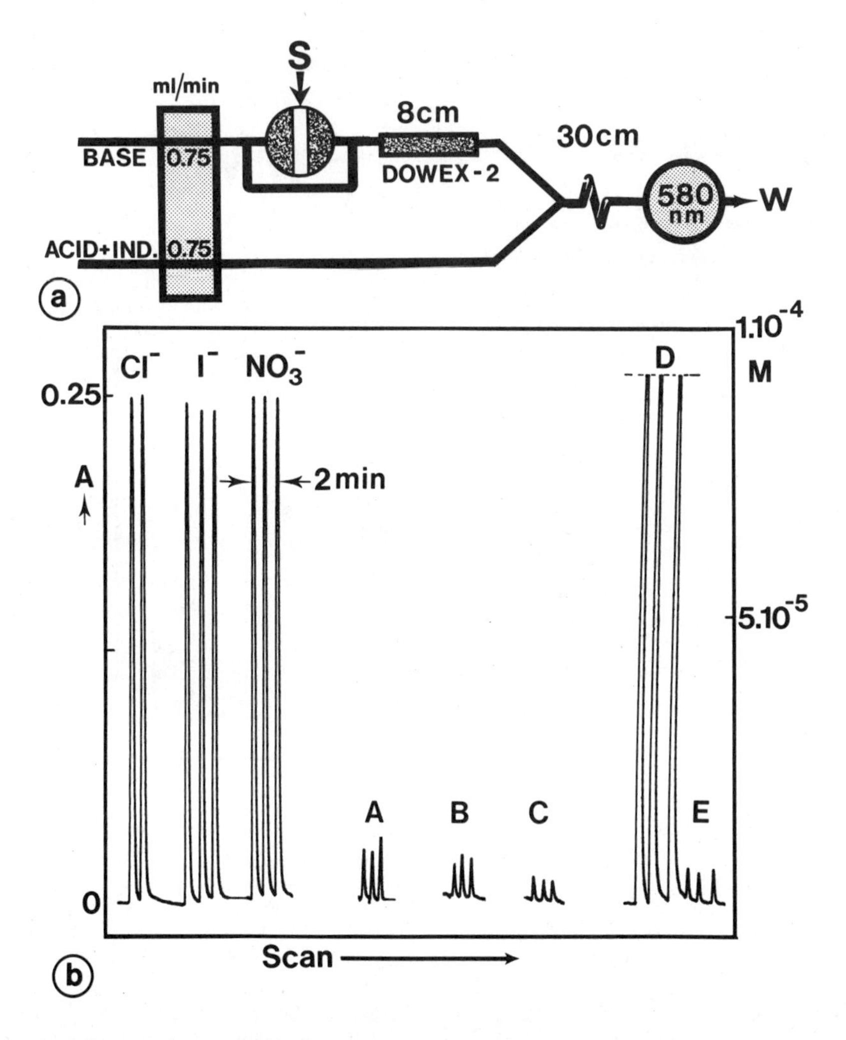

Figure 4.44. (*a*) FIA manifold used for the determination of low-level concentrations of anions. The sample is injected into an alkaline carrier stream (2×10^{-5} *M* NaOH), passed through the anion exchange column, and then merged with a stream of acid solution (2×10^{-5} *M* HNO_3). All anions in the sample solution are replaced in the ion-exchanger by OH^-, causing a change in pH of the combined stream, which is monitored spectrophotometrically via the acid–base indicator IND. (*b*) Response signals obtained by injecting equimolar solutions (8.5×10^{-5} *M*) of KCl, KI, and KNO_3, followed by outputs for samples of deionized water (*A*), distilled water (*B*), and Millipore-filtered water (*C*) (all samples injected in triplicate). *D* represents the peaks recorded for tap water (the dashed line indicating that the signal exceeded the scale of the recorder); *E* is a rerun of distilled water (*B*), showing the absence of carryover between a high and a low sample.

chloride was retained, before being guided through the thiocyanate column.

A disadvantage of the conversion techniques discussed above is the lack of selectivity, and therefore it is not surprising that considerable effort has been invested in designing systems that could remedy this. Selective conversion methods have recently been published for the determination of cyanide and sulfide [1010, 1247]. The principle of these approaches are illustrated in Fig. 4.45. In the determination of cyanide (Fig. 4.45*a*) advantage is taken of the fact that this anion with Cu(II) forms a complex ion of defined stoichiometric composition. Thus, injecting the cyanide-containing sample into a manifold comprising a column of CuS, the cyanide will form soluble tetracyanocuprate, the copper released from the

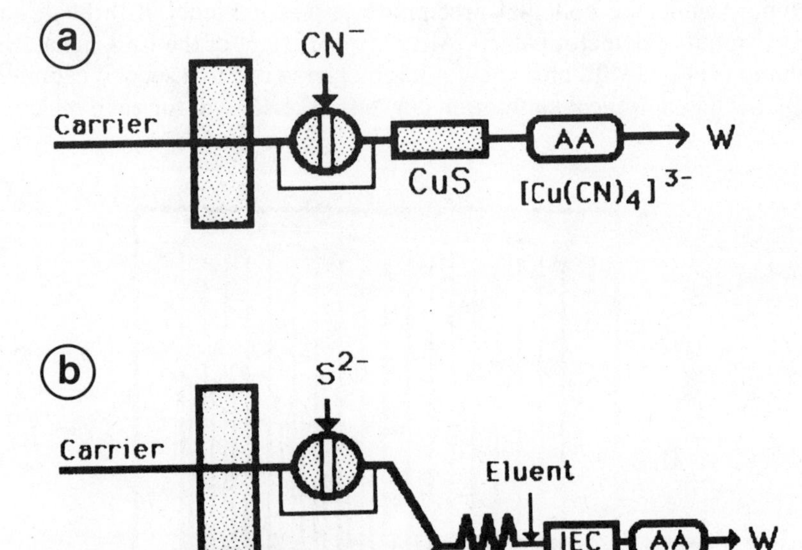

Figure 4.45. Principles of conversion methods in FIA. (*a*) A manifold in which the analyte (cyanide) in a packed column of copper(II) sulfide is converted to tetracyanocuprate(I), which is detected by atomic absorption spectrometry (AA). (*b*) A system in which the analyte (sulfide) is precipitated by cadmium(II) ions; the colloidal precipitate formed passes unhindered through the system and is detected by AA. Excess of cadmium(II) is retained by the ion-exchange column (IEC) and later eluted.

solid surface subsequently being detected by AAS. This approach allows determination of 1 μg/L of cyanide in the presence of a number of anionic and cationic species with an RSD value of 1.8%. Further increase of sensitivity, improvement of detection limit (down to 50 μg/L) as well as increase of selectivity has later been achieved by replacing the copper sulfide column with one containing silver sulfide [4.13]. In the procedure for determining sulfide [1247] the approach is somewhat different (Fig. 4.45*b*). Here the anion is converted to an insoluble compound by means of a cationic tag material. Sample (sulfide) and surplus of reagent [cadmium(II) ions] are injected simultaneously using the synchronous merging zones principle. When combined, colloidal cadmium sulfide is formed. Directing the sample zone toward a miniature cation-exchange column packed with a chelating resin excess of cadmium(II) is retained on the column, while the colloidal precipitate passes unhindered through and finally into the detector (AAS). After measurement of the CdS signal, the column is eluted with nitric acid, thereby giving rise to a second cadmium signal. Thus, a typical calibration curve (Fig. 4.46) will for each injection

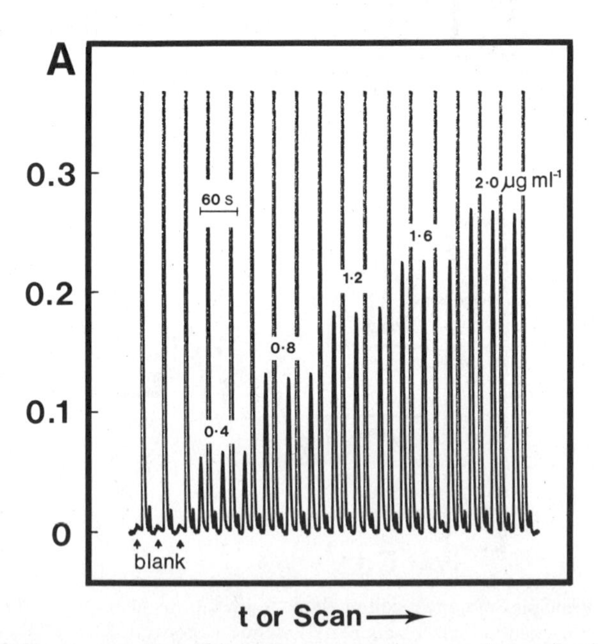

Figure 4.46. Calibration run for sulfide in the concentration range 0–2 mg/L with the system depicted in Fig. 4.45*b*. The cadmium(II) elution peaks were all out of range on the recorder and so they appear to be of equal heights.

of sample consist of two peaks: first, one reflecting the sulfide content and then a second (large) peak caused by the release of surplus cadmium(II) ions. Note that by this mode of operation the column material is not consumed (i.e., does not become exhausted as is the case with the cyanide method) because it is renewed during each cycle. By automating the FIA system shown in Fig. 4.45*b*, it was possible to detect sulfide down to 10 μg/L at a sampling rate of 100 samples/h with a typical standard deviation of 1.2%. This approach might very well be applied for the determination of other anions, such as fluoride or sulfate via precipitation of calcium fluoride or barium sulfate, which are notorious in gravimetric analysis for initial precipitation in colloidal form.

Therefore, the FIA conversion technique, which allows a full measurement cycle to be completed within less than 1 min, opens new avenues for detecting species normally nondetectable by conventional procedures such as spectrophotometry, AAS, or ICP. Using spectrophotometric detection for the determination of very low concentrations of calcium in chlor-alkali brines Wada et al. [1311] took advantage of the reaction between calcium ions and the zinc complex of EGTA in the presence of the chelate PAESPAP {2-(3,5-dibromo-2-pyridylazo)-5-[*N*-ethyl-*N*-(3-sulfopropyl)amino]phenol} which gives rise to the formation of the corresponding Zn-PAESPAP chelate. Having a very high molar absorptivity, this conversion reaction thereby allowed very low concentrations of calcium levels to be determined. Furthermore, for excessively low concentrations of calcium, the authors exploited the preconcentration and separation of calcium from sodium chloride with a chelating resin column, whereby concentrations as low as 0.25 mgCa/L in the presence of magnesium, iron(III), aluminum, copper(II), nickel, phosphate, chloride, sulfate, and carbonate could be determined at a sampling rate of 40 samples/h.

4.7.2 Immobilized Enzymes

Of all the FIA methods using packed reactors, systems with *immobilized enzymes* are far the most common (Table 7.1). They offer the selectivity of enzymatic reactions and the economy gained by immobilizing the often costly catalysts. To illustrate the versatility of this technique selected examples of published applications comprising use of immobilized enzyme reactors are given below.

Petersson et al. [1160] recently described a miniaturized flow-injection system for the determination of four different substrates, which each could be degraded enzymatically by means of appropriate oxidases according to the following reaction schemes:

(1) β-D-glucose + O_2 + H_2O $\xrightarrow{\text{D-glucose oxidase}}$ gluconic acid + H_2O_2

(2) Creatinine $\xrightarrow{\text{creatininase}}$ creatine

Creatine + H_2O $\xrightarrow{\text{creatinase}}$ sarcosine + urea

Sarcosine + O_2 + H_2O $\xrightarrow{\text{sarcosine oxidase}}$ glycine + formaldehyde + H_2O_2

(3) Cholesterol + O_2 + H_2O $\xrightarrow{\text{cholesterol oxidase}}$ Δ^4-cholestenone + H_2O_2

(4) L-lactate + O_2 + H_2O $\xrightarrow{\text{L-lactate oxidase}}$ $CH_3COCOOH$ + H_2O_2

In each case the ultimate reaction product is hydrogen peroxide, which subsequently is determined by chemiluminescence via reaction with luminol and hexacyanoferrate(III). Thus, the inherent lack of selectivity of the chemiluminescence reaction is compensated for by exploiting the substrate specificity of the enzymes. The manifold used for these procedures is depicted in Fig. 4.47*a*, while the actual FIA microconduit system is shown in Fig. 4.47*b*. The sample (*S*) is injected by a valve and then directed to the incorporated reactor containing the enzyme immobilized on controlled pore glass beads (the actual volume of the reactor is 80 μL—the authors having given a detailed description of the immobilization procedure employed), where hydrogen peroxide is generated. Then the sample zone is guided to the merging point where it is mixed with the luminol (*R*1) and the hexocyanoferrate reagent (*R*2) before entering the flow cell, in which two photo diodes record the generated light. As an example of the performance of this system, which was designed for the assay of serum samples, the output for a calibration series of lactate is shown in Fig. 4.48.

More elaborate systems incorporating a series of reactors have been described by a Swedish group in Lund, which has also conducted a number of theoretical studies on optimization of the design of packed reactors in FIA systems [344, 345]. Thus, Olsson et al. [1067] have devised a procedure for determination of sucrose based on the following sequence of reactions:

Sucrose + H_2O $\xrightarrow{\text{invertase}}$ α-D-glucose + D-fructose

α-D-glucose $\xrightarrow{\text{mutarotase}}$ β-D-glucose

β-D-glucose + O_2 + H_2O $\xrightarrow{\text{D-glucose oxidase}}$ gluconic acid + H_2O_2

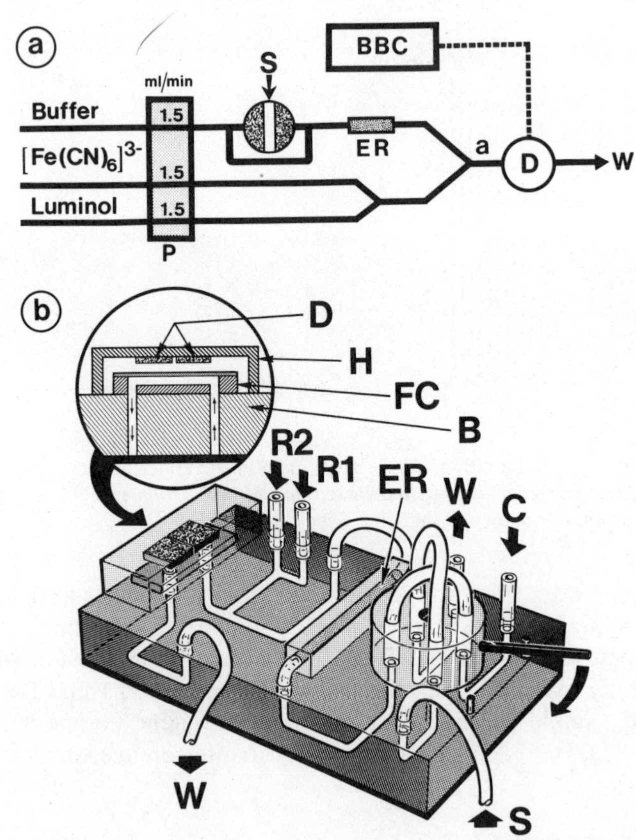

Figure 4.47. (*a*) Manifold for the determination of glucose, free cholesterol, creatinine, and lactic acid with detection by chemiluminescence. The sample (*S*) is injected into a carrier stream of buffer propelled forward by pump *P* and carried to a reactor (*ER*) containing immobilized enzyme (oxidase), in which the substrate of the sample is degraded, leading to the formation of hydrogen peroxide. Being confluenced with luminol and hexacyanoferrate(III), the sample zone is finally guided through a short channel *a* (2 cm, corresponding to 16 μL) into the light detector *D*, the output from which is led to a computer (*BBC*); *W* is waste. (*b*) Integrated FIA microconduit accommodating the manifold components shown for the system in (*a*). *C* represents the carrier stream of buffer, *R*1 is luminol, and *R*2 is hexacyanoferrate(III). The flow cell (*FC*) comprises two photo diodes (*D*) contained in a house (*H*) mounted on the microconduit base plate (*B*).

Using the manifold shown in Fig. 4.49, the injected sample of sucrose is first directed to a packed reactor (MGC) containing coimmobilized mutarotase, glucose oxidase, and catalase, which decomposes any glucose already present in the sample, and then on to a second reactor (IMG) containing coimmobilized invertase, mutarotase, and glucose oxidase in which the sucrose via the reactions described above is degraded to hy-

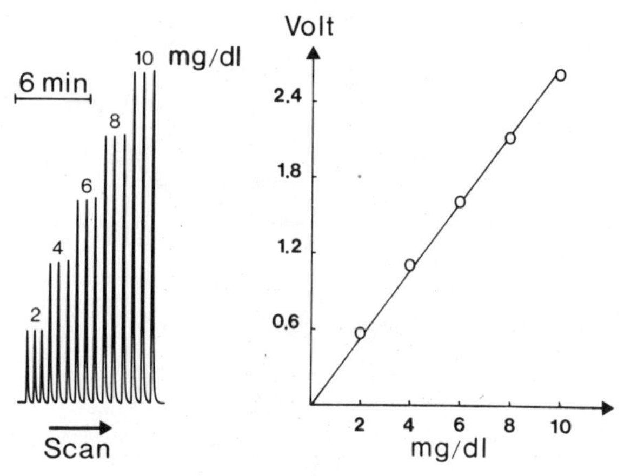

Figure 4.48. (*Left*) Calibration run for lactic acid in the concentration range 0–10 mg/dL using the FIA microconduit system depicted in Fig. 4.47. (*Right*) Calibration graph for the lactic acid standards shown at the left.

drogen peroxide, this product subsequently being measured spectrophotometrically after a chromogenic reaction employing a third reactor containing peroxidase (POD). Despite its complexity, this manifold was successfully used for determining sucrose down to 0.1 m*M* for injections of 80 μL sample solutions, the sampling frequency being 80 samples/h and the RSD 0.3%. Furthermore, glucose interference was so effectively

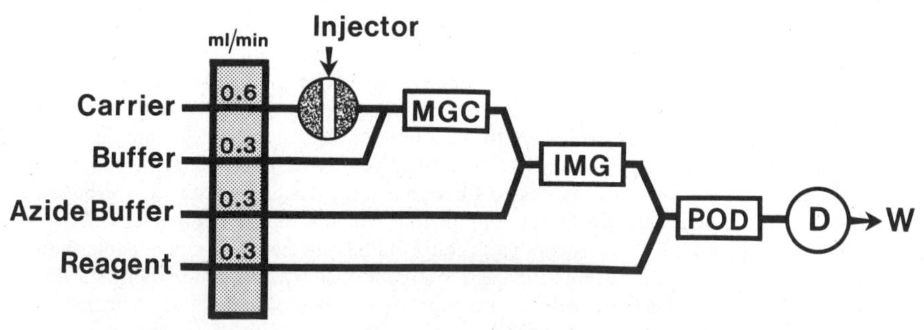

Figure 4.49. Diagram of the flow-injection system used for assay of sucrose. The sample is injected into the carrier stream, and, after being buffered, it is led to the first enzyme reactor (*MGC*) containing coimmobilized mutarotase, glucose oxidase, and catalase, which reactor eliminates any glucose possibly present in the sample. The sample zone is then guided to the second enzyme reactor housing coimmobilized invertase, mutarotase, and glucose oxidase. Here the sucrose is eventually converted to hydrogen peroxide, which subsequently is confluenced with a chromogenic reagent and via a third immobilized enzyme reactor containing peroxidase (*POD*) is led to the flow cell for spectrophotometric detection (*D*). (According to Ref 1067 by permission of Elsevier Scientific Publishing Co.)

eliminated that the response for this constituent was only 0.7% of that for sucrose.

The same Swedish group has designed a FIA system containing even more packed reactors and, indeed, a dialysis unit [758]. Aimed at the determination for galactose, the manifold for this system is shown in Fig. 4.50. Again, the final measurement is based on determination of hydrogen peroxide generated by an enzymatic reaction of an oxidase:

$$\text{D-galactose} + O_2 \xrightarrow{\text{galactose oxidase}} \text{D-galactohexodialdose} + H_2O_2$$

a fairly complex reaction, the details of which are only partially understood. The sample is dialyzed into an acceptor stream in order to eliminate unwanted matrix constituents, and then passed to a Bond-Elut-NH_2-Cu column (Cu^{2+}) for the removal of interferences and to a catalase reactor (Cat) for removal of hydrogen peroxide not only from the sample but produced by reactions in the Bond-Elut-NH_2-Cu reactor. After this pretreatment the sample zone is led into the galactose oxidase reactor (GalOD) where hydrogen peroxide is generated, which is finally measured spectrophotometrically through the color formed in an oxidative coupling reaction in the peroxidase reactor (POD). Fully automated and controlled by a computer, the linear range for the determination of galactose using an injected sample volume of 160 μL was from 10 μ*M* to 14 m*M*, the recovery from spiked serum samples, at low galactose levels, being close to 100%. The sampling frequency was 45 samples/h.

Most recently a Japanese group [1320] has described a rapid and sensitive FIA method for the enzymatic determination of adenosine and inosine in human blood plasma, which comprised the incorporation of five sequentially placed serial immobilized enzyme columns, the ultimate formation of hydrogen peroxide being measured fluorimetrically.

4.7.3 Generation of Reagents *in Statu Nascendi*

Many oxidizing or reducing species, which potentially might be applicable as reagents in chemical reactions, are thermodynamically unstable toward water in aqueous solutions and the reducing agents also toward oxygen. Thus the use of these reagents in classical analytical procedures is rather cumbersome owing to complications associated with storage and handling, and, therefore, their utilization has eventually led to their disappearance from the analytical scene. However, FIA can help in making possible the application of these reagents in quantitative analysis, by *generating* them in the flow system itself, starting from a stable oxidation state of the pertinent element. This avenue has in particular been pursued

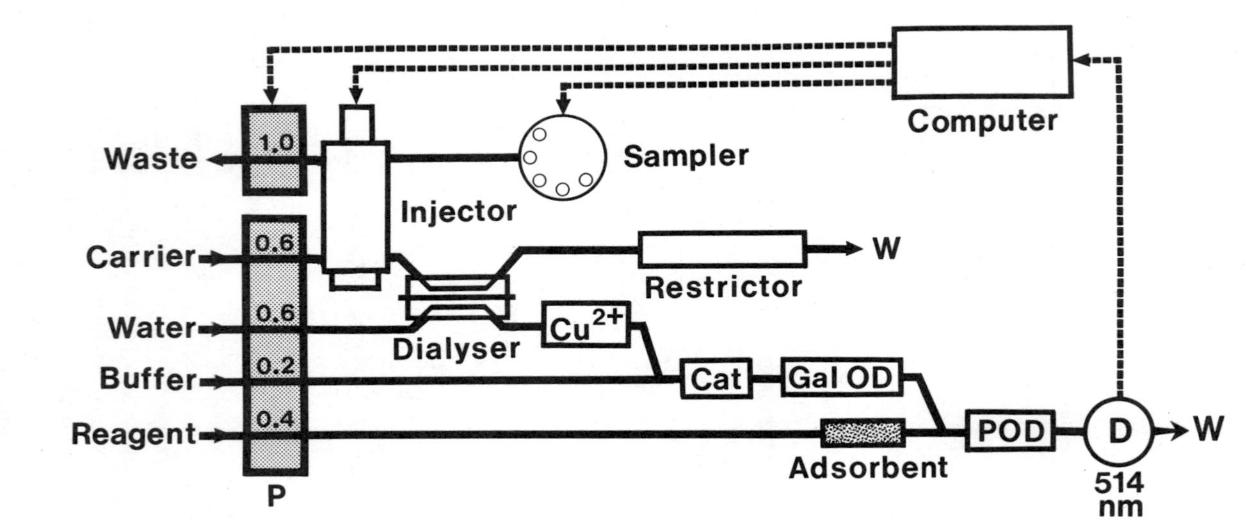

Figure 4.50. Schematic diagram showing the flow-injection system for galactose determination with spectrophotometric detection. The sample is dialyzed into an acceptor stream and then passed through two reactors (Cu^{2+}) and (Cat) for removal of interferences (for details, see text) before entering the reactor containing immobilized galactose oxidase (GalOD). The hydrogen peroxide generated there is combined with chromogenic reagent and measured spectrophotometrically at 514 nm, the operation of the system being fully automated by means of a computer. (From Ref. 758 by permission of Elsevier Scientific Publishing Co.)

by Schothorst and den Boef, who in a series of articles have described the use of strongly reducing [357, 459, 609, 637, 1112] or strongly oxidizing agents [774, 1077, 1112]. The principle is illustrated in Fig. 4.51, which shows a simple manifold comprising two reactors. A solution of a stable compound is introduced into the flowing system [such as Cr(III) for the generation of Cr(II) or U(VI) for U(III)]. The solution is pumped through Reactor I, which converts the starting solution into the unstable species. For the reducing agents a Jones reductor appears to be very effective and for the oxidizing agents an electrochemical reactor with a working electrode of gold powder kept at +2 V versus SCE is suitable. The sample to be analyzed is injected by means of the injection valve, the reaction between the reagent and the analyte occurring in Reactor II, which may be simply a coiled tube or a single-bead string reactor. After reaction, the detection is in the flowthrough cell. The great advantage of this method is that the reagent only needs to be sufficiently stable during its residence time in the flow system, which may be less than 1 min. Thus, for silver(II) [774] the half-life of the reagent solution under the experimental conditions was measured and appeared to be of the order of 1 min.

Also for reagents like Cr(II) and V(II), the application of a flowing system has great advantages. Although these reagents are relatively stable toward water, their storage is a problem because they are oxidized by air. When generated *in statu nascendi* in a flow system, oxygen is eliminated from the solution in Reactor I and the only ancillary operation to be done is to deaerate the sample solution before injection. This results in the simple applicability of these reagents.

The applicability of unstable reagents might not be restricted to reducing and oxidizing agents, but may well provide similar advantages for other types of reactions and in solvents other than water.

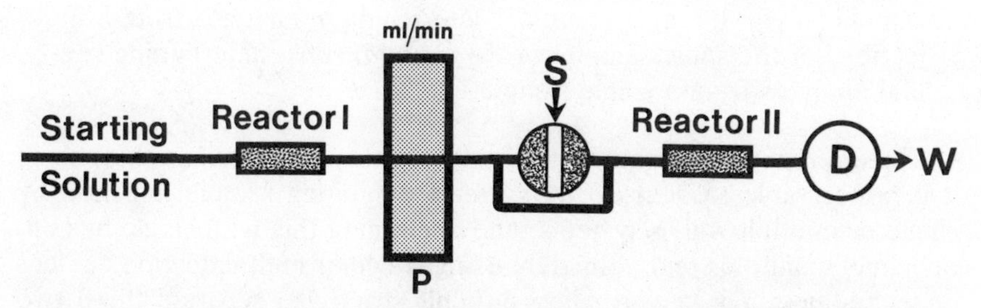

Figure 4.51. Simplest manifold for generation and use of reagents in unstable oxidation states. The starting solution is aspirated into Reactor I, which converts the stable starting solution into an unstable compound. Serving as carrier stream, the sample is then injected into this solution, the reaction between the reagent and the analyte taking place in Reactor II, the result of which is monitored by the detector.

4.8 FIA METHODS BASED ON MULTIDETECTION AND MULTIDETERMINATION

Reviewing the literature on FIA it is striking to observe that, although a large number of configurations and applications of FIA systems have been described, most methods are still based on the measurement of a single signal depending on the analyte concentration. However, as emphasized throughout this book the potential of FIA is much greater, allowing not only the exploitation of the concentration gradient for the collection of much analytical information, thereby inherently permitting the possibility of making several measurements per single sample injection, but furthermore facilitating the simultaneous determination of several species.

As rightly pointed out recently by Luque de Castro and Valcárcel [1044], the term "simultaneous" is not cearly defined in FIA because it has been indiscrimanately used to describe different alternatives. Hence, in their paper, they attempted to establish the difference between multidetection and multidetermination, and the sequential and simultaneous variants of each of these, arriving at the following definitions:

Multidetection. Two or more signals are obtained from a single injected sample. It can be performed with a single detector at different times (cf. Fig. 4.52*a*), at different instrumental conditions (wavelength, potential applied, etc.) or with several detection points located in series (cf. Fig. 4.52*b*) or in parallel. It may be (a) *sequential*, using one or several detection points to obtain several signals at different times per injected sample; or (b) *simultaneous*, using a single detection point performing simultaneous measurements at different instrumental conditions.

Multidetermination. Determination of two or more analytes on a sample. It can be (a) *sequential*, determining n analytes from n injections of the same sample; or (b) *simultaneous*, determining several analytes from a single sample injection.

Thus, although adhering to these definitions, it is obvious that, since it is conceivable to devise FIA systems exploiting sample injection by single or multiple valve systems, and combining this with single or multichannel manifolds and ultimately using single or multidetection devices or techniques, it is nevertheless difficult strictly to separate these two terms, because multidetermination might in fact be achieved via multidetection, although multidetection is not necessarily equivalent to multidetermination. No attempt is made here to provide an exhaustive review of multidetection and multidetermination, which have been treated in

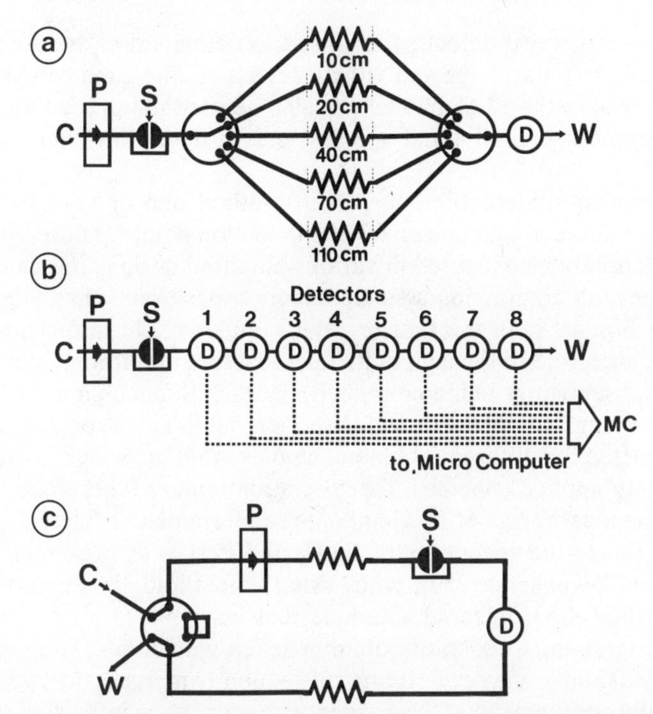

Figure 4.52. FIA manifold designs for sequential multidetection employing a single injection position. (*a*) Splitting of the sample into a number of subplugs, which are guided through individual reaction coils and finally routed to a common, single detector. (*b*) Use of multiple detectors located in series, the response from each of which is fed to a microcomputer. (*c*) Use of a closed-loop system where the injected sample is recycled around a number of times in a loop, which includes the detector, until a constant signal is emitted, whereupon the sample is directed to waste.

great detail by Luque de Castro and Valcárcel [540, 1044], although selected examples of these techniques of particular relevance to practical work will be illustrated in the following.

Turning initially to multidetection, and here first to simultaneous usage, an obvious application is to combine the gradient FIA techniques with the use of detectors that provide several signals at several values of the instrumental variables, which indeed gives FIA a doubly dynamic character. In these techniques, which have already been mentioned in Section 2.4, advantage can be taken by multidetectors, such as the fast-scan voltammetric detectors [288]; or by inductively coupled plasma atomic emission spectroscopy [808]; or by diode array detectors [1017, 1075, 1382]. Combined with the advantages offered by chemometrics, these multidetection procedures may in fact be extended to multideterminations, particularly if several detectors are used in combination, because, then, besides the outputs of the individual detectors, the time-concentration matrix formed in the FIA channel can also be used to improve the quality of measurement of individual species and to determine their chemical form.

Sequential multidetection, by which method one or several injected samples are subjected to one or several detection points at different times, might potentially be executed in various manifold designs, of which three simple ones, all comprising a single injection position, schematically are shown in Fig. 4.52. In the first (*a*), the sample is split into a number of subplugs, each routed through individual reactors of different lengths, the streams subsequently being combined via a confluence point so that the sample subplugs reach the detector sequentially. This type might for instance be used for differential kinetic studies [854] or, which is one of the most widely applied schemes, for speciation studies where two or more physicochemical forms of an element are determined (Table 7.3). An example of this is the speciation of Fe(II) and Fe(III) as presented by Faizullah and Townshend [759], who, using a manifold similar to that presented in Fig. 4.53, injected a sample that was split into two lines, one of them containing a reductor column packed with Jones reductor (amalgamated Zn) and a delay coil. Being subsequently merged, the two streams were finally confluenced with a reagent stream of phenanthroline in citrate buffer—which two components were premixed in the FIA system in an

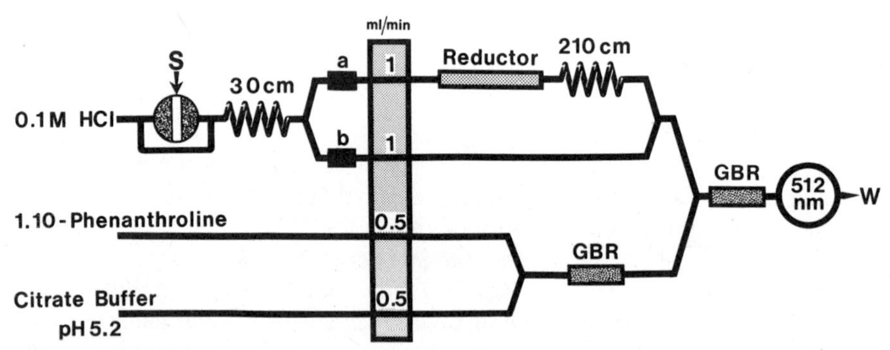

Figure 4.53. Manifold used for the simultaneous spectrophotometric determination of iron(II) and total iron. The sample (*S*) is injected into a carrier stream of acid and split into two channels: one comprising a column containing a (Jones) reductor and a delay coil, the other one simply being a short connecting tube to a common confluence point where the two streams are joined again. Being spaced in time, the two subplugs, where all iron now is present as iron(II), are then mixed with color-forming reagent (1,10-phenanthroline) and measured spectrophotometrically at 512 nm in a common detector. (According to Ref. 759 by permission of Elsevier Scientific Publishing Co.)

incorporated glass bead reactor (GBR)—before being passed via a second GBR mixing reactor into the detector. Hence, measuring Fe(II) plus Fe(III) by the subsample guided through the reductor-containing line, and Fe(II) from the signal obtained from the sample zone directed through the other line, Fe(III) could be calculated by the difference.

When several detection points are used to obtain sequential analytical signals, the detectors—either of the same or of different kind—can be located in parallel or in series (Fig. 4.52*b*). Thus with the series configuration of potentiometric–photometric detectors, the application of the FIA technique has resulted in methods for calculating ionization constants, which is especially useful for unstable compounds [834], and for multispeciation–numerical-calculations methods [1044], just as the use of this approach employing detectors of the same nature allows the carrying out of nonkinetic multidetermination as well as kinetic determinations [1044]. A special system of series detectors has been suggested by Hooley and Dessey [365] for kinetic determinations. It is based on the use of LEDs located in series along a quartz tube and in which the radiation, which impinges perpendicularly on the flow, is collected by a diametrically opposed photodetector. One type of sequential multidetection device with a single detector, which recently has attracted attention, is the so-called closed-loop or closed-open system [1049] (Fig. 4.52*c*), where the injected sample is recycled around a closed loop that includes the detector. On each circuit it passes the detector, the successive passage of the sample plug through the detector causing a series of atypical FIA peaks until a constant signal is emitted when physical (dilution) and chemical equilibria are attained, whereupon the sample is directed to waste. The maxima and minima profiles correspond to typical kinetic curves to which conventional kinetic considerations can be applied [840]. The recording allows one to discern physical and chemical kinetics. Its application to the determination of a single species [dichromate, iron(II)], simultaneous determinations, as well as the determination of physicochemical parameters such as reaction-rate constants and stoichiometries, and to the implementation of amplification and dilution methods, etc., are indicative of the applicability of this configuration [1044].

An example of simultaneous multidetermination executed via a single injection and employing several detectors has in fact already been mentioned in Section 4.1, where the simultaneous determination of pH and pCa [21] was presented (Fig. 4.4). Another similar application is the scheme presented by Virtanen [250] for the determination of Na^+, K^+, Ca^{2+}, and Cl^- in serum by the use of ion-selective electrodes in which system the flow impinged laterally on both the sensors and the reference electrode, which was located behind them and in the same position, an approach originally suggested in one of the early FIA works [8]. Betteridge and Fields [32,220] suggested an interesting mode of simultaneous determination by conventional FIA, involving a single detector and single injection, that is, creating a pH gradient within the carrier stream by using a reagent stream the pH of which was different from that of the sample solution, exploiting this gradient for multielement analysis. Thus, if the volume of the injected sample is sufficiently large when it reaches the detector, two zones of different characteristics exist that are close to the interphases with pH values different from that of the central zone of the plug. The characteristics of these regions can be used to determine several species in the same sample. An example of this type is the simultaneous hotometric determination of Pb(II) and V(V) with 4-(2-pyridylazo)-resorcinol (PAR) [32, 98]. As the complexes of this ligand exist in different pH ranges, the individual cations will, at the time the sample plug reaches the detector, only exist in discrete parts of the sample zones, consequently allowing their individual determination. An application of the multidetermination approach, which really does not fit any of the proposed definitions set forward in the beginning of this chapter, is the one described in the very early days of the FIA technique for the simultaneous determination of nitrogen and phosphorous [5]. Exploiting the fact that the chemistries employed both lead to the eventual formation of a blue color, a manifold was designed where the injected sample is diverted into two channels—to each of which the appropriate reagents are added by confluence—where the chemistries for nitrogen and phosphorous, respectively, are executed. The two streams are, however, not subsequently merged, but instead they are directed to two individual identical flowthrough cells, which are aligned in the optical path of the sample beam of a single spectrophotometer. Thus, provided that the two sample plugs are sufficiently spaced in time, discrete signals for each channel might be obtained. Interestingly, this configuration—which originally was conceived out of despair—was recently adopted and used by Ruz et al. [1321] in speciation studies of Cr(VI) and Cr(III).

Simultaneous determination with a single detector was first described by a Brazilian research group directed by Bergamin, the proposed approach actually constituting the very first attempt of speciation [81]. In this approach (Fig. 4.54), designed for the determination of nitrate and nitrite, the injector (commutator) is furnished with two loops, one of which is used for injection of sample solution for the nitrite assay, while the other one is used for the nitrite plus nitrate determination, the sample being routed through a reduction of column of cadmium, which reduces all nitrate to nitrite. Being subsequently merged, the two streams are, spaced in time, confluenced with color-forming reagent, and subsequently

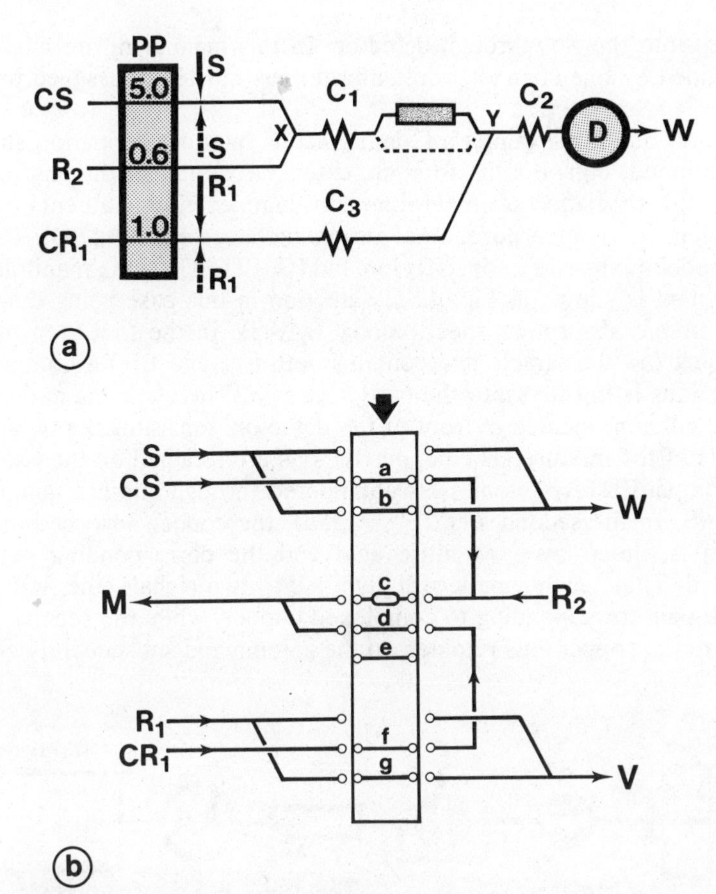

Figure 4.54. (*a*) FIA manifold for determination of nitrate and/or nitrite by the merging zones principle. The sample *S* is injected into a carrier solution of water *CS* to which at *X* is added buffer and masking agent R_2. After mixing in coil C_1 the sample zone either enters a cadmium-containing reduction column where nitrate is reduced to nitrite, or is—for nitrite determination alone—by-passed (dashed line), before being merged at *Y* with the color-forming reagent R_1 injected into the buffer stream CR_1. The resulting color is measured at 540 nm (see Fig. 4.5). (*b*) Schematic diagram of the injector commutator used in a two-channel system for determining nitrate and/or nitrite. The injector is shown in the nitrate plus nitrite position, where the sample, metered by loop *b* and carried by stream *CS*, is mixed with reagent R_2 before entering the reduction column *c*, and subsequent mixing in the manifold *M* with the color-forming reagent R_1, injected via loop *f* into CR_1. In the other position of the injector, the sample volume is metered by loop *a*, is flushed by *CS*, mixed with R_2, and led into the manifold via *d* to merge with R_1, metered by loop *g* and carried by stream CR_1 (in this position where the reduction column is by-passed, only nitrite is determined). While surplus sample is wasted (*W*), any uninjected reagent solution R_1 is collected in recovery vessel *V*. (From Ref. 81, by permission of Elsevier Scientific Publishing Co.)

directed into the flowthrough detector. In this manner nitrite might be determined by one channel, while nitrate plus nitrite are assayed by the other.

Finally, one last example of simultaneous multidetermination should be mentioned, constituting, to some extent, a variant of this approach, that is, the speciation of metal ions and complexed constituents of the same element via incorporation of an ion-exchange column [780]. Tested on a model mixture of copper(II) ion and $[CuEDTA]^{2-}$, the manifold for this system is shown in Fig. 4.55, detection in this case being done by flame atomic absorption spectrometry (fAAS). In the first step of the procedure (*a*), the sample that contains both free and EDTA-complexed copper ions is injected into the carrier stream. The chelating cation-exchange column, located in front of the detector, separates the two components of the mixture. The copper(II) ions are retained on the column, whereas $[CuEDTA]^{2-}$ ions pass unhindered through and are quantified by fAAS. In the second step (Fig. 4.55*b*), the copper adsorbed on the column is eluted by 2 *M* nitric acid and the corresponding peak is recorded. Thus, each sample will give rise to two signals, the first peak in each pair corresponding to complexed copper, while the second peak is due to the copper ions retained on the column and subsequently eluted

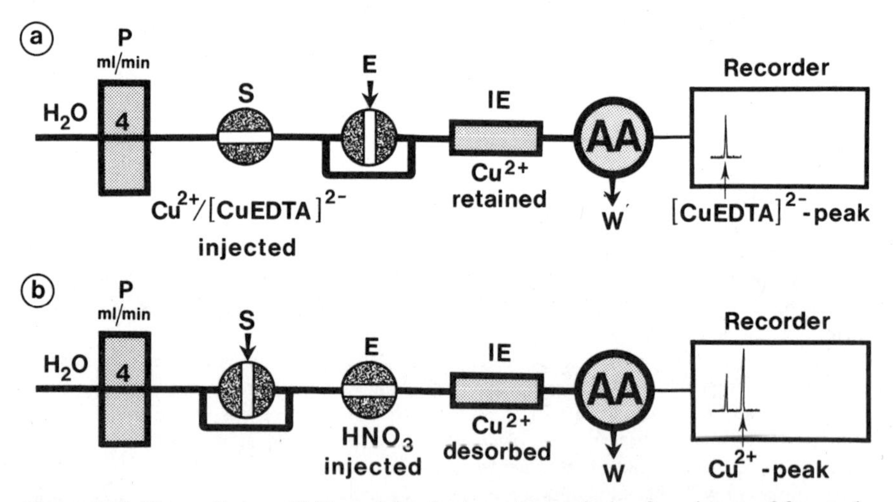

Figure 4.55. Two-valve manifold used for the two-step analysis of a mixture of free and EDTA-complexed copper(II) ions. *P*, peristaltic pump; *S* and *E*, sampling and eluting valves, respectively; *IE*, cation-exchange column; *AA*, atomic absorption spectrophotometer. (*a*) The sample is injected by valve *S*, Cu^{2+} is retained on the column, while $[CuEDTA]^{2-}$ is transported to the detector where the corresponding peak is recorded. (*b*) Eluent (2 *M* HNO_3) is injected by valve *E*; Cu^{2+} is desorbed from the column and the signal corresponding to Cu^{2+} is recorded.

by the nitric acid pulse. Because preconcentration occurs by elution into a smaller volume than the sample volume originally injected (cf. Section 4.7.1), and because the dispersion of the eluted zone is less than that for the complexed copper zone at the point of detection, the second peak is narrower and therefore comparatively higher than the first one. Typical relative standard deviations using this system were ($n = 10$) for the determination of $[CuEDTA]^{2-}$ 1.2–1.6%, while the corresponding values for copper(ii) ions were 1.0–1.2%. Most recently, Milosavljevic et al. [1310] utilized this approach for the simultaneous determination of Cr(VI) and Cr(III), applying an incorporated column of activated alumina, which has a much higher affinity for anionic Cr(VI) than for Cr(III) species, reaching standard deviations of the order of 0.5% at concentration levels of 100 μg/L.

4.9 FIA TITRATIONS AND PEAK WIDTH MEASUREMENTS

As detailed in Section 2.4.4, the FIA titration technique is a gradient method based on creation of a suitable concentration gradient of an injected sample (analyte) within a carrier stream of titrant, exploiting that both at the front and at the tail of the dispersed sample zone an element of fluid exists within which equivalence between analyte and titrant is obtained. These two equivalence points form a pair, having the same D value, and their physical distance, measured as Δt for a constant flow rate Q, will increase with increasing concentration of the injected sample, C_A^0, and decrease with increasing concentration of the titrant reagent, C_R^0, contained in the carrier stream (cf. Fig. 2.23). The relationship of the time span Δt to the concentrations of the analyte and the titrant depends on the concentration profile formed by the dispersion of the zone on its way from the injector to the detector. As shown in Section 2.4.4, Δt as a function of the concentrations C_A^0 and C_R^0 will, as long as the concentration profile conforms with the model known as "one mixing stage," be given by

$$\Delta t = (V_m/Q) \ln 10 \log[C_A^0/(C_R^0\, n)] + (V_m/Q) \ln 10 \log(S_v/V_m) \qquad (4.6)$$

where n is the stoichiometric factor of reacting components, S_v is the injected sample volume, and V_m is the volume of the mixing stage. As the last term for a given FIA system will be a constant, a plot of Δt versus $\log C_A^0$—as obtained by injecting a series of standard solutions—will thus yield a straight line with a slope of $(V_m/Q) \ln 10$. Hence, on the basis of this calibration curve, by measuring Δt of a given sample one can determine its initial concentration. Therefore, one of the inherent advantages of the FIA titration procedure is that, since the quantitation of the unknown solution is based on a relative measurement, that is, to the known standards, the absolute value of the reagent titrant solution need not be accurately ascertained nor indeed be known.

In order physically to fulfill the one-mixing-stage condition, the early FIA titration systems did in fact incorporate a small mixing chamber (equipped with a magnetic stirrer) the volume of which (ca. 1 mL) dominated the flow design. This approach is illustrated in Fig. 4.56, which

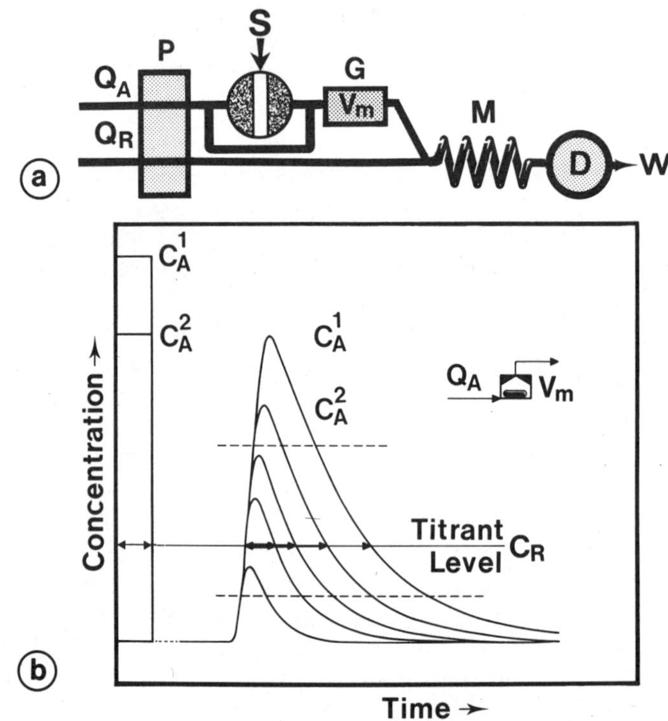

Figure 4.56. (*a*) Two-channel manifold for FIA titrations. The sample S is injected into an inert carrier stream, pumped at rate Q_A, mixed with the carrier in the gradient chamber G of volume V_m, and the controlled concentration gradient formed of the sample is then merged with titrant R, pumped at rate Q_R, before entering the flowthrough detector D via the reactor M. (*b*) Concentration gradients formed by injecting samples of different concentrations C_A^i. Equivalence between sample solution A and titrant R is encountered twice during each titration cycle, and the time between them, Δt, is for a fixed level of R proportional to C_A^i. Note that by increasing the concentration of R, Δt decreases, but that the limit of sensitivity simultaneously deteriorates, and vice versa. To the left is shown the Δt values that would be obtained if the sample did not undergo dispersion in the system (i.e., Δt would in such case merely reflect the length of the injected sample zone).

might serve as explanatory basis for the FIA titration procedure. The manifold (Fig. 4.56*a*) comprises two lines, *A* and *R*, referring to carrier stream and titrant reagent stream, respectively. The sample is injected into the inert carrier stream *A* and propelled into the mixing chamber *G*. Provided that the injected sample volume is much smaller than V_m and that the sample can be considered introduced into the mixing chamber as a plug and is mixed there instantaneously, a pure exponential concentration gradient of the sample within the inert carrier is formed. Corresponding to one stirred tank [$N = 1$, Eq. (3.9)], the concentration of *A* leaving the chamber at time *t* will thus be given by $C_A = C_{A0} \exp(-tQ_A/V_m)$, where C_{A0} is the concentration of *A* in the mixing chamber at $t = 0$, that is, $C_{A0} = S_v C_A^0 / V_m$. Downstream the concentration gradient is merged with the titrant solution *R*, the chemical reaction between the two species taking place in reaction coil *M*. This is via an appropriate indicator added to the titrant solution measured by detector *D* tuned to monitor the presence of either the analyte or one of the forms of the added indicator. Thus, although the concentration gradients formed in the system will have an exponential shape (Fig. 4.56*b*), the detector will during the passage of the dispersed sample zone indicate two abrupt changes of signal corresponding to those two elements of fluid on the titrant stream/sample zone boundaries where the equivalence conditions are fulfilled. The time span between these two equivalence points is the analytical readout, Δt, quantitatively expressed by Eq. (4.6). [If the two pumping rates Q_A and Q_B are different, an additional *constant* term is incorporated into the equation, cf. Section 2.4.4.] As seen from Fig. 4.56*b*, the magnitude of Δt will, for a fixed value of C_A, be a function of C_R (which trigger level thus might be prearranged by adjusting the concentration of the titrant solution), or for a fixed level of C_R be a function of C_A (which in turn will determine the lower limit of detection of analyte to be assayed by the titrant used).

Since the role of the mixing chamber solely is to create a controlled concentration gradient of the injected sample, the experimental setup in Fig. 4.56 can be simplified by using a single-line manifold such as that shown in Fig. 4.57*a*, where the sample to be titrated is added directly into the carrier stream of titrant and carried through a very short line into the well-mixed gradient chamber *G*, that is, $Q_A = Q_B$. The applicability of this manifold is demonstrated in Fig. 4.57*b*, which shows the titration of a series of strong acid solutions (HCl, in the range 0.008–0.1 *M*) with strong base (NaOH, 1×10^{-3} *M*) as monitored spectrophotometrically via an acido-basic indicator (bromothymol blue) added to the titrant stream, the detector being tuned to monitor the alkaline blue form of the indicator (cf. the titration exercise in Section 6.7). The color changes of

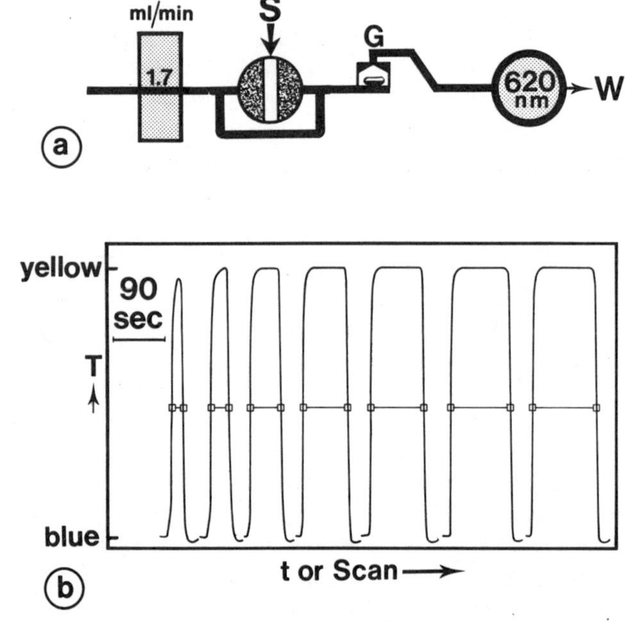

Figure 4.57. FIA titration of strong acid with strong base. The manifold (*a*) comprises a mixing chamber *G*, having a volume of 0.98 mL, where a well-defined concentration gradient is formed of the injected acid sample zone within the alkaline carrier solution, which also contains an acido-basic indicator. The flowthrough cell monitors colorimetrically the change of the indicator from blue to yellow as shown on the record in (*b*). As the color of the indicator is converted completely during each titration cycle, all curves have the same height, while their width, Δt, is a function of their individual concentration.

the indicator during the titration cycle (from blue base color to the yellow acid color and back to blue) are registered by the spectrophotometer as a "blue or no blue signal," that is, resulting in peaks of the same height, but of variable widths as a function of C_A^0. In practice, the Δt signal is measured as the peak width directly on the recorder paper at the same level, approximately halfway between the baseline and the top of the signal.

Referring to the discussion in Section 2.4.4, it is obvious that since the FIA titrations adhere entirely to the underlying principles of classical titrimetry, all the conventional titration procedures might be embodied in FIA titration systems. Therefore, an acid with two protons protolyzing at sufficiently different pK_a value ($pK_{a1} \geq pK_{a1} + 4$) will consequently be titrated in two steps. Such is the case with phosphoric acid (pK_{a1} = 2.15, pK_{a2} = 7.21, pK_{a3} = 12.36, so that $pH_{eq\ 1}$ = 4.7 and $pH_{eq\ 2}$ = 9.8); the titration of phosphoric acid by 0.001 M sodium hydroxide as carrier stream is shown in Fig. 4.58, right, the pH here being measured potentiometrically by means of a microquinhydrone electrode, yielding at the two prefixed pH_{eq} levels the respective titration values, Δt_1 and Δt_2 [338]. The concentrations of phosphoric acid in the three injected samples were in this case chosen to be in the ratio 1:2:4 (0.05 *M*, 0.10 *M*, and 0.20 *M*) and the carrier stream was 0.001 *M* NaOH containing 0.01 *M* quinhydrone so that according to Eq. (4.6) the horizontal distances between the equivalence points of curves 1 and 2, and 2 and 3, respectively, should be equal. As apparent from the titration curves, recorded from the same start (*S*), this is indeed the case.

Potentiometric titrations exploiting ion-selective electrodes to monitor the analyte are very attractive because ion-selective electrodes yield a logarithmic response and thus encompass a wider concentration range than do the color indicators. Yet, for the very same reason, the potentiometric titration curves are not flat at the top, but exhibit an S-shaped form. This is observed on the potentiometric titration of calcium with EDTA, which was performed in a single-line manifold similar to the one depicted in Fig. 4.57*a* using a gradient chamber (V_m = 1 mL), pumping the carrier solution of EDTA at a rate of 0.84 mL/min [10]. By injecting 200 μL aliquots of sample solution, containing calcium in the range 0.005–0.05 *M*, into the carrier stream containing 5×10^{-4} *M* EDTA, a series of titration curves was obtained (Fig. 4.59*a*). Each titration is initiated by an abrupt increase of the potential, followed by a typical S-shaped descending part, in which the inflection point marks the end of the titration. By choosing a fixed potential, Δt was read off and plotted versus the

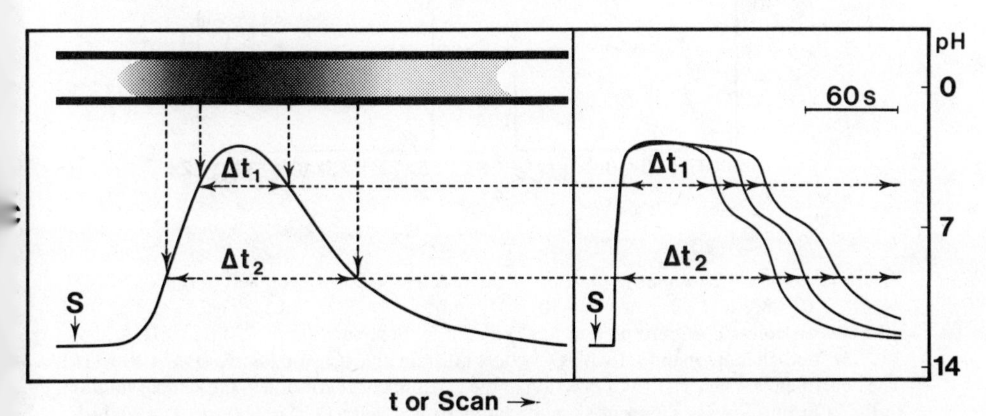

Figure 4.58. Gradient FIA titration, based on identifying pair(s) of segments with identical *D* values in the dispersed sample zone. The time span Δt between these segments is related to the total concentration of analyte. To the right is shown the stepwise potentiometric titration of three different concentrations of phosphoric acid by sodium hydroxide.

concentration of the injected sample (Fig. 4.59*b*). In agreement with Eq. (4.6), a straight line was obtained over the whole concentration range, its position in semilogarithmic coordinates depending on (1) the volume of the injected sample (compare curves *a–d*), because the S_v/V_m ratio changes, and (2) the concentration of EDTA. The slope of the calibration curve, as predicted by Eq. (4.6), depends on the pumping rate Q, but for three different pumping rates, all the lines intersect at approximately the same $\Delta t = 0$ (Fig. 4.59*c*), corresponding to the detection limit of the

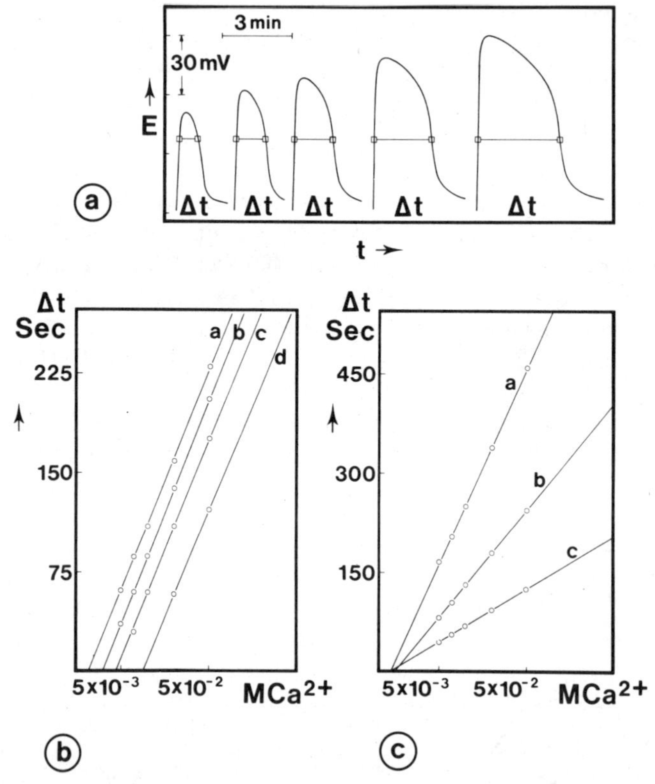

Figure 4.59. (*a*) Potentiometric titration of calcium with EDTA. From left to right 5×10^{-3} *M*, 7×10^{-3} *M*, 1×10^{-2} *M*, 2×10^{-2} *M*, and 5×10^{-2} *M* $CaCl_2$. The Δt values read off between points □–□ are plotted in (*b*) as line *c*. Sample volume, 200 μL; EDTA, 5×10^{-4} *M*; one-channel manifold with $Q = 0.84$ mL/min and $V_m = 0.98$ mL (Fig. 4.57*a*). (*b*) The equivalence time (Δt, s) as a function of the calcium concentration in the sample solution ($\log C_A^0$) in titrations of different sample volumes (*a* = 400 μL, *b* = 300 μL, *c* = 200 μL, and *d* = 100 μL). Titrant, 5×10^{-4} *M* EDTA; $Q = 0.84$ mL/min. (*c*) The influence of the flow rate Q on the slope of the calibration curve as obtained by titration of 300 μL Ca sample volumes by 5×10^{-4} *M* EDTA. The flow rates are *a* = 0.46 mL/min, *b* = 0.84 mL/min, and *c* = 1.60 mL/min; Δt (s) plotted versus $\log C_A^0$ (*M* Ca^{2+}).

titration, which is given by S_v/V_m and the concentration of titrant C_B. The chamber volume V_m, as calculated from the slopes, was found to be 0.98 mL.

Although the exponential concentration gradient obtained by the inclusion of a mixing chamber in the FIA titration manifold is desirable, use of a mixing stage of large volume has several drawbacks. Not only are the mechanics of a mixing chamber with a magnetic stirrer objectionable, but the large volume of the chamber is undesirable because it results in a large dispersion of the sample material causing loss of sensitivity as well as a low sample throughput, if a moderate reagent consumption is to be maintained. For example, with a typical $V_m = 1$ mL (as that used in the examples mentioned above) and a pumping rate of 1.5 mL/min, the purely physical process of decreasing C_A to one half requires 28 s ($t_{1/2}$) and reaching the baseline within 4% requires as much as 6 $t_{1/2}$, that is, 2.8 min (cf. Fig. 4.59*a*). This is, admittedly, the worst case possible, as during the titration a chemical reaction takes place, which depending on the C_A^0/C_R^0 ratio will decrease the above values perhaps as much as five times. Therefore, Ramsing et al. [183] undertook an investigation aimed at elimination of these drawbacks by designing a simpler flow channel. This was achieved by comparing the experimental data obtained by dispersion measurements with the theoretical models describing the dispersion of material during its movement through a tubular reactor. These data allowed the position and physical dimensions of a single mixing stage to be located, which is capable of providing the same function as a mixing chamber of the previous design. Demonstrating that the injection process formally conforms with the one-mixing-stage model, and that the crucial relationship in order to obtain N approximately equal to unity is that the ratio S_v/V_m is as close as possible to 0.5, that is, the system is operated at *limited to medium D* values, the authors designed a system useful for acido-basic, redox, and compleximetric titrations. It should be noted, however, that by replacing the mixing chamber with a gradient tube a concentration gradient is obtained that is not strictly exponential over a wide concentration range, and hence a semilogarithmic relationship between Δt and C_A is linear only over 2–3 decades.

The manifold used for acido-basic titrations of strong acid with strong base using bromothymol blue as indicator added directly to the alkaline carrier solution is shown in Fig. 4.60*a*, while the recorder output is shown in Fig. 4.60*b*. Here, all the five titration cycles made are recorded from the same starting point (S), showing the practical consequences of performing titrations at medium D values. First, the titration cycle, including the washout period, can be made as short as 12 s with an equivalence Δt of 2.2 s; hence, it is necessary to acquire the readouts by means of a

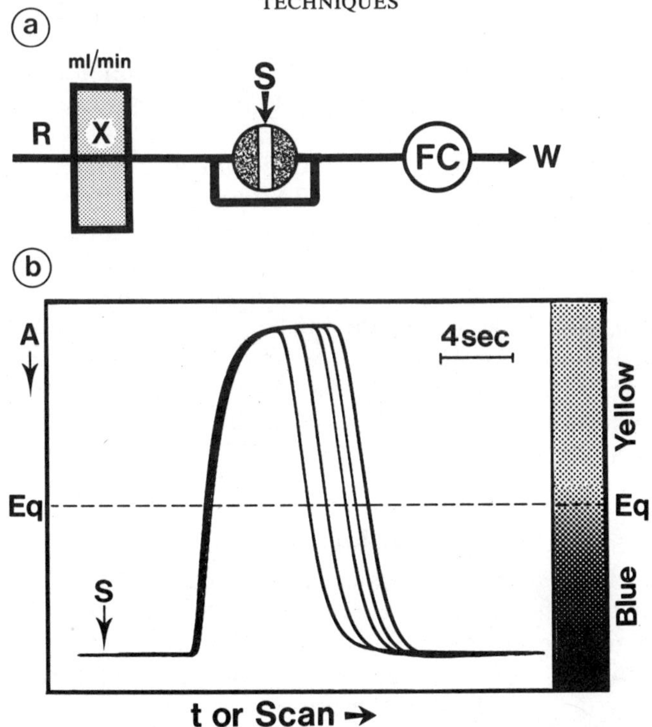

Figure 4.60. (*a*) Single-line flow-injection titration manifold used for high-speed acid–base, compleximetric, and redox titrations; the tube connecting the injecting port and the flow-through detector (*FC*) is 25 cm long, 0.5 mm ID, the injected sample volume (*S*) being 30 μL. (*b*) Readouts for titration of strong acid with strong base using the manifold shown in (*a*), and employing bromothymol blue as indicator added to the alkaline carrier stream (1×10^{-3} *M* NaOH) pumped at a rate of 1.46 mL/min. All samples (C^0_{HCl} = 0.01, 0.02, 0.04, 0.06, and 0.10 *M*, respectively) were injected at point *S*, and the Δt values were read at the Eq level.

computer because measurement of peak widths performed on the chart recorder paper are not sufficiently precise. Second, when medium dispersion is used, the sensitivity of titration is at least as good as that of conventional batch procedures. Third, the sample volume may be low (typically 50 μL or less) and therefore the reagent consumption is also low.

In this context a very important aspect of the FIA titration procedure should be emphasized [253]: From Eq. (4.6) it is apparent that Δt may be varied by changing the concentration of the titrant solution (C_B), that is, the dynamic measuring range of the procedure may be shifted as a function of C_B. This is readily demonstrated in Fig. 4.61, which shows a series of calibration runs of titration of strong acid (HCl) with strong base

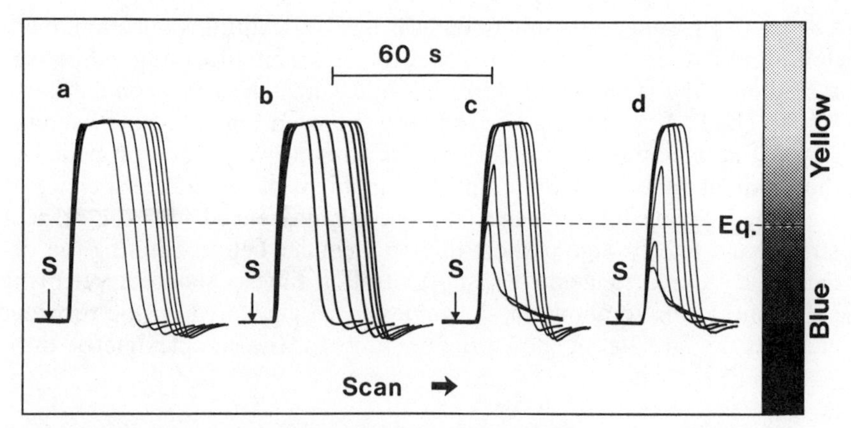

Figure 4.61. Recordings obtained in the titration of HCl (C^0_{HCl} = 5 × 10^{-3} *M*, 1 × 10^{-2} *M*, 2 × 10^{-2} *M*, 4 × 10^{-2} *M*, 6 × 10^{-2} *M*, 8 × 10^{-2} *M*, and 1 × 10^{-2} *M*) with (*a*) 5 × 10^{-4} *M* NaOH; (*b*) 1 × 10^{-3} *M* NaOH; (*c*) 5 × 10^{-3} *M* NaOH; and (*d*) 1 × 10^{-2} *M* NaOH. A decreasing concentration of the titrant in the carrier stream from 1 × 10^{-2} *M* to 1 × 10^{-3} *M* NaOH increases the limit of sensitivity of measurement (*d*, *c*, *b*), but beyond the latter level it only increases the time needed to complete the titration cycle (*a*). The manifold used was similar to that depicted in Fig. 4.64*e*, the volume of the gradient tube being 150 μL (2 mm ID) and the injected sample volume 75 μL.

(NaOH), using a manifold similar to the one described above, as obtained by adjusting the concentration levels of sodium hydroxide in the carrier stream. As seen on the figure, dilution of the titrant leads to longer Δt times, while increasing concentrations of titrant gradually reduce the peak widths (*b–d*). Yet, the *relative difference* in the Δt distances between the individual samples will remain constant, because the calibration curves for all titrant concentrations will be parallel [cf. Eq. (4.6)]. Therefore, varying C^0_B will *not*, per se, change the absolute precision of the titration procedure, but it may serve as a convenient means of adjusting the titration times for a series of samples to fall within a chosen time range, that is, changing the dynamic measuring range of the procedure.

Using the same manifold as that depicted in Fig. 4.60*a* Ramsing et al. [183] performed a compleximetric titration of zinc(II) by EDTA, using xylenol orange as indicator, and a redox titration of iodine by thiosulfate with starch added as indicating species. In both instances a plot of Δt versus log C^0 of the analyte yielded within 2 decades of concentration linear calibration curves of regression coefficients better than 0.995, the relative standard deviation of the individual measurements being of the order of 1%. These two experiments have in fact been adopted as part of the standard FIA exercises at this Department (cf. Chapter 6), being accommodated to the FIA system shown in Fig. 6.11. In this apparatus a series of FIA microconduit modules is used, constituting injection, manifold, and detection units. For titrations the manifold module comprises a gradient tube (volume ca. 150 μL) drilled into the microconduit baseblock. The FIA system is schematically shown in Fig. 4.62*a*. The sample (50 μL) is injected into an inert carrier stream (water) and propelled to the gradient tube, where a controlled gradient of medium dispersion is created. Leaving the tube the gradient is confluenced with the reagent stream and led through a short mixing meander before entering the detector. By this arrangement a short titration time is secured, yet being sufficiently long to allow the Δt values to be evaluated on the recorder chart. In Fig. 4.62*b* are shown the readouts for the compleximetric titra-

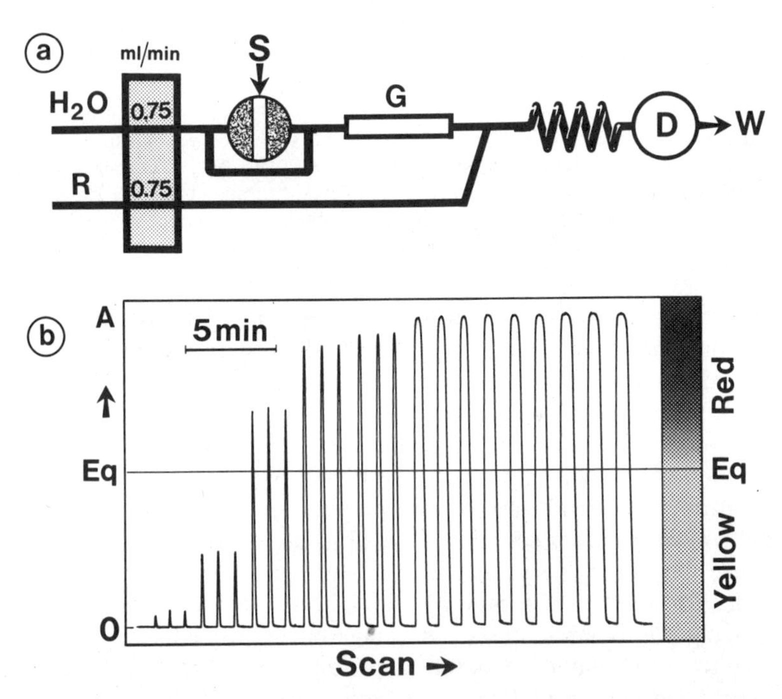

Figure 4.62. (*a*) Two-line titration manifold. The sample (50 μL) is injected into a carrier stream of water and then led to the gradient tube *G* (volume 150 μL), where a controlled gradient of the injected sample is formed. Being subsequently merged with reagent *R*, the solution is finally directed into the detector *D* and from there to waste, *W*. The pumping rate of each line is 0.75 mL/min. (*b*) Readouts for the compleximetric titration of zinc(II) by EDTA with xylenol orange added as indicator to the reagent stream *R* (1×10^{-3} *M* EDTA in 0.1 *M* acetate buffer, pH 4.7), using spectrophotometric detection at 575 nm. The samples (C^0 = 0.004, 0.005, 0.006, 0.008, 0.01, 0.04, 0.06, and 0.08 *M* $ZnSO_4$) were each injected in triplicate, the Δt values being read off at the Eq level, that is, the most dilute samples did not give rise to any Δt signal.

tion of Zn(II) with EDTA. The composition of the reagent stream was 1×10^{-3} *M* EDTA in 0.1 *M* acetate buffer (pH 4.7), containing 0.5% (v/v) of an aqueous xylenol orange indicator stock solution [0.5% (*w*/v)]. The carrier and reagent streams were each pumped at a rate of 0.75 mL/min, the combined streams being monitored by a spectrophotometer tuned to 575 nm, so that the recorder baseline was at zero balance when the yellow stream was passing through the flow cell. The equivalence level was selected approximately halfway between the baseline and the top of the highest signals, that is, the indicator transition was monitored when about 50% of the indicator was bound by zinc as the red complex. The figure shows the readouts obtained for injection of standards in the range 0.004–0.08 *M* $ZnSO_4$, each sample being injected in triplicate (in order to accommodate all the signals on the graph, the recorder was run at 1 min/cm, yet to improve the precision of the peak width measurements, it is recommended that a higher recorder speed be used in practice). A plot of Δt versus log C^0_{Zn} yielded—except for the two lowest standards—a straight line.

For the redox titration of iodine by thiosulfate with starch as indicator by means of the FIA system shown in Fig. 4.62*a*, the reagent stream consisted of 5×10^{-5} *M* thiosulfate containing 5% by volume of starch solution. The pumping rates of the aqueous carrier solution and of the reagent stream were 0.75 mL/min, the spectrophotometer being tuned to 610 nm. Since the combined streams of carrier and reagent solution were colorless, and hence exhibited zero absorbance, this signal served as the recorder baseline (Fig. 4.63). The equivalence level was set arbitrarily to absorbance 0.6, so that the presence of the blue starch–iodine complex was detected, and the titration curves were recorded in exactly the same manner as in the case of compleximetry. The iodine standards, injected in triplicate, were in the range 2.5×10^{-4}–1.0×10^{-2} *M* KI_3. Again, a plot of Δt versus the logarithm of analyte concentration was linear for those concentrations exceeding the selected equivalence level.

Voltammetric and amperometric titrations of metal ions by strong complexing agents have also been performed [288]. Thus, in Fig. 4.64 is shown the manifold and the readouts obtained for the amperometric titration of Cu(II) by EDTA, the sample solutions of Cu(II) being injected into a carrier stream of ammonia. When the metal ion forms a complex during such titrations, the half-wave potential or peak potential (depending on the technique used) is shifted to more negative values, the magnitude of the shift being dependent on the stability of the complex formed. Thus, with EDTA, if the potential of the indicator (mercury) electrode is set at −300 mV, the copper present as its ammine complex will be monitored, whereas the Cu–EDTA complex, which has a much higher stability con-

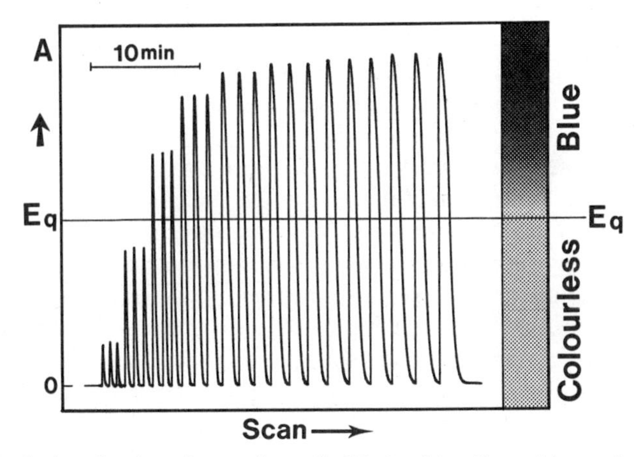

Figure 4.63. Redox titration of potassium triiodide by thiosulfate with starch added as indicator to the reagent stream R (5×10^{-5} M sodium thiosulfate) using the manifold depicted in Fig. 4.62*a* and the conditions stipulated there. All samples (C^0 = 0.25, 0.5, 1.0, 2.0, 4.0, 6.0, 8.0, and 10.0 mM KI_3) were each injected in triplicate and monitored spectrophotometrically at 610 nm. The Δt values were read off at the Eq level.

stant ($\log K_{CuEDTA} = 19.1$) than the ammine complexes, will become electroinactive.

In this manner a series of titration curves *b*, *c*, and *d* have been recorded using a carrier stream containing a selected level of EDTA in addition to ammonia buffer. In these experiments, the flow rate (Q) was kept strictly constant (1 mL/min) and the injected sample volume was 150 μL, while the gradient tube (G) had a volume of 300 μL. The copper(II) nitrate concentrations were 0.01 (*a*), 0.0075 (*b*), 0.005 (*c*), and 0.0025 (*d*) M. The carrier stream for the different experiments were as follows: (*a*) 0.1 M NH_4OH/0.1 M NH_4NO_3 buffer alone; (*b*) 0.001 M EDTA in the same buffer; (*c*) 0.0025 M EDTA in the same buffer; and (*d*) 0.005 M EDTA in the same buffer. While experiment *a* is a direct measurement and not a titration, because all copper is sensed by the electrode in the absence of EDTA, experiments *b*, *c*, and *d* clearly show the increasing influence of a particular concentration of EDTA as the copper(II) concentration decreases. As in all amperometric titrations, the end point is reached when the limiting current decreases to the level of the background current. Therefore, the time span Δt between the two endpoints of equivalence, which yields the analytical readout, is in this case located at the base of the recorded peak; however, in order to avoid uncertainty of readout caused by tailing at the baseline level, the time spans were read out at a slightly elevated and arbitrarily chosen level. Conforming to Eq. (4.6), it is was found that the value of the time span, as indicated by arrows on

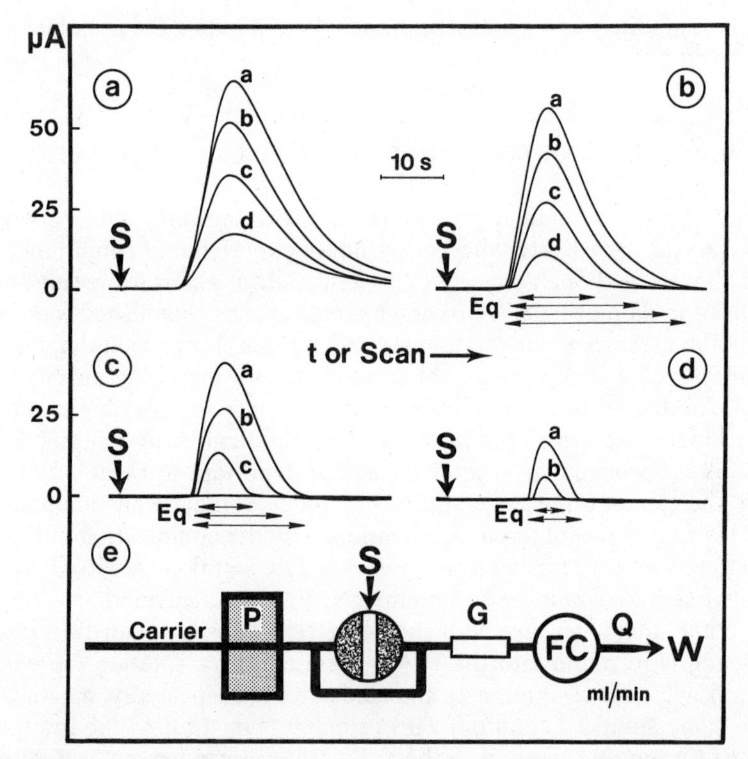

Figure 4.64. Flow-injection measurement (*a*) and flow-injection amperometric titrations (*b*, *c*, and *d*) of Cu(II) with EDTA. Executed at a potential of -300 mV (against a saturated Ag/AgCl reference electrode), the volume of the injected Cu(II) solution was in all cases 150 μL, while the concentrations were (*a*) 1×10^{-2} *M*, (*b*) 7.5×10^{-3} *M*, (*c*) 5×10^{-3} *M*, and (*d*) 2.5×10^{-3} *M*. The titrant carrier stream solution in the three experiments was (*b*) 1×10^{-3} *M* EDTA in 0.1 *M* NH_4OH/0.1 *M* NH_4NO_3 buffer, (*c*) 2.5×10^{-3} *M* EDTA in the same buffer, and (*d*) 5×10^{-3} *M* EDTA. In the manifold (*e*), *G* is the gradient tube.

Fig. 4.64, *b*, *c*, and *d*, indeed increased linearly with the logarithm of the concentration of analyte.

Referring to Figs. 4.62 and 4.63, it is evident that going from the solutions of lowest to those of the highest concentrations, first an increase in peak height—linearly related to the concentration of the injected sample solution—is observed, and later when the peak heights due to "saturation" have attained a constant level, the peak widths will increase as a (logarithmic) function of the concentration of analyte, because the volumetric method of titration differs from direct measurements by having, as the base for the analytical readout, the consumption of a reagent added into the reaction vessel until the equivalent point has been reached. However, as discussed in Section 2.4.4, measurement of peak widths may, in fact, be exploited as a general approach for extending the dynamic range of a given procedure.

4.10 FIA SCANNING METHODS AND SIMPLEX OPTIMIZATION

When adapting a manual procedure to a FIA system, it is commonly necessary to investigate whether optimum experimental conditions have been chosen. Although one would, as a rule, always start from well-known equilibrium data or from well-known conditions as established for a manual method, these might serve merely as a guide, because at the dyanmic conditions of a FIA system, the kinetic factors may play an important role. Whether trying to eliminate their influence, or possibly exploiting these kinetic factors as the basis for the analytical assay via kinetic discrimination procedures, it is nevertheless important to be able to investigate the execution of the actual assay under well-controlled conditions. Therefore, it is useful to change continuously the composition of the carrier stream while exposing it to a given analyte solution. A natural avenue to this exercise would be to employ the FIA gradient methods (Section 2.4), combining it perhaps with chemometric analysis, yet an even simpler approach is to pump into the FIA system a carrier solution the composition of which is continuously changed and then repeatedly injecting into it the same standard solution. This principle, constituting the first primitive *FIA scanning* method, is best illustrated by referring to Fig. 4.65, depicting the manifold and the readout obtained during the study of solvent extraction of metal dithizonates [90], where the pH of the carrier stream was continuously altered by means of a potentiostatic control. Thus by injecting a number of samples of identical analyte content [0.1 mM $Pb(NO_3)_2$ in the example shown in Fig. 4.65], and by recording on the ordinate, the absorbance of the chloroform-extracted species and simultaneously registering the pH of the analyte solution on the abscissa, one could, within less than half an hour, obtain the pH profile of the extraction procedure. This would otherwise require at least 10 times more chemicals and time to investigate by a manual method.

The catalytic activity of enzymes, the influence of pH on the response of an ion-selective electrode, and numerous other investigations can be performed in a similar manner, and conceivably faster, as they do not involve mixing and separation of immiscible phases [79]. As in all other FIA methods, in flow-injection scanning one does not measure the true equilibrium conditions, because the sample residence time in the system is so short, and because the solutions are not entirely mixed, but the measurement correctly reflects the conditions at the dynamic situation

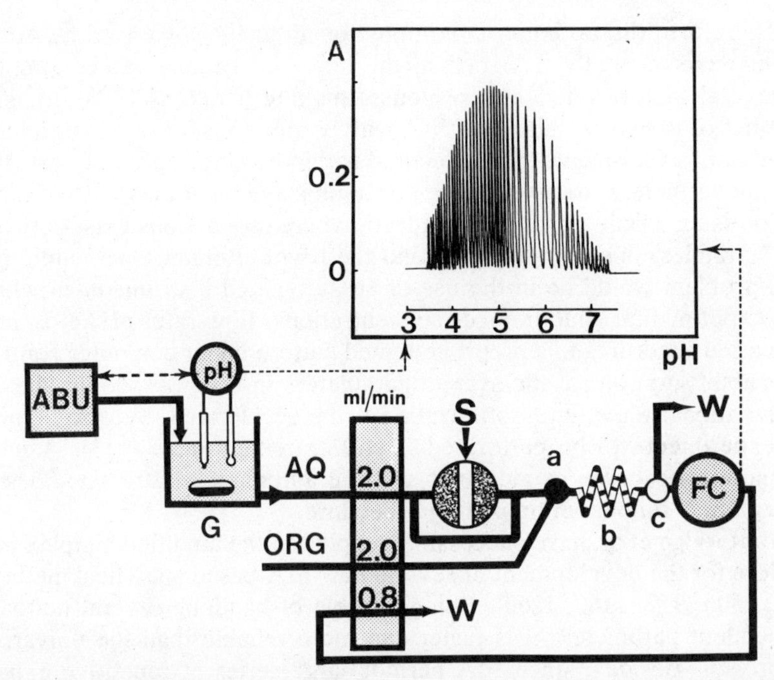

Figure 4.65. Simple FIA scanning method, as illustrated by the solvent extraction of lead by 0.1 m*M* dithizone solution in carbon tetrachloride, measured in a spectrophotometric flow cell at 520 nm. The pH in the gradient vessel *G*, originally containing 1 m*M* nitric acid, is gradually changed via a potentiostatically controlled pH electrode by adding 0.1 *M* NaOH from an autoburette (*ABU*). Thus an aqueous carrier stream (*AQ*) of gradually changing pH is pumped to the sampling valve *S*, where 30 μL of 0.1 m*M* lead nitrate solution is injected at regular intervals. After extraction with organic phase *ORG* and phase separation (point *c*) the absorbance *A* of the extracted lead dithizonate is continuously measured in the flow cell *FC*. The pH measured by the glass electrode and the absorbancy measured by the spectrophotometer are simultaneously fed to an *x-y* recorder, the output of which is shown at the top of the figure. It takes less than 25 min to make the 45 injections shown and to investigate the pH range between 3.5 and 7.5.

valid for that particular FIA manifold used, and therefore yields the optimum conditions for performing the automated determination.

FIA scanning, although easy to perform, is, admittedly, a rather inexpert way of optimizing a FIA method. Yet our present knowledge does not permit us to predict the optimum reaction conditions from equilibrium data. The difficulty is that the unknown factors involve not only the kinetics of chemical equilibria, but also the incompleteness of mixing, as a result of which the concentration and pH of the input solutions do *not* correspond to those under which the reaction takes place within the sample zone on its way toward the detector.

Therefore, the optimum conditions for an analysis have to be established experimentally. This task might, however, become rather complex if several factors affect the response, making it necessary to adjust a number of variables. Although FIA with its rapid response allows a large number of experiments to be executed within a limited space of time, this can, nevertheless, be a very time-consuming avenue if univariate optimization is undertaken manually; indeed, where interactions exist between the variables, one is unlikely to find the true optimum. One solution to this problem would lie in the use of an automated instrument, in which the experimental conditions of concentrations, flow rate, pH, etc., may be varied according to a preprogrammed pattern under computer control. The ability to control the system parameters in this way would allow a conventional univariate optimization to be accelerated, yet the number of experiments to be performed is still extremely large [413]. A more economical scheme would be to use the automatic control facilities to carry out a simplex optimization procedure.

Betteridge et al. have successfully exploited the modified-simplex procedure for the development of several new FIA-based analytical methods [413, 506, 972, 1166, 1260]. Being capable of handling several mutually dependent parameters it is faster and more reliable than the univariate approach. Besides, since FIA permits large series of repetitive experiments to be performed within a limited period of time, a number of different start simplexes can be used, thereby ensuring with high probability that a true optimum rather than a local one is located. However, while the modified-simplex procedure will identify the optimum reaction conditions, it will not per se implicate the significance of the individual factors and their interaction. Furthermore, if this method is used a number of times, the simplex applied might become severely deformed compared to a regular simplex, and, therefore, there is a risk that the simplex eventually will be so deformed that is becomes unusable. Therefore, modifications to the modified-simplex procedure have been devised (super- and super-super-modified simplex procedures [4.14]) that, yielding better estimates of the optimum by a reduced number of experiments, inevitably are characterized by increasing complexity. Therefore these algorithms require extensive computations that can only be resolved by the help of computers. Software for these tasks are, however, now commercially available [4.15].

4.11 REVERSED FIA AND FLOW REVERSAL FIA

In reversed FIA, which was first suggested by Johnson and Petty [279], the roles of "sample" and "reagent" are reversed, the method being based on injection of reagent(s) into a carrier stream of sample. Originally prompted specifically for the assay of phosphate by the molybdenum blue method in seawater samples aspirated continuously into an analyzer situated aboard an oceanographic vessel, this approach has proven ideally suited in cases where sample material is abundant, while reagents ought to be spared, injection of reagents improving the overall economy of the assay. Besides, in many applications the reversed FIA mode increases the sensitivity of measurement, partly by widening the dynamic concentration range attainable as compared to that achieved by conventional FIA, and partly by providing a lower range of concentrations to be reached, a feature that is of ultimate importance in microtrace analysis, thus compensating for the drawback of the comparatively higher consumption of sample material.

Reversed FIA has particularly been exploited by Valcárcel and coworkers in Spain, who have designed a number of ingenious systems [529, 540, 640, 757, 968, 1021, 1044]. Thus taking advantage of chemistries leading to the formation of products of approximately the same color, they have succeeded in performing multielement assay on wastewater samples by using a series of injection valves, each of which successively inserted into the carrier stream of sample solution suitable selective reagents for the individual analytes, using as a detection device a single colorimetric detector tuned to one wavelength (Fig. 4.66) [640, 1044]. Obviously, by employing a detector that is capable of monitoring several wavelengths, such as a diode array arrangement, this approach might readily be expanded. Alternatively, the sample stream might be split into different lines and subjected to individual treatment by appropriate reagents. Hence, the use of reversed FIA is a very useful tool in continuous monitoring and in process control.

Using a single detection device, the same Spanish group has employed reversed FIA for speciation studies [968, 1021]. Thus, in Figs. 4.67*a* and 4.67*b* are shown two configurations utilized for the assay of Cr(III) and Cr(VI), the quantification of both species ultimately being based on the reaction between Cr(VI) and 1,5-diphenylcarbazide (1,5-DPC). In the

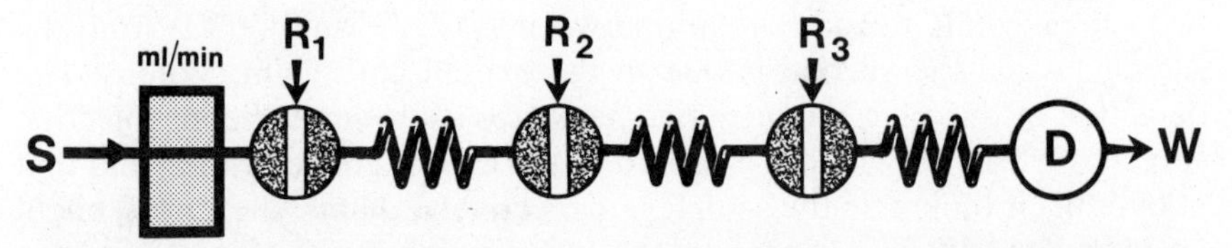

Figure 4.66. Manifold used for reversed FIA and sequential injection of reagents. The sample solution is pumped as carrier stream, selective reagents (R_1, R_2, and R_3) being injected by different valves for the different analytes, permitting time-based detection of the individual analytes by a single detector.

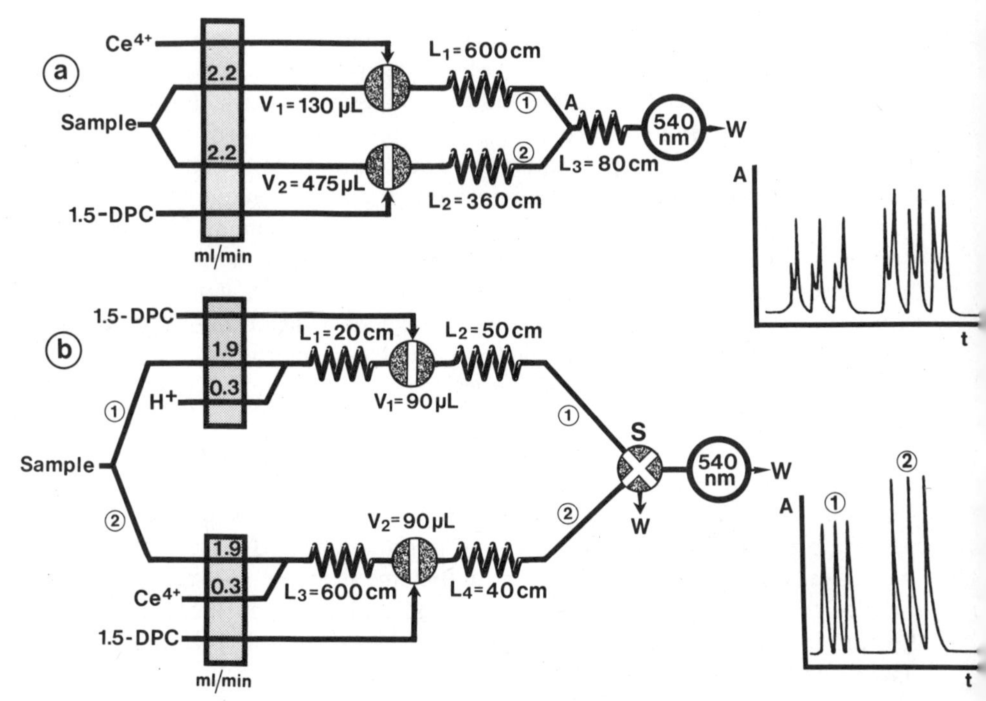

Figure 4.67. Manifolds designed for speciation analysis of Cr(VI) and Cr(III) by reversed FIA, both species being determined spectrophotometrically by a single detector. (*a*) Asynchronous merging zones mode for simultaneous determination of both species. (*b*) Sequential method for this speciation by use of a selecting valve for determination of Cr(VI) (channel 1) or overall chromium (channel 2). (From Ref. 968, by permission of Springer Verlag.)

manifold depicted in Fig. 4.67*a* the aspirated sample solution is split into two branches: in channel 1 Cr(III) is oxidized by injected Ce(IV) to Cr(VI), while in channel 2, 1,5-DCP is injected into the sample solution. Reactor lengths and injection volumes are optimized in such a manner that the confluence of the injected plugs at point *A* is asymmetric, the plug traveling along channel 1 merging with the tail of the plug propelled through channel 2. A plug is formed in reactor L_3 with two reaction zones. The first corresponds to the reaction between 1,5-DCP and Cr(VI) from the sample, while the second is due to the overall chromium content. The detection of this plug provides two peaks for each simultaneous injection (Fig. 4.67*a*, right), which are related to the Cr(VI) concentration and that of overall chromium in the samples, respectively, that is, the Cr(III) might be calculated by difference. In the manifold shown in Fig. 4.67*b*, the sample containing solution of Cr(III) and Cr(VI) is again split into two channels feeding two subconfigurations. However, in subconfiguration 1

Cr(VI) is determined by injecting the reagent (1,5-DCP) after merging with a stream of sulfuric acid, which ensures an acidic pH for the sample. In subconfiguration 2 the overall chromium content is determined, owing to the confluence of the sample channel with a stream of Ce(IV) in acidic medium prior to the injection of 1,5-DCP. Both streams are directed toward valve *S* located in front of the detector, which is a directional valve able to be open either to channel 1 or 2, while the content of the other channel simultaneously is directed to waste *W*. In this manner Cr(VI) and overall chromium in the samples can be determined sequentially depending on the position of valve *S* (Fig. 4.67*b*, right). Using this approach, a much higher sampling frequency than that attainable with the previously described system can be obtained.

Most recently, an approach termed *flow reversal FIA* has been proposed independently in this laboratory [4.16] and by Betteridge et al. [4.17], based on a strategy by which a sample plug in a FIA system might be subjected to pass through any of a wide range of effective coil lengths between the points of injection and detection, without modification of the reaction manifold, this being accomplished by reversing the flow in such a manner that either the sample plug is first directed to a dispersion coil and subsequently via flow reversal to a detector, or the sample plug, or parts of it, repeatedly is exposed to the detector by repeated flow reversals. Thus, by exploiting this feature one might treat the reaction coil length as a continuous variable in much the same way as has been possible for carrier stream composition, flow rate, and sample size. Flow reversal FIA thus makes more thorough automated optimizations a possibility. By use of several flow reversals, and hence sweeping the sample plug past the detector several times, different pH gradient profiles might for instance be created and exploited from a single sample injection, making multidetection or speciation feasible. Besides, being a significant departure from normal FIA practice, flow reversals should prove a useful tool to those who wish automatically to develop and optimize FIA methods or who need more verstile ways of controlling dispersion.

4.12 INTEGRATED MICROCONDUITS

When reviewing the development of FIA over the past 10 years, several patterns emerge, of which the trends toward miniaturization, exploitation of the possibilities inherent in the concept of controlled dispersion, and application and introduction of new flowthrough detectors are the most striking. While the first FIA system [1] used more than 10 mL of reagent and 0.5 mL of sample for a single measurement, contemporary designs require 10–20 times smaller volumes. This process of miniaturizing the FIA system has recently taken a great step forward with the introduction of integrated microconduits [608], which permit the integration of traditional manifold components, such as coils, connectors, injection valve, dialysis and gas-diffusion units as well as columns containing ion-exchangers or immobilized enzymes, into a single unit approximately of the size of a credit card. The purpose of miniaturization and integration has been twofold: (1) by reducing the overall volume of the FIA channel system, the economy of sample and reagent consumption is improved; and (2) by integrating the detector into the FIA channel itself, the detection can be executed exactly within that section of the channel where the dispersion and other conditions of an individual assay are optimal. This is in contrast to the conventional constructions of FIA manifolds, where the physical size and geometry of an available, external flowthrough detector limits the potential of the method, giving an undesired additional dispersion and, consequently, a loss of signal and sampling frequency. Although the conceptual idea of the integrated microconduits is simple, its practical implementation turned out to be a complex task. Yet, based on the theory of similarity (Section 2.6), it was possible to identify the scaling factors governing the characteristics of these circuits thereby optimizing their design.

The microconduits resulting from these investigations consist of plastic blocks (normally 70 × 45 × 10 mm in dimensions) into which grooves, forming the flow channel pattern, may be imprinted or engraved, which when closed, either by a flat plate with the aid of a pressure-sensitive glue or by means of an elastic base plate of a material such as silicone rubber, yield conduits of semicircular cross section, typically of an internal area of 0.8 mm^2. With these bonding techniques it is easy to reopen channels, for instance, in order to inspect, clean, or renew electrode surfaces, replace optical windows in flow cells (should they have become translucent), remove or renew incorporated membranes, ion exchangers or immobilized enzymes, or even to reroute channels in a previously imprinted channel pattern. Introduction of liquids into the channels and their withdrawal are effected through perpendicular holes drilled at appropriate positions and furnished with externally communicating tubes. The rigidity of such a structure not only ensures perfect repeatability of the dispersion of the sample zone, but the method of fabrication, engraving, or imprinting of the channels, and lamination of the layers secures perfect uniformity of units even produced on a large scale. The materials—plastics such as PVC (usually transparent, but for special applications opaque)—are inexpensive and can handle the same solutions as the PVC pump tubes used in the external pumps. If necessary, two or several microconduits can be interconnected, either in a layered structure or via short intercommunicating tubes (cf. Fig. 4.70).

In Fig. 4.68 is shown a microconduit incorporating two potentiometric pH electrodes and a common reference electrode, the introduction of sample solution being executed by means of an exteriorly placed injection port. The measurement of pH requires a system with limited dispersion coefficient, and no chemical reaction is needed in the flow channel. Consequently, a short residence time was chosen, and the two pH-sensitive PVC-based membrane electrodes, containing as electroactive material tri-*n*-dodecylamine [778], were placed in a single-line system and very close to the injection position (cf. Fig. 4.3), the Ag/AgCl wire reference electrode being situated in a side channel and connected to the main channel downstream from the indicator electrodes. The manifold construction is such that the reference solution and thus the liquid junction are renewed

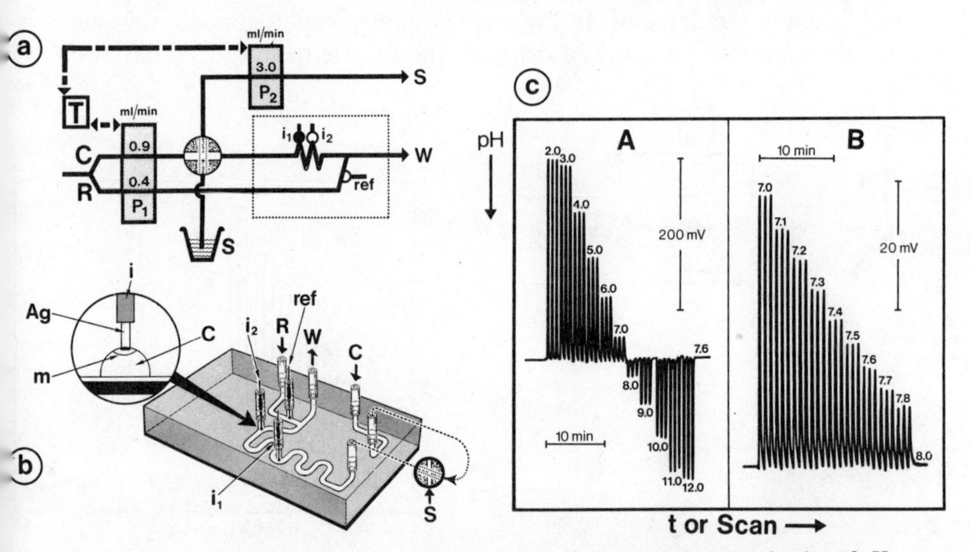

Figure 4.68. (*a*) Manifold and (*b*) microconduit unit for potentiometric determination of pH, the components within the boxed area of the manifold being those contained in the microconduit. Pumps P_1 and P_2 are operated sequentially, controlled by timer *T*, so that sample (volume 30 μL) is first aspirated by P_2 into the injection valve (during which period P_1 is stopped) and subsequently propelled forward by carrier stream *C* (diluted buffer) toward indicator electrode i_1 (or i_2), the construction details of which are shown in the inset (*m* is the pH-sensitive membrane). The reference electrode (ref) is placed in a side channel fed by stream *R*, and the combined streams are ultimately led to waste, *W*. (*c*) pH-response curves obtained with the microconduit system shown to the left. (*A*) A series of pH standards in the range 2.0–12.0; the carrier solution being 0.01 *M* phosphate buffer containing 0.1 *M* NaCl, at pH 7.6. (*B*) Performance of the system in the physiologically important pH range, the carrier solution here being adjusted to pH 8.0.

during each sampling cycle when carrier stream is pumped (by pump P_1) through both channels. This arrangement in turn results in a very high sampling frequency as seen on the recorder output (Fig. 4.68, *A* and *B*), without impairing the reproducibility of measurement, the calibration plot depicted in *A* being obtained by injecting buffers in the pH range 2.0–12.0 into a carrier stream consisting of phosphate buffer of pH 7.6 (RSD ± 0.01 pH), while the run obtained for the narrow pH range 7.0–7.8 (RSD ± 0.002 pH), corresponding to clinical applications, was recorded using a carrier stream of pH 8.0. Microconduits of this type have been used for pH measurements in the process control of a glucose isomerase reactor in a pilot plant [4.18]. Operated over a two-month period they performed satisfactorily without any need for servicing or replacement.

Spectrophotometric measurements require medium dispersion of the sample zone within the carrier stream containing a color-forming reagent, that is, sample and reagent must be mixed to a certain degree before being subjected to the detector. In Fig. 4.69 is shown a microconduit, designed for the colorimetric assay of calcium, the determination being based on

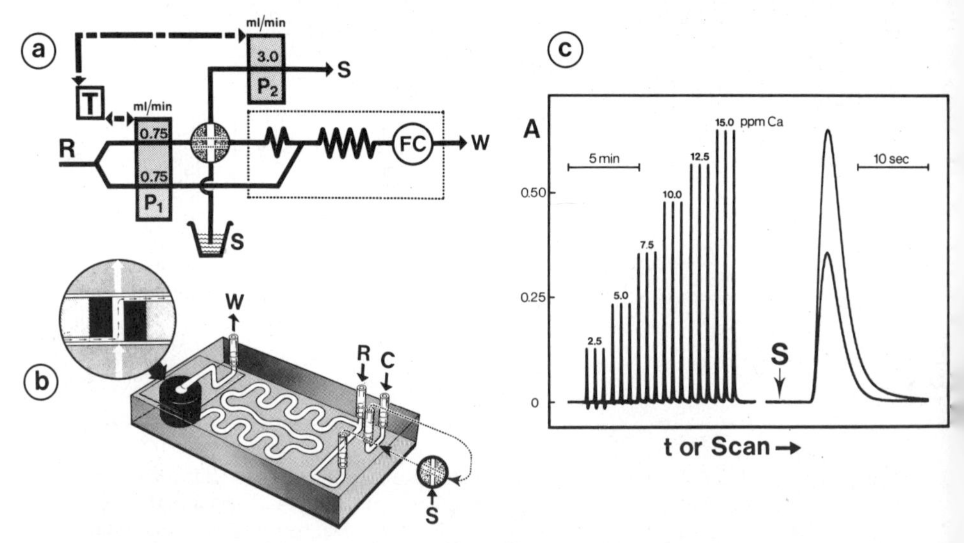

Figure 4.69. (*a*) Manifold and (*b*) microconduit integrated with optical flow cell (*FC*), details of which are shown in the inset. Light is piped via optical fibers. The pumping rates indicated are for the determination of calcium with this manifold. The two pumps are operated sequentially, controlled by the timer *T* (cf. Fig. 4.68). (*c*) Response curves for the determination of calcium obtained with the microconduit shown to the left (injected volume 10 μL), where the reagent stream of alkaline *o*-cresolphthaleine complexone is pumped to both inlets *C* and *R*, the stream being monitored at 580 nm. First is shown a series of standards in the range 2.5–15.0 ppm Ca, followed by the standards containing 7.5 and 15.0 ppm Ca recorded at high chart speed.

formation of the red–violet complex with *o*-cresolphthaleine complexone. Using again an externally placed injection port, this microconduit comprises an integrated optical flowthrough cell, optical fibers being used to interface the microconduit with a conventional spectrophotometer. Since the lengths of the communicating fibers, through which light is carried to and from the flow cell (shown on the inset; path length 10 mm) can be varied almost at will, this design illustrates the flexibility of the microconduit concept in optimizing the chemistry of a particular assay. Furthermore, as seen from the response curves the reproducibility of measurement is excellent (RSD better than 0.3%), being superior to that obtained in FIA systems built from "flexible" tubings.

In all the previous examples, the sample zone was injected into the microconduit channel from an external sample valve, yet ultimately this function, of course, ought to be integrated into the microconduit. Miniaturization of a rotary valve is one possibility (see below), while another is the use of the hydrodynamic injection principle ([338; cf. also Section 5.1.3], which involves a combination of hydrodynamic and hydrostatic forces to aspirate, meter, and inject the sample solution in the form of a well defined plug into the carrier stream.

To test this approach for sample injection and to demonstrate its flexibility, a manifold operated by two peristaltic pumps, controlled by a timer T (Fig. 4.70), was constructed; two microconduits were used, the basic unit furnished with a flow cell, and the injection unit with an imprinted volumetric channel and two sample cups S_1 and S_2. The volumetric channel (here having a volume of 17 μL) is shown in the manifold as the coil situated between points a and b, and is served by pump P_2. This channel forms part of two circuits: an open circuit starting in S_1 and leading through P_2 to S, and a closed circuit in which the inlets (R and C) and the outlet (to W) are hydrodynamically balanced so that the combined volumetric delivery of liquids to R and C equals the outflow to W. Thus when pump P_1 operates and P_2 does not, there is no movement of liquid between b and S_1 in either direction. The sampling cycle starts with P_1 in the stop position while P_2 is operating, as indicated by the solid lines in the upper corner of Fig. 4.71. During that delay period (DE1) the sample loop is washed and filled by sample solution from cup S_1. Next, pump P_2 is stopped and P_1 is activated, thus injecting the sample zone, located between a and b, into the coil and through the analytical channel. By using a colorless carrier stream (1×10^{-2} M sodium tetraborate) and bromothymol blue as "sample" solution, the smallest peak in Fig. 4.71 was recorded. [The reproducibility of injection is demonstrated on the right-hand side of the figure, which shows traces from repeated (seven) injections of the same concentration of dye.] In this mode of operation, the injected sample volume S_v corresponds exactly to the physical volume of the imprinted channel between a and b and therefore this approach may be termed a hydrostatic injection.

If the timing cycles of pumps P_1 and P_2 are allowed to overlap, the injected volume of sample can be increased at will. This can be understood by referring to Fig. 4.70a and Fig. 4.71 (top right). In this mode, the cycle starts with a period during which P_1 is stopped (DE1) and P_2 is activated, filling the sample loop, but instead of the stop–go pattern being reversed after the DE1 period, both pumps are operated simultaneously for an overlap period (marked P20), during which pure sample solution is directed at point b toward the detector at a volumetric rate corresponding to that of the carrier stream C, while stream C itself is directed at point a toward S (along with an amount of sample corresponding to the volumetric pumping rate of pump P_2 minus that corresponding to C). When pump P_2 is stopped, carrier stream C is redirected through the sample loop and carries its content of sample towards the detector. Thus, the

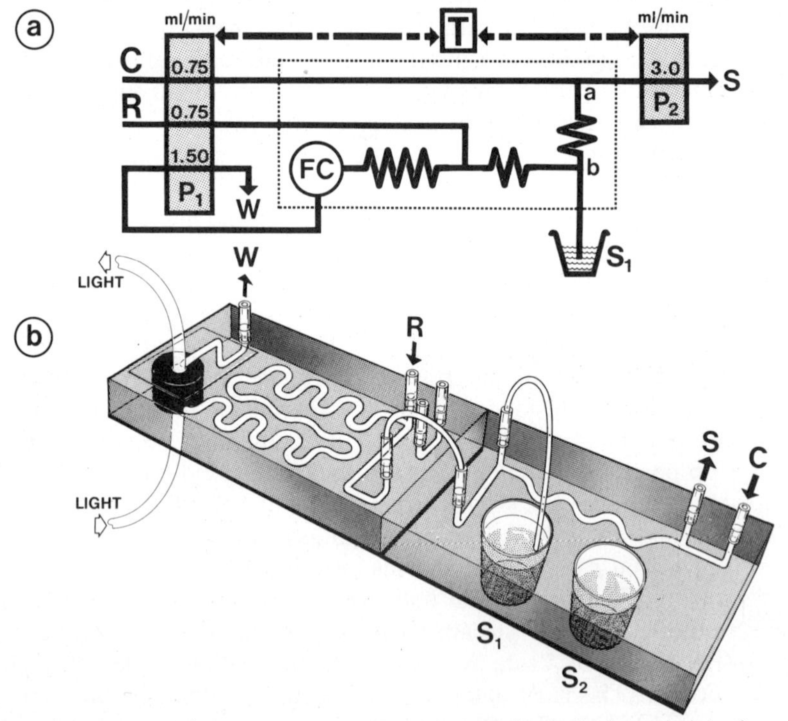

Figure 4.70. (a) Manifold and (b) microconduit incorporating hydrodynamic injection and integrated optical flow cell. For details of operation, see text.

amount of sample introduced is that aspirated directly during the overlap period, P20, *plus* the volume of the sample loop *a* to *b*; by increasing the overlap time, increasingly larger sample volumes can be injected, until the steady-state plateau is reached (Fig. 4.71). This is a true hydrodynamic injection as the sample zone is formed by a combination of hydrostatic and hydrodynamic forces exerted on the column of liquids by the two peristaltic pumps. As seen from the second series of seven injections on the right-hand side of Fig. 4.71, the reproducibility of the hydrodynamic injection (for an overlap time P20 of 3 s, corresponding to a total injected volume of 55 μL) is as good as that of the hydrostatically injected.

Recently, a novel type of integrated optical detectors based on the interaction of radiation with a surface situated in a flowing stream was introduced [848]. Exploiting absorbance, reflectance, and fluorescence of visible and/or UV light as it changes due to chemical reactions taking place at or in close proximity of a surface surrounded by a flowing stream,

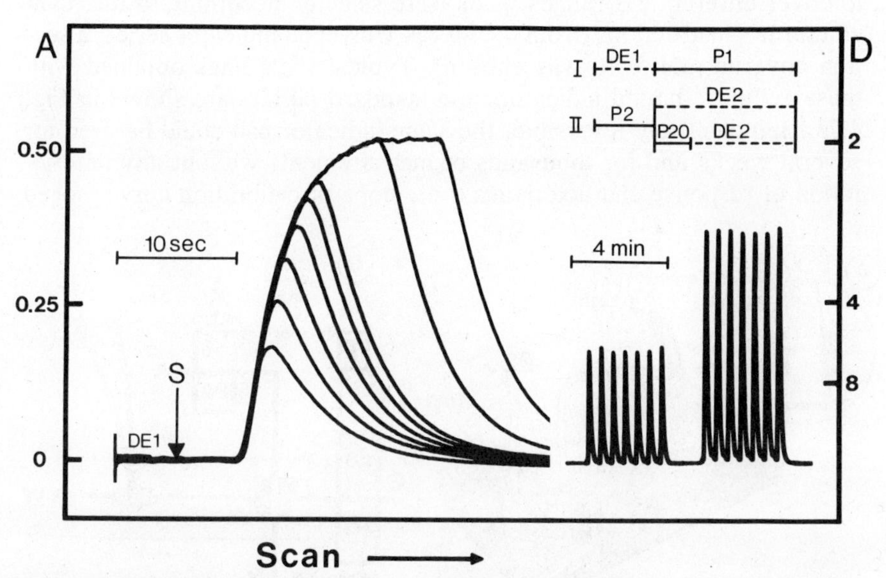

Figure 4.71. Hydrodynamic injection of increasing sample volumes of a dye solution (bromothymol blue) into a carrier stream (1×10^{-2} *M* sodium tetraborate), using the manifold depicted in Fig. 4.70, monitoring the color at 600 nm. To the left is shown a series of curves with increasing overlap times (P20) of the two pumps (0, 1, 2, 3, 4, 5, 10, and 15 s). Thus the smallest peak corresponds to an injected volume equal to the volume of conduit *a-b* in Fig. 4.70 (17 μL); for increasing overlap times, and, therefore, increasing injected sample volumes, progressively higher peaks are recorded until the steady-state plateau is reached. To the right is shown the reproducibility of measurement for signals corresponding to P20 values of 0 and 3 s.

such *optosensors* can be miniaturized and therefore they are ideally suited to be embodied into FIA microconduits. Thus in Fig. 4.72 is depicted a system used for the determination of pH, employing an immobilized pH indicator situated at the end of a bifurcated optical fiber. The sample is injected by means of the valve, integrated into the FIA microconduit, and then passed to the tip of the fiber at which position is situated a fibrous cellulose pad on the surface of which an acido-basic indicator is covalently bound. When the pH of the carrier stream due to the presence of the sample is shifted, the color of the indicator is altered, the ensuing change being sensed by the optical fiber that is bifurcated at the remote end to accept the incident light from an external light source and to accommodate a conventional spectrophotometer to monitor the light reflected from the fibrous pad. Using indicator pads cut from commercially available Merck nonbleeding indicator strips, which exhibit color changes from yellow (acid form) to blue (alkaline form), the reflectance was measured at 610 nm, that is, monitoring the alkaline blue form of the indicators. In order to cover different pH ranges, pads were selected according to the manufacturer's instructions (from the Merck Universal indicator series, a section covering pH 5–10 was chosen). Typical recordings obtained with these cellulose-bound indicators and standard buffers are shown in Fig. 4.73. Being entirely reversible, the same indicator pad could be used for several weeks and for thousands of measurements without any deterioration of response characteristics (i.e., slope of calibration curve, speed

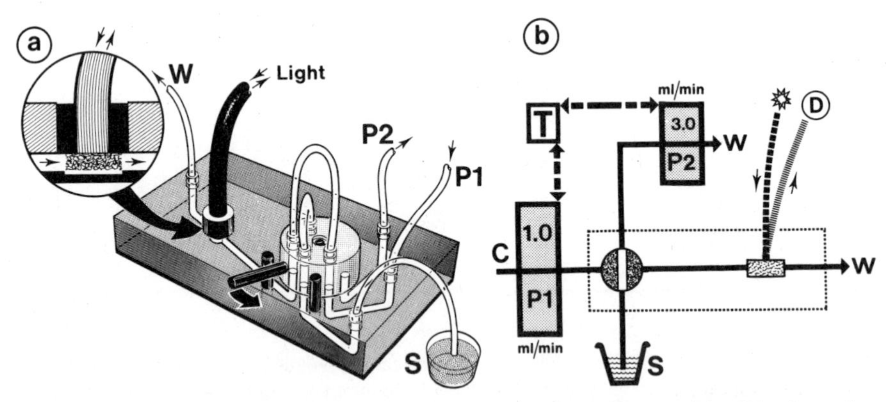

Figure 4.72. (*a*) Integrated micoconduit for measurement of pH comprising injection valve and optosensor. *S*, sample solution; *P*1 and *P*2, tubes leading to the peristaltic pumps [cf. (*b*)]; *W*, waste. The inset shows a cross section of the optosensor. (*b*) FIA manifold for pH measurement comprising two peristaltic pumps (*P*1 and *P*2), a timer (*T*) and the microconduit shown in (*a*) (embodying the components within the boxed area). *C* is the carrier stream. The flowthrough cell communicates with a light source and a spectrophotometer (*D*) through optical fibers.

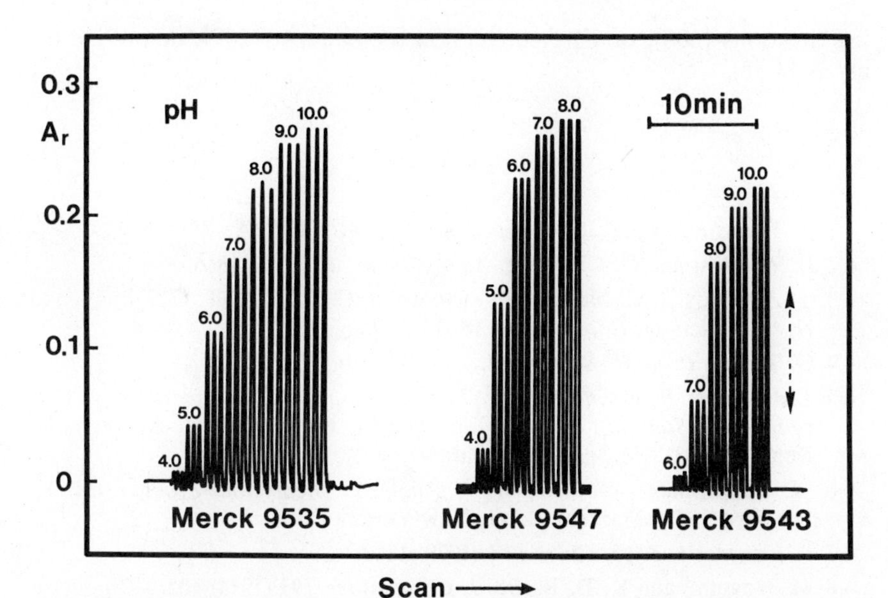

Figure 4.73. The pH responses of the optosensor furnished with the following immobilized indicators: Universal (Merck 9535), Spezial (Merck 9547), and Spezial (Merck 9543), recorded by injecting standard buffer solutions into a 5×10^{-4} *M* HCl carrier stream and monitoring the reflectance as increased absorbance (A_r) at 610 nm. The arrow at the far right indicates the pH range investigated for clinical applications.

of response). Because of the excellent homogeneity of the commercial materials and the reproducible geometry of the pad in the flow cell, a pad could be replaced, or an entirely new microconduit fabricated, with such repeatability that identical calibration curves were obtained with new and old devices [1232].

Various microconduit designs have, in fact, already been mentioned in the foregoing sections. Thus in Fig. 4.16 (Section 4.3) was described a system incorporating an optical flowthrough cell; in Fig. 4.35 (Section 4.6.2) was shown a microconduit including a gas diffusion unit used in conjunction with optosensing; in Fig. 4.40 (Section 4.7.1) was detailed a setup including an ion-exchange column used for preconcentration with ensuing detection by atomic absorption spectrometry; in Fig. 4.47 (Section 4.7.2) was depicted an integrated system comprising an immobilized enzyme reactor utilized in conjunction with determination by chemiluminescence; while in Section 6.8 is described an exercise where the manifold is based on the use of a FIA microconduit system. Other microconduit manifolds have been described in the literature, either comprising potentiometric or optical detection (Table 7.1), and facilities for performing titrations [608], or FIA conversion techniques [1247], or on-line dialysis [4.18].

REFERENCES

4.1 A. B. Crowther and R. S. Large, *Analyst,* **81** (1956) 64.

4.2 P. W. West and G. C. Gaeke, *Anal. Chem.,* **28** (1956) 1816.

4.3 C. A. Burtis, J. C. Mailen, W. F. Johnson, C. D. Scott, T. O. Tiffany, and N. G. Anderson, *Clin. Chem.,* **18** (1972) 753.

4.4 G. Taylor, *Proc. R. Soc. Lond., Ser. A,* **219** (1953) 186.

4.5 T. Guldager Petersen, *FIA Gradient Techniques. Development of a Procedure for Standard Addition* [in Danish]. Chem. Depart. A, Tech. U. Denm., (1983). (M.Sc. Thesis, Part II).

4.6 W. B. Furman, *Continuous Flow Analysis: Theory and Practice* (M. K. Schwartz, ed.), Marcel Dekker, New York, 1976.

4.7 P. Concus, *J. Phys. Chem.,* **74** (1979) 1818.

4.8 M. Novotny and K. D. Bartle, *J. Chromatogr.,* **93** (1974) 405.

4.9 M. G. Garn and C. Nystrup, *Fast Methods for the Determination of Trimethylamine and Dimethylamine in Aqueous Mixtures Using Flow Injection Analysis and Gas Diffusion* [in Danish]. Chem. Depart. A, Tech. U. Denm., (1984). (M.Sc. Thesis).

4.10 Z. Zhu and Z. Fang, *Anal. Chim. Acta* (in press).

4.11 E. J. Conway, *Microdiffusion Analysis and Volumetric Errors.* Crosby Lockwood, London, 1962.

4.12 J. W. Ross, J. H. Riseman, and J. Kreuger, Potentiometric Gas Sensing Elecrodes. In *Selective Ion-Sensitive Electrodes.* Proceedings of IUPAC Meeting, Cardiff, 1973. (G. J. Moody, ed.). IUPAC, Butterworths, London, 1973.

4.13 Bo Petersson, (priv. comm.)

4.14 H. Baekmark Andersen, *Optimization of FIA Procedures by Means of Statistical Methods* [in Danish]. Chem. Depart. A, Tech. U. Denm., (1986). (M.Sc. Thesis).

4.15 P. F. A. van der Wiel, B. G. M. Vandeginste, and G. Kateman, *Chemometrical Optimization by Simplex (COPS).* Elsevier Scientific Software, The Netherlands, (1986).

4.16 J. Růžička and E. H. Hansen, Eur. Patent. Appl. No. 86850367.3

4.17 D. Betteridge, P. B. Oates, and A. P. Wade, *Anal. Chem.* (in press).

4.18 M. de Bang and J. F. Gram, *Production of High Fructose Syrup.* Dept. Chem. Eng., Tech. U. Denm., (1986).

CHAPTER

5

COMPONENTS OF A FIA APPARATUS

Small is beautiful . . .

ERNST FRIEDRICH SCHUMACHER

Anything can be made smaller, never mind physics,
everything will be more expensive, never mind
common sense.

T. HIRSHFELD

In the first edition of this book the individual components, their manufacture and combination into a FIA apparatus were discussed in great detail, because at the time of publication the production of commercially available equipment was still in its infancy, the enterprising users of FIA being forced to make their own systems with whatever was at hand. In those days sometimes incompatible components were assembled into FIA systems, in spite of which surprisingly good results were obtained, this being a tribute to the robustness and versatility of the technique. As the optimization progressed, admittedly mostly by trial and error, more reliable and suitable FIA components were identified. Thus, Chapter 5 in the previous edition of this book reflected the state of the art of its time as well as the very limited availability of commercial instruments. Since then several commercial apparatuses have been marketed (cf. Section 5.5) and dedicated individual components needed for constructing FIA equipment can be purchased separately; therefore, there is no need to continue a detailed description. Consequently, this chapter is more devoted to general concepts with the emphasis on injection means and on propelling devices (pumps). At the end of the chapter the commercially available FIA instruments are briefly discussed, a summary of their concepts being presented rather than their technical parameters.

5.1 REVIEW OF INJECTION TECHNIQUES

Whether injecting a sample into a carrier stream of reagent or a reagent into a carrier stream of sample (reversed FIA) FIA is based on the creation of a zone, which is the starting point of each measuring cycle in terms of concentration, geometric form, and time. Hence, the successful operation of any FIA system requires injection of a well-defined zone into the analyzer channel, where the zone disperses in a controlled manner on its way toward and through the detector. Conceptually, the injection means and devices designed for this purpose can be divided into two categories: (1) volume-based injection; and (2) time-based injection, or a combination thereof. In category (1) the injection is based on the physical entrapment of sample solution into a geometrically well-defined volumetric cavity and subsequent transfer (injection) of the thus formed sample zone into the nonsegmented carrier stream. Category (2) is based on aspiration of sample solution at a constant flow rate for a fixed period of time into a well-defined section of a flowthrough channel, from where the metered sample solution is injected into the carrier stream by a combination of hydrostatic and hydrodynamic forces. The concepts and technical means of the two types of injection is described in more detail in the following.

5.1.1 Volume-Based Injection Procedures

The oldest volumetric-based FIA injection technique utilized a syringe furnished with a hypodermic needle, which by piercing the wall of the carrier stream tube allowed the introduction of the sample material [1]. This approach was later improved and simplified by combining the syringe with a flap-valve-type injection port [3]. These injection devices today are, of course, merely of historical interest, since they rely heavily on experimental skill, because the *volume* of the sample material introduced is adjusted by the position of the plunger, and because the *form* of the injected plug depends on the speed with which the plunger is being depressed. Besides, when disposable plastic syringes were used, hardly less than 200 μL could be reproducibly metered, and this was the main reason why early FIA systems were designed with large coils and high pumping rates. As the scaling down of the FIA system gained momentum, this type of injection was quickly, and inevitably, abandoned, being replaced by slider valves [208, 227] and especially by rotary valves (with four, six, or eight ports).

Although the rotary valves used in FIA might be designed and mechanically manufactured differently, the function of this type of valve is essentially that schematically shown in Fig. 5.1*a*. The valve has two po-

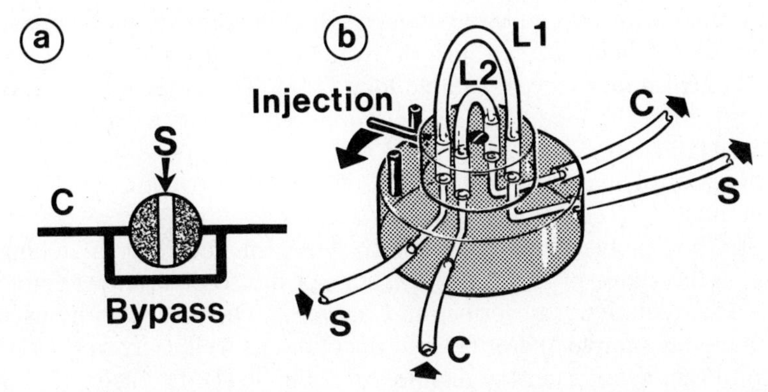

Figure 5.1. (*a*) Schematic drawing of FIA rotary injection valve. Sample (*S*) is introduced into the sample cavity the volume of which meters the sample solution. While in the sampling position, the carrier stream (*C*) is shunted through an external by-pass. When the valve is turned, the sample volume is swept by the carrier stream into the system. (*b*) Four-port rotary injection valve consisting of a rotor furnished with two loops (*L*1 serving as the sample cavity and *L*2 as the by-pass) and a stator accommodating matching pairs of ports. When the lever is turned from one position to the other, guided by the two stops fixed on the stator, the sample loop or the by-pass is made part of the carrier stream conduit. By changing the physical dimensions of the sample loop the injected volume of sample can readily be altered.

sitions, the sampling and injection position. During the first stage sample is filled into the volumetric cavity while the carrier stream is shunted via a by-pass in order to prevent build-up of back pressure. When the cavity is filled with sample—either effected manually by a syringe or automatically by aspiration of a pump—the valve is rapidly turned to the inject position and left there until the sample is completely swept by the carrier stream out of the volumetric cavity.* In Fig. 5.1*b* is depicted a typical injection valve consisting of a rotor and a stator. The rotor, preferably made of Teflon, has two pairs of holes, accommodating the sample loop (*L*1) and the internal by-pass loop (*L*2), respectively, drilled to match two pairs of ports made in the stator (which conveniently consists of clear PVC). This four-port valve has the advantage that the externally placed sample loop *L*1 (typically 0.5-mm-ID tubing) is interchangeable so that the sample volume can be altered at will. The minimum injected sample volume is, of course, restricted by the physical distance between the inlet and outlet holes drilled into the rotor. Alternatively, the valve might be

* Originally, it was believed that an external by-pass of higher hydrodynamic resistance than that of the sample cavity was needed to avoid pressure build-up *during* the manual switching of the valve between the sampling and inject positions. It was found, however, that the internal short by-pass loop *L*2 in Fig. 5.1*b* is sufficient provided that the switching of the valve is executed rapidly.

constructed with two stators between which a rotor is sandwiched, the sample cavity being drilled into the rotor itself, or contained in a rotor furnished with an external sample loop [153]. The latter approach is the one utilized in the ingenious *commutators* developed by Bergamin et al. in Brazil [1060; cf. Fig. 4.54]. Valves and commutators with a larger number of pairs of matching ports allow injection of two or several zones simultaneously (see below).

A six-port valve (Fig. 5.2) with six holes on the rotor matching six holes on the stator provides a wide range of the configurations employed in the FIA techniques described in Chapter 4. The valve can be used for simultaneous sample and/or reagent injection, as well as to *direct* streams into different sections of the manifold (cf. Fig. 4.41). Erickson et al. have recently reviewed and summarized these options [5.1], which are outlined in the following.

The six-port valve shown in Fig. 5.3 was incorporated into a FIA microconduit (Section 4.12; cf. also Fig. 4.47*b*), constructed by bolting a Telfon disk with six evenly spaced holes on a PVC plate, 1 cm thick. Matching holes are drilled through the plate and the disk, and channels

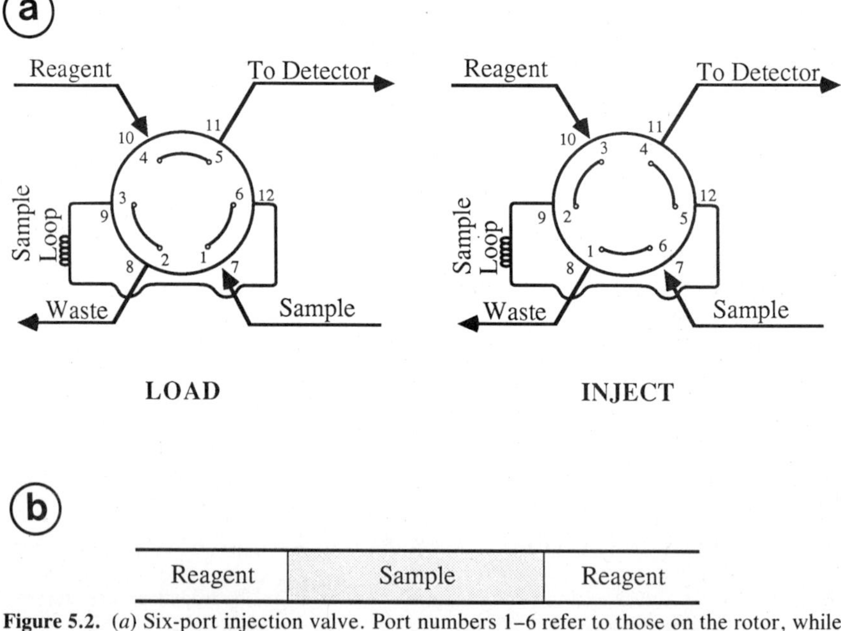

Figure 5.2. (*a*) Six-port injection valve. Port numbers 1–6 refer to those on the rotor, while port numbers 7–12 designate those on the stator. The valve configurations are shown in LOAD and INJECT positions, respectively, while the diagram depicted in (*b*) shows the resulting injection of a sample zone in the reagent carrier stream [5.1].

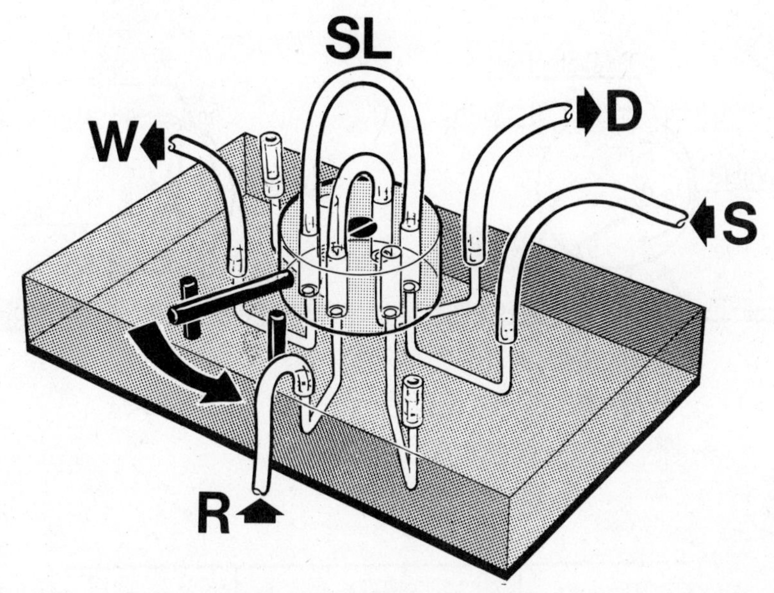

Figure 5.3. Six-port valve integrated into a FIA microconduit. The valve is turned by the lever, proper matching of the ports in the two positions being secured by the two stops fixed on the baseplate. The valve is shown in the sampling position where sample solution (*S*) is being aspirated into the sample loop (*SL*) and from here to waste (*W*). In this position the reagent carrier stream (*R*) is shunted by the small by-pass loop to the manifold and detector (*D*). For further details, see text [5.1].

are imprinted on the bottom of the plate leading to inlet and outlet holes situated at a distance from the rotating disk. Narrow plastic tubings are press fitted into the exposed holes on the top of the plate and in the disk. The bottom of the plate is sealed, leaving the impressed channels open for flow. A lever on the disk allows it to be turned from one position to the other, with stops on the plate ensuring that the holes in the disk and the plate are aligned in either the LOAD or INJECT position. The valve is normally operated manually, but might also been automated (if a stepper motor is used the stops on the plate are superfluous). For convenience, additional channels may be imprinted into the plate to be used for reaction coils, merging zones, sequential reagent additions, etc.

The standard six-port arrangement shown in Fig. 5.2 can be formed by simply connecting pairs of ports on the rotor (or on the plate). Alternatively for a simple injection the valve can be configured as shown in Fig. 5.4.

The valve scheme shown in Fig. 5.5 allows simultaneous sample/reagent injection after filling the two loops (merging zones). Note that the two carrier inlets may be propelled by a single pump tube, or by separate

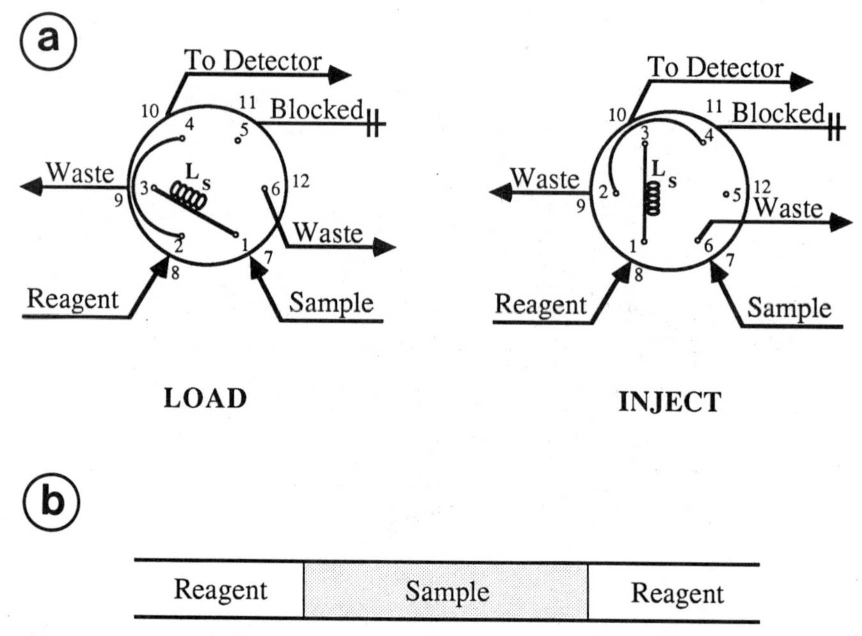

Figure 5.4. (*a*) Six-port valve configuration for simple alternate sample injection, L_S designating the sample loop while the line between ports 2 and 4 represents the by-pass. In (*b*) is depicted the resulting injection of a sample zone in the reagent carrier stream [5.1].

pump tubes, which is preferable, since flow rates through both loops do not need to be hydrodynamically balanced in order to ensure reproducible injections.

Even the injection of a plug of reagent (at a different pH than the reagent stream) in the center of the sample plug, exploiting the resulting pH gradient formed, might be accomplished with the single valve configured as shown in Fig. 5.6. As an alternative to this procedure Fig. 5.7 shows a manifold that simply allows a reagent injection followed by sample (or vice versa), which would create an information-rich asymmetric double pH gradient over the sample zone rather than an information-redundant symmetric double gradient when the reagent is injected in the center of the sample zone. The same valve configuration might also be used for the gradient scanning standard addition method proposed by Fang et al. (Section 4.5.2; Ref. 817) by replacing the reagent in the reagent loop with a standard solution of the analyte. By injecting a sample plug immediately followed by a plug of standard into the carrier stream (and then merging with a reagent stream if necessary), response can be observed at times corresponding to various sample/standard concentration ratios. This

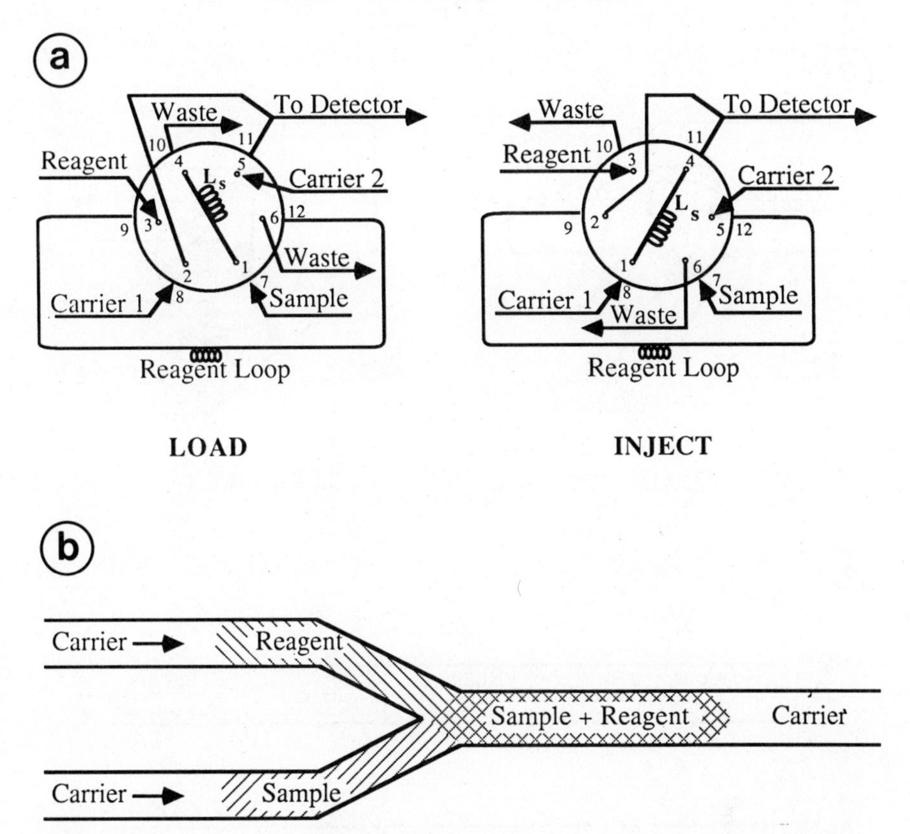

Figure 5.5. (*a*) Six-port valve configuration for metering of separate volumes of sample (L_S) and reagent, respectively, which two zones upon simultaneous injection into separate carrier streams are merged downstream (*b*). The merging might either be effected synchronously (as shown here) or asynchronously, depending on the path traveled by the zones before the merging point [5.1].

valve configuration can also be used for interference studies where the sample plug is immediately followed by a plug of potential interferent that partially merges with the sample plug while both mutually disperse in the carrier stream to which a reagent stream is later confluenced (Section 4.5.2; Ref. 378).

Another configuration is shown in Fig. 5.8, which results in a double injection separated by a plug of the reagent stream, which can result in two separate peaks. By using sample loops of different volumes, different goals can be accomplished. First the dynamic range of the system can be increased by calibrating high concentrations from the smaller injection volume and low concentrations by the larger injection volume. A readout

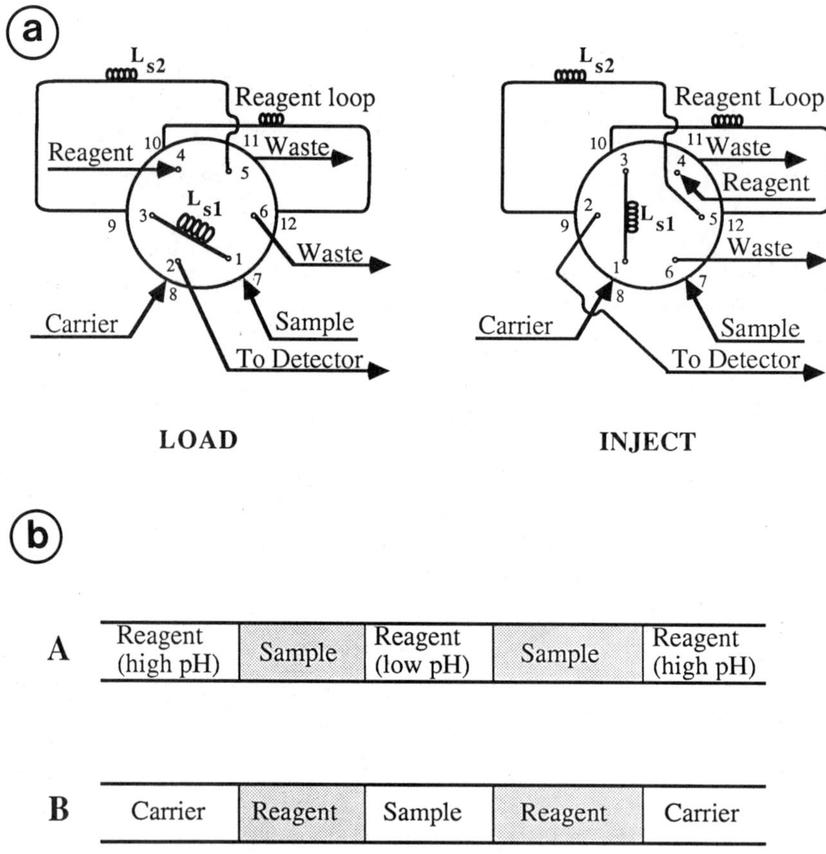

A	Reagent (high pH)	Sample	Reagent (low pH)	Sample	Reagent (high pH)
B	Carrier	Reagent	Sample	Reagent	Carrier

Figure 5.6. (*a*) Six-port valve configuration for triple zone injection. In (*b*), part (*A*), is shown an application of this approach for the injection of a reagent plug of low pH bracketed by zones of sample (L_{S1} and L_{S2}), the three zones being injected into a carrier stream of reagent of high pH. Ensuring adequate reagent supply throughout, the resulting pH gradient formed might thus be exploited analytically. By interchanging sample and reagent loops (*B*), sample can be injected between plugs of reagent in an unreactive carrier, thereby minimizing reagent consumption in a single-line system [5.1].

identical to that obtained with a bifurcated manifold [4] will be obtained, but since the present method does not rely on hydrodynamic balance of bifurcated streams, reproducible results over extended dynamic range and time periods will be obtained.

Next, since peak height is a known function of the injected sample volume, two-point calibration can be achieved by injecting the *same* standard solution by the two loops. By selecting a standard corresponding roughly to the highest expected sample and by selecting loop volumes

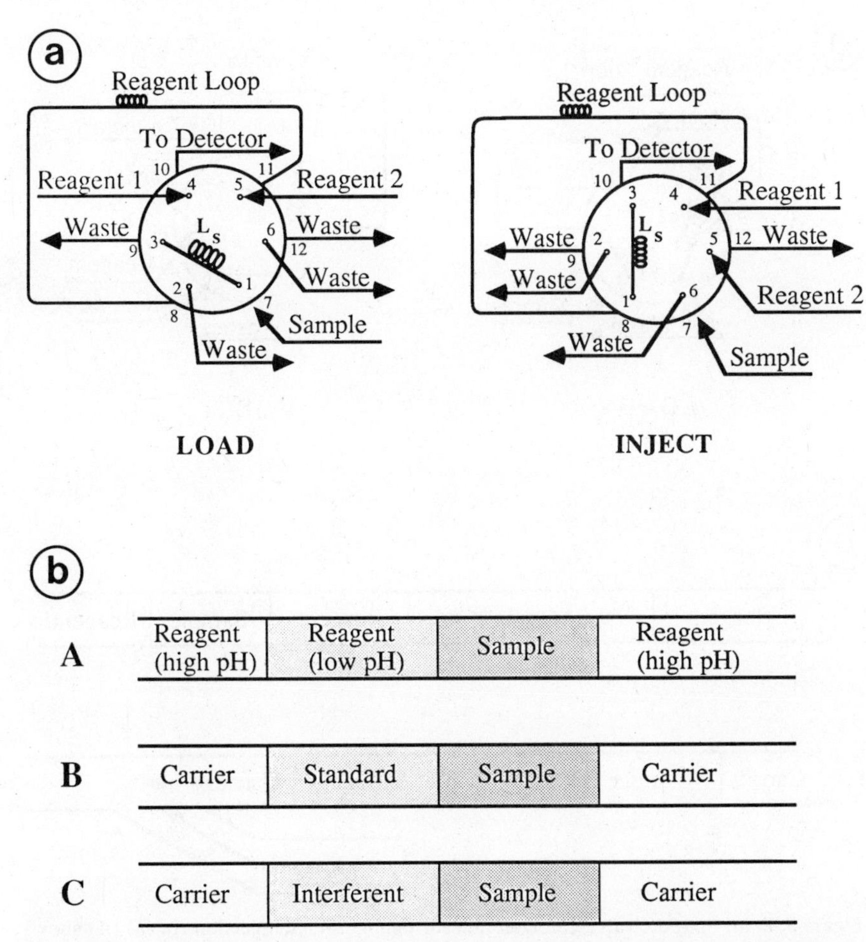

Figure 5.7. (*a*) Six-port valve configuration for injection of two adjacent zones, which in the manifold will chase and gradually merge into each other. (*b*) (*A*) Injection of sample (L_S) followed by a slug of reagent of low pH which, situated within the carrier stream of high pH, will give rise to an information-rich asymmetrical double pH gradient. (*B*) Injection of sample and a standard to be exploited for standard addition measurements. (*C*) Injection of sample and a zone of a potential interferent allowing interference studies [5.1].

corresponding to 50% and 100% of the $S_{1/2}$ value of the FIA system used, two peaks will be obtained for each injection, the first one being twice as high as the second one, thus allowing detector (or sensor) slope to be continuously rechecked, together with periodical recalibration of the whole system.

A third type of application of this double sample injection valve utilizes a packed reactor, placed in the reagent loop (cf. Fig. 5.10*d*). This con-

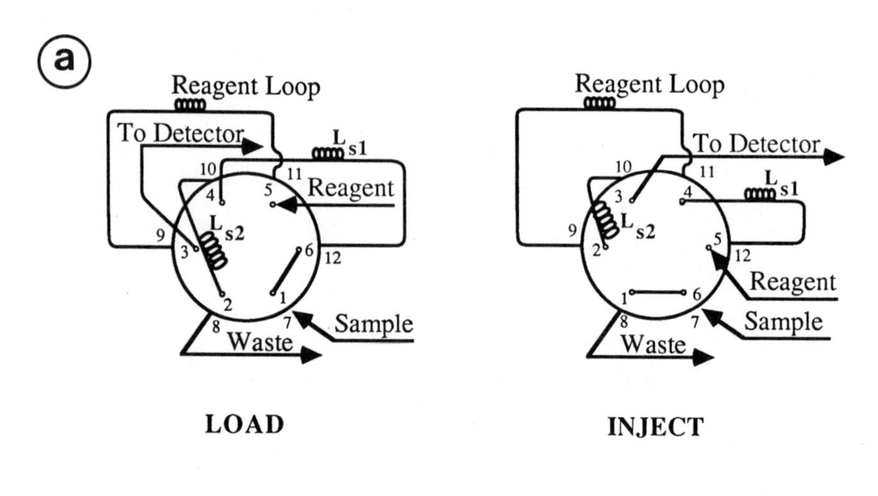

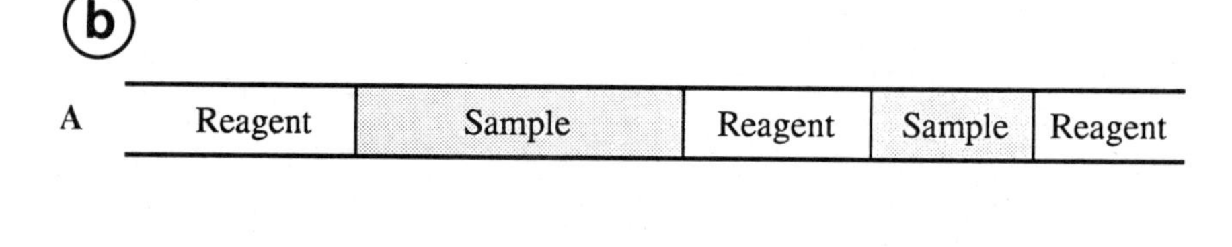

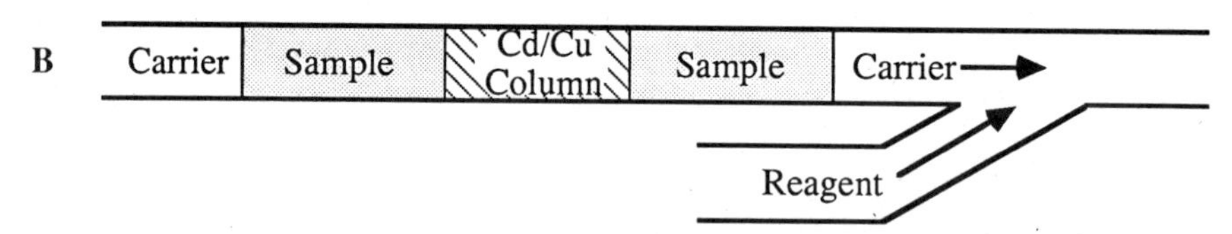

Figure 5.8. (*a*) Six-port valve configuration for double sample injection. In (*b*) (*A*) shows a system where zones of the same sample solution yet of different volume (L_{S1} and L_{S2}) are separated by a plug of reagent, the ensuing chemical reaction resulting in a readout comprising two separate peaks. This approach might be exploited for extending the dynamic measuring range of a given analytical procedure. In (*B*) also two individual sample zones are injected, yet one of these is allowed to pass through a packed reactor for pretreatment (allowing, for instance, as implied on the drawing, nitrate to be reduced to nitrite on a Cd/Cu reductor column), reagent being added further down the manifold [5.1].

figuration would be useful for determination of nitrite (Section 4.7). The reagent loop is turned into a copper/cadmium reaction column and the two loops are filled with the sample. When injected, the first sample passes directly to the reaction coil where nitrite is detected. The second, however, first passes through the copperized cadmium reduction column converting nitrate to nitrite allowing measurement of total nitrate plus nitrite.

Nitrate concentration can then be calculated by difference. Note, however, that each peak position must be calibrated separately, since the second peak travels a greater distance causing additional dispersion.

The final configuration (Fig. 5.9) allows double sample injection, which can be used for similar purposes as the previous configuration or for injection of two *different* reagents. Thus a triple zone, very rich on information (which advantageously might be decoded with the aid of chemometrics) is obtained by a mutual dispersion of sample and reagents in an inert carrier stream.

The most important technological feature of this approach is that the sample–reagent zone sequences from all the above configurations are after injection propelled by the same drive (i.e., single pump tube) through the FIA channel and detector. Thus, unlike the traditional merging zone approach, where reproducible synchronization of merging zones is difficult to maintain during longer experimental periods, here high reproducibility of mutual zone penetration is guaranteed. It is also noteworthy that only two peristaltic pumps (equipped with several pump tubes) are

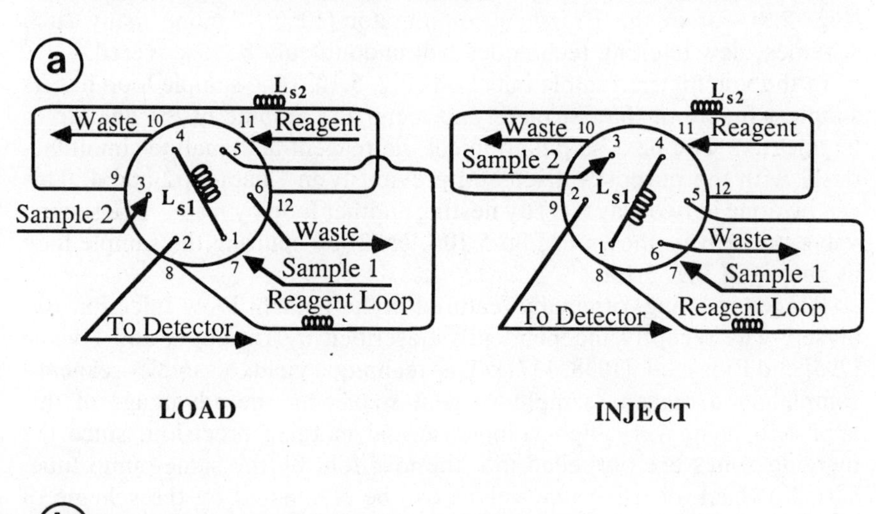

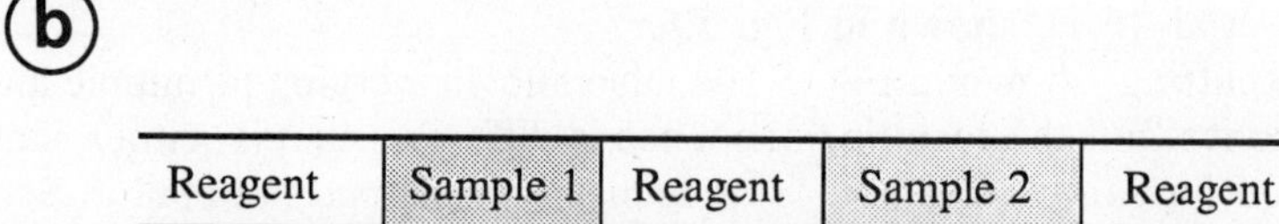

Figure 5.9. (*a*) Six-port valve configuration for triple zone injection, allowing either a double injection of sample solution (L_{S1} and L_{S2}) spaced by a reagent zone (*b*) or the injection of a sample plug sandwiched between two different reagents. Alternatively, a sample in L_{S2} may be injected with reagent, standard, or interferent (similar to Fig. 5.7) in L_{S1} separated by a small plug of the carrier [5.1].

needed to operate any of the above combinations, one pump being run continuously, the other one being in GO essentially only for the loading postition, since it might be stopped for the inject position, should sample and/or reagent material be scarce or expensive.

5.1.2 Split and Nested Sample Loops

The external sample loop of a rotary valve may, beside serving as a simple volumetric cavity accommodating a selected sample volume, be split or nested in order to perform more sophisticated functions (Fig. 5.10). So far only individual components, such as columns, dialyzers, and gas-diffusion units, have been nested in the loop, but it is likely that in the future a series of components, perhaps even an entire FIA microsystem, will become nested in a loop to perform a complex assay. Although a four-port rotary valve scheme was used in this section to review the techniques, it should be kept in mind that a sample loop can be split or nested equally well when mounted in a six-port valve—as implied in the previous section (Fig. 5.8)—or in the Brazilian commutator [1060]. Among many such schemes, new exciting techniques will undoubtedly be discovered.

In the volumetric sample injection (Fig. 5.10*a*) the sample loop has its simplest function, that is, merely to meter the volume of the analyte to be injected. The next step is to inject the reagent and analyte simultaneously with the purposes discussed previously in Chapters 2 and 4. This can be done in two ways: (1) by nesting another loop by means of a second valve in the way shown in Fig. 5.10*b*, or (2) by splitting the sample loop as shown in Fig. 5.10*c*.

Reagent nesting, originally featured in the Hitachi Flow Injection Analyser, was recently independently described by Dasgupta and Hwang [795] and Rios et al. [1038, 1371]. This technique yields a sample–reagent–sample, or a reagent–sample–reagent sequence, the advantage of this approach being very high volumetric and merging precision, since the merging zones are propelled into the manifold by the same pump tube. The drawback of using two valves can be eliminated by the scheme of Erickson et al. [5.1], shown in Fig. 5.6.

Loop splitting, shown in Fig. 5.10*c*, also allows merging of sample and reagent zones, but the resulting sequence is different, that is, either sample–reagent or sample–(sample + reagent) as explained in detail in Section 4.5.2 (Fig. 4.28; Fig. 4.36). As schematically shown here, in the load position the two sections of the split loop are filled with sample and reagent by aspiration via a common outlet (*W*). When the valve is turned, the two zones are swept down the manifold, chasing and gradually penetrating into each other.

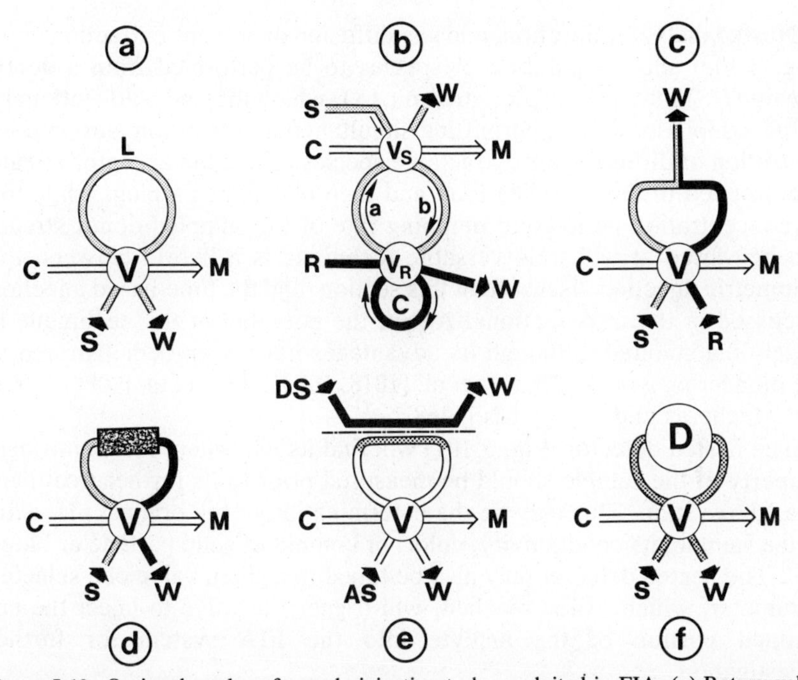

Figure 5.10. Optional modes of sample injection to be exploited in FIA. (*a*) Rotary valve (*V*) furnished with an externally mounted sample loop (*L*), the volume of which determines the amount of sample solution injected into the carrier stream (*C*) and transported into the manifold (*M*); *W* is waste. (*b*) Nested loop injection, where a two-valve system is used to entrap a zone of reagent (*C*) between two zones of sample (*a* and *b*). (*c*) Split loop configuration where the external loop is divided into two sections accommodating sample and reagent solutions, which upon injection chase and merge into each other. (*d*) Sample loop comprising a packed reactor (containing, for instance, an oxidant or a reductant) allowing pretreatment of sample solution prior to injection, or a column packed with an ion-exchanger permitting preconcentration of the sample material prior to injection. (*e*) Sample loop constituting the acceptor part of a separating unit (dialysis, gas diffusion, or extraction device) where sample material is transferred from the donor (*DS*) to the acceptor stream (*AS*), thereby allowing pretreatment of sample solution and possible preconcentration. (*f*) Sample loop comprising a detector (*D*) to measure a selected species or parameter of an analyte prior to further treatment in the FIA system.

The nested packed reactor (Fig. 5.10*d*) allows sample pretreatment *prior* to injection by means of solid oxidants, reductants, ion exchangers, immobilized enzymes, or suitable surface-active sorbents. The potential of this approach is largely unexploited, since so far such sample pretreatment has been used only to remove unwanted matrix, which is not retained on the column, for sample preconcentration, and for analyte conversion in connection with AAS and ICP (cf. Chapter 4.7).

Nested dialyzer, filtration, and gas-diffusion or solvent-extraction units (Fig. 5.10*e*) allow separation of species to be performed from a donor stream (*DS*) into an acceptor stream (*AS*), which may be held stationary in the sample loop, thus permitting simultaneous separation and preconcentration of diffusable or extractable species. Since the acceptor stream is entrapped in the sampling loop and then injected as a plug, while the preconcentration period and pumping rate of the sampled donor stream can be varied at will, this versatile technique is a hybrid between the volumetric injection discussed in this section, and the time-based injection discussed in the next sections. Again, the potential of this technique is largely unexploited, although its advantages have been demonstrated in the pioneering works of Pacey et al. [1018, 1073], Fang et al. [1080, 1257], and Macheras and Koupparis [1268].

The nested detector (Fig. 5.10*f*) will find its use whenever an intrinsic property of the sample should be measured prior to its further treatment in a FIA system. This may be the determination of the original pH value of the sample, its conductivity, color, or content of solid particle or blood cell. The nested detector may also be tuned to a given value of a selected parameter, which, when reached, will trigger the valve to inject the entrapped portion of the analyte into the FIA system for further investigation.

5.1.3 Hydrodynamic Injection

Inherently, all injection valves have one thing in common, that is, they have moving parts that might eventually become worn. Yet, the recent generation of FIA valves, made of a suitable combination of polymers, have proven very reliable in continuous use—in fact, van der Linden [1059] has reported that rotary valves subjected to exhaustive tests even after 50,000 switches still functioned properly, and no wear or tear was noticeable provided that the sample and carrier stream were virtually free from particles. Even better performance is feasible with valves if powered by stepper motors, which results in superior reproducibility of injection and excellent duration. Nevertheless, it would, particularly for special applications, be beneficial to be able to meter the sample volume and subsequently transfer it reproducibly into the carrier stream by means requiring no mechanically moving parts.

This feature is possible by applying the hydrodynamic injection technique [338], the principle of which is best explained by referring to Fig. 5.11. By this approach a column of liquid can be exactly metered into a geometrically well-defined conduit (of length L and internal radius R) which is at all times open to other channels, provided that these channels

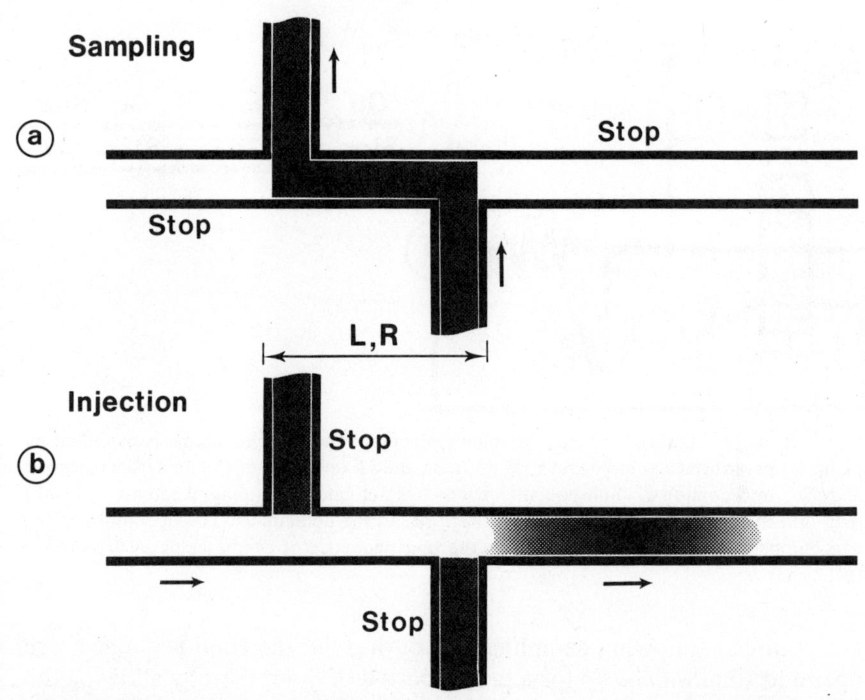

Figure 5.11. The principle of hydrodynamic injection. A fixed volume of sample solution is metered into a conduit, of length L and internal radius R (*a*—Sampling), and this volume is subsequently propelled downstream by the carrier stream (*b*—Injection). During the sampling cycle, the carrier stream circuit is stopped, and vice versa. When aspirating the next sample, the column of carrier stream solution contained within the common conduit L is emptied to waste along with surplus of sample solution (see also Fig. 5.12).

are filled by a stagnant liquid (Sampling, Fig. 5.11*a*). These columns of liquid exert a hydrostatic force, which serves as a lock while the sample volume $S_v = \pi R^2 L$ is being filled. Thus by an alternate and intermittent pumping of carrier and sample solutions, a well-defined sample zone can be formed and then inserted into the carrier stream (Injection, Fig. 5.11*b*). The manifold serving for this purpose (Fig. 5.12*a*) uses two peristaltic pumps to control the movement of sample and carrier solutions. It is obvious that if $x = z$ and if the pumps follow the sequence indicated in Fig. 5.12*b*, then the sample conduit L will be filled by sample solution from sample reservoir S when pump 1 is in motion, and during the next cycle, when pump 1 is stopped and pump 2 operates, the carrier stream will empty this exactly metered volume of sample from conduit L and will carry it further toward the detector (D). Various other schemes employing combinations of hydrodynamic and hydrostatic forces, where

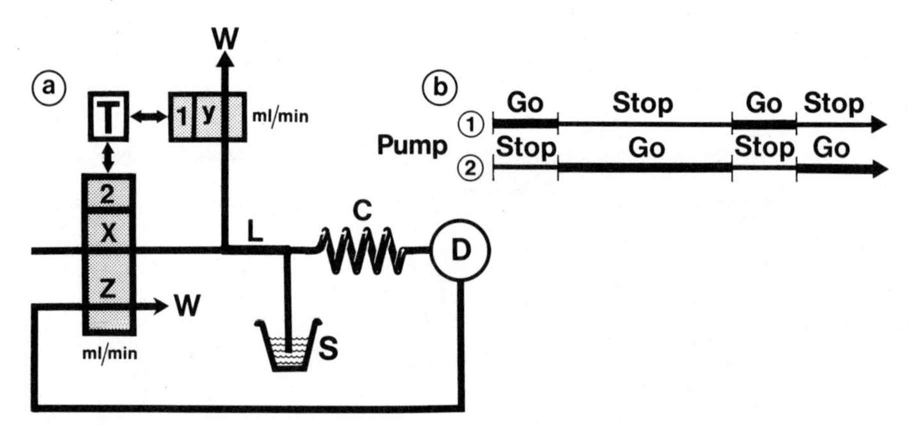

Figure 5.12. (*a*) Manifold for hydrodynamic injection. The sample volume is aspirated by pump 1, operating at a pumping rate of y mL/min, and a fixed volume of sample from reservoir S passes into conduit L. Subsequently pump 2 is activated, pumping at rates $x = z$ mL/min, and the sample is flushed through reactor C to the detector D. The operations of the two pumps is controlled by the timer T, the time sequence of events being as depicted in (*b*).

both pumps, following sampling and during the injection sequence, are operated *simultaneously* for a preset period of time, thereby allowing the volumetrically entrapped sampled volume in the common conduit to be supplemented by a time-based metered amount of sample solution, have been implemented (see Sections 5.1.4 and 4.12).

The reproducibility of measurement when using hydrodynamic injection is apparent from Fig. 5.13, where the colorimetric method for chloride was used for testing [338]. The relative standard deviation for a series of 10 injections of one of the chloride standards was better than 0.5%. This value compares well with that obtained by means of the normally used FIA rotary valves (RSD $\leq$ 0.3%). The hydrodynamic injection approach is especially attractive for monitoring purposes (cf. Fig. 5.13*c*), permitting not only frequent sampling of the analyte to be assayed but, equally important, periodic checking of the analytical instrumental setup by intermittent injections of standard sample materials. In Section 6.8 is detailed a FIA exercise illustrating this approach for an analytical procedure imitating industrial process control, the purpose of the exercise being to investigate the course of the oxidation of sulfite to sulfate with air under different reaction conditions.

5.1.4 Time-Based Injection Procedures

Common for the *time-based* injection procedures is that the sample volume is metered as a function of time, the most important feature of this

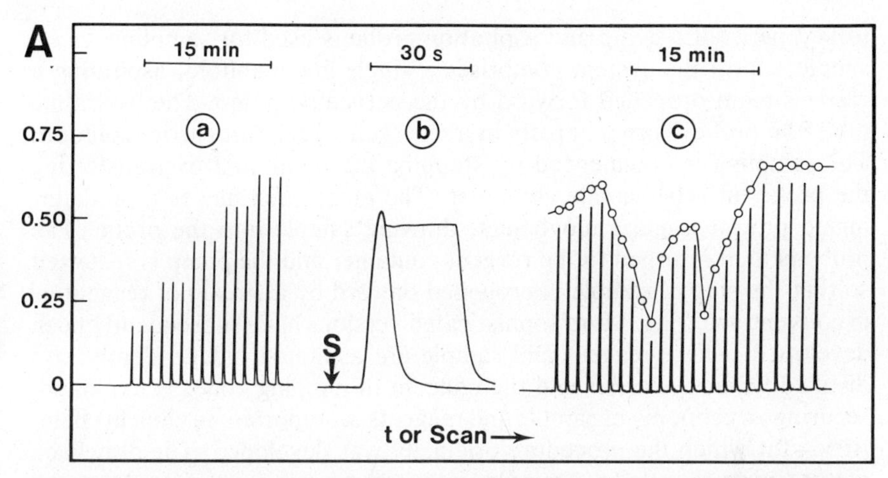

Figure 5.13. Determination of chloride by the mercury thiocyanate method with the hydrodynamic injection manifold in Fig. 5.12*a*, where *C* was 100 cm long and the volume of conduit *L* was 25 μL. The pumping rate $x = z$ was 1.1 mL/min and the aspiration rate *y* was 3.0 mL/min, operated for 12 s. The detector was tuned at 490 nm. (*a*) Standard calibration run of samples in the range 10–50 ppm Cl; (*b*) stopped-flow experiment with the 40 ppm Cl standard recorded at high paper speed to demonstrate the fast rate of reaction; and (*c*) monitoring of the content of chloride in a solution of NaCl in which the analyte concentration was changed intermittently and measured at fixed time intervals by the system.

approach being the control of the injected sample volume by means of a selected time interval. Thus the injected sample volume can be readily computer controlled, while in the volumetric methods the sample volume is changed by physically changing the volume of the dimensions of the sample loop. Admittedly, the latter methods have the advantage of being independent of volumetric flow rates and therefore are inherently more precise, because the precision of the time-based injection relies on maintenance in constancy of flow rates. Fortunately, any changes caused by wear of pump tubes and pump parts are usually gradual and might therefore be corrected.

Time-based injection might be accomplished simply by partially emptying a cavity of a sample conduit. This avenue was actually used by the U.S. company FIAtron in their early versions of FIA systems, the metering of the injected sample volume being executed by sequencing a series of open–closed magnetic-driven valves. FIAtron has recently resorted to the more reliable injection of sample volume by means of an eight-port rotary valve (FIAtron FIA-Valve 2000) through which sample solution is pumped for a selected period of time, user-programmable via the panel of the valve unit.

Riley et al. [394] in 1983 introduced an injection technique that they termed "controlled-dispersion flow analysis." In this, a computer-controlled peristaltic pump and aspiration probe is used for sampling. In its simplest form the system comprises a single-line manifold, aspirating a carier stream propelled forward by the peristaltic pump. The aspirating tube (the probe) normally rests in a reservoir of reagent carrier solution. The sampling is commenced by stopping the pump and by transferring the probe into the sample container. The pump then makes a predetermined precise angular movement, drawing sample into the probe. The probe is then returned to the reagent container and the pump is restarted so that the slug of sample is propelled onward by a stream of reagent as in conventional FIA. More sophisticated versions have subsequently been developed, where reagent and sample are aspirated independently into distilled water as carrier and the concept of merging zones is exploited, securing an economy of sample and reagents so important in clinical chemistry—for which the procedure originally was developed. The drawback of this approach is that the sample and reagent zones are aspirated through the peristaltic pump before entering into the manifold, which invariably makes control of the dispersion more difficult to achieve.

In the previous section the hydrodynamic principle was discussed for volumetric sample injection. In Section 4.12 it is explained how this approach allows the injected volume of sample to be increased at will by overlapping the operation of the sample aspiration pump with that of the carrier solution propulsion pump, that is, combining the volumetric sample metering with that of time-based injection. However, a disadvantage of these two versions of the hydronamic injection procedure is that they require an absolute balancing of the in- and outgoing pumping rates ($x = z$, Fig. 5.12). Hence, alternative variants eliminating this drawback have been suggested.

All of these exploit the use of a confluence point at which a well-defined sample zone is formed by means of the alternate motion of sample and carrier stream. The principle of the first alternative is best explained by referring to Fig. 5.14, which shows a system comprising two pumps (1 and 2) pumping at volumetric pumping rates x and y, respectively, where $x > y$. Initially, that is, prior to injection (*a*), pump 1 propels carrier stream solution toward the confluence point at a higher volumetric pumping rate than pump 2, which by its mode of action aspirates the liquid of the analyzer channel, that is, pumps away from the confluence point, resulting in a positive outflow of carrier stream through the inlet channel (L) and into waste, while the section from the confluence point and onward, including the detector (D), concurrently are filled with carrier stream flowing through pump 2 and toward waste (W). In the second step (*b*), sample aspiration, pump 1 is stopped and the column of carrier stream solution in the carrier stream channel is held still, while pump 2 continues to

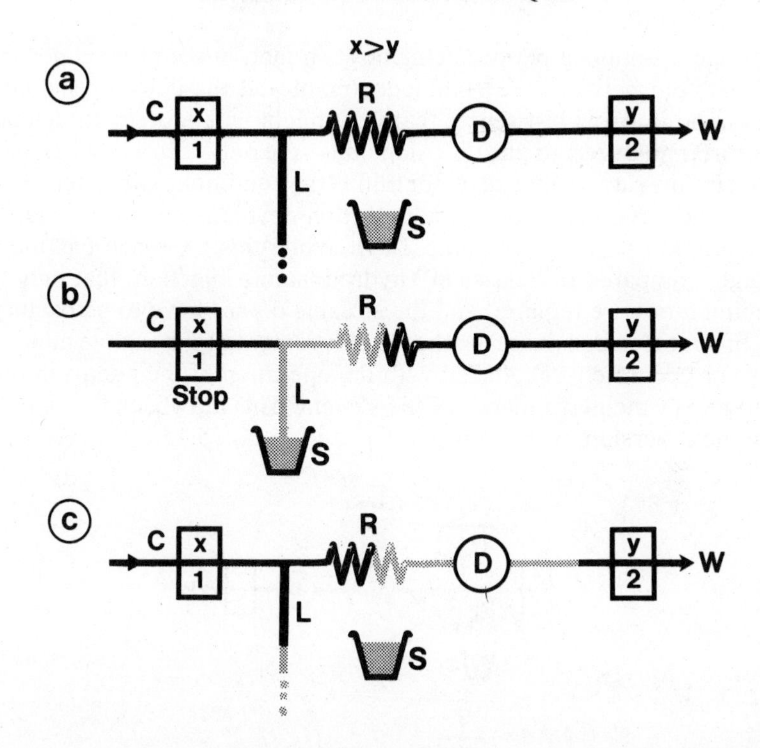

Figure 5.14. Time-based injection FIA system comprising two pumps, 1 and 2, operating at volumetric flow rates x and y, respectively, where $x > y$. In the standby position (*a*) both pumps are activated, propelling carrier stream solution (*C*) through the manifold, excess of carrier being expelled via channel *L*. For sampling (*b*), the sample container is contacted to *L*, pump 1 is stopped while the action of pump 2 is maintained, whereby a defined zone of sample solution is aspirated into the manifold. In the injection step (*c*), the sample cup is removed and pump 1 is restarted, the carrier stream thereby forcing the sample zone through the FIA manifold and into the detector, while surplus of carrier is wasted through channel *L* simultaneously emptying it for sample solution.

operate thereby aspirating sample solution from a container, which now has been moved into contact with the inlet. In the sample injection step (*c*), the pumping of pump 2 is maintained, while action of pump 1 is resumed and the sample source is withdrawn. Thus all sample solution in the inlet channel is forced in countercurrent fashion toward waste, while the sample solution to the right of the confluence point in the form of an injected sample zone is aspirated, followed by carrier stream solution, through the analyzer channel and into the manifold for further treatment and subsequent measurement in the detector. As both pumps $P1$ and $P2$ continue pumping, all sample material is eventually expelled from the system either via the inlet or the outlet and the system is thus reestablished for the next sampling period, being now, in fact, in the original position. The manifold between the confluence point and the detector should be understood in its widest sense, that is, it might be a system of channels, detector(s), dialyzer, or gas-diffusion unit—the only restriction being that the net inflow rate into the detector fulfills the conditions stipulated above, that is, the flow rate through the detection device is governed solely by the aspiration flow rate of pump 2 (y). While this approach has the advantage, compared to "classical" hydrodynamic injection, that only two pumping tubes are required and the streams do *not* have to be balanced, it is, of course, a disadvantage that the movement of the sample container has to be accurately sequenced with the operation of the pumps in order not to impair the performance of the system. This drawback is eliminated in the next version of the system, depicted in Fig. 5.15.

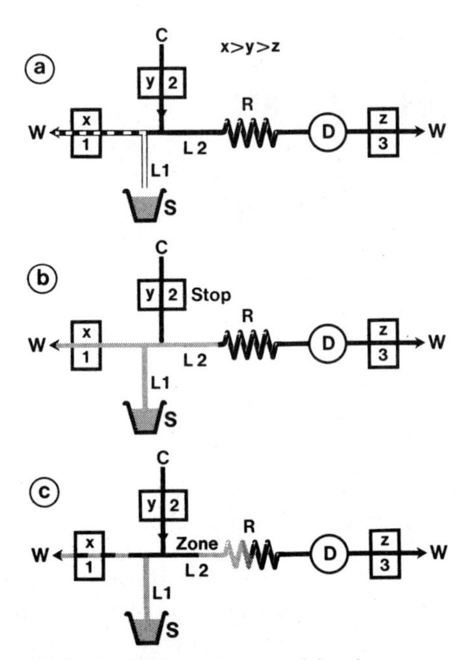

Figure 5.15. Time-based injection FIA system comprising three pumps, 1, 2, and 3, operating at volumetric flow rates x, y, and z, respectively, where $x > y > z$. Pumps 1 and 3 are in action continuously, while pump 2 propelling carrier stream solution (C) is sequenced in a go–stop mode. In the standby position (a), all three pumps are operating whereby the FIA manifold is filled with carrier solution, the surplus being wasted through pump 1 along with air (aspirated through inlet $L1$). During sampling (b), pump 2 is stopped and sample solution is aspirated via inlet channel $L1$, being directed partly into conduit $L2$ and partly toward pump 1. In the injection step (c), the action of pump 2 is resumed, which will propel forward and into the manifold that zone of sample solution in $L2$ entrapped to the right of the junction point of the manifold and pump 2.

Here the system essentially comprises three pumping stations (1, 2, and 3), furnished with pump tubes of volumetric flow rates of x, y, and z, respectively, these flow rates fulfilling the conditions of $x > y > z$. Pump 2, which delivers carrier stream solution, is operated sequentially in stop–go modes, while the two others continuously are in go; hence, pump stations 1 and 3 might in reality be combined into a single unit accommodating both tubes x and z, yet for the purpose of clarity they are shown as separate units on Fig. 5.15. In (*a*) the system is shown in the standby position. During this situation all three pumping stations are in go position, the carrier stream being furnished by pump 2. The carrier stream is partly aspirated through the FIA manifold (by pump 3) and is partly directed to waste trough pump 1, which additionally aspirates air via the sampling line $L1$. Note, however, that the sample cup container very well might be brought into contact with $L1$, without any sample solution thereby entering the manifold part of the system, because the sample solution will in this situation be exclusively directed toward waste of pump 1, thus only impairing the economy of sample solution. When the actual sampling sequence is initiated (*b*), the sample container is brought into contact with $L1$, and after this conduit has been filled with sample solution, pump 2 is stopped. Sample solution will now proceed from the junction toward both pumps 1 and 3, sample solution thereby filling conduit $L2$ (during this sequence pump 1 might actually be stopped as soon as the sample solution has reached the junction). Injection of sample is initiated by resuming of pumping of pump 2 (*c*). The carrier stream will now propel forward and into the manifold that zone of sample solution entrapped in $L2$ in front of the second junction point, while part of the carrier stream solution will be directed toward pump 1. Note, that the sample cup does not need to be removed during this operation, because the volumetric flow rate x is larger than that of y, while by removing the cup any loss of sample material is prevented and the standby position is resumed.

An interesting variation of this approach, the reverse injection, is presented in Fig. 5.16. Again, three pumping stations, 1, 2, and 3, are required, accommodating tubes of volumetric flow rates x, y, and z. As before, pumps 1 and 3 are operated continuously and might therefore be combined, while pump 2 is sequenced in stop–go, and the volumetric flow rates are adjusted so that $x > y > z$. However, in this approach pump 2, which delivers the carrier stream, is placed *behind* the FIA manifold. In standby postion (*a*) all three pumps are operating, carrier solution being furnished to the manifold, while either air or sample solution (depending on the position of the sample container) is directed through conduit L and toward waste via pump 1. Sampling is initiated by contacting the sample

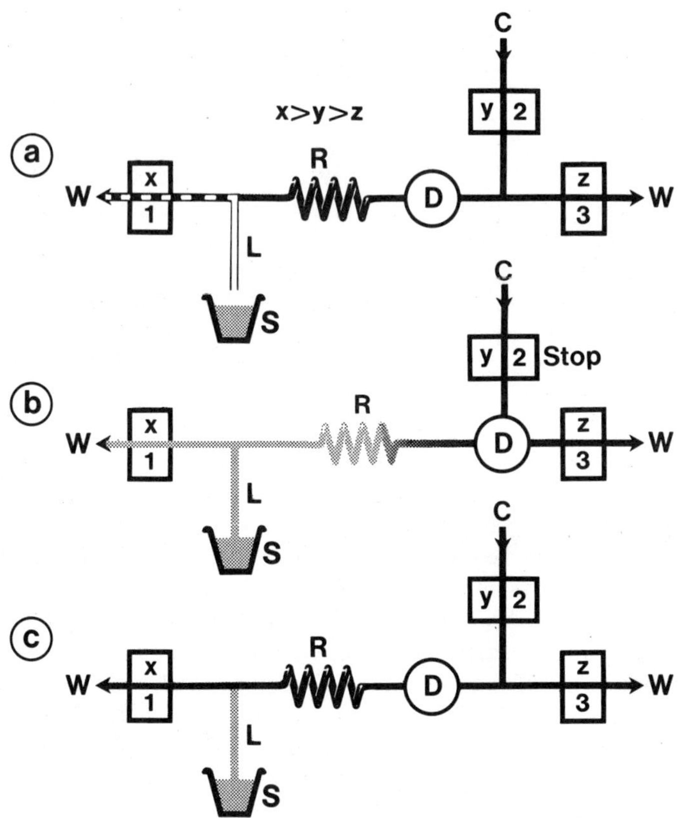

Figure 5.16. Reverse-injection FIA system comprising three pumps, 1, 2, and 3, operating at volumetric flow rates x, y, and z, respectively, where $x > y > z$. Pumps 1 and 3 are in action continuously while pump 2 propelling carrier stream solution (C) is sequenced in a go–stop mode. In the standby position (a), all three pumps are operating, whereby the FIA manifold is filled with carrier solution, the surplus being wasted through pump 1 along with air (aspirated through inlet L1). For sampling (b), pump 2 is stopped and sample solution is aspirated via inlet channel L, being directed partly into the manifold and partly toward pump 1. In the manifold the sample solution will chase and be mixed with the carrier reagent solution. After the dispersed zone has reached the detector and a suitable signal has been recorded, the action of pump 2 is resumed (c), which countercurrently will direct the sample/carrier solution toward pump 1.

cup to L, filling this conduit, whereupon pump 2 is stopped (b). Sample solution will now enter the manifold chasing and being mixed with the carrier solution, eventually reaching the detector. After a suitable signal has been recorded, the action of pump 2 is revoked, which *countercurrently* will direct the sample/carrier solution toward pump 1 (c). Hence, the sample will by this approach "peek" into the detector. Consequently, almost symmetrical peaks will be recorded. This reversed injection has numerous advantages, one of which is that the detector is readily cleansed. Compared to conventional FIA, where the detector is the last station of the FIA system and where the disappearance of the analytical signals signifies the cleansing of the entire FIA manifold, the disappearance of the peak here merely signifies the clearing of the detector, and therefore it is necessary to operate pump 2 for a sufficiently long period of time until the whole manifold is washed thoroughly with carrier solution.

Being suitable for conventional FIA procedures, as well as for gradient techniques and stopped-flow measurements, the hydrodynamic injection variations presented above are also ideally suited for microminiaturization.

Bergamin et al. [5.2] have recently described an original and elegant time-based injection procedure. When analyzing metal samples from a smelter, they used a specially constructed electrolytic cell arrangement in which the spherical metal specimen was made the anode. Using a 1 A current for a preselected period of time (5 s was generally used), a defined amount of the metal became electrolytically dissolved, forming a sample zone that subsequently was carried into the FIA manifold where the species of interest (Cr and Ni) were colorimetrically determined. Many years ago such anodic dissolution was used in connection with spot tests for qualitative and semiquantitative assay (developed by a fellow Brazilian, Feigel), yet Bergamin's approach is quantitative and characterized by a very high reproducibility of measurement.

5.2 PUMPS AND OTHER DRIVES

Multiroller peristaltic pumps are, in spite of their shortcomings, the most suitable means of propelling the carrier and reagents streams in FIA systems, because they may accommodate several channels, whereby, according to individual tube diameters, equal or different volumetric pumping rates may be obtained. Being, in principle, capable of maintaining constant flow rates and corresponding constant residence times, they are independent of minor changes in viscosity or variations in backpressure due to restriction changes in the remainder of the FIA system. As each tube might be used for either aspirating or delivering solution, a propelling as well as a withdrawing motion may be executed by the same pump, the latter function being needed in automated sampling, solvent extraction, and sample splitting. A well-constructed pump stops and starts instantaneously, thus allowing precise control of the movement of all streams for stopped-flow or intermittent pumping functions. A modern pump also has a very small inner holdup volume, permitting rapid startup and short washout periods. If an aggressive liquid is pumped, the only part likely to be damaged is the pumping tube, which may easily be replaced. Additionally, peristaltic pumps are sufficiently robust and reliable for use in industrial process analysis. The vital parts can easily be protected against corrosive environmental attack. Several manufacturers have even marketed pumps for use in explosive environments, where special precautions have to be taken to meet existing safety requirements.

The main drawback of peristaltic pumps is that the stream is never completely pulse-free, although the slight fluctuations that may occur are dampened to some extent by the use of flexible and somewhat elastic tubing. However, a well-designed pump with many closely spaced rollers rotates sufficiently rapidly so that the pump tubes are compressed frequently for short periods of time (e.g., eight rollers on a wheel rotating at 40 rpm), thus generating a pulsation of high frequency and low amplitude. This is true for the Ismatec Minipump (Model S-840), which, capable of accommodating up to four channels, has been used extensively by the authors. Equally satisfactory experience has been gained with Alitea pumps. Most pump manufacturers offer a wide array of models featuring continuous or stepwise variation of rotational speed of the pump head, thereby allowing regulation of the flow rates, yet it should be borne in mind that the adverse influence of pulsation will increase with decrease in rotation. Recently, pumps operated by stepper motors (e.g., Alitea and Watson–Marlow) have been marketed, and they are very attractive for FIA, particularly in conjunction with sample introduction procedures such as hydrodynamic injection (see Section 5.1). As a rule of thumb, it can be said that most commercially available multiroller peristaltic pumps might be used in FIA, yet besides generating a nearly pulse-free flow as possible (if necessary, an additional pulse-dampening device can be used), it is absolutely imperative that they do not have any inherent inertia preventing them from being stopped and restarted precisely.

A minor drawback of peristaltic pumps is that all models generate, to a greater or lesser extent, pulses of static electricity that might disturb measurements with ion-selective electrodes. This problem has been known for some time, as it is especially troublesome in systems based on air-segmented streams, where the air bubbles serve as insulators [5.3]. Its origin is difficult to diagnose, as the frequency of disturbances is synchronized with the movement of the pump rollers, and, therefore, it is easily confused with the pulsing of the liquid. If the disturbances decrease when the pump tubes are made conductable (by wetting with water or antistatic liquid), static electricity is indeed to blame. A convenient, and permanent, remedy is obtained by short-circuiting the inlet and outlet sections of each pump tube so that there is electric contact between the incoming and outcoming streams (see Section 6.9). Also, all streams

Table 5.1. Characteristics of Commercially Available Pump Tubes

Pump Tube Material	Applicable for	Unsuitable for
PVC, Tygon (e.g., Technicon Clear Standard; Ismatec ENE)	Aqueous solutions; dilute acid and basic reagents; formaldehyde; actaldehyde; dilute alcoholic solutions	Concentrated acids; pure organic solvents
Modified PVC (e.g., Technicon Solvaflex)	Alcohols; aliphatic hydrocarbons; cyclohexane; carbon tetrachloride; dilute methyl cellosolve	Esters; aldehydes, ketones; aromatic hydrocarbons; chloroform; ethers; acids; bases
Silicon rubber[a]	Lower alcohols (including butanol); acetone; dilute acids and bases; acetic acid; acetic acid anhydride	Higher alcohols (e.g., isoamyl and isopropyl alcohol); esters; ethers; strong acids and bases; aliphatic and aromatic hydrocarbons; chlorinated solvents
Fluoroplasts (e.g., Technicon Acidflex; Viton; Isoversinic)	Concentrated (mineral) acids; chlorinated solvents; aromatic hydrocarbons (benzene, toluene, xylene); (hydrocarbons)	Ketones; methyl alcohol; dioxane; ethers; aldehydes; tetrahydrofuran
Marprene (Watson–Marlow)	Medium to strong acids and bases; alcohols; aldehydes; perchloric acid; pickling solutions; turpentine; oils	Acetone; benzene; ketones; chlorinated solvents; ethers; cyclohexane

[a] Often porous, allowing microbubbles to be aspirated into the FIA system via the tube wall

should be made sufficiently conductable by addition of a well-dissociated compound [60].

Pump tubes are available in different materials, according to use (Table 5.1*), and are often color coded to designate the delivery rate. These delivery rates are, however, only approximate and may vary up to 20%, even if the tubes are used in the pump for which they are designed. For an *exact* description of experimental conditions, the delivery rate should

* This is a very abridged list. For specific information the reader is advised to consult the catalogs supplied by various manufacturers of pumping tubes. The British company Watson–Marlow has issued a highly recommendable and very detailed *Tubing Selection Guide*.

always be checked by pumping the carrier stream for a few minutes and by measuring the delivered volume. The delivery rates also tend to change somewhat with time, depending on the quality of the tubing, on whether the tube is new or used, and on the composition of the carrier stream. Fortunately, the changes of delivery rates are slow and monotonous and therefore do not affect the analytical results, provided that the system is regularly recalibrated by standard solutions. This is easily accomplished in FIA, where the high sampling capacity and instantaneous availability of the readout allows, within minutes, checking not only the position but also the slope of the calibration curve. Thus a change of the respective peak height, which may be caused not only by change of delivery rates, but also of temperature or of sensitivity of detector and associated electronics, can be readily corrected. Although certain modern pump tubes (such as Marprene) are compatible with a wide array of organic solvents, it should be mentioned that delivery of corroding solvents in FIA might be accomplished by using the displacement technique where water is pumped into a sealed bottle containing the solvent reagent.

Propulsion of liquids might also be accomplished using piston or reciprocating positive displacement pumps. The most frequently quoted advantage of these devices is the precision of delivery both in terms of actual value and time stability, which characteristics should make them superior to peristaltic pumps for the FIA gradient techniques where a high reproducibility in flow rates is imperative. Depending on construction, these pumps are capable of generating very high pressures, up to several hundred atmospheres, which often may be required for the HPLC applications for which they are originally designed. Low-pressure pumps delivering up to 50 atm are less expensive, but still cost at least 10 times as much per channel as the peristaltic pumps. From the viewpoint of constructing a FIA apparatus, their drawback is that each channel requires an individual pump, that they cannot perform a withdrawing action, and that the change from one analytical method to another, when a different reagent is to be used, is more time consuming. The stream generated by reciprocating pumps is, however, not entirely pulse-free and therefore dampers and restrictors have to be used in FIA applications [74, 167]. Syringe pumps are a less expensive alternative [1063], yet, owing to their generally limited capacity, they have the disadvantage that the pumping action periodically must be interrupted for filling of the syringe.

For simple FIA systems use of gas-pressurized reservoirs or of constant-head vessels ostensibly offer inexpensive alternatives for propelling of reagent and carrier streams, free from interfering electric disturbances [102, 253]. Apart from the benefit of an almost completely pulse-free fluid flow, their main advantages are simplicity and the lack of any moving parts. As long as the geometry of the whole setup is fixed and clogging or narrowing of the conduits is avoided, a continuous and very constant flow can be guaranteed; yet, in manifolds requiring two or several lines, it becomes cumbersome to adjust the flow rates, and therefore their use in practice is limited.

5.3 REACTORS, CONNECTORS, AND OTHER MANIFOLD COMPONENTS

The most frequently used *reactors* are made of plastic tubing, which can be coiled, knitted, or knotted. The purpose of such geometric deformation is to decrease zone dispersion (cf. Chapters 2 and 3, and Fig. 2.8) and its importance cannot be overemphasized. The coils are made of suitable lengths of plastic tubing tightly wound around a core of uniform diameter (approx. 1 cm). Knitting is much more effective than coiling, but requires skill and use of tubing that, being sufficiently flexible, has walls thick enough to prevent the tube from collapsing or narrowing at tight bends. Up to 30-m-long knitted reactors made of Teflon tubing (0.8 mm ID), thermostated in a water bath at 55°C, have been used for the routine assay of catecholamine and its derivatives [5.4]. Knotting, which does not require any skill, is most convenient for short reactors, being very useful to promote intense mixing downstream from confluence points. When used for sample loops, it improves the washout, notably for long loops.

The most frequently used internal tube diameter in FIA applications is 0.5 mm, but 0.8 mm is useful either to increase dispersion or to increase the holdup volume and decrease the flow resistance of extremely long reaction coils. To decrease the dispersion in short straight lines, often necessary to interconnect a manifold and a bulky detector (such as an AAS instrument), narrow tubing (0.3 mm ID) is useful. The most suitable tube material is Teflon, which, besides being chemically resistant, adsorbs the least solutes on its surface. A proprietary material, produced by Thermoplastic Inc. for making Micro-Line tube, is clearly transparent, easy to form (and restore to shape), and easy to fit into all connectors because it is tractable. Polyethylene or polypropylene tubing is inexpense and easy to flange. For subminiaturization experiments, the glass capillaries of 20–125 μm ID, commercially available as material for HPLC, without inner coating, are very suitable, since they are flexible and extremely resilient due to their outside layer of plastic.

Packed reactors are readily made from pieces of suitable tubing into which the solid material is placed and where, if necessary, at both ends a small tuft of glass wool, fine plastic grid, or frit is placed, to keep the packing material within the column. Typical column dimensions are 1–3 cm in length and max. 5 mm ID. The packing material should be noncompressible, should not swell during use, and should be coarse enough to create minimum backpressure. Silica particles of 150 μm average size are ideal for this purpose, because they can be coated with a wide variety of active materials.

The chemically inactive single-bead string reactor (Chapters 2 and 3) is made by filling a suitable length of plastic tubing with glass beads (packed into the tube by aspirating a glass bead/ethanol slurry). A 3-m-long reactor of 0.85 mm ID carefully packed with approximately 1250 glass beads (≅0.6 mm) can be operated at usual flow rates by a peristaltic pump.

It is very important that reactors and connecting pieces of tubing be well attached on a solid support, so that the flow path remains unchanged during each set of experiments. Loose pieces of tubing may easily change position, altering the radius of the bent parts. This will alter the flow pattern, peak shape, and ultimately the calibration curve. It is therefore not only neat, but also necessary to have an arrangement whereby all manifold components are well fixed in a suitable way. This is why the first FIA systems built in Brazil incorporated the Lego toy modular blocks donated to us [3]. Although the Lego is still part of some of our student exercise setups, its proper place is in FIA history. In the commercial instruments such as the TECATOR/BIFOK FIAstar 5020 Analyzer, a system of module blocks, CHEMIFOLDS, is used. These modules have a system of channels that serve as inlets, outlets, confluence points and to accommodate reactor coils, gas-diffusion, and dialyzer units. All confluence points should always be carefully machined to minimize dispersion and to promote radial mixing, being designed in the geometrical configuration discussed in Section 3.2.3 (cf. also Fig. 4.9).

The *connectors* can be divided into three groups, (1) push fitted, (2) threaded (Omnifit or Swagelok type), and (3) permanently glued. Push fitted are most common; when properly made, they are reliable and have no dead volume. Threaded connectors are easy to reassemble, reliable but somewhat costly, being available as HPLC fittings, furnished with a standard thread. They are used in commercial setups, such as the CHEMIFOLDS in the TECATOR/BIFOK FIAstar Analyzer.

For solvent extraction (Section 4.6.1) Teflon coils of 0.5 mm ID are used, together with two special components: a segmentor and a separator for the organic phase [19, 116, 699, 1079]. The function of the *segmentor* (Fig. 4.30*a*) is to create a regular pattern of organic and aqueous segments, which will equilibrate during their passage through the mixing coil. By adjusting the segmentor, shorter or longer segments of organic and aqueous phases can be formed.

The *separator* can be made in a simpler (Fig. 4.30*b*) or in a more sophisticated (Fig. 4.30*c*) version, the latter construction that employs a hydrophobic membrane being preferable, since it is capable of handling a wide variety of solvents. The simple version is made of a standard A4 fitting (Technicon) into which either a Teflon fiber (or a thread of fibers), or a stripe of Whatman IPS phase-separating paper is inserted. The organic phase is separated by its adhesion to these hydrophobic materials, and this process is further assisted by differential pumping so that the flow rate of the organic phase drawn from the T piece (and into the flow cell) is only 60–80% of the one entering the T piece (from the left, as shown in Fig. 4.30*b*). Thus, some of the organic phase and all the aqueous solution leaves for waste (*W*). The membrane phase separator (Fig. 4.30*c*) uses a microporous Teflon membrane with polyethylene backing, and is commercially available (Tecator/BIFOK). At flow rates of 0.5–1.0 mL/min, this device is capable of separating alcohols, alkanes, chlorinated hydrocarbons, and aromatic solvents from the aqueous phase with up to 95% efficiency.

The *gas-diffusion unit* (Fig. 4.33*c*) allows an efficient, rapid, and reliable transfer of a gaseous species from a donor to a recipent stream. The streams are pumped in parallel at equal or different rates, and during their passage through the unit they are separated by a gas-diffusion membrane. Shallow and long grooves allow efficient gas diffusion, without undue increase of dispersion and sample zone broadening. The choice of the membrane material is critical, Teflon and especially Cellgard being preferable. A gas-diffusion unit furnished with a microporous Teflon membrane is commercially available (TECATOR/BIFOK).

Integrated microconduits accommodate injection valve, connectors, confluence points, diffusion units, solid reactors, and detectors in a carefully designed geometry around a meandering channel imprinted into a rigid plastic plate, so that all dimensions together with the geometry of flow channel are selected to accommodate a given solution-handling task in the most economical way. Details are given in Section 4.12 (cf. also Section 5.1).

5.4 DETECTORS

It is beyond the scope of this book to review all detectors applicable to FIA and to specify in any detail the requirements for their performance characteristics. They are indeed many (cf. Table 7.1) and their number is steadily increasing. Most recently, besides detectors that we know from classical instrumental analysis, a new breed of sensors, such as transistors, piezoelectric and optical-fiber-based devices, are being incorporated into FIA systems either for sensing of analyte species or for testing of the performance of these sensing devices (cf. Chapter 8). Much can be learned from past experience with detectors used in flow analysis and even more from the practice of high-performance liquid chromatography [5.5–5.7]. Therefore, in this section emphasis is placed on specific requirements of FIA detectors so that readers may keep them in mind when choosing a detector for their particular task.

Similar to chromatographic detectors, the linearity, noise level, and peak broadening effects are important criteria valid to FIA detectors. Without quoting limits or actual values, one may say that the requirements for HPLC are as stringent as for FIA, and therefore a well-constructed HPLC detector will have a noise level so low that it will be suitable for high-sensitivity FIA measurements. Also, the degree of linearity of response will easily meet FIA requirements. The peak broadening effect has to be considered more carefully, as it is the result (a) of the flow velocity in, and (b) the holdup volume of the detector; (c) of the speed of detector response; and (d) of the time constant of the associated electronics [5.5–5.7]. As discussed in detail in Chapter 3, it is desirable that the peak broadening caused by the detector, expressed as its variance $\sigma^2_{\text{detector}}$, is no more than 5% of the overall peak width. Therefore, if a maximum sampling frequency of, say, 180 samples/h is to be obtained, σ_{overall} should not be more than 3.3 s, allowing σ_{detector} to be of maximum 1 s. Thus for a pumping rate of 1 mL/min and assuming that the detector cell behaves like a small mixing chamber, the volume of the detector must not, for $\sigma_{\text{detector}} = 1$ s, exceed 17 μL. {Assuming, rather optimistically, that plug flow prevails in the cell, the volume of the detector might be up to $\sqrt{12}$ as large (i.e., 57 μL) [5.7]}. As the associated electronics usually have a response time much shorter than 1 s, it is the cell holdup volume and its geometry that have the greatest influence on the peak broadening contributed by the detector. Generally, chromatographic detectors with cell volumes below 20 μL are well suited for FIA work. Alternatively, conventional detectors such as spectrophotometers and fluorimeters can be used, provided that their optical system operates with sufficiently coherent beams, thus being capable to accommodate flowthrough cells of sufficiently small apertures. If such an optical detector also has a sufficiently powerful light source, it can be equipped with optical fibers, which will guide the lightbeam into the FIA system and back into the instrument, thus allowing miniaturization of FIA manifolds. Prior to trying this technology the reader is advised to consult a suitable text [5.8] for potential, limitations, tools, and components of fiber optics. The investment of time will be amply returned since the technique has tremendous advantages for FIA applications. As one of the future innovations, the availability of superminiaturized, low-cost, solid-state lasers, as marketed by Hitachi, will bring further changes in the flow cell and fiber optic design.

Atomic absorption and inductively coupled plasma spectroscopy rely on the use of a nebulizer to introduce the analyte into a flame or a plasma. Presently, the nebulizers are clumsy devices with large internal volumes through which gases flow at high velocities while liquid samples are nebulized with a very low efficiency. The marriage of FIA with these techniques is both a gunshot and one of convenience, and in spite of certain incompatibility, a working compromise can be found, if guidance is taken along the lines discussed recently by Tyson [790]. As far as ICP is concerned, the micronebulizer recently developed by Fassel's group [1105, 1375] is a first step toward future harmony in performance parameters of the two partners.

Generally speaking, the main difference between chromatographic and FIA detectors is that the latter technique preferably uses selective detectors, whereas HPLC, where the separation occurs on the column, relies on nonselective detectors that should yield a readout for as many species (solutes) as possible, preferably with the same sensitivity. Thus, taking an example from the field of electrochemical sensors, a conductivity detector is ideal for ion chromatography, for example, whereas for FIA applications its usefulness is limited, as its response to anions and cations is nonselective. On the other hand, ion-selective electrodes have been successfully applied in FIA, for selective assays of individual cations or anions in mixtures, but they find no real application in chromatography. An exception to this rule is a case when a nonselective detector is coupled to a highly selective chemical reaction. Thus the evolution of light, caused by the nonselective chemical reaction between luminol and H_2O_2, will, when coupled to the selective enzymatic production of H_2O_2 from substrates by selective oxidases, result in selective determination of clinically important species. Scanning detectors, especially UV-VIS diode array devices, fulfill equally well the requirements of both HPLC and FIA, since their detection capability is both broad and selective.

It is important to realize that owing to the specific feature of FIA, that is, the formation of a concentration gradient of the dispersed sample zone, the function of the detector has to be considered from a somewhat different angle than for all other measurements. From this viewpoint all detectors fall into two groups depending on the way of probing the dispersed sample zone, that is, by bulk sensing or by surface sensing, the typical representatives of these two categories being optical detectors and electrochemical detectors. In most optical sensors the signal approximates the mean composition of the flowing stream present in the detector cell, and thus reflects the composition of the bulk of the solution. This is because in spectrometry the beam penetrates the sample zone [either radially, or more often axially (Fig. 5.17*a*)]. The microstructure of the stratified sample zone is integrated over the optical path length and the detector yields an instant response. In the theory of flow such a detector function is referred to as a mixing cup detector. Note that for this reason, the D, σ, or T values refer, in fact, to a nonexistent element of fluid, as they involve integration of a certain portion of the flowing stream by the beam—and therefore these values are affected by the geometry of the beam as it penetrates the flowing stream. Similarly, fluorescence and chemiluminescence (Fig. 5.17*b*) reflect the composition of the bulk of the solution, and so does atomic absorption flame photometry and ICP (Fig. 5.17*c*), where the sample is effectively nebulized into the flame. Among electrochemical methods conductimetry is the only one probing the bulk of the flowing stream.

Electrochemical methods, on the other hand, rely on transport of an electroactive species toward an electrochemically active surface. If the species is not effectively transferred from the bulk solution to the diffusion layer and across the diffusion layer to the sensing surface, it cannot be sensed. This is why in FIA, where the sample zone is stratified, the species

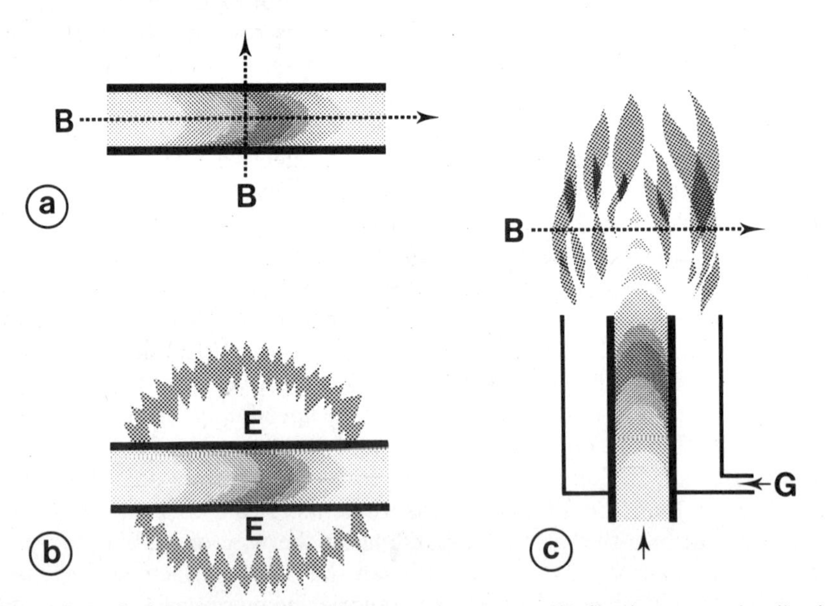

Figure 5.17. Optical detectors for (*a*) spectrophotometry, (*b*) fluorimetry or chemiluminescence, and (*c*) atomic absorption flame photometry and ICP: *B*, light beam; *E*, emitted light; *G*, gas inlet.

to be measured may in extreme cases pass the ion-sensitive surface unmonitored. This is most likely to happen when the measuring channel is straight (Fig. 5.18*a*), and is nearly certain if the sensitive surface (*S*) is, because of a faulty construction, recessed in the wall. Therefore, a wall-jet construction, well known from polarography and voltammetry, yields fast response and high sensitivity. This is best achieved by suddenly changing the direction of flow so that the stream impinges on the sensitive surface, which may be either integrated into the outer conduit wall, or attached perpendicularly to the end of the tube. A cascade type of flow cell (Fig. 5.18*b*) is a further extension of this principle [21]. A wire-type detector is the most effective to probe the center of the zone (Fig. 5.18*c*) but also the most awkward to construct.

Microminiaturization of the flow through cells has been, within the last few years, the most important development, which is best documented on practical examples. Thus for the most frequently used detector, a colorimeter or spectrophotometer, the glass made Hellma flow cell, Model OS 178.12, or the corresponding quartz made Hellma type flow cell (Fig. 5.19*a*) is still the most popular one; yet since it is expensive, and since it has a relatively large inner volume as well as an awkward inlet, it is being replaced by the so-called Z cell (Fig. 5.19*b*), which has been originally designed for HPLC. This is an easily adaptable construction, where the transparent windows (*A*), made of glass, quartz, or sapphire, enclose the optical cavity at both ends, forming an optical path (usually 10 mm long and 1.5 mm wide) within a Teflon body (*B*). It is advisable to mount this cell in such a way that the carrier stream enters it from below, as this facilitates the removal of entrapped air bubbles. The commercial cells

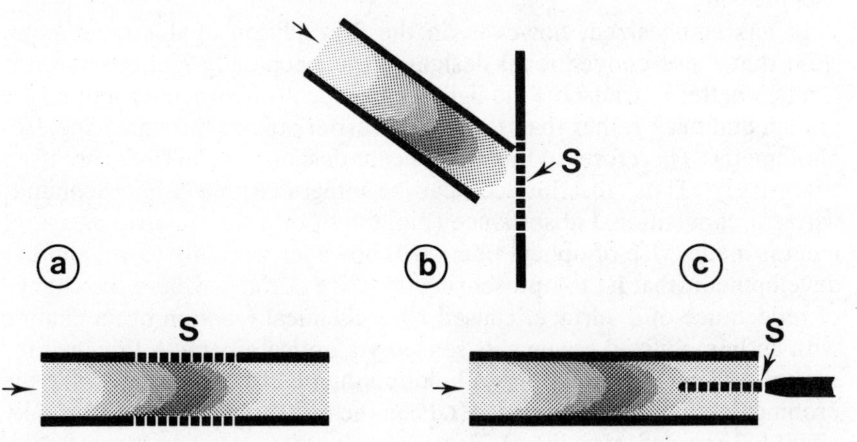

Figure 5.18. Electrochemical detectors (*S*, sensitive surface): (*a*), annular sensor; (*b*) cascade-type sensor (cf. Fig. 5.20*b*); (*c*) wire-type sensor.

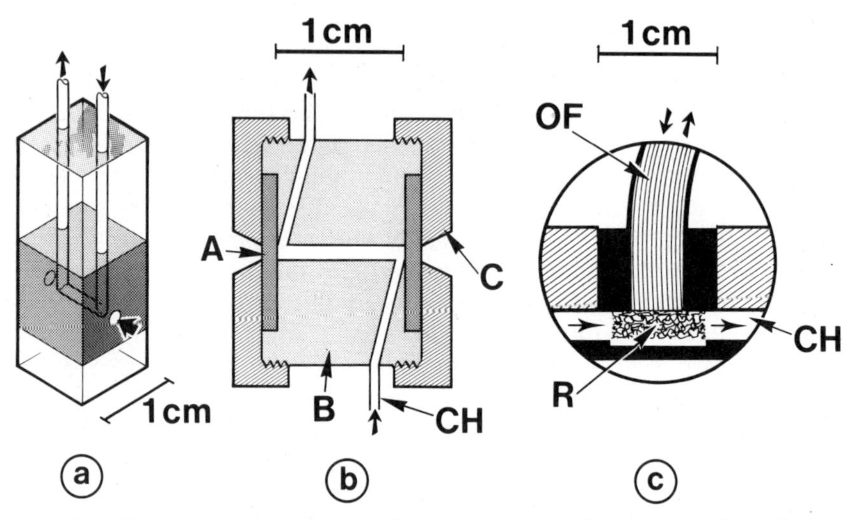

Figure 5.19. Progress in miniaturization of spectrophotometric flowthrough cells. (*a*) Hellma cuvette, which fits into most commercial photometers. (*b*) Z-cell, where *A* are the transparent windows; *B*, Teflon made body; *C*, cell house, and *CH*, the inlet channel. (*c*) A fiber optic reflectance cell for optosensing at active surfaces [848], where *CH* is the channel of a microconduit, *R* is the chemically active reflecting material, OF is the optical fiber.

designed for HPLC monitoring are always furnished with very narrow steel made connecting tubes (0.1 mm ID), which, for FIA use, have to be replaced (by 0.4-mm-ID tubes), as otherwise the low-pressure peristaltic pump and the FIA system will operate under unnecessary backpressure.

It was emphasized, however, in the first edition of this monograph [153] that these conventional designs are conceptually inappropriate; it is much better to transport the light from a spectrophotometer into a FIA system and back rather than to pump a carrier stream through a spectrophotometer. Therefore, in the most recent designs optical fibers are used extensively. Thus, the flow cell can be integrated into a microconduit, where a conventional absorbance (Fig. 4.69), or a fluorimetric measurement is made. Use of optical fibers led, however, recently to yet another development, that is, to optosensing at active surfaces where the change of reflectance of a surface, caused by a chemical reaction of an analyte with an immobilized reagent, is sensed via optical fibers. A flow cell for such a design (Fig. 5.19*c*) has a holdup volume of barely 1 μL since the probing depth of the fiber optics (*OF*) in the reflecting medium (*R*) is only about 0.3 mm (cf. also Fig. 4.73).

For electrochemical sensors (e.g., ion-selective electrodes) the miniaturization is easier to achieve, provided that the sensing area can be miniaturized. For electrodes with larger surfaces, this problem may be circumvented either by a wall-jet design (Fig. 5.20*a*) or by a casacade flow design (Fig. 5.20*b*). Although the holdup volume of these cells may be as large as 1 mL, the effective volume is very small (approx. 10 μL), as it constitutes only the very thin layer of the carrier stream that impinges onto the sensitive surface or cascades over the sensitive window of the electrode (marked *S* in the inset shown in Fig. 5.20*b*). For ion-selective electrodes furnished with PVC membranes, microminiaturization is easily achieved by implanting a silver wire (Ag) into a PVC made block furnished with a channel (Fig. 5.20*c*). Such a wire is furnished first with a reference layer, and then a PVC membrane is cast over it (m) so that it slightly protrudes into the channel (CH). The concept of coated wire electrodes, originally introduced by Freiser et al. [5.9], is thus fully exploited with the aid of the technology of integrated microconduits, where pH [608], potassium, nitrate, lithium [1200, 1251, 1327], and other ion-selective electrodes were successfully operated in miniaturized FIA systems. Obviously other electrode types and materials may be incorporated into FIA microconduits with the aim to perform conductimetry, voltammetry, or coulommetry in the microscale.

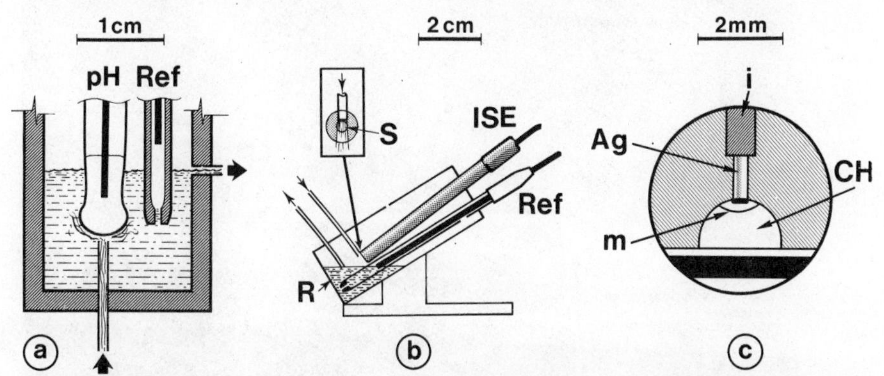

Figure 5.20. Progress in miniaturization of electrochemical sensors shown for potentiometric flowthrough cell designs. (*a*) Wall-jet configuration showing a glass electrode (pH) and a reference electrode (REF) in a larger reservoir furnished with a narrow jet from which the sample streams at the sensitive surface. The level of the liquid in the reservoir is maintained constant in the same way as in the next configuration. (*b*) Cascade-type cell housing an ion-selective electrode (ISE) and a reference electrode (REF). The carrier solution is pumped over the ion-sensitive surface *S* of the electrode, the level of liquid in the reservoir *R* is maintained constant by differential pumping. (*c*) Coated silver wire electrode (Ag) integrated into the channel (CH) of a microconduit [608], the PVC made membrane (m) containing an electroactive material (cf. Fig. 4.68).

5.5 COMMERCIALLY AVAILABLE FLOW-INJECTION ANALYZERS

Much has changed since the first commercial FIA analyzer was introduced by Bifok. This FIA 05 instrument (Ref. 153, Fig. 5.1), developed by the Swedish group of R. Lundin, T. Anfält, and B. Karlberg in a then typical "garage" enterprise (actually an abandoned milk store) had a motorized valve, modular flow system but no computer control of the analyzer functions. Later, FIAtron, Lachat and Control Equipment (all U.S.), Hitachi, Jasco, and Soma Kazaku (Japan) as well as Micronal (Brazil) entered the market with different designs. Meanwhile Bifok developed a new generation of FIA instruments and in a joint venture with Tecator (Sweden) marketed these analyzers worldwide.

Most recently a trend toward marketing of specialized FIA apparatus has begun, indicating that the technology is maturing. It is not surprising that such types of equipment are being introduced by those manufacturers who gained the experience by producing and marketing the first and second generations of FIA instruments. Thus, Tecator/Bifok produces, besides an all purpose laboratory instrument with an open structure (FIAstar 5020, Fig. 5.21), a specialized instrument for analysis of water and soil

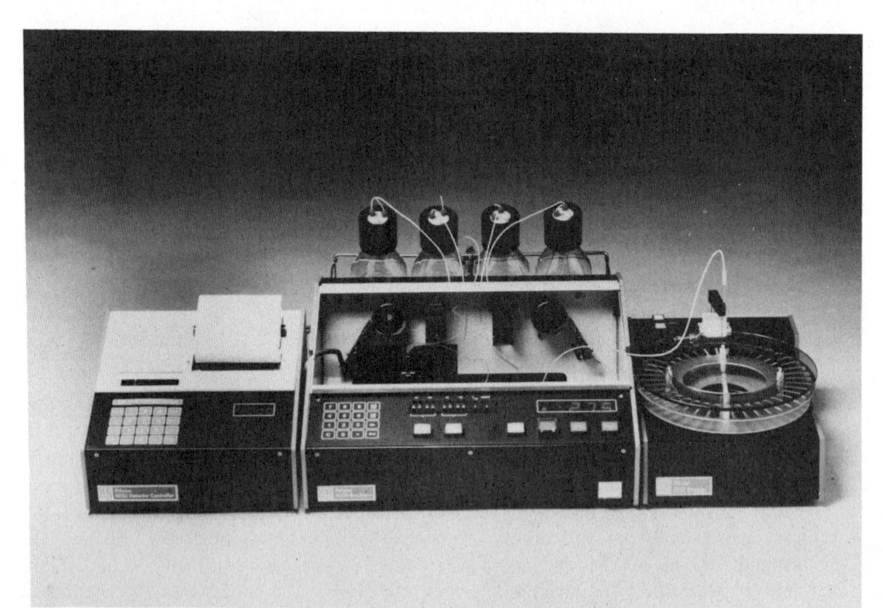

Figure 5.21. Bifok/Tecator FIAstar 5020 Analyzer (center) shown with the 5032 Detector System (left) and the 5007 Sampler (right). Comprising two independently programmable peristaltic pumps and injection valves, the analyzer applies small dedicated, ready-to-use chemical manifolds with fixed designs (Chemifolds), which are easily interchangeable. The compartment of the analyzer can be thermostated.

extracts (Aquatec, Fig. 5.22). Similarly FIAtron has introduced a very flexible line of modular laboratory equipment (valve module, all purpose detector module, peristaltic pump module with an ingenious magnetic clutch), along with a specialized FIA process control analyzer characterized by a robust design and pneumatic drive. For the reader's convenience the features of some of the currently commercially available FIA equipment, as kindly communicated to us by the manufacturers at the 1987 Pittsburgh Conference, are summarized below.

Bifok/Tecator (Bifok AB, P.O. Box 124, S-191 22 Sollentuna, Sweden, Tecator AB, P.O. Box 70, S-263 01 Höganäs, Sweden, or P.O. Box 405, Herndon, VA 22070, USA). The FIAstar 5020 Analyzer is a modular microprocessor controlled FIA system comprising two independently programmable peristaltic pumps, a motorized single or dual-channel injector, microprocessor-controlled photometer or spectrophotometer, printer and optional sample changer. The instrument allows all FIA modes, including stopped-flow, hydrodynamic injections, gas diffusion, dialysis, solvent extraction and preconcentration, to be performed by means of Chemifolds adaptable for a variety of flow designs. Besides

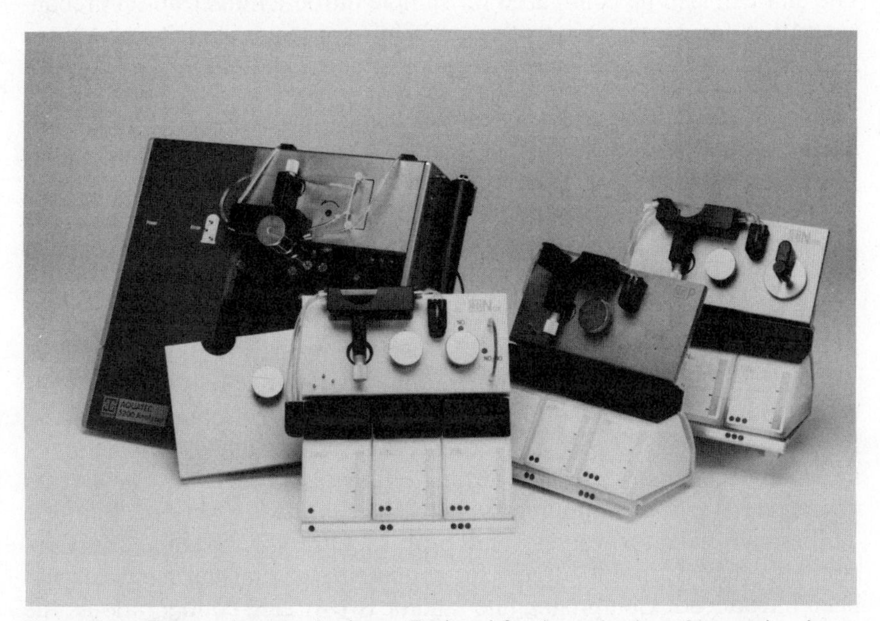

Figure 5.22. Tecator Aquatec Analyzer. Designed for determination of ammonia, nitrate, nitrite, and phosphate in aqueous samples, the microprocessor controlled instrument (left) employs dedicated modules comprising microconduit based Chemifolds and furnished with prepacked reagent solutions (right), facilitating rapid change of function and simplifying the operation of the system.

research in advanced FIA techniques, the FIAstar 5020 is suited for a number of routine assays described in application notes for industrial, pharmaceutical, and agricultural applications. Aquatec is a specialized analyzer for determination of ammonia, nitrate, nitrite, and phosphate in water and aqueous samples such as soil extracts. It is a compact microprocessor-controlled instrument with a minimum of operational requirements. Being based on microconduit technology, it has a good economy in terms of reagent and time consumption. Its modular design and prepacked reagents allow simple and rapid change of function, thus making the instrument economical even for small sample series.

Control Equipment Corporation (171 Lincoln Street, Lowell, MA 01851, USA). The AMI-103 FIA Analyzer is designed to automate a wide variety of chemical analyses for the environmental, pharmaceutical, food and beverage, and biotechnical industries. Utilizing a chemically inert gas displacement pumping system and a unique injection-valve arrangement, the system can process up to 120 samples/h with no constraints on system chemical compatibility. Detectors are available for UV/VIS, fluorimetric, electrochemical, and polarographic analyses for most sample matrices. The unit can also be configured for sample introduction/clean-up/preconcentration for atomic absorption/emission spectrophotometers. An automatic sample changer is an integral part of the system.

FIAtron Systems, Inc. (510 S. Worthington Street, Oconomowoc, WI 53066, USA). The FIAtron laboratory equipment is a line of modular microprocessor-based pumps, sample inject and selector valves as well as a range of flowthrough detectors. Each detector consists of a master module and a detector drawer. The FIA-DUCT (conductivity), FIA-TRODE (potentiometry), FIA-ZYME (enzyme-amperometry), and FIA-LITE (colorimetry) detector drawers are interchangable and automatically adapt the master to become a specific analyzer. The sophisticated software and hardware allows integration of external valves with the FIAtron pump and interfacing of the modules to external computers via an RS-232 and serial I/O port. Application notes are available for a series of automated chemistries. The FIAtron on-line process control analyzers comprise sample conditioning systems, that is, heated process dilutors, multistream selector valves, and process filtration systems. Most recently, a dual-endpoint titrator for processing of unusual materials has been introduced. Comprising one- and/or two-stream configurations, the on-line FIA analyzer is a gas-driven, microprocessor-controlled instrument contained within a rugged, enclosed, gas-purgeable cabinet. The process applications range from determination of free fatty acid and iodine values in edible oils to the determination of ammonia and peroxides in organic solvents.

Hitachi Ltd. (Instrument Div., Shin Maru Bld. 5-1, Marunouchi Chiyoda-Ku, Tokyo, Japan 100). The company offers a laboratory instrument for serial assays of industrial samples in a single unit. The carrier stream is propelled by a piston pump, samples are aspirated into the injector by means of a peristaltic pump. A sophisticated 16-port valve allows combination of sample and reagent solutions to be introduced by the merging zone technique. General purpose instrument for industrial and environmental analysis. Detector systems have to be acquired separately.

Lachat Instruments (10500 N. Port Washington Road, Mequon, WI 53092, USA). The QuikChem Automated Ion Analyzer is an integrated system of modular components that operates on the principle of flow injection analysis. The flexibility of the technology enables the system to determine the diversity of analytes in both simple and complex matrices from the ppb to percent levels. The system is characterized by rapid startup and shutdown (5 min each), short analysis times (<1 min), high sample throughput (60–360 samples/h), fast method switching (<10 min), complete intersample washout, and simultaneous determination of up to four analytes in a single sample.

REFERENCES

5.1 B. C. Erickson, J. Ruzicka, and B. Kowalski, *Anal. Chem.*, **59** (1987) 1246.

5.2 H. Bergamin F° (private comm.).

5.3 J. Růžička and J. C. Tjell, *Anal. Chim. Acta,* **47** (1969) 475.

5.4 B. Neidhart (private comm.).

5.5 W. B. Furman, *Continuous Flow Analysis: Theory and Practices* (M. K. Schwartz, ed.). Marcel Dekker, New York, 1976.

5.6 R. P. W. Scott, Liquid Chromatography Detectors, *Journal of Chromatography Library,* Vol. 11, Elsevier, Amsterdam, 1977.

5.7 J. F. K. Huber (ed.), Instrumentation for High-Performance Liquid Chromatography, *Journal of Chromatography Library,* Vol. 13, Elsevier, Amsterdam, 1978.

5.8 *Fiber Optics Handbook.* Hewlett-Packard GmbH, Boeblingen Instrument Division, FRG, 1983.

5.9 H. Freiser, *Ion-Selective Electrodes in Analytical Chemistry.* Plenum Press, New York, 1978.

CHAPTER

6

EXPERIMENTAL TECHNIQUES AND FIA EXERCISES

If anything can go wrong, it will.

MURPHY

Within the last few years FIA has been incorporated into numerous undergraduate and graduate analytical chemistry courses. Furthermore, the literature records an increasing number of papers that have become the quintessence of many dissertations. Thus, FIA has found its way to academia at all levels, and it is the purpose of this chapter to support this development by summarizing the existing material, by illustrating it with selected exercizes, by offering practical hints to those who are about to use FIA for the first time, or to help those who wish to introduce FIA into their courses, and to exploit the most recent development in this area, the microconduit-based pedagogical FIA system. Additional information on the educational aspects of FIA can be found in the works referenced in Chapter 7 [59, 139, 253, 299, 333, 446, 727, 1224, 1265].

The first part of this chapter deals with practical details of preparing a FIA apparatus for analysis; the second part deals with six colorimetric exercises, four of which have for several years been part of an undergraduate laboratory course in instrumental analysis at our department [59] (in this reference, applications of the exercises on practical problems are detailed). The third part of the chapter contains a brief description of a method for diagnosing the types of errors and malfunctions most frequently encountered in practical work with FIA systems.

6.1 PRACTICAL CONSIDERATIONS

Before any chemical procedure is adapted to FIA, there are various practical, often trivial, considerations to be made, such as whether the reagents to be used will attack or degrade the material of one or several components of the analyzer, or whether the samples to be assayed need to be pretreated (filtered, diluted, neutralized) to render them suitable for injection. Ignoring these basic requirements may cause serious problems.

In both commercial and home-built FIA instruments, the injection valve is made of PVC and/or Teflon, the coils and connecting tubes of polyethylene, polypropylene, or Teflon, and the pump tubes of PVC. All these materials are fairly resistant to polar solvents, and as long as aqueous solutions are used throughout, no problems will be encountered. Yet if nonpolar solvents have to be handled, such as in solvent extraction, it is necessary to use pump tubes resistant to these solvents [e.g., Acidflex or Viton tubes made of black flurorplast rubber, or the recently introduced Marprene tubes (see Table 5.1)], Teflon coils, and to construct the system so that the aqueous samples are injected into an aqueous carrier stream (i.e., the Perspex or PVC made parts of the injection port do not contact the organic solvent). For FIA procedures performed exclusively in non-aqueous media, compatible materials have to be chosen carefully and the injection valve must be made entirely of Teflon or a stainless-steel–Teflon combination, such as that used in chromatographic valves.

As long as flowthrough detectors made of glass, silica, or Teflon are used, organic solvents are harmless. However, some flowthrough cells are constructed, either partly or entirely, of plastic-type materials and they might be attacked by organic solvents. In this context special attention should be paid to potentiometric sensors furnished with PVC membranes. Even small amounts of organic solvents (e.g., chloroform added to wastewater samples as preservative) might lead to deterioration of the function of the electrode.

The reagent solutions, especially when prepared from distilled water supplied from a central unit via pressurized pipelines, may contain considerable amounts of dissolved air or gases, which can become liberated as microbubbles in the FIA system, causing noisy signals. Although the signal from an air bubble is readily detectable [see Fig. 6.10(2*b*)], it is nevertheless a source of irritation, and frequent bubbles will render the measurement impossible. Thus, it is advisable to degas all reagent solutions prior to use. This is readily achieved within a few minutes by *stirring* the solutions, using a magnetic stirrer, in a slightly evacuated Erlenmeyer flask, the vacuum being provided by a water pump or a simple hand-operated pump.

In this context, attention should be drawn to those procedures where formation of gases might take place in the FIA system itself. This will happen whenever the partial pressure of the gas exceeds that corresponding to the solubility of the gas under the prevailing conditions, thus leading to formation of microbubbles. Thus without protection from the ambient air, alkaline reagent solutions will absorb carbon dioxide, and if the reagent stream in the analyzer is subsequently acidified, carbon dioxide may be released.

6.2 STARTING A FIA SYSTEM

Before any analysis the FIA apparatus should be thoroughly checked. After it has been ensured that all components have been assembled correctly, distilled water should be pumped through all tubes and a visual inspection carried out to establish whether there are any leaks. A practical hint: if small air bubbles are trapped in the system—and this might happen if the system has been "dry" or if the tubes are new—they can easily be swept out by introducing several larger air segments into the system, either by means of the injection valve or via one of the reagent pump tubes, by removing its supply end from the reagent vessel for short periods of time.

It is always advisable to perform a number of functional tests and dispersion measurements on a newly designed manifold. This is helpful if the user is not familiar with the technique. The time so spent will result in a better understanding of the physical functions of the system so that problems, connected with the chemical reactions on which the analytical procedure is based, will be identified and readily solved.

6.3 FUNCTIONAL TESTS AND MEASUREMENT OF DISPERSION

To operate the FIA system successfully, the injection valve must function smoothly, without any leakage, thus allowing reproducible injection. This is easily checked by pumping a suitable colorless buffer solution through the system and then repeatedly injecting an exact volume of a colored dye—as metered by the injection valve—and recording the peaks as detected by means of a flowthrough spectrophotometer (see Fig. 6.1, left). By recording the peaks at high paper speed (Fig. 6.1, right) one may simultaneously determine the residence time T, the dispersion coefficient D^{max}, the axial dispersion σ_t, and the dispersion coefficient $\beta_{1/2}$. This allows computation of the time needed to flush the sample out of the system and the maximum frequency with which the samples may be injected into the system without intermixing [cf. Eqs. (2.8)–(2.11) as well as Eq. (3.5)].

As repeatedly emphasized in this book, the dispersion coefficient D^{max} or D is the key parameter in FIA, as it indicates the degree of dilution of the original sample solution and thus the extent to which the injected

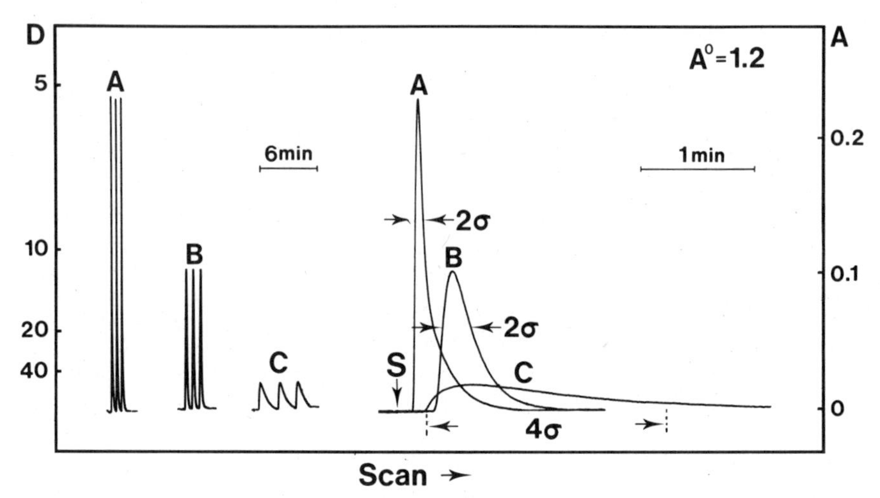

Figure 6.1. FIA response curves, as obtained with the manifold in Fig. 6.2 by injecting 30 μL aliquots of a dye "sample" solution of absorbance $A^0 = 1.2$ into an inert carrier stream, recorded at low (left) and high (right) paper speed. Peak series *A*, *B*, and *C* correspond to the manifold *A*, *B*, *C* in Fig. 6.2.

sample is mixed with the carrier stream (reagent). In conjunction with knowledge of the residence time *T*, this parameter will readily allow comparison of conditions in a FIA system with the dilutions and reaction times of conventional manual or automated procedures. Indeed, the *D* value should always be determined for any type of FIA manifold and reported with the analytical results obtained with it. Therefore, the first exercise in this chapter is devoted to measurement of *D* (and *T* and $\beta_{1/2}$) in three manifolds. It is hoped that these experiments may give an intuitive feeling for the magnitude of these parameters in manifolds of different types and assist in a better understanding of the more abstract concepts discussed in Chapter 3.

The dispersion coefficient *D* is defined as the ratio of the original analyte concentration C^0 to the concentration of analyte in that element of fluid that yields the analytical readout (cf. Chapter 2). Restricting ourselves in these exercises to the element that corresponds to the maximum of the peak, C^{max}, $D = D^{max}$ might be determined simply by comparing the peak signal obtained by injecting a sample of colored dye solution, and the signal recorded when the flow cell has been completely filled with undiluted dye solution. (This may be achieved either by pumping the dye solution through all tubes, or by stopping the FIA pump, disconnecting the tube leading to the flow cell, and filling it by a syringe with the dye solution.)

The manifold for all three experiments (Fig. 6.2, *A–C*) uses the same pumping rate (1.5 mL/min), the same carrier stream solution (borate buffer), and the same dye solution (of bromothymol blue). The "reactor" in experiment *A* is simply a short (20-cm) tube (0.5 mm ID) connecting the injection valve and the flow cell (volume 18 μL) in the spectrophotometer, which is adjusted to 620 nm. (Actually, in the student exercises [59], a simple Corning 252 filter colorimeter was used.) For experiment *B*, the reactor consists of a 30-cm coil (0.5-mm-ID) which has been inserted into the previous manifold, yielding a total length of connecting tube equal to 50 cm. For experiment *C* the 30-cm-long coil is replaced by a miniaturized mixing chamber (volume 0.98 mL—cf. Fig. 6.7*c*), equipped with a small magnetic stirrer, placed on a magnetic stirring table.

Carrier Stream Solution. 1×10^{-2} *M* aqueous borax solution.

Dye Solution. A dye stock solution is prepared by dissolving 0.400 g of bromothymol blue in 25 mL of 96% ethanol, making the final volume up to 100 mL with the 1×10^{-2} *M* borax solution. The dye solution used is then prepared by mixing 1 mL of the stock dye solution with 199 mL of the 1×10^{-2} *M* borax solution (the absorbancy of this solution in a 10-mm-long cell is 1.2 A).

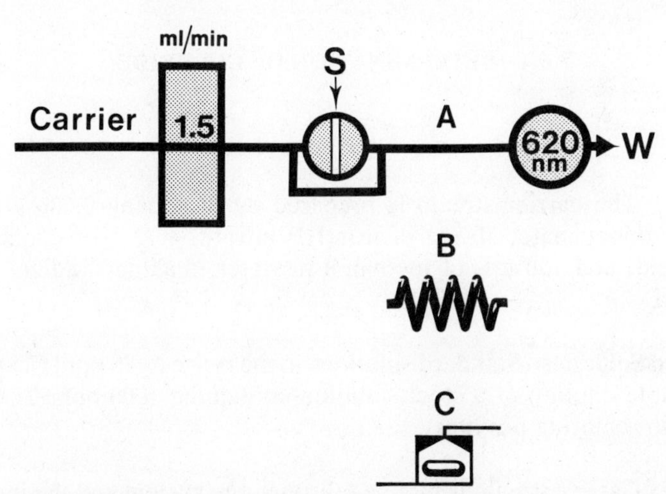

Figure 6.2. The manifolds for dispersion experiments. As carrier stream serves a 1×10^{-2} *M* borate buffer solution into which 30 μL aliquots of dyed sample solutions are injected. The reactor is (*A*) a 20-cm, 0.5-mm-ID tube; (*B*) a 0.5-mm-ID coil of total length 50 cm; and (*C*) a mixing chamber of volume 0.98 mL, equipped with a small magnetic stirrer, placed on a magnetic stirring table (cf. Fig. 6.7*c*).

Exercise. The FIA system, assembled according to Fig. 6.2(*A*), is used to record the curves shown in Fig. 6.1(*A*), first as repetitive injections recorded at a paper speed of 1 cm/min; then at a higher paper speed the peak is recorded to allow measurement of σ_t or $\beta_{1/2}$. Experiments *B* and *C* are performed in the same manner, with the coil and mixing chamber, respectively, included in the manifold. Values of D^{max}, *T*, and S_{max} obtained from the parameters of the recorded curves might then be computed and compared. For experiment *A* the standard deviation of peak heights should not be greater than $\pm 1\%$, and for experiment *B* not more than $\pm 0.5\%$. When measuring D^{max}, note that $D^{max} = C^0/C^{max} = A^0/A^{max}$, where A^0 is the absorbance measured with the cell filled by the undiluted dye solution ($A^0 = 1.2$ for the recording shown in Fig. 6.1).

Additional experiments can be made by (a) repeatedly injecting dye solutions of various concentrations in manifold *B* and computing standard deviations for the peak heights in the thus-simulated calibration curve; (b) varying the injected sample volume and computing $\beta_{1/2}$ and S_{max} and verifying Eq. (2.3) [see Fig. 2.6*a* as well as Eq. (3.51) and Fig. 3.15]; (c) investigating the stopped-flow method (see Figs. 2.7 and 4.11); and (d) changing the coil length *L* and geometry (straight, coiled, or knitted tube, cf. Figs. 2.8 and 2.10) and computing and comparing $\beta_{1/2}$ [Eq. (2.8)], *N*, and *H* (Section 3.7).

6.4 DETERMINATION OF CHLORIDE

Manifold: Fig. 2.1a.

Reagent: The carrier stream is prepared by dissolving 0.626 g of mercury(II) thiocyanate, 30.3 g of iron(III) nitrate, 4.72 g of concentrated nitric acid, and 150 mL of methanol in water, making the final volume up to 1 L.

Standard Solutions: Standard solutions in the range 5–75 ppm Cl are made by suitable dilution of a stock solution containing 1000 ppm Cl (1.648 g of sodium chloride per liter).

Exercise: Carrier stream is pumped through the system and the individual Cl standards are injected successively in quadruplicate, thus yielding a recording such as that shown in Fig. 2.1*b* (left). (Note that the waste is toxic and should be collected, not discharged into the sink.)

The analytical procedure is based on the following reactions:

$$Hg(SCN)_2 + 2Cl^- \rightarrow HgCl_2 + 2SCN^-$$
$$2SCN^- + Fe^{3+} \rightarrow Fe(SCN)_2^+$$

The carrier stream contains $Hg(SCN)_2$ and Fe(III). The chloride of the injected sample reacts with $Hg(SCN)_2$, liberating SCN^-, which in turn forms with Fe(III) the red-colored complex ion $Fe(SCN)_2^+$, the intensity of which is measured spectrophotometrically at 480 nm. The height of the recorded absorbance peak is then proportional to the concentration of chloride in the sample (see Fig. 2.1*b*). Besides $Fe(SCN)_2^+$, other (higher) complex ions might be formed; thus the calibration curve cannot be expected to be linear over wide range of concentrations.

The reproducibility of the procedure might be estimated by calculating the standard deviation on the peaks obtained by injecting one of the standards 10 times. Note that the coil length is the same as that used in the exercise in Section 6.3, experiment *B*; that is, the dispersion coefficient D^{max} can be estimated directly from the previous exercise, although the pumping rate has been halved. Also compare σ_t values and the resulting S_{max} and consider the differences.

6.5 DETERMINATION OF PHOSPHATE

Manifold: Fig. 6.3*a*

Reagents: The carrier stream consists of a mixture of two solutions pumped at equal rates; that is, (a) 0.005 *M* ammonium heptamolybdate (6.1793 g/liter) in 0.4 *M* nitric acid; and (b) 0.7% (w/v) aqueous solution of ascorbic acid to which is added 1% (v/v) glycerine (to minimize precipitation of reaction products on the walls of the flow cell).

Standard Solutions: Standard solutions in the range 0–40 ppm $P\text{–}PO_4$ is prepared by successive dilutions of a 100 ppm phosphate stock solution (0.4390 g of anhydrous potassium dihydrogen phosphate per liter).

Exercise: The analytical procedure is based on the following reactions:

$$H_3PO_4 + 12H_2MoO_4 \rightarrow H_3P(Mo_{12}O_{40}) + 12H_2O$$
$$Mo(VI) \xrightarrow{\text{ascorbic acid}} Mo(V)$$

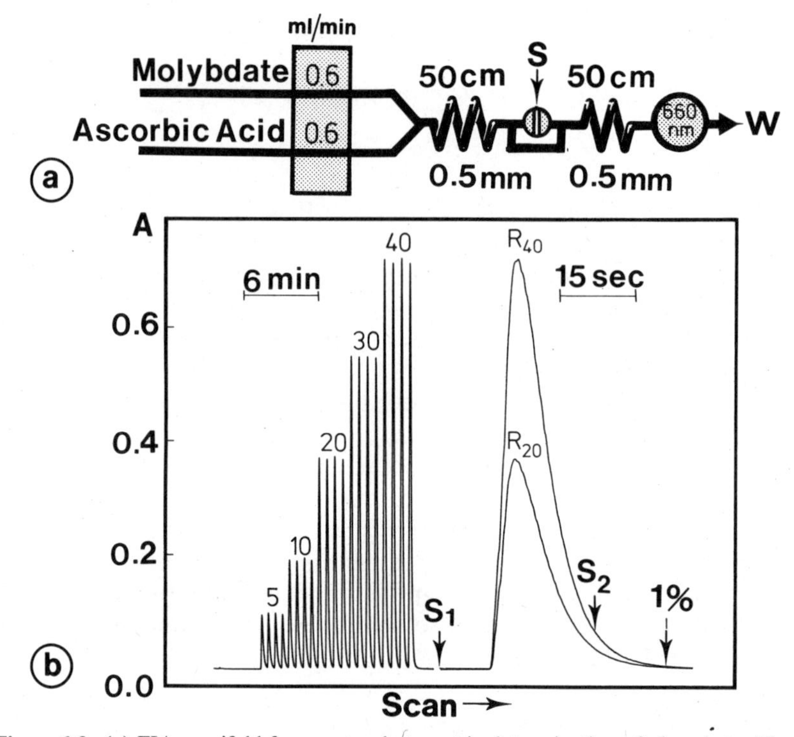

Figure 6.3. (*a*) FIA manifold for spectrophotometric determination of phosphate. The two reagents are premixed in the first coil, whereupon sample is injected (30 μL). All tubes are 0.5 mm ID. (*b*) Left: record obtained by injecting standards in quadruplicate, containing 5–40 ppm P–PO_4; the record to the right shows a scan where the time scale is expanded to show the peak shape when injecting 20 and 40 ppm solutions. Note that it takes only 15 s between sample injection *S* and peak maximum readout *R*, and another 15 s until the next sample (S_2) can be injected. Hence, the signal will be below the 1% level before the next readout will be taken, and therefore there is no carryover even at a rate of 120 samples/h.

The carrier stream contains molybdate and ascorbic acid. Since a mixture of these two components is not stable, they are mixed in the system ahead of the sample injection valve. Phosphate forms with molybdate a heteropolyacid in which molybdenum can be reduced from oxidation state 6 to 5 with ascorbic acid forming an intensely blue complex which can be measured spectrophotometrically at 660 nm. Although the first reaction is fast, the second one is relatively slow; however, the precise timing of the FIA system ensures that the same fraction of the heteropolyacid is reduced. Thus the recorded absorbance peak is proportional to the concentration of phosphate present in the sample (Fig. 6.3*b*).

The reproducibility of the method is determined as previously described—by injecting a standard solution 10 times and calculating the standard deviation.

Finally, note that in this case the coil length is also 50 cm, as in experiment *B* of the exercises in Sections 6.3 and 6.4; that is, D^{max} will be very similar in all three cases, despite the slightly different pumping rates. Thus the phosphate manifold may actually be used for determining chloride simply by pumping the pertinent carrier stream through both pump tubes (and changing the filter in the photometer), with the result of higher S_{max} (because of lower σ_t).

An inherent drawback of the system depicted in Fig. 6.3*a* is that with a sample volume of 30 μL, the maximum sensitivity reached is only about 5 ppm P–PO_4. Yet as derived from Rule 1 in Chapter 2, the sensitivity of measurements is, within a certain range, proportional to the injected sample volume S_v. Thus by increasing S_v above 30 μL, the detection limit and sensitivity of the assay will be further increased. This will be demonstrated in the next exercise.

6.6 HIGH-SENSITIVITY DETERMINATION: RELATIONSHIP OF SAMPLE VOLUME TO PEAK HEIGHT

Manifold: Fig. 6.4a, where the injection valve is the type furnished with an external, interchangeable sample loop (Section 5.1). The following sample volumes are used (the number in parantheses denoting the corresponding lengths in cm of 1.14 mm tubing except for the smallest volume, where a 0.7-mm-ID tube is used): 23 μL (6.0); 70 μL (6.9); 115 μL (11.3); 180 μL (17.6); and 320 μL (31.3).

Reagents: Molybdate solution: 0.0025 *M* ammonium heptamolybdate (3.0897 g/liter) in 0.2 *M* nitric acid. Ascorbic acid solution: 0.25% (w/v) in 10% (v/v) aqueous glycerine solution.

Standard Solutions: Standard solutions in the range 0–2 ppm P–PO_4 are prepared by successive dilutions of the 100 ppm stock solution described in the preceding exercise.

Exercise: The analytical procedure is identical to that described in Section 6.5. However, as the manifold here is designed for high-sensitivity measurements of phosphate, large sample volumes are injected. Therefore, a manifold such as that shown in Fig. 6.3*a* cannot be used, as this would result in insufficient mixing and formation of double peaks (see Fig. 6.6).

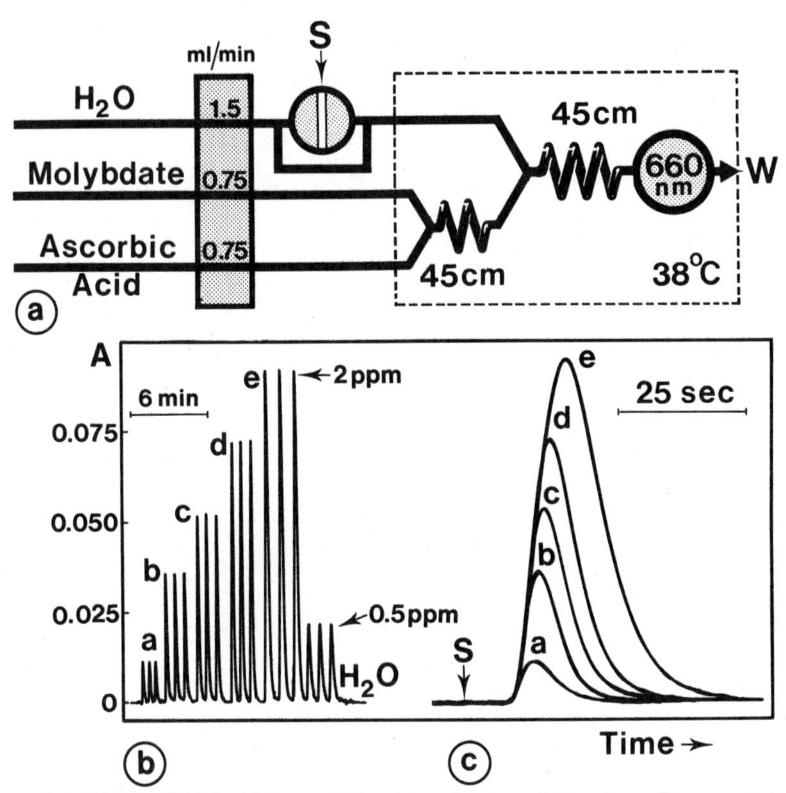

Figure 6.4. (*a*) Manifold for high-sensitivity determination of phosphate. The sample is injected into an inert carrier stream (water) merged with the reagent solution, and, after mixing in the reaction coil, it is measured spectrophotometrically. To obtain effective mixing without excessive dispersion of the sample solution, the confluence point is fashioned as shown in Fig. 4.9. All tubes are 0.5 mm ID. To increase the rate of reaction, the system is thermostated at 38°C. (*b*) At the left are the records obtained by injecting increasing volumes of sample of a 2 ppm P–PO_4 standard solution (S_v being: *a*, 23 μL; *b*, 70 μL; *c*, 115 μL; *d*, 180 μL; and *e*, 320 μL), followed by the output for a 0.5 ppm P–PO_4 sample solution, where S_v = 320 μL. To the right are shown peaks *a–e*, recorded at high paper speed. Miniaturization of this exercise is achieved by using the system depicted in Figs. 6.11 and 6.12.

To avoid this phenomenon, the sample is injected into a stream of water, which is then *merged* with the combined reagent stream of molybdate and ascorbic acid. Furthermore, to obtain effective mixing without excessive dispersion of the sample solution, either the streams are confluenced as shown in Fig. 4.9 or an imprinted meander (in a microconduit) is inserted immediately following the confluence point, permitting the use of a short reactor before the flowthrough cell.

In Fig. 6.4*b* are shown the readouts obtained by triplicate injections of a 2 ppm standard solution with gradually increasing sample volumes. Note that the peak height (absorbance) increases with increasing sample volume (tending toward a maximum value corresponding to an absorbance $A^{max} = A^0 \cong 0.105$; Fig. 6.5) and that for lower S_v values there is an almost linear relationship between S_v and peak height (absorbance), whereas S_v versus log $(1 - A^{max}/A^0)$ yields a linear relationship over the entire range [cf. Eq. (2.3)]. Yet, while the sensitivity is increased with increasing sample volume, the residence time T of the sample increases as well (Fig. 6.4*c*), leading to a decrease in the overall sampling frequency. These combined effects might be observed by comparing the first three peaks of Fig. 6.4*b* (2 ppm, S_v = 23 μL, T = 13 s) with the last three peaks (0.5 ppm, S_v = 320 μL, T = 20.5 s), from which it appears that although the injected volume is increased almost 14 times, leading to an eightfold relative increase in the recorded signal, the residence time T is increased only about 60%, while σ_t is more than doubled (see Fig. 6.4*c*).

The exercise can be further extended by investigating the relation be-

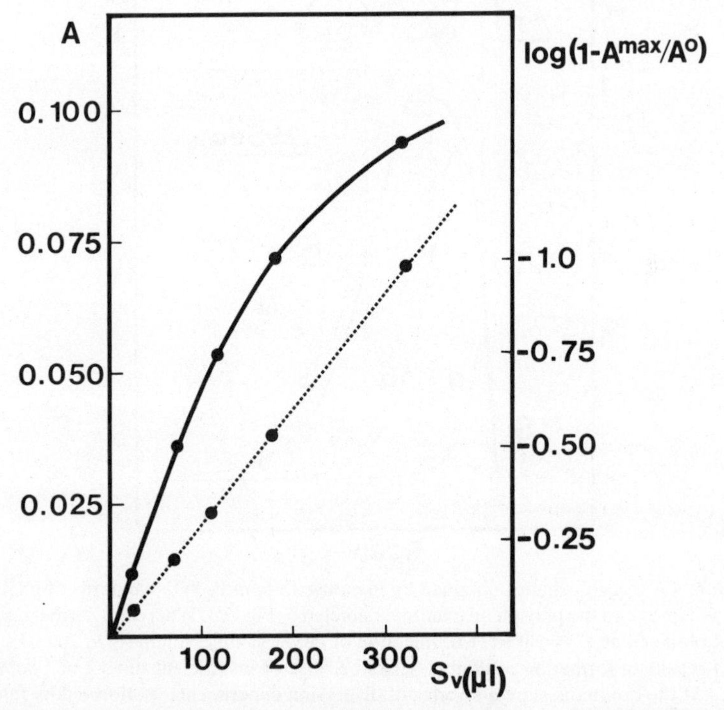

Figure 6.5. Plot of the peak height values (A^{max}) for samples *a–e* of Fig. 6.4 versus injected sample volume, S_v (solid line). In accordance with Eq. (2.3), a plot of log$(1 - A^{max}/A^0)$ versus S_v yields a linear relationship (dashed line).

tween dispersion of the sample zone and the sensitivity of measurement, that is, by showing how insufficient mixing will adversely affect the measured signal. As explained previously (Sections 2.4, 2.5, and 4.2.1) the reagent has to penetrate, in sufficient concentration, into the *center* of the sample zone, where the peak height is actually measured. Therefore, the confluence manifold (Fig. 6.4*a*) has to be used when sample volumes exceeding 30 μL are injected. To investigate the influence of insufficient mixing, the manifold shown in Fig. 6.3*a* is used, and sample volumes of

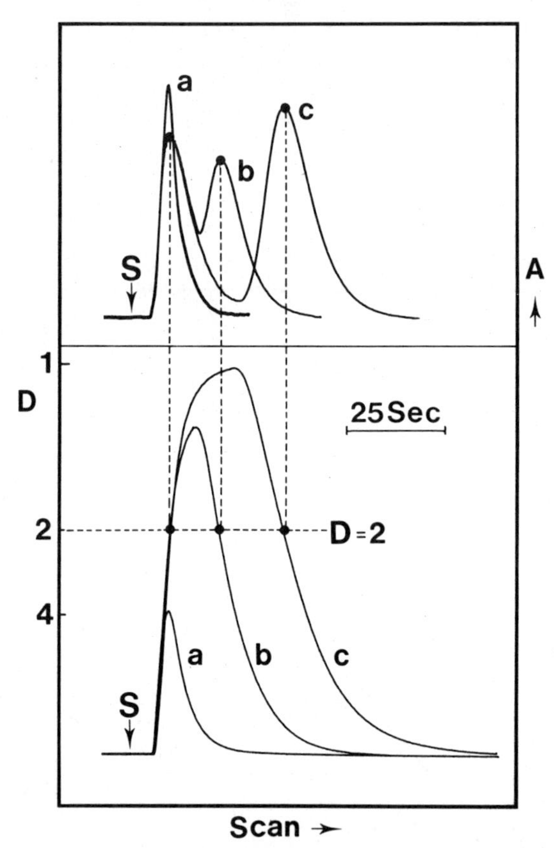

Figure 6.6. (*Top*) Peak shapes obtained by injecting 15 ppm P–PO_4 standards of increasing sample volumes into the phosphate manifold depicted in Fig. 6.3. Whereas a normal response curve is obtained at S_v = 30 μL (*a*), injection of larger sample volumes (*b*, 200 μL; and *c* 400 μL) results in formation of double peaks because of insufficient mixing of sample and reagent. At the bottom is shown a series of dispersion experiments, performed by injecting dye solution into an inert carrier stream, using the same manifold and injecting the same three sample volumes. By comparing these two figures, it is apparent that lack of reagent occurs for dispersion coefficient values (*D*) equal to or smaller than 2.

30, 200, and 400 μL containing 15 ppm of P–PO_4 are injected. Thus, while a satisfactory response will be obtained for the 30 μL sample (Fig. 6.6 top, curve *a*), larger sample volumes will result in the formation of double peaks (curves *b* and *c*). It is interesting to note that (a) the leading edges of all three curves are identical; (b) the larger sample volume gives a lower peak; and (c) the concentration gradients formed are highly reproducible, yielding identical shapes of certain curve sections, which even partially coincide for the falling edges of curves *b* and *c*. The proof that insufficient, although highly reproducible, mixing is the reason for these peculiar peak shapes is obtained through dispersion experiments (Fig. 6.6 bottom) executed in the same manifold in the usual way, showing that for given initial sample and reagent concentrations the lack of reagent becomes apparent for D values equal to or smaller than 2 (cf. Section 2.4).

6.7 TITRATION OF STRONG ACID WITH STRONG BASE

Manifold: Fig. 6.7a. Note that in this exercise, the injection valve is furnished with an external sample loop of a volume of 125 μL (12.1 cm of 1.14-mm-ID polyethylene or polypropylene tubing). In the latter part of the exercise the sample loop is replaced by loops of different volumes (i.e., 75, 100, 150, and 200 μL, corresponding to 7.3, 9.8, 14.7, and 19.6 cm, respectively, of 1.14-mm-ID tubing).

Reagent: The carrier stream (1.0×10^{-3} *M* NaOH, carbonate free) is prepared by mixing 5 mL of 0.1 *M* sodium hydroxide (it is convenient to use, for example, Titrisol standard solution; Merck, Germany) with 50 mL of 96% ethanol and 2 mL bromothymol blue indicator stock solution (see Section 6.3), making the final volume up to 500 mL with distilled water.

Standard Acid Solutions: Standard acid solutions in the range 8×10^{-3}–1×10^{-1} *M* are made by appropriate dilutions of a 0.1 *M* hydrochloric acid stock solution (e.g., Titrisol).

Exercise: The quantitative determination of the acid concentration is based on titration of the acid (HCl) with base (NaOH), the acid samples being injected into a FIA system comprising a mixing chamber (Fig. 6.7*a*) in which a controlled dilution gradient of the samples can be created (cf. Sections 2.4.4 and 4.9). As carrier stream is used a 1.0×10^{-3} *M* NaOH solution containing the indicator bromothymol blue.

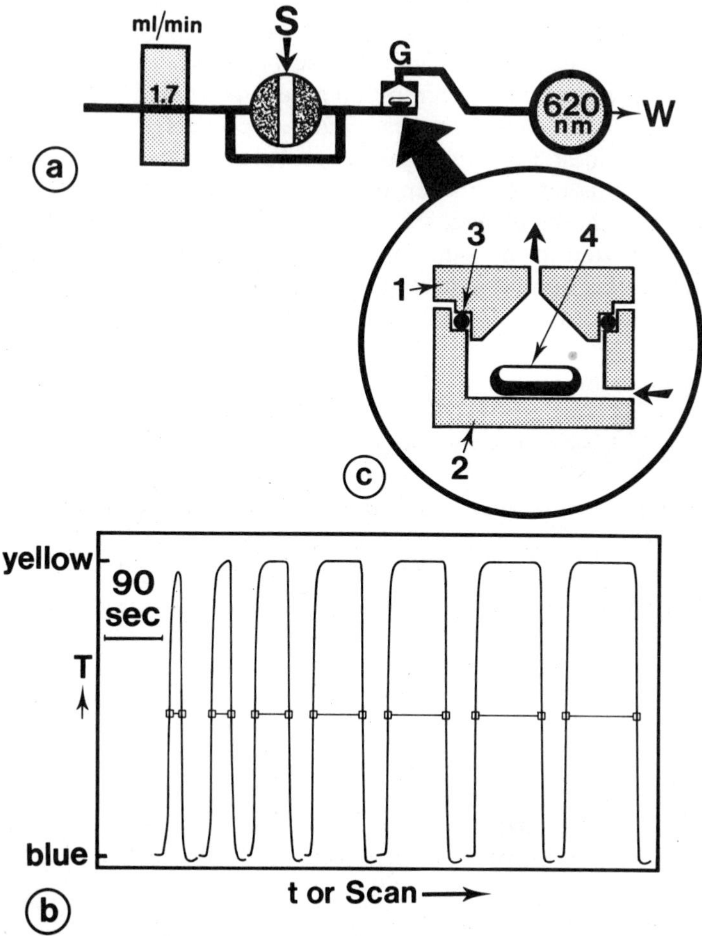

Figure 6.7. FIA titration of acid with base. The manifold (*a*) comprises a mixing chamber *G*, having a volume of 0.98 mL, where a well-defined concentration gradient is formed of the injected acid sample zone within the alkaline carrier solution, which also contains an acid–base indicator. The flowthrough cell monitors the change of the color indicator from blue to yellow as shown on the record in (*b*). The time interval Δt, which elapses between the two square points on the ascending and descending parts of each curve, is the measure of the concentration of the injected sample. (*c*) Construction of the mixing chamber, consisting of two parts, a lower circular unit housing a Teflon-covered magnetic stirring bar, and an upper part with a dome-shaped inner cavity to avoid entrapment of air. When assembled, a rubber gasket is placed between the two parts to ensure complete tightness.

When an acid sample of sufficiently high concentration is injected, the indicator in the mixing chamber will undergo a color change from blue to yellow. Yet by continuous addition of carrier stream, the acid sample will be continuously titrated, and when its concentration has decreased to the level of the carrier stream, the color will change to blue again; that is, the color change will indicate the equivalence point; and the time elapsed between the two color changes, Δt, will yield a measure of the concentration of the acid (Fig. 6.7*b*). The actual detection of the color changes is effected by continuous monitoring of the carrier stream in a flowthrough cell at 620 nm. Thus in this particular case it is the width of the recorded peak that yields the analytical readout, and as previously derived (Section 4.9), it can be shown that Δt (min) is given by the expression

$$\Delta t = (V_m/Q) \ln 10 \log C_{A0} - (V_m/Q) \ln 10 \log C_{NaOH} \qquad (6.1)$$

where C_{A0} is the concentration of acid in the mixing chamber at time $t = 0$ (i.e., at that point when all of the injected sample is situated in the mixing chamber), Q is the volumetric flow velocity (mL/min), V_m the volume of the mixing chamber (mL), and C_{NaOH} the concentration of the carrier stream. A plot of Δt versus log C_{A0}—as obtained by injecting a series of acid standards—will thus yield a straight line with a slope of V_m/Q (ln 10). Hence, on the basis of this calibration curve, one can by measuring Δt of a given sample determine its initial concentration C_s^0, which is related to C_{A0} (concentration of the sample in the chamber at t = 0) according to

$$C_{A0} = S_v C_s^0 / V_m \qquad (6.2)$$

where S_v is the injected sample volume. Substituting Eq. (6.2) into Eq. (6.1) yields

$$\Delta t = (V_m/Q) \ln 10 \log C_s^0 - (V_m/Q) \ln 10 \log C_{NaOH} + (V_m/Q) \ln 10 \log(S_v/V_m) \qquad (6.3)$$

Extrapolated to $\Delta t = 0$, Eq. (6.1) or (6.3) will directly yield the detection limit of the procedure.

From Eq. (6.3) it also appears that if samples of constant concentration (C_s^0) but different volume S_v are injected into a carrier stream of fixed concentration C_{NaOH}, a plot of Δt versus log S_v will yield a straight line, also having a slope of V_m/Q (ln 10).

First a calibration curve is run by injecting 125 μL of each of the standard solutions. Whether measuring transmittance or absorbance, the color changes of the indicator during the titration cycle—which will be from blue base color to the yellow acid color and back to blue again—will be registered by the spectrophotometer as a "blue or no blue signal"; that is, all peaks will be of the same height, but their widths will vary as a function of C_{A0} (and for constant S_v, of C_s^0) (Fig. 6.7*b*).

In the calibration curve, Δt is plotted on the ordinate versus log C_{A0} or log C_s^0 on the abscissa; Δt (min) is directly read off the recorder paper as the peak width b (mm)—measuring the peak width of all peaks at the same level, approximately halfway between the baseline and the top of the peak—that is, $b = \Delta t \cdot u_r$, where u_r is the recorder chart speed (mm/min).

If the flow rate Q is accurately determined (i.e., by collecting the waste over a fixed period of time), it is possible to calculate the volume of the mixing chamber V_m from the slope of the calibration curve, using Eq. (6.1) or (6.3). Furthermore, the detection limit might be determined (for $\Delta t = 0$) and compared with the calculated value [Eq. (6.1)–(6.3)].

By replacing the 125 μL sample loop with loops of identical internal diameter but of different lengths, that is, volumes (e.g., 75, 100, 150, and 200 μL), and in each case injecting samples of the same HCl standard, a plot of the obtained Δt versus log S_v should, according to Eq. (6.3), yield a straight line of slope (V_m/Q) ln 10. Thus V_m might equally well be calculated from this curve and compared to the previously found value.

Note that the rotor of an injection valve furnished with external sample loops itself has a small internal volume (i.e., the holes drilled in the rotor, cf. Fig. 5.1*b*). Thus when making this graphical representation, S_v for each sample loop in fact means the nominal loop volume $S_{v,\text{nom}}$ plus the internal volume ΔS_v. If ΔS_v is not known, this experiment might, however, be used to ascertain it, as the slope of the curve is known to be (V_m/Q) ln 10 which value is already determined by the calibration curve. Thus by measuring the Δt values for two nominal sample loop values (Δt_1 and Δt_2), Eq. (6.3) yields

$$\begin{aligned}\Delta(\Delta t) &= \Delta t_1 - \Delta t_2 \\ &= (V_m/Q)\ln 10 \log[(S_{v,\text{nom}(1)} + \Delta S_v)/(S_{v,\text{nom}(2)} + \Delta S_v)]\end{aligned} \qquad (6.4)$$

from which ΔS_v may be calculated.

Additional experiments with compleximetric, redox, or precipitation-based titrations have been designed using the same concepts as described above, employing colorimetric indication of the end point. Pertinent details can be found in Ref. 183, in Chapter 2.4.4 (Fig. 2.23), Chapter 4.9

(Figs. 4.62 and 4.63), and in textbooks on classical titrimetry. Needless to say that other means of indicating the equivalence point, that is, ion-selective electrodes, (Fig. 4.59), amperometric (Fig. 4.64), or dead-stop endpoint electrochemical detection, would be well suited for FIA titrations and that practically all titrimetric procedures can be automated by the FIA technique. Titrations without a magnetically stirred chamber, using a gradient device (Figs. 4.60—4.64), were the first important step toward miniaturization of the FIA apparatus [183]. An updated version of this concept has been adopted in the pedagogical version of the FIA system utilizing integrated microconduits (cf. Fig. 6.11*d*).

6.8 CONTINUOUS MONITORING WITH HYDRODYNAMIC INJECTION: INVESTIGATION OF THE CATALYZED OXIDATION PROCESS OF SULFITE WITH AIR

Manifold: Fig. 6.8. In this exercise the injection of sample solution is accomplished by hydrodynamic injection (Chapter 5.1) by means of the manifold components depicted within the boxed area and accommodated in a FIA microconduit (Chapter 4.12). The cross-sectional area of the channel of this module is approximately 0.8 mm^2, that is, the reactor R being ca. 15 mm long corresponding to a volume of 120 μL, while the sample loop—which for convenience is attached externally—is made to accommodate 20 μL of sample solution (10-cm, 0.5-mm-ID tube). The connecting line between the microconduit and the flowthrough cell placed in a photometer consists of 20-cm, 0.5-mm-ID Tygon tubing.

Reagents: The carrier stream is prepared by first dissolving 0.3 g of starch (Amidon soluble, Merck) in 250 mL of water while being heated. After cooling the starch solution is combined with a solution of 2.14 g of potassium iodate in 150 mL of water to which is added 200 mL of 1 *M* sulfuric acid, whereupon the volume is made to 1 L with distilled water.

Additionally, the following aqueous stock solutions are required: (a) 0.1 *M* phosphate buffer of pH 6.8, which per liter solution requires the dissolution of 6.805 g of potassium dihydrogen phosphate and 17.908 g of disodium hydrogen phosphate dodecahydrate; (b) 0.005 *M* copper(II) solution [1.208 g of copper(II) nitrate trihydrate per liter]; and (c) 0.01 *M* ethylene diamine tetraacetic acid (EDTA) (3.7224 g of the disodium salt per liter).

Standard Solutions: Standard solutions of thiosulfate in the range 1–2.0 m*M* in 1×10^{-2} *M* phosphate buffer (pH 6.8) are made by suitable di-

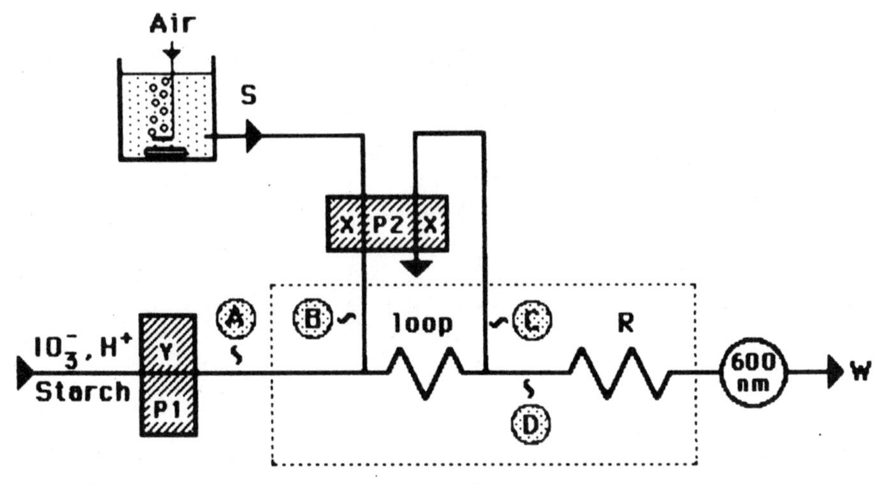

Figure 6.8. Manifold for the determination of sulfite comprising two separate pumps *P*1 and *P*2, where *P*1 is used for propelling carrier solution while *P*2 aspirates the sample solution. Sampling is accomplished by the hydrodynamic injection principle, which rests on operating the two pumps alternately: At the initiation of the analytical cycle, *P*1 is stopped while pump *P*2 is activated. Thereby sample (*S*) is drawn from the reservoir (*X* = 1.2 mL/min) and into line *B*. Since line *A* is filled with carrier solution the sample is forced through the sample loop (20 μL) and into line *C*, where the volumetric pumping rate should be ideally identical to that of line *B*, but for practical reasons is made a fraction larger than that of *B*. Hence, the solution in line *D* will run slowly backward, whereby the volume of sample to be injected will be limited to that entrapped within the sample loop. When *P*2 is stopped and *P*1 is activated, lines *B* and *C* will be blocked and the sample solution contained within the loop will be propelled forward by the carrier stream (*Y* = 1.2 mL/min) into line *D* and the reactor *R* (125 μL), and via a connecting line to the detector and finally to waste (*W*). For a sampling rate of 1 sample/min, and allowing for adequate aspiration of fresh sample to the loop and subsequently securing complete wash-out of the analyzed material, the sequential operation of the two pumps during the analytical cycle is:

	$t = 0$ to 25 s	25 s to 60 s
Pump 1	Stop	Activated
Pump 2	Activated	Stop

lutions of an aqueous stock solution containing 0.1 *M* sodium thiosulphate (24.817 g per liter).

Exercise: In aqueous solution sulfite ions are oxidized by dissolved oxygen, which process is catalyzed by copper(II) ions:

$$2SO_3^{2-} + O_2 \xrightarrow{Cu^{2+}} 2SO_4^{2-} \quad (6.5)$$

The purpose of the exercise is to investigate the course of this oxidation process under different reaction conditions. This is implemented by three separate experiments where the chemical reaction is monitored with and without a catalyst and, in the last experiment, with an inhibitor present. In each experiment 1 L of 1×10^{-2} *M* phosphate buffer of pH 6.8 is added to a beaker and air is bubbled through the solution until it is saturated with dissolved oxygen (this requires only a few minutes of bubbling). While maintaining the addition of air, a weighed amount of sodium sulfite is added quantitatively and the concentration of sulfite is followed as a function of time by aspirating samples from the reservoir at fixed time intervals (here once per minute) and injecting them into the FIA manifold. The sample is injected hydrodynamically by means of pumps sequentially controlled by a computer (Fig. 6.8).

The continuous aeration secures a constant concentration of dissolved oxygen during each experiment, that is, the transport of oxygen in the liquid is not the limiting factor of the reaction rate, which is consequently defined as:

$$r = -\left(\frac{dC_{SO_3^{2-}}}{\mathrm{dt}}\right) = \frac{dC_{SO_4^{2-}}}{dt} \qquad (\mathrm{mol/L} \cdot t) \qquad (6.6)$$

The measurement of the concentration of sulfite is based on the following analytical reactions:

$$2IO_3^- + 5SO_3^{2-} + 2H^+ \rightarrow I_2 + 5SO_4^{2-} + H_2O \qquad (6.7)$$

$$I_2 + \text{starch} \rightarrow I_2\text{–starch complex} \qquad (6.8)$$

The FIA system used is shown in Fig. 6.8, showing the sample injection by means of the hydrodynamic forces, the principle being explained in Fig. 6.8 (cf. Section 5.1.3). The carrier stream contains iodate and dissolved starch in diluted sulfuric acid. In the acidic solution the sulfite of the sample reacts with iodate leading to the formation of free iodine that with the starch forms a strongly colored blue complex which is measured spectrophotometrically at 600 nm. Since the formation of the end product is a very complex reaction which proceeds in several steps depending on the concentration of iodine, a linear calibration curve cannot be expected.

Because standard solutions of sulfite are unstable, the system has to be calibrated by standards of sodium thiosulfate, reaction (6.7) thus being replaced by

$$2I\,O_3^- + 10\,S_2O_3^{2-} + 12\,H^+ \rightarrow I_2 + 5\,S_4O_6^{2-} + 6\,H_2O \qquad (6.9)$$

that is, 2 moles of thiosulphate correspond to 1 mole of sulfite in the calibration procedure.

In the actual exercise a calibration curve is first obtained by operating the FIA system in the manner described in Fig. 6.8, by aspirating successively thiosulfate standards in the range 0–2 m*M* (all prepared in 1×10^{-2} *M* phosphate buffer), Fig. 6.9*a*. Next, a beaker is filled with 1 L of 1×10^{-2} *M* phosphate buffer. The beaker is placed on a magnetic stirrer and by means of a fritte air is bubbled through the solution at a rate of ca. 10 L/min. After having measured 4–5 blank values a weighed amount of ca. 125 mg of sodium sulfite (1 m*M*) is dissolved in a few milliliters of water and quantitatively transferred to the beaker. The oxidation process is monitored by aspirating samples regularly at 1 min intervals (Fig. 6.9*b*). After 15 min, 100 μL of the 0.005 *M* copper(II) nitrate solution is added to the solution and the oxidation process is followed to completetion, which typically requires less than 10 min (cf. Fig. 6.9*b*).

The experiment is repeated with a new batch of phosphate buffer, the only modification now being addition of 100 μL of the copper(II) nitrate solution at the beginning of the experiment, that is, together with the addition of the ca. 1 m*M* sodium sulfite. The oxidation procedure, cat-

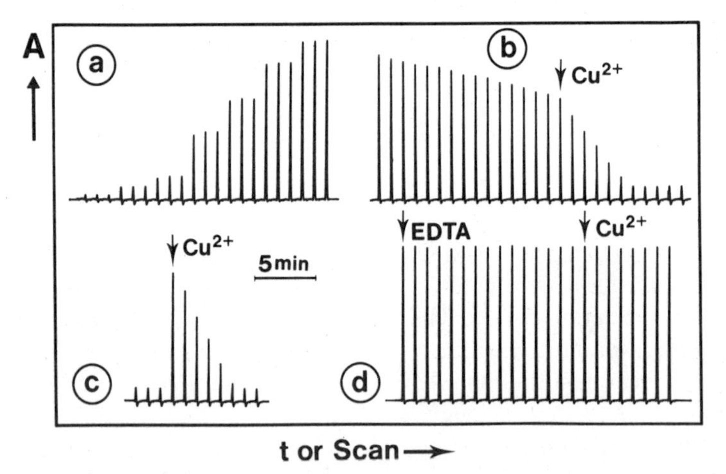

Figure 6.9. Readouts for the determination of sulfite in a dilute phosphate buffer solution (0.01 *M*, pH 6.8) using the manifold system shown in Fig. 6.8. (*a*) Calibration curve obtained by aspirating thiosulfate standards of 0, 0.25, 0.50, 0.75, 1.00, 1.50, and 2.00 m*M* prepared in the same buffer solution. (*b* and *c*) Response signals obtained when monitoring the oxidation of a 1 m*M* sulfite solution saturated with air, aspirating a sample each minute, to which a catalyst (Cu^{2+}) is added after (*b*) a delay period of 15 min, and (*c*) at the beginning of the experiment. (*d*) Repetition of the experiment outlined in (*b*), yet adding 0.01 *M* EDTA to the sulfite solution at the outset.

alyzed by the presence of cooper(II) ions, is now progressing rapidly. The monitoring is continued until the recorded peak has almost disappeared (Fig. 6.9*c*). In the third and last experimental run, a fresh 1 L portion of phosphate buffer is metered into the beaker to which 10 mL of 0.01 *M* EDTA solution is added. After dissolving the 1 m*M* sodium sulfite the solution is monitored for ca. 15 min, whereupon 100 μL of the copper(II) solution is added. As seen from Fig. 6.9*d* the recorded signal remains uneffected by the presence of complexed copper(II), thus showing that only the free copper(II) ions catalyze the oxidation reaction.

Concluding the exercise one of the thiosulfate standards is injected 10 times and the standard deviation, obtained by peak height measurements, is computed. Based on the calibration graph the sulfite concentrations of the individually measured signals in the three runs are calculated, and the concentrations for each experiment are plotted versus time. From these graphs the initial reaction rate r [Eq. (6.6)] of the oxidation process can be computed (from the slope of a tangent drawn to the graph), and from the shape of the reaction curve the reaction order might be estimated. Since the order of reaction without the presence of catalyst obviously is zero (cf. Fig. 6.9, *b* or *d*, from which it appears that dC/dt is constant), while of a higher order when copper(II) has been added, the analysis of the curve in the latter case might possibly be supplemented by a log $C_{SO_3^{2-}}$ versus time graph, which clearly will reveal the order of reaction to be 1 in this instance. Additionally, from the peak heights recorded in experiment *d* (or *b*) as compared to those expected from the calibration graph, the purity of the sodium sulfite used might be calculated.

6.9 DIAGNOSIS OF ERRORS

In experimental work, three sources of errors may be encountered: those having their origin in the nature of the injected sample material, those due to malfunction of the FIA apparatus, and those due to poorly designed chemistry.

Sample pretreatment, like dilution, neutralization, and filtration, may sometimes be necessary prior to injection of a sample into a FIA system, although highly concentrated, very acid (basic) or viscous samples may be directly diluted by injecting microliter volumes into confluence-type manifolds. Filtration cannot be avoided if suspended solids are present initially, yet precipitates may also be formed during the analytical procedure. Not only can the presence of solid particles lead to possible blocking of the tubes, but they can equally well interfere with the transducers, especially in optical systems. Although sample cleanliness is usually easy to maintain by filtration or centrifugation, particle formation as a result of chemical reactions can be more difficult to overcome. Often, however, the addition of a suitable surfactant or detergent may hinder formation of deposits (e.g., the addition of glycerine in exercise 6.5). Chemical binding was found to be effective when determining the total SO_2 in red wine by a spectrophotometric procedure (see Fig. 4.14), where it was initially found that after the red wine had been pretreated with base and then acidified (manipulations required by the analytical method), a reddish, tannin-containing precipitate was formed, leading to an ever-increasing baseline signal. However, by adding Fe(III) to the acidified sample the precipitation was effectively hindered.

6.9.1 Malfunction of the Apparatus

This can be diagnosed from the form of the recorded peak, either when actual chemical analysis is performed or from dispersion measurements:

1. *Poor reproducibility for repeated injections.* Carryover. This is easily checked by injecting alternately high and low sample concentrations (see Fig. 4.5). Carryover is eliminated either by decreasing the sampling frequency or increasing the pumping rate of the carrier stream(s) or both. Also check the valve for possible leakage and—when an automated FIA system is used—the level of sample liquid in the cups.

2. *Peak response is sluggish in reaching baseline (large σ_t value).* The system has a dead volume (bad connectors, large flow cell volume, large volume of the terminals of the flow cell) that acts as one or several small "mixing chambers." These dead volumes should be eliminated or minimized.

3. *Baseline drift* [Fig. 6.10(*1*)]. In optical detectors this is due to deposition of material on the windows of the flow cell. Such depositions may often be removed by injecting a reagent that dissolves the material, or by washing the system with a suitable wash liquid or detergent. In potentiometric detector systems, the drift may be caused by a change in the standard cell potential (either drift in the E^0 value of the indicator electrode or the reference electrode or both) or by a change in junction potentials.

4. *Air bubbles.* Carrier stream not deaerated; gases are formed as a result of a chemical reaction (e.g., CO_2; see Section 6.2) or due to a Venturi effect in the flow cell, caused by the sudden decrease of pressure in the flow cell, owing to a larger inner diameter of the flow cell terminals than of that of the tube used. This effect can be eliminated by inserting the connecting tubes all the way into the terminals, so that the orifices of the tubes are as close as possible to the actual cell chamber, and by

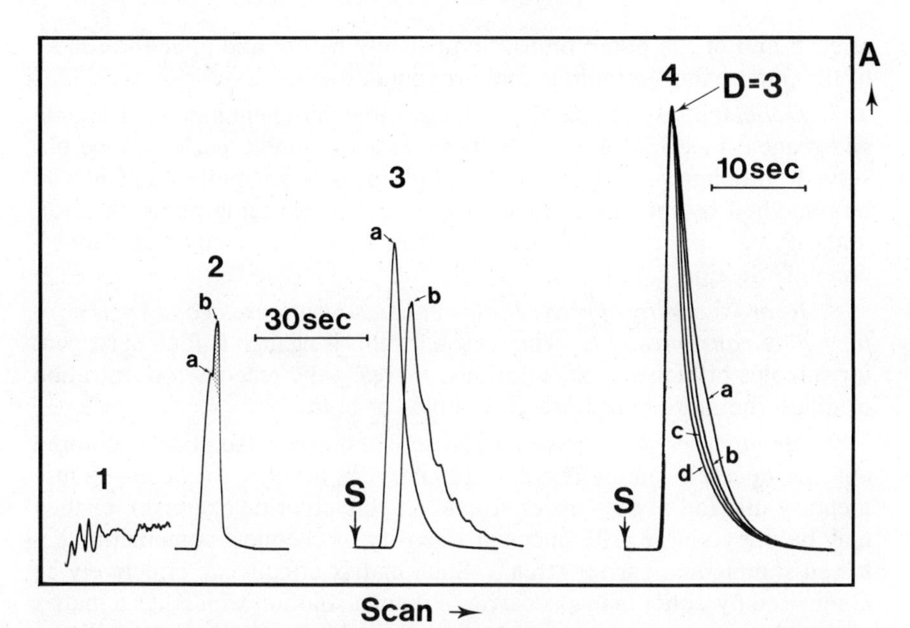

Figure 6.10. Diagnosis of errors in FIA. (*1*) Baseline drift. (*2*) *b*, Air bubble passing through the detector at the time of recording a normal peak (*a*). (*3*) Noisy signal output as caused by a faulty pump or worn pump tubes. Instead of the normal signal (*a*), the recorded peak is lower (*b*) and has a sawtooth shape and a longer residence time. (*4*) Peak shapes obtained by emptying the volumetric bore of the injecting valve over different periods of time: *a*, 5 s; *b*, 2.8 s; *c*, 1.8 s; and *d*, 1.3 s at a pumping rate of 1.5 mL/min. Note that the peak height is the same in all instances; yet further reduction of the injected fraction will not only lead to a narrower peak width but also to a decrease in peak height and hence a loss of reproducibility.

connecting to the outlet of the cell approximately 20 cm of 0.5-mm-ID tubing through which the stream is led to the rest of the waste tube.

5. *Abrupt noise on peaks* [Fig. 6.10(2*b*)]. Small air bubbles passing through the detector. The presence of air may either be due to the reasons mentioned above, or be a result of imcomplete filling of the sample loop of the injection port (check sample valve).

6. *Noisy signal.* Baseline is stable, but the movement of the recorder pen is not smooth, so that the peaks are recorded in a sawtooth fashion [Fig. 6.10(3*b*)]: The pump pulsates excessively, either due to a fault in construction or because the tubes are not pressed hard enough against the moving roller. Also, the pump tubes might be worn and should be replaced. (When the FIA system is not in use, the pump tubes should always be released, since this will prolong their lifetime.) If a potentiometric flow cell is used, a noisy signal might also be due to static electricity. This can be eliminated by inserting a small piece of metal tubing

at each end of the pump tubes, immediately before and after the rollers of the pump, short-circuiting and grounding them.

7. *Doublet peaks*. Lack of reagent due to insufficient mixing of sample and reagent (see Fig. 6.6). In extreme cases a double peak may be observed, although more often noise atop a peak is encountered. This can be remedied by increasing residence time T (decreasing pumping rate), intensifying mixing (confluence), or decreasing the injected sample volume (cf. Section 2.4.5).

8. *Reproducibility of peaks for lower concentrations is good but is poor for higher concentrations*. The reason for this is again insufficient reagent for samples of higher concentrations. Increase the reagent concentration or dilute the more concentrated samples or both.

9. *Negative peaks*. These may arise if the carrier solution is colored and the injected sample is colorless and very diluted, thus causing momentary dilution of the carrier stream (cf. the chloride exercise), or they may be due to large differences in viscosity or chemical composition between sample and carrier stream. Such matrix effects can effectively be eliminated by either using as carrier stream a solution which has a matrix composition very similar to that of the samples, or, better, employing a system such as that depicted in Fig. 6.4, that is, injecting the sample in an inert carrier stream which then is confluenced with the reagent stream.

Finally, it should be added that *automated injection* with a motorized valve can be used to reduce peak width (and increase the sampling frequency) if the valve can be returned from inject to fill position after a chosen period of time. Thus the volumetric sample cavity can either be emptied completely by leaving the valve in the inject position for a sufficiently long time [5 s, Fig. 6.10(4*a*)] or partly by letting the valve remain in the inject position for shorter periods of time (curve *b*, 2.8 s; *c*, 1.8 s; *d*, 1.3 s). Shorter injection periods, however, will lead to decreasing peak heights and may result in loss of reproducibility (cf. also Chapter 5.1).

6.10 CLOSING DOWN AN FIA SYSTEM

As a standard precaution, a FIA system should always be thoroughly rinsed if stopped for more than half a day, by pumping distilled water through it. Because of its small overall internal volume (usually only a few hundred microliters), washing can be done in a few minutes. Failure to do so may lead to clogging of the tubes. When washing the system, it is important to turn the injection valve a couple of times to rinse the bypass.

If the chemistry performed has resulted in the deposition of reaction products on the walls of the tubes or the flow cell, the system should be washed, first with an appropriate rinsing solution (acid, base, alcohol, or nonionic detergent) and then with water.

After the washing procedure has been completed, the pump tubes should be released, as this will prolong their lifetimes and preserve their delivery rates.

It should be added that if the FIA instrument is to be used only intermittently, yet always be ready for use (as for standby or stat operation), it is not necessary to wash it between each series of assays. The system should, however, be kept "wet"; i.e., completely filled with reagent solutions (or, preferably, water) in all lines.

6.11 FIA LABORATORY—THE MICROCONDUIT-BASED PEDAGOGICAL SYSTEM

Miniaturization of the FIA apparatus and integration of the manifold components leads to reduced consumption of sample and reagents and therefore less production of chemical waste per assay. Indeed, FIA allows up

Figure 6.11. FIA teaching system based on the concept of integrated microconduits, showing (from right to left) a four-channel peristaltic pump, an injection valve furnished with a bypass and an external sample loop, a reactor conduit, and a flowthrough cell (detector) with optical fibers, by means of which light is communicated from a light source and to a photometer (not shown). The photograph presents the setup for the colorimetric determination of phosphate (cf. Fig. 6.4) using microconduits *a* and *c* in Fig. 6.12, the sample being injected into a carrier stream of water which is subsequently merged with the reagent solutions, supplied from the containers placed in the rear compartment of the FIA lab instrument.

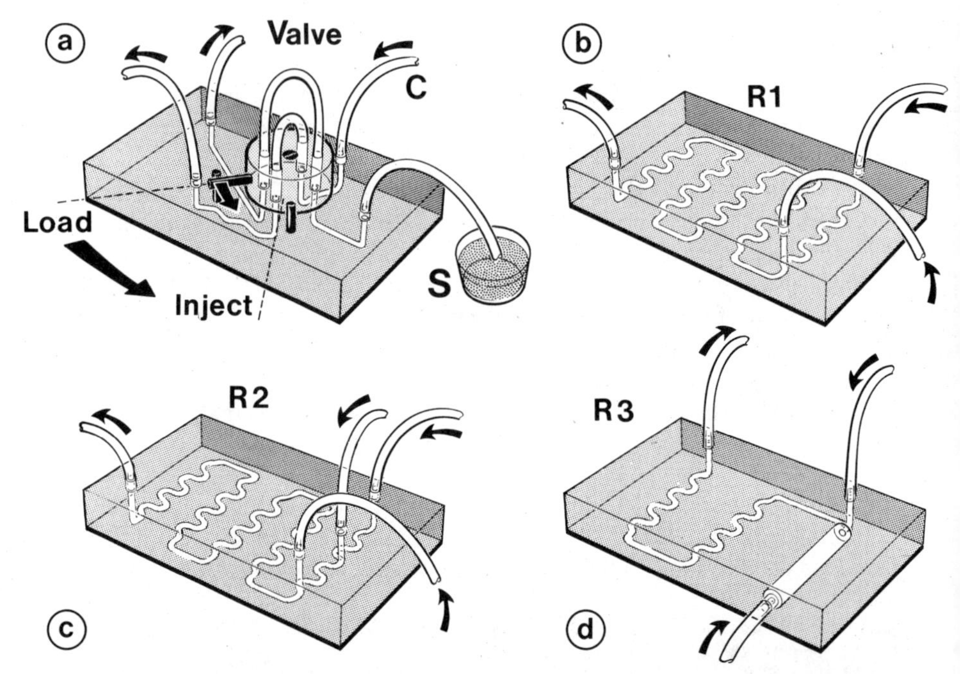

Figure 6.12. Detailed drawings of: (*a*) The injection valve module; (*b*) reactor module R1 comprising one mixing tee; (*c*) reactor module R2 comprising two mixing tees; and (*d*) reactor module R3, which, designed for titrations, comprises a gradient tube and and a mixing tee (cf. Figs. 4.62 and 4.63). The reactors, valve, and flow cell modules, as well as additional electrochemical detector modules, are interchangeably accommodated on an elastic back-plate so that the versatility of the miniaturized FIA lab system can be further expanded.

to 20 times reduction of these volumes per assay compared to traditional wet chemistry analysis as performed in university courses. Considering the amount of students and the number of chemistry courses affected, such enormous reduction of material consumption represents a substantial saving of funds, time, and waste disposal, as well as an improvement of laboratory safety and environment.

This observation led recently to a cooperative project between B. Kowalski, G. Christian, and the authors of this monograph with the aim to develop and test a robust miniaturized pedagogical FIA laboratory system (Fig. 6.11 and Fig. 6.12). Its modular concept allows dispersion experiments, colorimetric procedures, titrations, and other exercises, which are typically performed in undergraduate and graduate courses, to be made using a single FIA instrument. The prototypes have been successfully tested in courses at the Chemistry Department A at the Technical University of Denmark in Lyngby, Denmark and at the Department of Chemistry, University of Washington, Seattle, USA.

CHAPTER 7

REVIEW OF THE FLOW-INJECTION LITERATURE

Literature references are most useful if relatively recent or very old.

T. HIRSCHFELD

Whenever you set out to do something, something else must be done first.

MURPHY

7.1 FIA THROUGHOUT ITS FIRST DECADE

To aid the readers who want "to do something" with FIA themselves, we have done the "something else [which] must be done first" by attempting in this chapter to categorize all the FIA papers published from 1975 until the end of 1986—to the extent that they have been accessible to us. The papers have, according to their contents, been arranged into three tables, where Table 7.1 makes reference to theory, methods, and techniques used in conjunction with FIA, Table 7.2 lists the species that have been determined by FIA, and Table 7.3. divides the cited references according to areas of application. All three tables refer to the List of FIA References at the end of the book, wherein the titles of the individual papers are included. The references are—as far as it has been practically feasible—arranged chronologically; thus, in cases where more than one reference is cited for an individual species or technique the number might readily serve as a key to evaluating the "state of the art" and "prior art."

In the previous edition of this book, the classification of the FIA references into a similar set of tables was supplemented by a series of briefs in which a condensed description of the procedure(s) used for assaying a particular species, including their sensitivity and selectivity, and of the requirements as to apparatus and chemicals, was particularized. However, as the first edition comprised merely 100 references—and hence only dealt with a limited number of species determined by FIA—while this one exceeds this number by a factor of almost 14, it is evident that such a task not only would be excessively time and page consuming, but necessarily also uncritical. Therefore, we have refrained from even attempting this endeavor; yet, in order to compensate for this, the contents of the tables have been prepared in greater detail, which should facilitate the manual retrieval. Thus, a practical guideline to use these tables is first to locate the species of interest in Table 7.2 and then for the thus identified references to consult Table 7.3 for the matrix and/or area of application. Then Table 7.1 will render information as to the detection principle used and further details as to methods and techniques applied. In this systematic manner the number of useful references might readily become confined. It is to be noted that in Table 7.3 a large number of references is classified under "General." This group refers to papers where no specific area of application has been assigned, that is, the authors have mainly restricted themselves to testing the procedures or techniques recounted on aqueous standard solutions. Under this heading, however, are also collected papers dealing with the description and/or developments of FIA apparatus, or particular parts of it.

The primary sources for acquiring the FIA references have been the CHEMABS base of the ESA-IRS (European Space Agency–Information Retrieval Service) *Current Awareness Service*, and the Physical Chemical, and Earth Sciences section of *Current Contents* published by the Institute for Scientific Information, Inc. These sources have been supplemented by our own scrutiny of available periodicals of the analytical chemical literature, by the FIA bibliography listed in the recently launched Japanese *Journal of Flow Injection Analysis*, and by reprints supplied kindly to us by authors themselves.

The criterion for including the papers into the List of References has been the authors very own choice, that is, whether they themselves have regarded their work to be related to FIA. If they did, then the title, summary, or text contained the key word flow injection analysis. Every effort has been made by us to locate such papers and to include them into the List, but the overwhelming flood of scientific literature has made our task progressively more difficult. Doubtless, we may be guilty of having omitted material of importance, and therefore any additions which the reader may care to suggest will be much appreciated. We doubt, however, that we shall be able in a foreseeable future to face the task of again reviewing the new material. Anyone who has tried to maintain a comprehensive review of a rapidly expanding field would be aware our predicament, augmented by two major problems: (1) it has become an increasingly more difficult task to keep track of the FIA articles because authors often do not include Flow Injection or FIA in the titles of their papers nor indeed in the abstracts; and (2) FIA papers are now being published at increased frequency in periodicals not covered by the international retrieval services. Although the papers of the second category mainly comprise those published in Chinese, Japanese, or other for us rather somewhat inaccessible languages, they do nevertheless often appear in periodicals which are important scientific sources of information in their respective countries and hence these papers deserve recognition. Whatever the reason for the problems, these FIA publications are inevitably identified at great delays and therefore it is becoming arduous or close to impossible to maintain the strict chronology of the List of FIA References.

7.2 PREHISTORY OF FIA

There is universal agreement that the paper entitled "Flow Injection Analysis. Part 1. A New Concept of Continuous Flow Analysis" [1], which appeared in *Analytica Chimica Acta* in the spring of 1975, is the very first publication in which the name of the method was introduced, and for this reason alone this reference rightfully appears as the first one in the List of References. Any reader interested in the history of FIA will find, however, by further scrutiny of this paper that, besides coining the new name, it also contains the concepts and experiments outlining the principle and the scope of this new method, and that its immediate sequels, published with our co-workers and associates from Brazil, Denmark, England, India, and Sweden [1–20], mapped the novel area and plotted the course of future research.

There have been, however, expressed other views and claims as to the discovery of FIA and they, besides offering interesting reading [216, 607, 1056, 1352], confirm that the history of human endeavour, and scientific discoveries in particular, tells many stories, often confused by conflicting claims. Valcárcel and Luque de Castro have in Chapter 2 of their FIA monograph [665] reviewed material pertinent to this topic published prior to 1975. The reason for selecting this year is that the principles and experiments described in the first FIA publication were summarized in an earlier Danish Patent Application filed in September 1974 [1].

Considering that FIA is based on sample injection, controlled dispersion of the injected sample zone, and reproducible timing and that it is physically embodied in the system shown in Fig. 2.4, it becomes evident that none of the papers published until 1975 conforms with that description. Closest to this principle was undoubtedly the work of Nagy et al. [7.1], who suggested a novel approach to hydrodynamic voltammetry, based on the injection of a small sample volume into a supporting electrolyte streaming continuously at a constant flow rate. *No* chemical reaction took place in the flowing stream, and the authors found that:

Reproducible results were obtained only if rapid and total *homogenization* of the solution to be analyzed and the supporting electtrolyte was ensured. Preliminary experiments and visual observations of the streamlines by a suitable technique showed that the test solution injected into the supporting electrolyte did not mix with the latter to the desired extent, even at high flow rates. *The reproducibility of the measurement was impeded by a concentration gradient perpendicular to the direction of flow* [our emphasis].

Hence the whole system in fact was a large mixing chamber while its inlet and outlet connectors were rudimentary pieces of tubing. Subsequent papers published within the next 9 years by the same group [7.2–7.5] were limited to the same scheme where the injected samples constituted 1–2% of the volume of the mixing unit (typically V_m = 5 mL, Q = 8 mL/min), all systems using amperometric detection, eventually yielding S_{max} = 40 samples/h (see Fig. 10, Ref. 7.5). This description of performance should not be misunderstood as a harsh criticism of a pioneering research effort, which, at its onset in 1970, was far ahead of any other design of nonsegmented continuous flow. It is important to realize that what was faced at that time was a conceptual dilemma: on one hand, a desire to mix the sample solution *homogeneously* with the carrier stream; on the other hand, the resulting unavoidable loss of sensitivity and sampling frequency due to the use of the mixing chamber.

Thus it is not incidental that the independently designed instrument of Eswara-Dutt and Mottola from 1975 [7.6], as well as its later versions [7.7, 7.8], were again based on the use of a *large mixing chamber* into which the sample was *directly injected*. The chamber dominated the system, so that not only the chemical reaction but also the detection took place within its volume (5 mL). A high sampling frequency was sustained by high pumping rates, about 80 mL/min [7.7]. Therefore, the carrier stream had to be circulated to save reagents, yet this approach, originally introduced by Bergmeyer and Hagen [7.9], is a dead end solution which did not find wider acceptance, because accumulation of reaction products and matrix materials in the reagent reservoir impairs the function of the analyzer.

Frantz and Hare [7.10] described in an internal report in 1973 a system where a sample was inserted into a nonsegmented stream by manually breaking a capillary of a defined volume. A colorimeter was used to measure reaction product. Hare later joined Beecher and Stewart who replaced the awkward sampling device by a chromatographic injection valve of standard design. In their first paper published in 1975 [379] they conclude:

> Closely associated with the problem of dilution is the restriction that thus far this system has been successfully operated with *only a single reaction stream*. To date, attempts to introduce two or more merging streams in the reaction flow system have met with only limited success [our emphasis].

Therefore, the designs of this group employed only a single reagent line for the next four years [62, 167] and the authors themselves decided to identify their work as FIA as late as in 1979 [62].

To conclude, a paper always can be found which *in retrospect* may be viewed to be *prior art* to a discovery. Thus H. L. Daneel's experiments on electrochemistry, performed under Nernst's guidance, might be considered *prior art* to Heyrovsky's polarography. Also Stowe's carbon dioxide sensor was, strangely enough, declared by U.S. patent examiners to be a *prior art* to Clark's oxygen sensor [7.11]. The absurdity of such notions can, however, be obviated by a simple criterion. If Daneel or Stowe were farsighted enough to appreciate that their experiment would lead to a further significant discovery, they would have immediately pursued the idea energetically and published their findings with no delay, since no true scientist can resist the challenge of verifying a new discovery and sharing the joy of it with his peers. The material summarized in this chapter gives an opportunity to apply this criterion to FIA.

7.3 GROWTH OF PUBLISHED PAPERS ON FIA

The number of FIA publications has increased exponentially since 1975. Depicted in a semilogarithmic diagram, showing the cumulative number of publications as a function of time (Fig. 7.1), the doubling time, T_d (i.e., the time span required for the literature to double in size), can be read out as the slope of the curve. Initially, this doubling time is seen to be less than 1 year, yet at that time the absolute number of FIA publications was modest. With the doubling time increasing to 1.2 years, it nevertheless required six years to reach Ref. 153, the first monograph on FIA, while it has taken only an additional five years to reach the ca. 1400 papers summarized here, the doubling time now having attained a value of 1.9 years. Compared to other analytical techniques, for which similar type of analysis has been made, the doubling time for the FIA literature is very short, the T_d values for other subsets of analytical chemistry being: amperometry, 8.0 years; conductometry, 6.4 years; general electroanalysis,

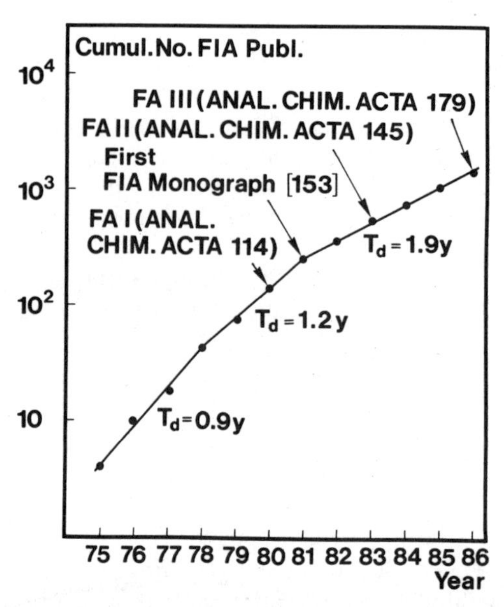

Figure 7.1. Growth of publications on FIA from 1975 to 1986. *FA I, II and III* refer to the three international conference on flow analysis held in Amsterdam, Lund, and Birmingham, respectively.

4.6 years; potentiometry, 4.0 years; and voltammetry, 2.8 years. Interestingly, the doubling times of the entire world chemical and analytical literature are even longer: 13.9 and 14.5 years, respectively [7.12].

This broad acceptance of FIA is undoubtedly due to its versatility, which allows the method to be used in conjunction with a wide variety of detectors and analytical techniques (Table 7.1), and for the assay of a multitude of organic and inorganic substances (Table 7.2). A closer look at the variety of detection principles used in conjunction with FIA (Fig. 7.2) reveals that optical methods (and particularly visible spectrophotometry) predominate. This is not surprising considering that spectrophotometry generally accounts for approximately 50% of all detection principles used in analytical chemistry. Yet, a statistical scrutiny of the data for the two time periods of 1975–1980 and 1981–1986 clearly reveals that the highest *relative increases* in detectors used with FIA actually are encountered in the areas of electrochemistry and optical methods (excluding spectrophotometry), and in the latter domain notably atomic absorption (AAS) and inductively coupled plasma spectrometry (ICP) dominate, followed by chemiluminescence and fluorometry. This trend is likely to continue, because flow-injection systems enhance the performance of AAS and ICP, by increasing the sensitivity over 100 times [684,

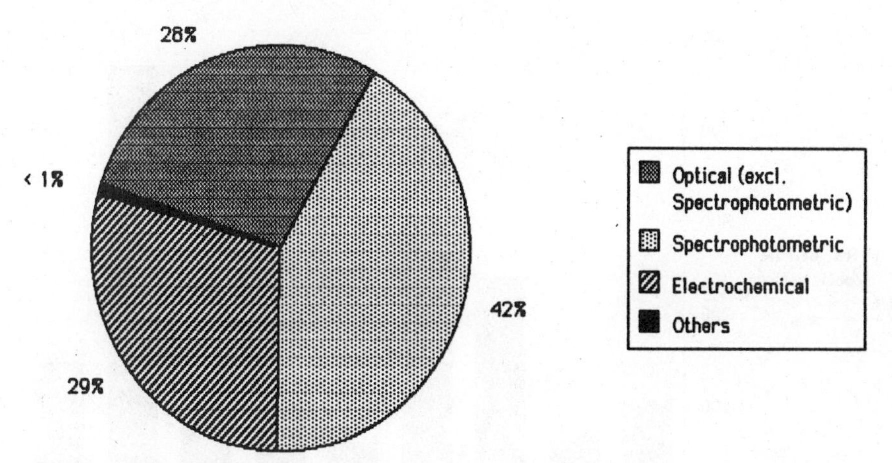

Figure 7.2. Distribution of the detection principles used in conjunction with FIA (cf. Table 7.1).

712, 738], by allowing speciation, and by removing matrix effects. For kinetic reasons, FIA is an ideal vehicle for chemiluminescence [330, 731, 1160].

As far as applications are concerned, several interesting trends appear when the papers are closely inspected. In Fig. 7.3 is shown the distribution of areas of application for all the FIA references, the papers classified under "General" being those where the authors have not specified any particular application, but in most cases have restrained themselves to use of aqueous standard solutions. While most of the earlier FIA papers were indeed centered on exploring the potential of the FIA approach per se, there has in recent years, however, been a marked increase in dedicated applications. This is illustrated in Figs. 7.4 and 7.5 where dedicated FIA applications for the time spans 1975–1980 and 1981–1986 are considered in terms of the absolute number of publications (Fig. 7.4) and in relation to the total number of FIA papers (Fig. 7.5) published during these two time periods. When comparing these two figures, it becomes apparent that the relative weight of agricultural applications in recent years have decreased considerably (although it has more than doubled in terms of absolute numbers), whereas pharmaceutical and environmental applications have remained practically at the same percentage level, and there has been a marked increase in food and clinical but particularly industrial and biochemical (including biotechnological) uses of FIA (Table 7.3). One of the reasons for these extensions of FIA outside research laboratories is likely to be attributed to the large number of international seminars, workshops, and conferences, such as the Flow Analysis Meet-

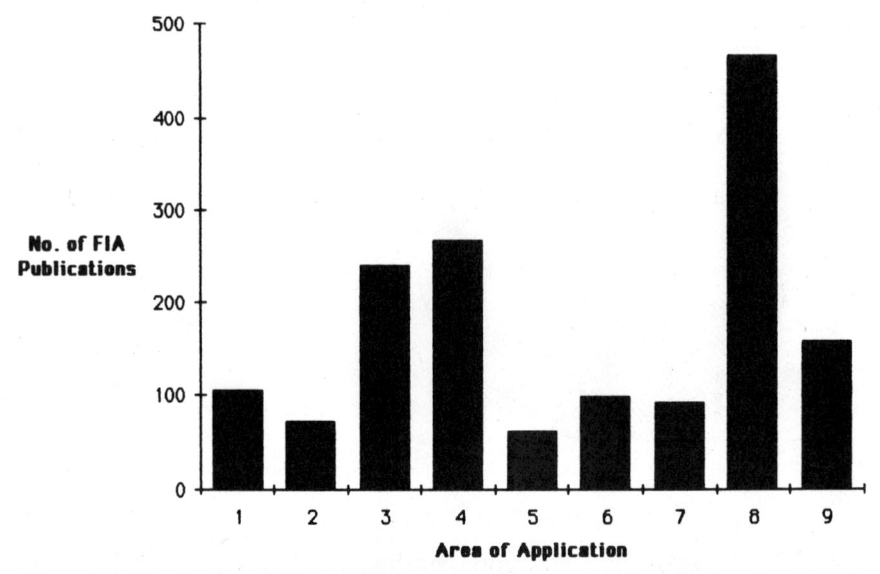

Figure 7.3. Distribution of FIA publications according to area of application. 1—Agricultural; 2—Biochemical; 3—Clinical; 4—Environmental; 5—Food; 6—Industrial; 7—Pharmaceutical; 8—General; and 9—Review articles, books, and educational papers.

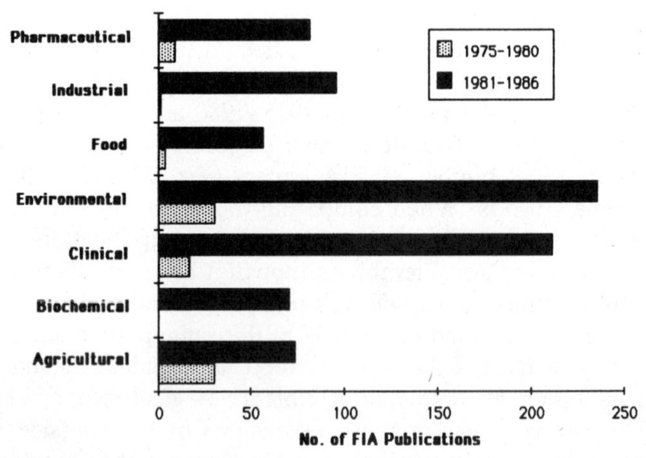

Figure 7.4. Distribution of FIA publications on dedicated applications in the time spans 1975–1980 and 1981–1986.

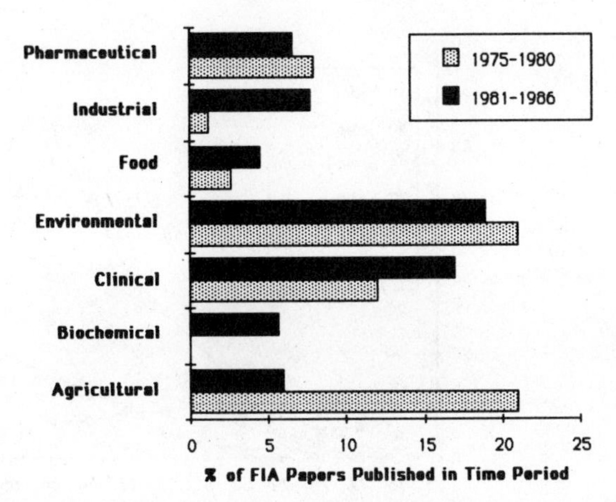

Figure 7.5. Relative distribution of dedicated FIA applications in the time spans 1975–1980 and 1981–1986.

ings (cf. Section 7.4), at which FIA has been the focus of attention. Another reason may be the increasing availability of commercial instrumentation for FIA, which is now produced in Sweden (Bifok-Tecator), United States (FIAtron, Lachat, Control Equipment), Japan (Hitachi), and Brazil (Micronal), the manufacturers being arranged in the same sequence in which they introduced their equipment (cf. Chapter 5.5).

When the FIA publications are arranged according to their country of origin (Fig. 7.6), the story of FIA is highlighted. Overall, 35 countries are represented; the United States and Japan clearly dominate the field at present—each accounting for 20–25% of all the papers published annually over the past five years—closely followed by Great Britain. Obviously, in these countries, there are various groups working independently on FIA. In fact, according to Ishibashi [7.13] over 60 groups are conducting FIA research in Japan where even a FIA Society has been established, publishing a periodical exclusively devoted to FIA, the *Journal of Flow Injection Analysis*. Significant contributors to the growth of FIA are also countries like Holland, Brazil, and Sweden, which all were fairly early to embark on FIA, and where the absolute number of publications in recent years has amounted to approximately 10–20 per year. In most other countries represented on the FIA list of contributors, the absolute number of publications is of the order of a few papers per year (thus, the very first paper originating in the USSR was published in a Russian periodical in 1985, followed by two more in 1986), notable exceptions being, however, China, Spain, and Germany, where the growth curve within the last few years, after a long incubation time, has exhibited a sharp increase (Fig. 7.6).

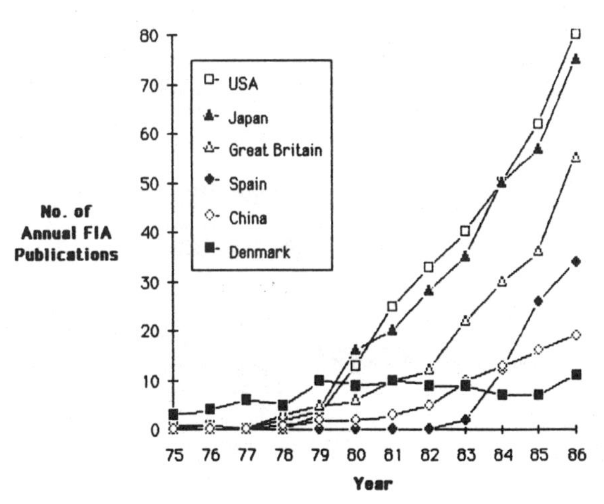

Figure 7.6. Growth of the number of FIA papers published annually in six different countries, where United States, Japan, and England represent the three most prolific countries, while China and Spain are examples of nations where the number of publications, after a long delay time, is now increasing significantly.

It is interesting to note that the overall increase in the volume of FIA publications gained momentum in 1979, and it is therefore relevant to ask what triggered the interest in this new technique. It takes one to two years from the time when a research project is conceived to the time when a paper is published, thus 1978 seems to be the year in or before which the "trigger" papers on FIA were published. Until then, 60% of *all* published FIA papers originated from three research teams: one in the United States (headed by K. K. Stewart), one in Brazil, and the Danish group, the latter two working in a close cooperation and publishing a number of joint papers [2–5, 13, 14] (Fig. 7.7). By 1979, the versatility of FIA had been proven by the development of a number of multiline and two-channel systems [1–23], and the feasibility of separations like dialysis [6], solvent extraction [19], and gas diffusion [57] had been demonstrated. Importantly, FIA was shown to be based on the combination of sample injection, controlled dispersion, and reproducible timing, and the concept of limited, medium, and large dispersion was thus established in Part X of the series of the papers originating from the Danish group [20], thereby allowing the rational design of any FIA manifold. Since at that time the Stewart team used only a single-line system and was only to adopt the FIA term for the first time one year later (AMFIA, [62]), it was mainly the experimental evidence published outside the United States, summarized in an excellent *Anal. Chem.* review written by Betteridge in 1978 [33], which was the triggering factor for future FIA proliferation.

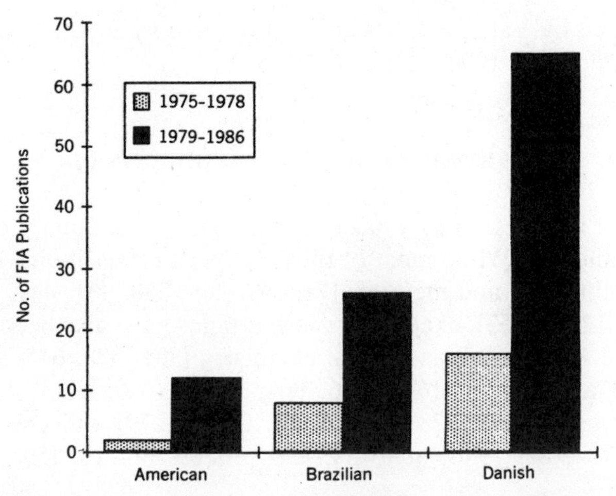

Figure 7.7. Distribution of FIA publications amongst the three original research teams in 1978 and in 1986.

Since then, numerous significant advances have been made in the concepts and technology of FIA (Table 7.1): The concept of controlled dispersion led to the development of gradient techniques (Sections 2.4, 2.5, and 4.5; Table 7.1). Reversed FIA, in which the sample rather than reagent serves as carrier stream, was suggested by Johnson and Petty [279] and in particular exploiteded by Valcarcel and co-workers (Section 4.11). The reagent-saving merging zones approach (Mindegaard [42] and Bergamin et al. [23]) became the first step toward truly economical Flow Injection systems, and various ingenious intermittent sampling modes, based on an original design of the Brazilian sampling device, were reported (Section 4.5). The group at CENA in Piracicaba (Brazil) further demonstrated the practicability of the FIA approach by routinely analyzing in 1982 over 300,000 samples of agricultural and environmental origin in their laboratory on a routine basis [440]. New research groups discovered special ways of exploiting FIA. Thus, for example, biochemical and biotechnological applications using immobilized reagents and enzymes (Section 4.7.2), formation of active reagents *in situ* (Section 4.7.3), kinetic advantages of FIA (Section 2.7.3), FIA–AAS using hydride generation (Section 4.6.2) as well as Karl Fisher methods [638, 1143], and comprehensive studies of liquid–liquid extraction mechanisms in FIA systems (Section 4.6.1) have been reported by different groups. Clinical applications have been actively pursued in several countries, especially England and Japan (Table 7.3). Advanced electrochemical applications have appeared from two American research groups [639, 618, 1239]. A very active Spanish group, headed by Valcárcel, has contributed to both theoretical and practical aspects of FIA (Chapters 3 and 4).

7.4 MONOGRAPHS AND REVIEWS ON FIA

A number of reviews on FIA has been published, a summary of which is given in Table 7.3. While most of these surveys primarily have been devoted to techniques and methods [71, 141, 268, 338, 431, 441, 808, 1055, 1205, 1231, 1257, 1353], excellent review articles have also been published in dedicated areas, such as clinical chemistry [301, 588, 645, 999, 1111], agricultural applications [673, 876, 964, 1060, 1080, 1114, 1191, 1257, 1296], pharmaceutical [328, 798, 1179], food [1279] and environmental [384, 749, 784, 822, 946] analysis, and process control [598, 718, 961, 1393].

The first monograph on FIA, which was the first edition of this book, was published in 1981 [153]. A Japanese translation of it (by N. Ishibashi and N. Yoza) became available in 1983 [663], while most recently Fang Zhaolun has completed a Chinese version [1223], in which the translated text was supplemented by a short chapter on recent developments in FIA including gradient techniques and the application of microconduits. In 1983 K. Ueno and K. Kina authored their *Introduction to Flow Injection Analysis. Experiments and Applications* [664], which, as the title implies, emphasizes the practical aspects of FIA and therefore is a most useful guide to the practitioner. Unfortunately, this book has so far only been printed in Japanese. A few years ago (1984), an excellent, comprehensive monograph, written by M. Valcárcel and M. Luque de Castro, *Análisis por Inyección en Flujo* [665], was published in Spain, and this book most recently appeared in an English translation [7.14]. Besides, to our knowledge preparations for two additional (dedicated) monographs on FIA are in progress.

7.5 FLOW ANALYSIS I, II, AND III

Throughout the past decade, the flow-injection method has grown in scope, both in its concept and in the number of applications. The recognition of FIA as an important, independent discipline in analytical chemistry is amply reflected in the number of international conferences either centered on FIA or where special sessions have been devoted to this subject. In recent years the most significant of these have been the Flow Analysis meetings, from which the proceedings have been published in special volumes of *Analytica Chimica Acta* (volumes 114, 145, and 179, respectively). Flow Analysis I took place in Amsterdam, Holland in 1979 [71–88], the second one in Lund, Sweden (1982) [338–361], while Flow Analysis III was situated in Birmingham, England (1985) [1055–1102]. Of conferences without formal proceedings, yet with sessions or symposia devoted to FIA, should be mentioned the annual SAC (Society for Analytical Chemistry, England) and ACS (American Chemical Society) meetings, including the Pittsburgh Conferences, the biannual congress of FACSS (Federation of American Chemical and Spectroscopy Societies), the triannual International Symposia on Kinetics in Analytical Chemistry, and the Tecator-sponsored International Symposium on Flow Injection Analysis in Örenäs, Sweden in 1985.

REFERENCES

7.1 G. Nagy, Zs. Feher, and E. Pungor, *Anal. Chim. Acta,* **52** (1970) 47.

7.2 Zs. Feher and E. Pungor, *Anal. Chim. Acta,* **71** (1974) 425.

7.3 Zs. Feher, G. Nagy, K. Toth, and E. Pungor, *Analyst*, **99** (1974) 699.

7.4 E. Pungor, G. Nagy, and Zs. Feher, *J. Electroanal. Chem.,* **75** (1977) 241.

7.5 E. Pungor, Zs. Feher, G. Nagy, K. Toth, G. Horvai, and M. Gratze, *Anal. Chim. Acta,* **109** (1979) 1.

7.6 V. V. S. Eswara-Dutt and H. A. Mottola, *Anal. Chem.,* **47** (1975) 357.

7.7 V. V. S. Eswara-Dutt, A. E. Hanna, and H. M. Mottola, *Anal. Chem.,* **48** (1976) 1207.

7.8 V. V. S. Eswara-Dutt and H. A. Mottola, *Anal. Chem.,* **49** (1977) 776.

7.9 H. U. Bergmeyer and A. Hagen, *Fresenius Z. Anal. Chem.,* **261** (1972) 333.

7.10 J. D. Frantz and P. E. Hare, *Ann. Rep. Director Geophys. Lab.,* **1630** (1973) 704.

7.11 P. Astrup and J. W. Severinghaus, *The History of Blood Gases, Acids and Bases*. Munksgaard, Copenhagen, Denmark (1986).

7.12 T. Braun and E. Bujdosó, *CRC. Crit. Reviews Anal. Chem.,* **13** (1982) 233.

7.13 N. Ishibashi (private comm.).

7.14 M. Valcárcel and M. D. Luque de Castro, *Flow-Injection Analysis. Principles and Applications,* Ellis Horwood Series in Analytical Chemistry. Ellis Horwood, Ltd., Chichester, England (1987).

Table 7.1 Theory, Methods, and Techniques Used in FIA

Category	References
Theory	1, 10, 14, 20, 33, 39, 65, 71, 72, 73, 74, 75, 142, 146, 150, 151, 153, 155, 156, 176, 181, 183, 197, 257, 296, 339, 343, 345, 353, 363, 372, 376, 378, 422, 430, 500, 501, 526, 541, 554, 566, 608, 639, 663, 664, 665, 667, 688, 701, 702, 714, 802, 819, 852, 864, 872, 905, 1000, 1006, 1022, 1053, 1059, 1061, 1062, 1063, 1064, 1065, 1089, 1103, 1118, 1124, 1143, 1155, 1198, 1223, 1226, 1248, 1256, 1273, 1329, 1354, 1374
Spectrometric methods	
Spectrophotometry in solution	1, 2, 3, 4, 5, 6, 7, 9, 15, 17, 18, 19, 21, 22, 25, 26, 27, 28, 30, 32, 34, 37, 38, 42, 48, 49, 51, 53, 54, 57, 59, 64, 69, 71, 78, 79, 81, 85, 89, 90, 91, 92, 93, 94, 95, 96, 97, 98, 101, 104, 106, 109, 110, 114, 116, 117, 118, 119, 120, 121, 122, 123, 124, 125, 126, 128, 129, 130, 132, 133, 135, 136, 138, 139, 145, 147, 148, 152, 154, 157, 158, 160, 161, 162, 163, 164, 166, 167, 169, 171, 174, 175, 177, 179, 180, 182, 183, 186, 189, 190, 191, 196, 197, 199, 202, 203, 207, 212, 220, 221, 224, 226, 228, 232, 233, 234, 235, 236, 237, 238, 239, 240, 241, 243, 245, 246, 252, 253, 256, 258, 259, 261, 262, 264, 266, 267, 268, 272, 273, 274, 275, 277, 279, 281, 284, 286, 293, 294, 295, 299, 303, 304, 305, 311, 312, 319, 320, 321, 325, 326, 328, 334, 339, 341, 346, 348, 349, 350, 351, 357, 361, 362, 363, 364, 365, 369, 371, 375, 378, 379, 384, 387, 388, 390, 391, 394, 398, 400, 401, 402, 403, 408, 409, 413, 415, 420, 422, 423, 424, 425, 430, 432, 433, 442, 443, 444, 447, 454, 457, 459, 461, 462, 463, 467, 470, 474, 477, 481, 485, 486, 493, 494, 495, 496, 499, 508, 511, 514, 517, 520, 522, 524, 533, 544, 546, 548, 549, 551, 555, 556, 557, 560, 564, 567, 568, 570, 573, 575, 577, 579, 581, 585, 591, 596, 597, 600, 601, 602, 603, 604, 608, 609, 610, 614, 615, 619, 623, 624, 625, 626, 627, 629, 631, 632, 633, 634, 635, 636, 637, 638, 640, 641, 643, 647, 648, 649, 650, 662, 669, 670, 671, 672, 673, 679, 681, 682, 686, 688, 691, 695, 696, 697,

Table 7.1 (Continued)

Category	References
	699, 701, 710, 711, 721, 722, 723, 725, 727, 729, 730, 733, 734, 737, 740, 741, 742, 743, 749, 750, 751, 753, 755, 756, 757, 758, 759, 762, 766, 767, 768, 769, 771, 773, 774, 775, 782, 786, 787, 796, 801, 802, 805, 811, 813, 814, 815, 816, 822, 824, 826, 828, 829, 830, 834, 838, 839, 840, 846, 847, 849, 859, 864, 866, 868, 870, 871, 876, 879, 880, 884, 885, 891, 892, 893, 896, 897, 898, 900, 901, 902, 904, 909, 910, 911, 912, 913, 914, 915, 918, 919, 922, 926, 932, 938, 940, 945, 947, 950, 952, 960, 961, 964, 968, 970, 971, 972, 974, 976, 977, 978, 979, 986, 987, 990, 991, 992, 993, 995, 998, 1000, 1001, 1003, 1004, 1008, 1009, 1012, 1016, 1028, 1029, 1033, 1035, 1040, 1042, 1045, 1046, 1047, 1048, 1049, 1056, 1065, 1066, 1067, 1068, 1070, 1071, 1072, 1073, 1074, 1076, 1077, 1079, 1092, 1093, 1094, 1096, 1108, 1109, 1110, 1111, 1112, 1114, 1122, 1127, 1132, 1136, 1137, 1138, 1143, 1145, 1147, 1152, 1156, 1158, 1159, 1164, 1166, 1170, 1171, 1172, 1180, 1184, 1185, 1186, 1188, 1191, 1194, 1203, 1204, 1206, 1207, 1216, 1218, 1220, 1224, 1225, 1227, 1229, 1230, 1234, 1237, 1242, 1244, 1250, 1256, 1260, 1265, 1267, 1268, 1269, 1272, 1277, 1279, 1280, 1283, 1285, 1286, 1287, 1290, 1291, 1292, 1293, 1297, 1298, 1299, 1300, 1305, 1309, 1311, 1321, 1323, 1324, 1326, 1329, 1330, 1336, 1338, 1339, 1340, 1343, 1344, 1346, 1348, 1351, 1353, 1357, 1372, 1373, 1378, 1379, 1380, 1383, 1384, 1388, 1389, 1390, 1392, 1393, 1394, 1395, 1396, 1404, 1405, 1406
Diode array detection	1017, 1075, 1382
Atomic absorption spectrometry	43, 62, 66, 108, 114, 127, 137, 145, 184, 195, 208, 223, 227, 263, 331, 339, 352, 353, 372, 383, 395, 396, 397, 414, 421, 428, 436, 439, 445, 458, 471, 472, 489, 497, 513, 526, 527, 535, 543, 550, 562, 576, 579, 583, 614, 617, 622, 628, 651, 683, 684, 693, 707, 715, 734, 735, 736, 738, 739, 754, 777, 780, 790, 791, 792, 793, 808, 812, 820, 821, 842, 849, 858, 864, 890, 906, 907, 924, 967, 969, 998, 1000, 1010, 1018, 1022, 1023, 1024, 1025, 1080, 1081, 1082, 1097, 1099, 1101, 1111, 1114,

Table 7.1 (Continued)

Category	References
	1127, 1139, 1142, 1169, 1189, 1190, 1191, 1217, 1231, 1245, 1246, 1247, 1257, 1261, 1263, 1264, 1281, 1294, 1296, 1307, 1310, 1314, 1322, 1324, 1342, 1362, 1369
Atomic emission spectrometry	43, 114, 318, 817, 836
Atomic fluorescence spectrometry	475
Chemi- or bioluminescence	58, 83, 99, 140, 172, 178, 193, 265, 280, 302, 315, 327, 330, 337, 358, 374, 399, 406, 484, 488, 498, 523, 545, 553, 571, 587, 660, 674, 675, 678, 686, 705, 731, 760, 761, 783, 788, 797, 823, 835, 863, 886, 901, 903, 933, 934, 994, 1000, 1007, 1011, 1039, 1073, 1078, 1104, 1119, 1122, 1129, 1134, 1135, 1160, 1162, 1174, 1202, 1209, 1213, 1214, 1252, 1274, 1315, 1324, 1337, 1350, 1353, 1356, 1363, 1391, 1400, 1402
Fluorimetry	31, 47, 80, 88, 185, 194, 214, 231, 260, 269, 278, 287, 310, 317, 318, 322, 329, 339, 347, 426, 444, 468, 509, 529, 611, 666, 677, 700, 724, 726, 748, 776, 781, 795, 800, 810, 822, 823, 825, 851, 854, 860, 887, 916, 920, 955, 959, 999, 1002, 1041, 1058, 1069, 1095, 1146, 1147, 1163, 1173, 1201, 1221, 1238, 1240, 1255, 1266, 1308, 1320, 1324, 1325, 1332, 1349, 1353, 1364, 1371
Inductively-coupled plasma atomic emission spectrometry	105, 188, 206, 354, 385, 518, 528, 565, 612, 656, 661, 712, 736, 738, 739, 763, 789, 790, 808, 861, 908, 963, 1005, 1080, 1098, 1105, 1140, 1195, 1228, 1231, 1249, 1254, 1257, 1304, 1324, 1328, 1375
Inductively-coupled plasma mass spectrometry	1303
Infrared spectrometry	942, 958, 1243, 1259
Laser-based detection	317, 642, 644, 685, 873, 989, 1115
Molecular emission cavity analysis	538, 809, 1030, 1100
Phosphorimetry	448, 698
Reflectometry/optosensing	848, 942, 1000, 1055, 1232, 1302
Refractometry	27, 162, 164, 232, 362, 1109, 1367

Table 7.1 (Continued)

Category	References
Thermal lens spectrometry	642, 644, 685, 873, 989, 1115
Turbidimetry/nephelometry	11,18, 37, 52, 87, 160, 254, 298, 355, 525, 530, 568, 706, 729, 935, 1183, 1212, 1324
Electrochemical methods	16, 29, 61, 67, 82, 92, 149, 159, 165, 168, 174, 187, 192, 198, 209, 233, 285, 309, 332, 339, 360, 404, 412, 483, 504, 510, 539, 595, 618, 673, 680, 732, 744, 745, 746, 752, 779, 828, 921, 996, 997, 1015, 1107, 1109, 1111, 1114, 1199, 1210, 1256, 1279, 1284, 1285, 1290, 1291,1356
Amperometry	61, 86, 102, 103, 115, 143, 155, 218, 219, 221, 231, 242, 247, 249, 255, 283, 290, 316, 356, 376, 377, 381, 386, 419, 452, 453, 455, 460, 465, 476, 478, 490, 491, 507, 531, 534, 546, 563, 592, 597, 606, 609, 630, 637, 657, 658, 659, 676, 703, 704, 708, 713, 716, 719, 721, 747, 764, 765, 770, 785, 799, 818, 819, 831, 833, 844, 853, 862, 867, 869, 872, 875, 883, 888, 923, 930, 931, 937, 975, 981, 982, 983, 985, 1013, 1014, 1027, 1031, 1034, 1050, 1051, 1052, 1054, 1064, 1084, 1087, 1091, 1103, 1113, 1153, 1161, 1175, 1187, 1196, 1211, 1233, 1276, 1278, 1301, 1316, 1331, 1333, 1335, 1365, 1366, 1374, 1376, 1381, 1386, 1387
Conductometry	318, 362, 363, 479, 605, 973, 1106, 1147, 1181, 1319, 1338, 1341, 1345
Coulometry	61, 113, 156, 187, 487, 507, 521, 536, 542, 594, 1360
Polarography	144, 159, 210, 289, 341, 443, 459, 506, 593, 1085, 1276, 1377
Potentiometry	1, 3, 152, 173, 196, 199, 213, 238, 262, 271, 332, 336, 416, 537, 547, 556, 572, 592, 616, 659, 717, 757, 778, 794, 827, 832, 844, 874, 923, 941, 948, 956, 985, 1000, 1037, 1059, 1133, 1147, 1165, 1176, 1276, 1338, 1347, 1396, 1399
Chronopotentiometry/ potentiometric stripping	215, 366, 521, 877, 1086, 1263, 1403
Ion selective electrodes/ ISFETS	8, 12, 13, 20, 21, 24, 38, 50, 60, 68, 111, 112, 134, 170, 204, 205, 229, 248, 250, 251, 268, 270, 282, 323, 359, 405, 407, 429, 446, 473,

Table 7.1 (Continued)

Category	References
	479, 480, 537, 552, 561, 578, 586, 608, 621, 652, 653, 687, 709, 749, 881, 882, 894, 900, 922, 936, 940, 957, 962, 988, 1026, 1035, 1036, 1083, 1088, 1101, 1102, 1121, 1144, 1191, 1193, 1200, 1235, 1251, 1258, 1260, 1270, 1271, 1289, 1295, 1306, 1312, 1313, 1317, 1318, 1327, 1334, 1355, 1358, 1370, 1385
Voltammetry	20, 40, 67, 131, 200, 217, 225, 276, 288, 300, 324, 370, 389, 410, 449, 464, 466, 483, 492, 503, 512, 515, 558, 574, 580, 582, 613, 620, 639, 646, 668, 720, 747, 845, 928, 939, 980, 1053, 1090, 1157, 1236, 1239, 1275, 1368, 1377, 1387, 1397
Enthalpimetry	516, 532, 837, 917, 1197
Photoacustic detection	965
Photoelectrochemistry	841
Piezoelectric quartz crystal	1361
Radiochemistry	211
Enzymatic methods	9, 20, 46, 60, 65, 79, 154, 167, 194, 199, 226, 231, 233, 234, 247, 262, 264, 265, 278, 280, 283, 287, 322, 330, 340, 344, 345, 346, 348, 360, 369, 386, 387, 420, 422, 426, 432, 444, 460, 465, 478, 488, 493, 504, 510, 516, 530, 542, 546, 547, 557, 573, 585, 589, 592, 596, 597, 615, 632, 660, 674, 678, 679, 680, 681, 686, 694, 703, 704, 719, 744, 746, 747, 750, 752, 758, 768, 779, 781, 786, 794, 824, 833, 835, 862, 863, 867, 869, 874, 875, 883, 897, 901, 916, 917, 921, 923, 926, 930, 933, 953, 955, 960, 961, 966, 976, 977, 985, 988, 996, 999, 1000, 1001, 1011, 1014, 1015, 1048, 1067, 1068, 1069, 1078, 1084, 1087, 1103, 1107, 1113, 1119, 1129, 1146, 1153, 1158, 1160, 1161, 1173, 1176, 1202, 1211, 1225, 1232, 1234, 1256, 1266, 1267, 1308, 1316, 1319, 1320, 1331, 1335, 1337, 1346, 1350, 1353, 1363, 1366, 1379, 1388, 1389
Immunoassay	80, 193, 231, 322, 348, 358, 530, 674, 681, 706, 760, 855, 856, 954, 965, 966, 999, 1014, 1058, 1119, 1129

Table 7.1 (Continued)

Category	References
Kinetic methods (nonenzymatic)	39, 45, 49, 56, 85, 96, 128, 129, 133, 150, 161, 174, 181, 191, 220, 232, 235, 252, 262, 302, 315, 349, 351, 364, 365, 371, 391, 411, 484, 501, 517, 522, 529, 533, 540, 541, 549, 553, 554, 561, 564, 566, 574, 587, 603, 604, 618, 631, 670, 695, 700, 701, 706, 713, 725, 748, 766, 770, 776, 788, 811, 814, 825, 838, 840, 850, 851, 854, 909, 915, 1009, 1033, 1040, 1041, 1065, 1073, 1094, 1121, 1122, 1134, 1135, 1159, 1172, 1186, 1203, 1207, 1216, 1242, 1252, 1315, 1325, 1353, 1368, 1376, 1384, 1386, 1400
Gradient techniques	32, 42, 65, 87, 135, 140, 150, 151, 154, 162, 196, 197, 201, 220, 223, 232, 264, 301, 318, 373, 418, 435, 441, 482, 540, 598, 608, 660, 692, 791, 812, 836, 838, 848, 849, 887, 889, 919, 998, 1000, 1017, 1019, 1038, 1058, 1060, 1063, 1064, 1065, 1117, 1121, 1124, 1127, 1177, 1198, 1242, 1253, 1256, 1260, 1275, 1285, 1367, 1371, 1382, 1393
Gradient calibration	227, 264, 338, 353, 372, 385, 441, 489, 526, 693, 707, 790, 812, 889, 1000, 1097, 1127, 1231, 1245, 1257
Gradient dilution	264, 275, 338, 441, 449, 707, 889, 1000, 1245, 1257, 1362, 1394
Gradient scanning	275, 288, 338, 441, 449, 464, 492, 817, 889, 980, 1000, 1080, 1305
Merging zones	23, 42, 43, 46, 48, 64, 65, 71, 79, 80, 126, 137, 166, 202, 206, 268, 286, 369, 393, 394, 530, 532, 610, 645, 681, 706, 827, 835, 926, 935, 1000, 1060, 1061, 1069, 1080, 1170, 1202, 1247, 1257, 1262, 1363, 1388
Peak width (non-titration)	151, 152, 223, 318, 526, 1019, 1056, 1062, 1064, 1097, 1124, 1127, 1145, 1147, 1237, 1338, 1362, 1374
Sample splitting	3, 4, 701, 759, 776, 795, 922, 923, 930, 968, 1071, 1095, 1117, 1206, 1308, 1321, 1362
Selectivity studies	253, 338, 378, 441, 889, 1000, 1073
Standard addition	354, 372, 385, 496, 513, 565, 817, 836, 849, 882, 1000, 1060, 1080, 1116, 1257
Stopped flow	20, 46, 64, 65, 71, 79, 80, 119, 154, 177, 191, 226, 253, 264, 266, 268, 275, 321, 338, 348, 349, 369, 393, 394, 414, 420, 422, 530, 540,

Table 7.1 (Continued)

Category	References
	559, 597, 611, 613, 681, 700, 706, 755, 834, 838, 848, 868, 889, 926, 935, 942, 1000, 1056, 1060, 1069, 1070, 1073, 1109, 1136, 1232, 1251, 1326, 1371, 1382, 1388
Titrations	10, 44, 59, 64, 71, 76, 82, 158, 183, 211, 213, 221, 253, 275, 288, 312, 318, 338, 373, 393, 414, 427, 559, 608, 671, 807, 814, 887, 889, 940, 951, 1000, 1028, 1032, 1056, 1109, 1124, 1171, 1226, 1280, 1326, 1338, 1374
Zone Sampling	145, 166, 202, 415, 496, 540, 1060, 1117
Separation methods	
Chromatography (liquid/gas)	89, 163, 195, 290, 304, 460, 514, 560, 574, 604, 721, 732, 751, 801, 805, 929, 1149, 1150, 1243, 1303
Dialysis	6, 9, 160, 199, 262, 393, 660, 686, 719, 743, 747, 758, 771, 852, 867, 923, 1000, 1011, 1107, 1173, 1200, 1219, 1240, 1268, 1313, 1317, 1318, 1326, 1332
Extraction	19, 20, 22, 31, 41, 55, 63, 64, 71, 77, 84, 90, 101, 116, 124, 136, 185, 189, 211, 214, 268, 304, 306, 307, 320, 352, 393, 437, 445, 447, 457, 462, 508, 528, 559, 584, 649, 651, 688, 697, 699, 714, 715, 729, 772, 777, 782, 791, 793, 823, 864, 871, 899, 922, 929, 932, 944, 972, 1000, 1025, 1079, 1081, 1093, 1117, 1191, 1205, 1257, 1261, 1272, 1309, 1324, 1326, 1329, 1330, 1372, 1373
Filtration	298, 308, 1225
Gas diffusion	57, 58, 124, 173, 196, 273, 402, 430, 451, 471, 475, 494, 500, 586, 608, 609, 629, 816, 839, 848, 881, 901, 924, 947, 1000, 1018, 1035, 1037, 1050, 1051, 1052, 1059, 1073, 1080, 1109, 1122, 1156, 1173, 1174, 1191, 1192, 1220, 1225, 1230, 1237, 1242, 1326, 1332, 1402
Ion exchange	103, 110, 172, 237, 412, 428, 439, 497, 511, 576, 608, 628, 647, 683, 684, 712, 758, 766, 780, 907, 913, 967, 998, 1000, 1040, 1080, 1102, 1126, 1140, 1146, 1169, 1173, 1191, 1217, 1231, 1228, 1247, 1257, 1264, 1280, 1307, 1310, 1311, 1323, 1324, 1326, 1328, 1353, 1401
Isothermal distillation	50, 238, 268

Table 7.1 (Continued)

Category	References
Micellar formation/micro-emulsions	571, 903, 1134, 1135, 1163, 1201, 1349, 1364
Preconcentration	412, 428, 439, 503, 576, 583, 608, 628, 683, 684, 712, 754, 763, 773, 939, 998, 1000, 1080, 1098, 1102, 1126, 1140, 1169, 1191, 1217, 1228, 1231, 1257, 1264, 1268, 1307, 1310, 1311, 1317, 1318, 1324, 1326, 1328, 1329, 1353, 1401
Other techniques	
Chemometrics	413, 506, 843, 880, 942, 972, 980, 1016, 1017, 1075, 1151, 1166, 1231, 1259, 1260, 1276
Closed-loop configuration	174, 212, 493, 840, 1094, 1224, 1266, 1353
Conversion	1010, 1025, 1081, 1126, 1231, 1247, 1261, 1281, 1311, 1363, 1391
Hydride generation	208, 421, 612, 617, 735, 820, 906, 907, 908, 924, 1018, 1030, 1073, 1080, 1189, 1190, 1191, 1249, 1257
Hydrodynamic injection	338, 373, 394, 559, 608, 645, 828, 1000, 1060, 1109, 1139
Impulse response	853, 1050, 1051, 1055, 1251, 1393
Intermittent pumping	48, 71, 132, 355, 1000, 1060
Nested/Split-loop injection	795, 848, 1000, 1018, 1038, 1073, 1080, 1232, 1257, 1268, 1371
Viscosimetry	7, 162, 224, 362, 363
Microprocessor control	76, 91, 111, 118, 224, 260, 261, 286, 305, 312, 336, 346, 349, 361, 362, 365, 366, 373, 382, 418, 434, 501, 516, 530, 556, 588, 597, 635, 707, 726, 753, 756, 758, 814, 826, 834, 838, 839, 851, 919, 960, 961, 965, 974, 992, 1016, 1019, 1028, 1047, 1069, 1073, 1074, 1075, 1090, 1094, 1097, 1160, 1234, 1247, 1249, 1254, 1260, 1268, 1272, 1305, 1339, 1369, 1373, 1386, 1388, 1389, 1393
Incorporated column reactor (chemically active/nonactive)	71, 146, 176, 199, 212, 233, 234, 280, 283, 284, 296, 340, 343, 344, 345, 348, 360, 386, 387, 422, 428, 439, 459, 460, 478, 490, 497, 501, 502, 511, 542, 546, 554, 566, 573, 576, 589, 597, 608, 609, 638, 647, 648, 650, 678,

Table 7.1 (Continued)

Category	References
	679, 680, 683, 684, 686, 694, 697, 703, 704, 712, 719, 747, 752, 754, 758, 759, 761, 763, 773, 774, 779, 781, 786, 789, 794, 795, 801, 833, 846, 860, 862, 863, 867, 869, 870, 883, 897, 907, 913, 916, 917, 922, 923, 930, 967, 976, 977, 998, 1000, 1005, 1011, 1015, 1016, 1040, 1048, 1058, 1060, 1067, 1068, 1069, 1071, 1078, 1080, 1087, 1095, 1098, 1102, 1103, 1104, 1107, 1112, 1113, 1126, 1129, 1137, 1143, 1146, 1160, 1173, 1176, 1191, 1211, 1217, 1225, 1228, 1231, 1247, 1252, 1256, 1257, 1264, 1266, 1280, 1300, 1307, 1308, 1310, 1311, 1316, 1319, 1320, 1323, 1324, 1328, 1330, 1331, 1335, 1337, 1346, 1350, 1353, 1386, 1401
Microconduits	441, 608, 684, 838, 848, 889, 936, 998, 1000, 1055, 1118, 1160, 1177, 1193, 1200, 1232, 1251, 1272, 1327
Special reactor geometry	567, 568, 608, 795, 1146, 1173, 1250, 1332

Table 7.2 Species Determined by FIA

Species	References
Acetaminophen	464, 491, 558, 853, 995, 1351
Acetic acid/acetate	1199, 1355
Acetoacetate	596
Acetone	1051
Acetylcholinesterase	1069
Acid/base	44, 59 253, 414, 532, 671, 716, 717, 814, 900, 940, 951, 1032, 1096, 1145, 1147, 1171, 1237, 1338
Acidity constants	782, 834
Acids, volatile	312
Adenosine	1320
Adenosine-5-triphosphate sulphurylase	1363
Adenosine phosphate	835, 746, 1078, 1129
Alanine	596, 788
Albumin	42, 47, 80, 95, 138, 262, 369, 553, 681, 713, 868, 948, 911, 974, 1119, 1213, 1356
Alcohols	131, 218, 226, 247, 387, 510, 680, 774, 863, 875, 955, 1051, 1266, 1267
Aldehydes	829, 841, 1267
Aldoses	1406
Alkaline phosphatase	1069
Alkalinity, total	619, 749
α-Amylase	557
Amyloglucosidase	1234
Aluminium	25, 48, 166, 206, 245, 259, 293, 295, 520, 656, 757, 769, 945, 956, 979, 1147, 1257, 1269, 1281, 1304, 1307, 1328, 1338, 1395
Amines	255, 310, 324, 448, 832, 1201, 1329, 1364
Aminoacylase	615
Amino acids	255, 290, 310, 604, 618, 732, 783, 788, 822, 814, 832, 900, 1355, 1356

Table 7.2 (Continued)

Species	References
Ammonia/ammonium	1, 52, 91, 110, 173, 238, 245, 246, 266, 273, 415, 451, 494, 580, 629, 632, 737, 741, 745, 749, 848, 900, 901, 920, 1003, 1032, 1225, 1237, 1244, 1295, 1361, 1371, 1378
Amoxillin	536
Anilide	217
Anilines	410, 1104, 1329
Antimony	458, 820
Arabinose	941
Arginine	788
Aromatic hydrocarbons	214, 306
Arsenic	102, 103, 225, 384, 518, 612, 617, 735, 820, 906, 924, 1005, 1012, 1018, 1030, 1249, 1368
Ascorbic acid	20, 29, 61, 478, 536, 563, 764, 814, 1028, 1029, 1250, 1387
Aspartate	863, 788
Aspirin	410
Barium	518, 572, 656, 712, 956, 1228, 1254
Benzoic acid	774, 1077, 1137
Benzoquinone	143, 316
Beryllium	241, 712, 1228
Bile acid	677, 823
Bilirubin	623, 733
Bismuth	135, 208, 414, 421, 458, 558, 820, 1254, 1257
Bittering compounds (beer)	1373
Boron	38, 105, 169, 206, 245, 401, 468, 1005, 1132, 1257
Bromamine	277
Bromate	1133
Bromide	767, 1016, 1092, 1391
Bromine	129, 235
Brucine	388

Table 7.2 (Continued)

Species	References
Buprenorphine	1274
Butylated hydroxyanisole	503
Cadmium	37, 90, 172, 215, 366, 370, 377, 428, 462, 464, 508, 515, 518, 521, 527, 558, 639, 683, 684, 712, 793, 877, 908, 956, 1080, 1085, 1140, 1169, 1217, 1228, 1254, 1365, 1403
Caffeic acid	464, 491, 595
Caffeine	19, 1079, 1272
Calcium	10, 21, 26, 43, 49, 54, 56, 85, 108, 112, 114, 195, 202, 206, 223, 244, 245, 248, 250, 253, 258, 270, 349, 350, 373, 383, 397, 425, 473, 497, 528, 543, 550, 570, 572, 608, 622, 634, 656, 717, 792, 814, 817, 830, 836, 879, 900, 913, 922, 956, 957, 1024, 1032, 1101, 1227, 1228, 1254, 1257, 1294, 1295, 1311, 1338
Carbohydrates	219, 514, 618, 704, 747, 758, 833, 867, 923, 941, 973, 1011, 1039, 1048, 1067, 1084, 1103, 1106, 1211, 1225, 1331, 1335, 1347, 1350, 1367, 1380
Carbonate (total)	451, 1199
Carbon dioxide	57, 196, 430, 451, 881, 901
Catalase	1379
Catecholamine	770, 799, 1360
Cellulose dehydrogenase	1389
Cephalosporines	516, 679
Cerium	774, 969, 1077, 1095, 1137
Chemical oxygen demand	104, 117, 120, 123, 125, 175, 180, 261, 274, 294, 303, 305, 384, 912, 1074, 1180
Chloramine	277, 983, 1157
Chloranilines	448
Chlorate	755, 986, 1073
Chlorhexidine	1392
Chloride	4, 6, 59, 91, 148, 171, 204, 207, 245, 250, 251, 253, 282, 355, 394, 398, 401, 409, 416, 525, 616, 662, 666, 749, 756, 767, 818, 859, 878, 880, 922, 1003, 1047, 1066, 1071, 1088, 1244, 1313, 1334, 1391

Table 7.2 (Continued)

Species	References
Chlorine	129, 235, 271, 272, 400, 646, 896, 983, 1037, 1242
Chlorine dioxide	947, 1073, 1122, 1174, 1242, 1402
Chlorite	1073
Chloro-organic compounds	1258
Chlorophenoles	448, 563, 490, 721
8-Chlorotheophylline	447
Chlorpromazine	412, 464, 558, 1042
Cholestorol (total/free)	386, 625, 869, 883, 923, 1160
Choline	386
Cholinesterase	460, 602
Chromium	30, 69, 253, 353, 388, 391, 442, 513, 518, 555, 579, 656, 707, 711, 763, 774, 885, 893, 900, 968, 1005, 1013, 1049, 1097, 1098, 1137, 1139, 1187, 1206, 1283, 1310, 1321, 1328
Chymotrypsin	1069, 1204
Citric acid	605
Cobalt	27, 83, 140, 191, 203, 220, 232, 252, 302, 371, 411, 423, 484, 498, 518, 527, 549, 564, 572, 712, 774, 793, 861, 915, 956, 1033, 1094, 1142, 1214, 1228, 1304
Codeine	41
Colouring matter (food)	200, 582
Complex stability constants	522, 604, 1017
Concanavelin A	530
Copper	24, 27, 37, 62, 122, 127, 135, 140, 162, 174, 203, 206, 215, 220, 232, 282, 315, 331, 352, 354, 366, 377, 436, 442, 464, 470, 485, 515, 527, 528, 529, 535, 571, 572, 576, 583, 622, 651, 683, 684, 700, 712, 720, 722, 748, 754, 757, 780, 792, 793, 797, 846, 849, 913, 938, 956, 967, 1008, 1023, 1024, 1075, 1082, 1086, 1102, 1134, 1135, 1147, 1228, 1254, 1257, 1264, 1304, 1338, 1342, 1365
Corticosteroids	78, 239, 1039
Creatine	632
Creatine kinase isoenzymes	1001

Table 7.2 (Continued)

Species	References
Creatine phosphate	1078
Creatine phosphokinase	1078
Creatinine	408, 1015, 1160
Cyanide	251, 452, 455, 561, 578, 586, 611, 616, 621, 640, 669, 726, 818, 922, 1010, 1192, 1209, 1271, 1315, 1399, 1400
Cyclophosphamide	1002
Cysteine	448, 490, 563, 832, 850
Cystine	490
Dicrotophos	1100
Diffusion coefficients	142, 162, 224, 325, 363, 519
Digoxin	1014
Dihydroxyacetone	867
Dihydroxyphenylalanine	61, 285
Diltiazem	1196
Dimethoate	1100
Dioctylsulfosuccinate	942, 1243
Diphenylhydramine	447
Disaccharides	1347
Dithionite	632
Dopamine	61, 285, 464, 492, 536, 613, 764, 832, 980, 1053
Doxorubicin	939
Drugs	41, 193, 304, 347, 361, 488, 740, 937, 939, 1007, 1016, 1042, 1155, 1162, 1212, 1274, 1405
Drug-protein binding	193, 887, 1058, 1240, 1268
Elements (multi-)	105, 162, 188, 206, 354, 385, 518, 528, 572, 656, 661, 980, 1008, 1227, 1228, 1254, 1303, 1328
Enalapril	932
Epinephrine	61, 536

Table 7.2 (Continued)

Species	References
Ergonovine maleate	1054
Ethanol	131, 207, 218, 226, 247, 387, 510, 680, 863, 875, 955, 1051, 1266, 1267
Ethylenediamine	832, 1076
Ethylenediaminetetraacetic acid	825, 832, 888, 900
Europium	1163, 1349
Ferroxidase	768
α-Fetoproteins	674, 760
Flavin mononucleotide	1337
Flocculants	884
Fluoride	111, 282, 359, 652, 653, 725, 729, 749, 847, 913, 1003, 1083, 1244, 1270, 1289, 1295, 1312, 1370, 1385, 1390
5-Fluorouracil	1236
Fluphenazine	937
Formaldehyde	459, 637, 650, 1359
Formate	774, 982, 1077, 1137
Formate dehydrogenase	960, 961
Free fatty acids	556, 649
Fructose	514, 941, 1039, 1331
Galactose	386, 747, 758, 867
Gallium	88, 185, 260
Germanate	225
Globulins	322, 530, 553, 706, 935, 948, 965, 1213
Glucose	9, 20, 46, 64, 71, 199, 231, 269, 280, 287, 330, 345, 348, 360, 386, 422, 493, 542, 547, 573, 597, 626, 659, 660, 678, 686, 703, 704, 719, 743, 779, 833, 863, 901, 917, 923, 926, 941, 1039, 1048, 1084, 1103, 1136, 1153, 1160, 1161, 1202, 1211, 1225, 1308, 1335, 1350, 1366
Glucose oxidase	231, 247, 287

Table 7.2 (Continued)

Species	References
Glutamate	680, 788, 863
Glutamine	788
Glutamyl transferase	394
Glycerol	7, 23, 1308, 1388
Glycine	47, 788, 832, 900
Glutalic oxalacetic transaminase	420
Glutamic pyruvic transaminase	420
Hardness (water)	524, 900, 1032
Heterocycyclic nitrogen compounds	448
Hexacyanoferrate (II/III)	459, 487, 491, 492, 536, 559, 609, 637, 1387
Histidine	788, 832
Humic substances	1188, 1238
Hydrazine	28, 563, 574, 580, 609, 632, 637, 705, 1371
Hydrogen peroxide	58, 194, 247, 265, 356, 345, 546, 603, 643, 747, 761, 781, 795, 810, 976, 981, 1146, 1147, 1173, 1202, 1233, 1256, 1338, 1376
Hydrogen sulfide, see sulphide	
Hydroquinone	487, 536, 1034
Hydroxyacetanilide	217
Hydroxybutyrate	596
17-α-Hydroxyprogesterone	760
Hydroxyproline	1344
Hydroxy compounds (aliph./arom.)	448
Hypohalites	580, 1073, 1233
Hypoxanthine	746
Imidazole	832
Immuno-γ-globulin (IgG)	322, 530, 706, 935, 965, 999, 1129
Indium	845
Indole derivatives	523
Inosine	916, 746, 1320
Inosine phosphate	746

Table 7.2 (Continued)

Species	References
Iodate	357, 459, 609, 637, 1077, 1096, 1137
Iodide	102, 282, 364, 376, 381, 453, 534, 561, 616, 618, 767, 796, 811, 922, 1041, 1064, 1090, 1096, 1306
Iodine	327, 545, 670, 873, 1233
Iodine value	556
Insulin	760
Iron	29, 86, 166, 206, 243, 245, 281, 293, 395, 423, 442, 461, 474, 508, 527, 528, 579, 581, 594, 614, 622, 627, 636, 656, 658, 682, 749, 757, 759, 773, 774, 776, 819, 825, 861, 866, 902, 913, 922, 971, 1013, 1024, 1075, 1077, 1082, 1094, 1137, 1147, 1159, 1191, 1197, 1216, 1229, 1237, 1252, 1254, 1257, 1328, 1338, 1342
Isoleucine	788
Isoniazid	410, 476
Isoprenaline	413, 443, 972, 1166
Isosorbide	144
Ketones	829, 841, 1051
Ketone Bodies	402
β-Lactams	516
Lactic acid/lactate	278, 283, 369, 373, 426, 585, 596, 680, 694, 867, 897, 917, 923, 930, 941, 1160, 1308, 1316, 1355
Lactate dehydrogenase	264, 369, 785, 1316
Lanthanum	969
Lead	32, 37, 90, 215, 232, 334, 352, 366, 370, 428, 457, 458, 464, 515, 518, 521, 527, 558, 572, 683, 684, 712, 793, 845, 877, 928, 956, 967, 980, 1038, 1085, 1086, 1099, 1228, 1257, 1264, 1322, 1358, 1403
Levamisole	1212
Leucine	788
Leucine dehydrogenase	960, 961

Table 7.2 (Continued)

Species	References
Lipase	1069, 1364
Lipids	624
Lithium	263, 528, 792, 817, 891, 936, 1026, 1144, 1200, 1251, 1257, 1327
Lysine	788
α-Macroglobulin	369
Magnesium	43, 45, 49, 56, 66, 85, 195, 206, 349, 383, 397, 403, 518, 526, 528, 543, 550, 570, 572, 622, 656, 707, 792, 842, 900, 913, 922, 1024, 1032, 1097, 1101, 1227, 1228, 1246, 1254, 1257
Malathion	1100
Maleic acid	459, 637, 1039
Maltose	514, 941
Manganese	53, 162, 206, 220, 232, 245, 378, 518, 572, 622, 695, 712, 774, 776, 849, 886, 893, 956, 967, 1009, 1040, 1137, 1142, 1195, 1228, 1257, 1348
Meptazinol	67
Mercury	165, 381, 453, 471, 475, 518, 748, 991, 1080
Metals (traces)	20, 115, 127, 184, 188, 189, 220, 352, 354, 366, 384, 428, 439, 442, 445, 515, 527, 528, 551, 651, 683, 712, 861, 876, 909, 915, 977, 998, 1009, 1040, 1086, 1098, 1115, 1203, 1217, 1216, 1228, 1229, 1254, 1263, 1264, 1304, 1328, 1401
Methanethiol	1332
Methionine	788
Molybdenum	22, 55, 351, 561, 610, 631, 656, 1005, 1121, 1386
Morphine	1007, 1162
Nicotineamide adenine dinucleotide	493, 589, 630, 1337
Nicotineamide adenine dinucleotide phosphate	1337

Table 7.2 (Continued)

Species	References
Nickel	203, 220, 232, 352, 354, 371, 527, 564, 572, 628, 683, 707, 712, 793, 956, 1008, 1097, 1228, 1304, 1339
Nitrate	8, 12, 34, 51, 71, 81, 91, 148, 157, 170, 245, 284, 326, 357, 373, 389, 424, 459, 481, 496, 609, 637, 640, 687, 737, 749, 822, 922, 962, 1003, 1025, 1071, 1081, 1185, 1191, 1244, 1300, 1323, 1378, 1391
Nitrite	15, 51, 71, 81, 132, 148, 245, 266, 284, 300, 326, 357, 424, 459, 481, 499, 512, 609, 637, 640, 708, 757, 822, 922, 931, 774, 1003, 1025, 1050, 1075, 1077, 1081, 1104, 1137, 1191, 1218, 1244, 1255, 1301, 1317, 1391
Nitrofurantoin	1027
Nitrogen/total nitrogen	3, 5, 13, 38, 50, 68, 109, 126, 228, 245, 258, 268, 425, 827, 990, 1220, 1287, 1383
Nitrogen oxides	1317
Nitrogenase	632
o-Nitrophenol	459
Nitroprusside	765
Norepinephrine	558
Nucleotides	746, 921
Ochratoxin A	800
Ovalbumins	948, 1213
Oxalacetate	863
Oxalic acid/oxalate	900, 902, 1031, 1221
Oxygen	345, 346
Ozone	575, 839, 1073, 1242
Palladium	1008
Parathion	1100
Penicillins	516, 723, 794, 1176
Penicilloic Acid	289
Perchlorate	777, 899
Peroxidase	953

Table 7.2 (Continued)

Species	References
Perphenazine	937
pH	21, 64, 112, 244, 248, 270 429, 608, 640, 757, 778, 848, 881, 882, 1165, 1191, 1302, 1334, 1355
Pheniramine maleate	1330
Phenols	249, 316, 410, 483, 558, 646, 853, 1034, 1052, 1278
Phenothiazine	937, 1042
Phenylephrine hydrochloride	1330
Phenyl isocyanate	958
Phosphate/phosphates	1, 6, 23, 59, 64, 89, 106, 113, 133, 163, 182, 225, 237, 266, 276, 279, 320, 375, 466, 495, 512, 567, 568, 601, 606, 616, 632, 666, 737, 749, 751, 822, 896, 904, 918, 978, 1003, 1005, 1012, 1036, 1045, 1046, 1080, 1152, 1172, 1184, 1199, 1218, 1225, 1244, 1292, 1297, 1299, 1325, 1338, 1363, 1378
Phosphoenol pyrovate	1363
Phospholipids	465
Phosphonate	601, 751
Phosphorous	2, 5, 13, 38, 94, 109, 126, 190, 206, 245, 258, 267, 268, 425, 433, 544, 548, 560, 696, 789, 809, 978, 1191, 1254, 1257, 1383
Phosvitins	948
Polycyclic aromatic hydrocarbons	214, 306
Potassium	8, 13, 31, 38, 43, 112, 114, 145, 206, 245, 248, 250, 323, 446, 472, 528, 622, 656, 817, 836, 842, 882, 1024, 1083, 1235, 1254, 1257, 1309
Procyclidine	304
Proline	788
Propantheline bromide	635
Prostatic acid phosphatase	1070
Protease	750, 961
Protein(s)	17, 118, 406, 425, 553, 587, 591, 674, 706, 713, 826, 948, 974, 1004, 1213, 1356
Pullulanase	750, 961

Table 7.2 (Continued)

Species	References
Pyridoxal	724, 854
Pyridoxal phosphate	854
Pyrocatechol	1034
Pyrogallol	1034
Pyrophosphatase	432, 824, 1068
Pyrophosphate	89, 106, 133, 182, 904, 1152, 1172, 1363
Pyrovate	596
Pyrovate kinase	1363
Reaction rate constants	517, 541, 838, 840, 982, 1065, 1224, 1259
Refractive Index	27, 98, 442
Resorcinol	1034, 1052
Rubredoxin	861
Salicylamide	635
Salicylic acid	1318
Selenium	518, 617, 734, 820, 821, 907, 1005, 1080, 1189, 1190, 1191, 1357, 1384
Silicate/silicon	177, 225, 319, 466, 477, 565, 656, 697, 710, 822, 813, 860, 910, 914, 959, 1072, 1325
Silver	381, 419, 453, 458, 508, 1164, 1217, 1257
Sodium	8, 114, 250, 472, 528, 622, 656, 817, 836, 842, 1024, 1254, 1257
Sodium chloride	1181, 1280, 1341, 1345
Sodium dioctylsulphosuccinate	1079
Sodium dodecyl sulphate	1079
Soil organic matter	1188, 1238
Solubility product	1193
Sorbose	941
Starch	1113
Steroids	823, 1039
Stoichiometry	1224
Strontium	45, 85, 518, 922
Succinic acid	1355

Table 7.2 (Continued)

Species	References
Sucrose	833, 923, 941, 1011, 1039, 1067, 1225, 1367
Sugars	219, 514, 704, 747, 758, 833, 867, 923, 941, 973, 1011, 1039, 1048, 1067, 1084, 1103, 1106, 1211, 1225, 1331, 1335, 1347, 1350, 1367, 1380
Sulfamethizole	635
Sulfanilamides	618, 740, 1268
Sulfur	866, 1254
Sulfur organics	448, 618, 850, 1104
Sulphate	11, 87, 160, 162, 171, 186, 245, 254, 256, 298, 355, 508, 511, 533, 538, 647, 672, 729, 730, 816, 922, 1036, 1071, 1183, 1358, 1391
Sulphide	83, 134, 179, 381, 453, 454, 467, 538, 640, 672, 676, 709, 850, 922, 1059, 1072, 1247, 1283, 1332, 1340
Sulphite	16, 321, 374, 407, 451, 538, 616, 672, 816, 831, 898, 903, 1087, 1104, 1156, 1230, 1332
Sulphite oxidase	1087
Sulphonyl haloamines (aromat.)	277
Sulphur dioxide	71, 147, 212, 266, 1091, 1184
Surfactants	63, 101, 136, 437, 531, 762, 892, 1079, 1093, 1261, 1272, 1383
Tartaric acid	1194
Tellurium	820
Terbium	329, 545
Terbutaline sulphate	361
Terodiline	929
Tetracycline	675
Thallium	877, 980
Theophylline	681
Thiamine	77, 509
Thiocyanate	616, 618, 811, 922
Thiols	134, 1332
Thiosulphate	850, 1147

Table 7.2 (Continued)

Species	References
Thiourea	618, 850, 1210, 1381
Thorium	135
Threonine	788
Thyroid hormones	1119
Thyroxine	358
Tin	877, 1403
Titanium	311, 656, 851, 1257, 1328
Triethylamine	99
Triethylenetetramine	832
Triglycerides	119, 231, 681, 1388
Trypsin	167, 379
Tryptophan	788
Tungsten	561, 656, 1121
Tyrosine	788
Uracil derivatives	1236
Uranium	390, 459, 577, 637, 775, 858, 956, 970, 972
Urea	60, 64, 71, 79, 199, 360, 744, 848, 923, 1080, 1110, 1158, 1232, 1319
Uric Acid	633, 678, 752, 853, 1346
Valine	788
Valinomycin	539
Vanadium	32, 128, 161, 232, 459, 637, 656, 766, 774, 1005, 1038, 1077, 1137, 1186, 1203, 1328
Viscosity	7, 162, 224, 362, 363
Warfarin	975
Water	82, 221, 414, 537, 638, 844, 870, 1143
Xylose	941, 1039

Table 7.2 (Continued)

Species	References
Zinc	62, 172, 203, 206, 245, 331, 352, 354, 377, 428, 436, 445, 515, 521, 527, 528, 558, 562, 572, 622, 684, 715, 720, 771, 792, 842, 871, 890, 956, 977, 1024, 1082, 1086, 1191, 1217, 1228, 1257, 1314, 1336, 1342, 1369
Zirconium	656

Table 7.3 Flow-Injection References Listed According to Area of Application

Area of Application	References
Agricultural	93, 268, 673, 978, 1257, 1296
Feed	26, 258, 425, 1380
Fertilizers	12, 13, 38, 190, 964, 971, 1080
Plant material	2, 3, 5, 11, 17, 22, 23, 25, 43, 48, 50, 53, 55, 92, 109, 126, 145, 166, 169, 202, 206, 238, 245, 281, 310, 355, 457, 496, 612, 617, 622, 631, 820, 876, 964, 978, 1030, 1135, 1185, 1189, 1190, 1292, 1299, 1380
Soil	8, 12, 30, 48, 52, 68, 69, 81, 92, 94, 170, 202, 204, 228, 267, 401, 424, 429, 520, 621, 622, 627, 710, 769, 778, 817, 827, 871, 876, 945, 964, 1045, 1046, 1080, 1114, 1188, 1189, 1190, 1191, 1220,1287, 1295, 1379, 1386
Biochemical	167, 219, 255, 301, 309, 345, 347, 348, 379, 402, 404, 406, 436, 478, 488, 516, 539, 587, 591, 597, 615, 630, 632, 633, 770, 787, 788, 799, 861, 863, 887, 906, 908, 921, 933, 948, 953, 996, 1001, 1024, 1039, 1078, 1086, 1129, 1169, 1195, 1204, 1213, 1337, 1344, 1346, 1356, 1360, 1403
Biotechnology	546, 679, 680, 723, 743, 750, 770, 771, 826, 852, 862, 867, 960, 961, 1113, 1225, 1256, 1389
Clinical	58, 95, 108, 118, 138, 154, 178, 193, 194, 195, 231, 250, 262, 269, 270, 301, 323, 358, 369, 387, 394, 396, 436, 474, 488, 530, 588, 645, 674, 692, 694, 706, 734, 855, 856, 901, 954, 965, 966, 999, 1024, 1057, 1069, 1111, 1119, 1123, 1256, 1298, 1342, 1388
Blood/blood plasma	42, 57, 173, 226, 244, 273, 348, 426, 460, 623, 629, 675, 686, 719, 744, 745, 800, 845, 881, 882, 916, 955, 1014, 1015, 1082, 1107, 1116, 1153, 1320
Blood serum	6, 8, 9, 21, 47, 60, 71, 80, 95, 112, 131, 174, 199, 248, 251, 263, 278, 280, 287, 322, 330, 331, 347, 360, 369, 383, 386, 394, 395, 397, 408, 420, 446, 465, 470, 472, 485, 493, 528,

Table 7.3 (Continued)

Area of Application	References
Blood serum (cont.)	543, 550, 553, 562, 570, 573, 596, 602, 623, 624, 625, 626, 629, 633, 656, 660, 677, 678, 681, 682, 700, 703, 704, 713, 719, 733, 740, 741, 744, 745, 758, 760, 768, 777, 785, 792, 854, 868, 869, 875, 883, 891, 901, 911, 913, 923, 929, 935, 938, 974, 1016, 1026, 1058, 1069, 1070, 1083, 1142, 1144, 1160, 1200, 1211, 1213, 1227, 1232, 1268, 1307, 1318, 1319, 1369
Urine	160, 359, 408, 412, 462, 606, 629, 633, 703, 752, 777, 842, 853, 867, 901, 939, 1015, 1029, 1098, 1221, 1319, 1322, 1385, 1403
Environmental	90, 205, 384, 471, 646, 673, 1338
Air	147, 212, 235, 374, 409, 560, 745, 1156, 1359, 1361
Water	4, 7, 11, 12, 15, 21, 24, 25, 28, 34, 52, 53, 63, 69, 81, 87, 89, 91, 101, 104, 106, 110, 113, 114, 117, 120, 123, 125, 132, 136, 157, 170, 171, 175, 180, 182, 186, 202, 210, 215, 236, 237, 245, 246, 254, 256, 261, 265, 272, 274, 279, 281, 284, 294, 295, 298, 303, 305, 310, 319, 320, 326, 327, 349, 350, 351, 355, 359, 364, 366, 375, 388, 391, 398, 400, 403, 407, 409, 411, 415, 416, 419, 424, 428, 432, 433, 437, 439, 451, 452, 454, 455, 457, 461, 467, 468, 473, 475, 477, 480, 481, 483, 494, 495, 499, 508, 511, 515, 520, 524, 525, 527, 531, 570, 575, 576, 579, 586, 612, 617, 619, 621, 628, 634, 640, 641, 643, 647, 650, 652, 666, 669, 676, 683, 684, 696, 709, 711, 725, 729, 730, 735, 737, 741, 745, 754, 757, 762, 766, 810, 814, 817, 822, 839, 847, 859, 866, 878, 884, 892, 896, 900, 902, 909, 910, 912, 913, 915, 920, 922, 924, 945, 946, 959, 960, 968, 970, 977, 979, 983, 990, 992, 1003, 1012, 1016, 1032, 1037, 1040, 1071, 1072, 1074, 1083, 1086, 1088, 1092, 1101, 1122, 1134, 1140, 1157, 1159, 1164, 1172, 1173, 1174, 1180, 1183, 1189, 1190, 1192, 1209, 1218, 1222, 1228, 1229, 1238, 1242, 1244, 1255, 1261, 1269, 1278, 1283, 1289, 1291, 1300, 1309, 1312, 1323, 1325, 1332, 1336, 1340, 1348, 1378, 1383, 1385, 1402, 1404
Speciation	474, 555, 579, 581, 594, 614, 636, 711, 759,

Table 7.3 (Continued)

Area of Application	References
	763, 780, 922, 968, 1021, 1049, 1187, 1206, 1252, 1277, 1310, 1321, 1357
Food	16, 47, 54, 62, 200, 556, 582, 605, 617, 680, 700, 701, 708, 746, 747, 807, 820, 883, 977, 1009, 1011, 1081, 1161, 1181, 1230, 1250, 1279, 1306, 1331, 1341, 1345, 1403
Beer	312, 510, 816, 1373
Beverages	226, 402, 451, 488, 503, 653, 704, 816, 833, 867, 874, 973, 1011, 1091, 1106, 1110, 1158, 1184, 1194, 1230, 1250, 1267, 1279, 1313, 1355, 1385, 1403, 1405
Industrial	207, 214, 243, 356, 385, 393, 409, 466, 508, 557, 580, 649, 682, 717, 723, 747, 844, 849, 866, 967, 986, 1035, 1074, 1098, 1183, 1226, 1254, 1271, 1280, 1297, 1311, 1314, 1359, 1378, 1399, 1403
Geological/mineralogical	86, 223, 293, 311, 329, 390, 533, 548, 578, 579, 581, 670, 775, 813, 820, 821, 830, 849, 910, 970, 991, 1197, 1246, 1290, 1294
Metallurgical	122, 203, 241, 259, 353, 354, 458, 513, 544, 586, 610, 722, 789, 820, 906, 907, 1085, 1139, 1290, 1336, 1339, 1395
Process control	261, 338, 393, 546, 556, 578, 598, 680, 718, 743, 826, 852, 857, 1055, 1059, 1086, 1208, 1393, 1394
Pharmaceutical	19, 29, 37, 41, 56, 57, 61, 67, 77, 78, 144, 193, 221, 239, 247, 249, 264, 276, 277, 283, 285, 289, 304, 307, 316, 325, 328, 341, 349, 361, 399, 409, 410, 413, 447, 448, 464, 476, 490, 491, 492, 503, 509, 514, 516, 523, 536, 563, 569, 589, 635, 638, 740, 767, 783, 788, 794, 798, 814, 815, 819, 823, 870, 887, 932, 937, 942, 951, 975, 995, 1002, 1007, 1016, 1042, 1054, 1076, 1138, 1155, 1162, 1179, 1196, 1212, 1236, 1240, 1243, 1249, 1250, 1268, 1274, 1330, 1336, 1351, 1392, 1405
General	1, 10, 20, 44, 45, 49, 51, 66, 79, 82, 83, 92, 98, 102, 105, 111, 116, 124, 127, 128, 140, 148, 152, 156, 158, 159, 161, 163, 164, 165,

Table 7.3 (Continued)

Area of Application	References
General (cont.)	168, 172, 177, 179, 185, 191, 208, 209, 210, 211, 213, 220, 223, 224, 225, 242, 252, 257, 265, 282, 285, 286, 288, 291, 300, 302, 306, 308, 313, 314, 315, 317, 318, 321, 324, 329, 332, 334, 339, 340, 342, 346, 352, 357, 362, 363, 365, 367, 368, 370, 371, 372, 373, 376, 377, 378, 379, 381, 389, 392, 396, 405, 414, 422, 423, 430, 434, 435, 440, 442, 443, 444, 449, 453, 459, 463, 469, 479, 482, 486, 487, 495, 497, 498, 500, 502, 504, 506, 507, 512, 517, 518, 519, 521, 522, 526, 529, 532, 534, 535, 537, 538, 541, 542, 547, 549, 551, 552, 554, 558, 562, 565, 566, 567, 568, 571, 572, 574, 577, 583, 584, 593, 594, 595, 599, 600, 601, 603, 604, 607, 609, 611, 613, 614, 616, 618, 620, 637, 639, 642, 644, 648, 651, 657, 658, 659, 661, 662, 667, 671, 672, 682, 685, 687, 688, 689, 690, 691, 693, 695, 697, 698, 702, 705, 707, 712, 714, 715, 716, 720, 721, 724, 726, 732, 736, 738, 739, 748, 751, 753, 755, 756, 761, 764, 765, 773, 774, 776, 779, 781, 782, 791, 793, 795, 796, 797, 801, 802, 803, 804, 809, 811, 818, 824, 825, 829, 831, 832, 834, 835, 836, 838, 840, 841, 843, 846, 848, 850, 851, 858, 864, 865, 870, 872, 873, 877, 879, 880, 885, 886, 888, 890, 893, 894, 897, 898, 899, 900, 903, 904, 914, 917, 918, 919, 925, 927, 928, 930, 931, 936, 940, 941, 947, 949, 956, 957, 958, 962, 969, 972, 976, 980, 981, 982, 987, 989, 993, 997, 1004, 1005, 1008, 1010, 1013, 1017, 1018, 1019, 1020, 1022, 1023, 1025, 1027, 1028, 1031, 1033, 1034, 1036, 1038, 1041, 1047, 1048, 1050, 1051, 1052, 1053, 1061, 1062, 1063, 1064, 1065, 1066, 1067, 1068, 1073, 1075, 1077, 1079, 1084, 1087, 1089, 1090, 1092, 1094, 1095, 1096, 1097, 1100, 1102, 1103, 1104, 1105, 1108, 1112, 1115, 1118, 1120, 1121, 1124, 1125, 1127, 1130, 1131, 1132, 1133, 1136, 1137, 1141, 1143, 1145, 1146, 1147, 1148, 1149, 1150, 1151, 1152, 1154, 1163, 1165, 1166, 1167, 1168, 1170, 1175, 1176, 1182, 1186, 1193, 1198, 1199, 1201, 1202, 1203, 1207, 1210, 1214, 1215, 1216, 1217, 1219, 1233, 1235, 1237, 1239, 1245, 1247, 1248, 1251, 1252, 1258, 1259, 1260, 1262, 1263, 1264, 1266, 1270, 1272, 1275,

Table 7.3 (Continued)

Area of Application	References
General (cont.)	1276, 1281, 1284, 1286, 1293, 1301, 1302, 1303, 1304, 1305, 1308, 1315, 1316, 1317, 1327, 1328, 1329, 1333, 1334, 1335, 1343, 1347, 1349, 1350, 1352, 1357, 1358, 1362, 1363, 1364, 1365, 1366, 1367, 1368, 1370, 1371, 1372, 1374, 1375, 1376, 1377, 1381, 1382, 1384, 1387, 1390, 1391, 1396, 1397, 1398, 1400, 1406
Educational	59, 139, 253, 299, 333, 446, 727, 1224, 1265
Review articles	20, 33, 35, 36, 64, 65, 71, 93, 96, 97, 100, 107, 121, 124, 130, 137, 139, 141, 152, 162, 178, 184, 188, 196, 197, 198, 201, 205, 216, 222, 227, 229, 230, 232, 233, 234, 240, 262, 266, 268, 275, 292, 296, 297, 301, 328, 335, 337, 338, 373, 382, 384, 393, 418, 427, 431, 438, 440, 441, 450, 456, 488, 489, 501, 505, 540, 545, 559, 569, 588, 590, 592, 598, 608, 645, 654, 655, 673, 692, 699, 718, 731, 742, 749, 784, 790, 798, 806, 808, 812, 815, 816, 822, 828, 837, 876, 889, 895, 922, 923, 933, 934, 943, 946, 961, 963, 964, 984, 985, 988, 994, 998, 999, 1021, 1035, 1043, 1044, 1055, 1056, 1060, 1080, 1109, 1111, 1114, 1117, 1126, 1128, 1177, 1178, 1179, 1191, 1205, 1231, 1241, 1253, 1257, 1279, 1282, 1285, 1288, 1296, 1324, 1326, 1353, 1393
Books	153, 663, 664, 665, 1000, 1223

CHAPTER

8

FLOW INJECTION ANALYSIS NOW AND IN THE FUTURE

The spirits I summoned, now I can't get rid of them!

W. GOETHE, *The Sorcerer's Apprentice*

There are very few things one may predict with any certainty, and our past expectations as to the future of FIA confirm this observation. Originally designed as a means to automate serial assays based on colorimetric detection, FIA has undergone an amazing development as documented by comparing the volume of this monograph with that of its first edition. It seems in retrospect as if the spirits have released a tidal wave of FIA publications far beyond imagination, the flooding power of which is rapidly undermining our ability to write about it.

It is the versatility of FIA that is the decisive factor behind the widespread application of the technique. Replacement of the established practice of homogeneous mixing by the exploitation of concentration gradients, well defined in time and space, has provided a new approach to solution handling and information gathering on chemical systems. It allows the use of numerous detectors, incorporation of a variety of separation techniques, and miniaturization of the apparatus so that serial assays may be performed in microscale, on small sample volumes and with limited reagent consumption. Such an enclosed, miniaturized automated instrumental technique contributes to improvement of the working environment.

These are the main factors behind the growth of the FIA publications recorded in this monograph. It would be futile to try to predict all future trends, in view of the research effort and the capacity of our colleagues in Japan, United States, Europe, and Brazil. There is, however, growing evidence that FIA will find a much wider range of applications than described so far. Our present work, conducted in cooperation between research teams in Seattle and Copenhagen, indicates that in the future FIA will be used as:

A tool for *serial assay*
A tool for *miniaturization* and *integration*
A means of *enhancing* detector performance
A *link* between chemistry and instruments
A *hyphen* yielding new information via multidimensional readouts
A tool for *continuous monitoring* and *process control*; and
An *impulse response* technique

8.1 SERIAL ASSAYS OF BLOOD, RANDOM ACCESS, AND REVERSED FLOW FIA

Highly viscous, nonaqueous, or other liquid samples of unusual properties are difficult and often unpleasant to handle manually, and it is only a question of time and some investment in research resources before the obvious advantages of FIA for these types of samples matrices will be exploited. Examples of such materials are oils, concentrated chemicals such as dye stuffs, sugar solutions, polymers, and reactants for polymer production. However, if one matrix should be singled out it is, of course, blood, which undoubtedly is the most frequently analyzed sample material.

Determination of glucose, urea, ammonia, electrolytes, and other clinically important species in full blood without pretreatment would be a major breakthrough in clinical chemistry, where these assays are traditionally performed on plasma and/or serum. This would avoid the necessity of separation of blood cells by centrifugation and other manual operations. Such manual handling of thousands of samples is not only time and labor consuming but also objectionable since the personnel are at increased risk to be infected by contagious disease, among which AIDS is the most recent threat. A stat instrument capable of yielding the analytical result on full blood within a minute after sample injection would eliminate these risks, and could be used at bedside and perhaps even with less skilled personnel, but certainly with less expense per assay than needed today with traditional equipment operating on large series of serum samples in a central clinical laboratory.

The most recent experimental evidence obtained as a part of a graduate programme at the Chemistry Department A, in cooperation with Ib Andersen of Herlev County Hospital, confirms that determination of urea in full blood yields results which are in agreement with the more laborious urea assay performed on plasma samples [8.1]. The FIA manifold used incorporated a flow cell combining gas diffusion and optosensing (cf. Fig. 4.35), the separating barrier between the donor and accepting streams consisting of a sandwich of a hydrophobic gas-permeable membrane and a hydrophilic membrane between which was contained a glutaraldehyde–albumin gel on which urease was covalently immobilized. The urea of the sample solution, injected into an aqueous carrier stream of tris buffer, was determined via the color change of an acid–base indicator contained within the acceptor stream, the change in reflectance being a function of the ammonia diffused through the gas-permeable membrane and thus the sample urea content. The correlation coefficient between assay of the full blood samples and the plasma samples (analyzed at Herlev County Hospital) was 0.9931 (RSD 1.8%), the response being independent of the pH, the buffering capacity, and the hematocrite content of the blood sample. The injected sample volume was 20 μL and the analysis frequency 60 samples/h.

A *random access FIA* is a logical extension of the flow analysis concept for applications when several components should be analyzed in the same sample or in a series of samples (cf. Chapter 1.3). Although its use in the clinical field is the most obvious, a random access FIA analyzer should, due to its simplicity, find a much wider range of applications. Its principle is best seen from Fig. 8.1, showing a double barrelled valve by means of which sample and reagents are injected into a suitable (inert) carrier stream. Using in fact a six-port valve as that shown in Fig. 5.3 (in the configuration depicted in Fig. 5.7), the injected zones will during the transport through the manifold chase and penetrate into each other (cf. Sections 2.5 and 4.5.2), their components by reaction yielding a composite peak from which sample blank, product signal and reagent blank may be read out. Reagents may be changed at will as well as samples, stopped flow kinetic measurements can be incorporated if required, and the readout of this stat system will be available within a minute after sample injection. The function of the carrier stream is merely to transport the sample and reagent zones reproducibly through the system and to provide a proper washout. Such a simple concept and the construction of a random access FIA analyzer contrasts with the complexity of the clinical continuous-flow random access analyzer discussed in Section 1.3, where an intricate combination of oil films and air bubbles are used in order to secure homogenization of the sample and reagent solutions on their long (and time consuming) travel through the conduits of the system and to prevent sample crosscontamination.

Reversed flow FIA is based on the injection principle depicted and explained in Fig. 5.16. Note that reversed flow injection has the following attractive features: (a) countercurrent flow in the merging tee, reactor, and detector; (b) flexibility of selecting the sample/reagent ratio by appropriate adjustment of the flow rates and by choosing the delay stop

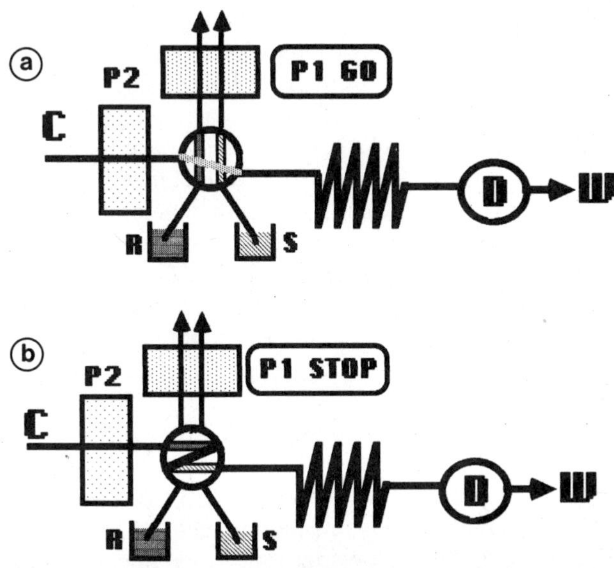

Figure 8.1. Random access FIA system furnished with a double barrelled injection valve. In the load position (*top*) sample (S) and reagent (R) are by means of pump $P2$ aspirated into the injection valve. During this sequence the carrier stream (C) propelled by pump $P1$ is by-passing the valve. In the inject position (*bottom*) the two zones are by means of stream C carried into the FIA manifold, chasing and penetrating into each before reaching the detector (D). By this approach a composite readout due to sample alone, reaction product, and reagent blank is obtained.

period of the reagent pump, since this allows choice of the element of fluid to be selected as readout; and (c) possibility of reaction rate measurement by stopped flow. The advantage of countercurrent operation of the whole system is that it allows filtration to be incorporated thereby avoiding clogging of the system should the sample material contain particulate matter, and that the detector is exposed to any unwanted matrix material only during the measurement interval, since the sample zone only "peeks" into the flow cell.

8.2 ENHANCEMENT OF INSTRUMENT RESPONSE

Enhancement of detector performance by FIA is exemplified by increase in sensitivity, ability to distinguish between different species and by conversion of nondetectable species into detectable ones.

The on-line microcolumn technique allowing improvement of the detection limit of atomic absorption and atomic emission spectroscopy by a factor of hundred has been discussed in detail in Section 4.7.1. With an inductively coupled plasma atomic emission spectroscopy instrument the same approach allowed the following typical detection limits (and linear range) in parts per billion: Ba 0.01 (0.1–100); Be 0.009 (0.1–100); Ca 0.6 (1.0–100); Cd 0.04 (0.1–100); Co 0.1 (0.1–100); Cu 0.08 (0.1–100); Mg 0.007 (0.1–100); Mn 0.06 (0.1–100); Ni 0.09 (0.1–100); Pb 2 (5.0–100); Zn 0.2 (0.1–100). Besides, the use of on-line microcolumns permit speciation as well as conversion (cf. Section 4.7.1), and it is conceivable that other techniques than AAS, AES, and ICP would benefit from similar enhancement.

Speciation allows one to distinguish between different chemical forms of an element. Since retention of each element on an ion-exchange column depends on its chemical form, preconcentration on a microcolumn situated in a FIA channel also depends on the type of chemical form of the analyte and on the column material [780, 1310]. By selecting suitable column materials (anion exchanger, cation exchanger, chelating agents, or C-18 bonded silica) a systematic scheme may be devised which should allow multielement speciation in complex matrices using AAS or ICP. While a manual approach would not be sufficiently reproducible, and certainly would be too labor and time demanding, automated solution handling makes use of several exchange columns on-line feasible.

The resonance and emission lines of most nonmetals occur in the vacuum UV region and therefore these elements are difficult to determine by AAS and ICP. Indirect methods, based on heterogeneous equilibria (solvent extraction, precipitation, or ion exchange) have been proposed, yet they have not found wider application, being time and labor consuming and also because their manual execution lack reproducibility. The FIA conversion technique (cf. Section 4.7.1) utilizes the fact that in the highly reproducible time-concentration domain of the dispersed sample zone, kinetic chemical exchange between any two phases (e.g., a liquid sample and a solid surface) can be executed reproducibly. In this manner a nondetectable species can be converted through a heterogenous chemical reaction into a detectable species. Thus the determination of cyanide using an exchange column with CuS, relying on the detection of the copper released from the solid surface as the water-soluble copper(I)cyanate complex, allowed measurement down to 1 ppm of cyanide in the presence of a number of anionic and cationic species with an RSD value of 1.8% [1010].

Indirect determination of sulfate by AAS is an example of instrument enhancement achieved in a novel way. When developing a conversion method for sulfate, based on the formation of $BaSO_4$, A. J. Rasmussen, a graduate student at Chemistry Department A, observed that the quantities of Ba detected by AAS depended on whether Ba was in the ionic or in the insoluble sulfate form. Closer examination showed that at the low temperature of the acetylene/air flame used, the $BaSO_4$, being refractory, does not dissociate in the flame, leaving the Ba atoms undetected. This led to the development of an indirect method for sulfate, where an reagent zone of $BaCl_2$ is allowed to react with a sample zone of a soluble sulfate, the excess of unreacted barium being measured by AAS [8.2]. It will be interesting to further develop this method, which is very simple, since it does not make use of an incorporated column, and to consider its potential for determination of species that form refractory compounds.

Instrument enhancement may in many cases cleverly be attained by exploiting the possibilities available through kinetic discrimination. Thus in enzyme electrodes advantage might be taken of the difference in diffusion rates of analyte and interfering species within the membrane layer (Section 4.1). Pacey et al. [947] utilized a similar approach in a system comprising a gas-diffusion cell furnished with Teflon membrane exploiting the selectivity of the membrane to discriminate between chlorine dioxide and chlorine, ultimately achieving an overall selectivity factor of 1500 in favor of chlorine dioxide (Section 4.6.2). Another example of instrument enhancement using the kinetic approach is encountered in voltammetry. Contrary to normal practice, where the sample has to be deaerated prior to analysis, the oxygen in the test solution does not have to be removed if the differential pulse technique is performed in the FIA mode [1239]. This is due to the irreversibility of oxygen reduction so that kinetic advantage can be taken by comparing the readouts of two electrodes sequentially monitoring the same sample zone.

Miniaturization of flow systems and integration of detectors into the flow channel of FIA system may be viewed as yet another example of instrument performance enhancement since it improves sample, reagent, and time economy (cf. Section 4.12). These were also the objectives behind a graduate program recently undertaken at this laboratory aimed at devising an immunoassay procedure at minimum consumption of expensive antibody material [8.3]. This task was solved by immobilizing the antibody on a porous support material placed in a small packed reactor within the FIA manifold. Since antibody–antigen reactions are diffusion dependent at low concentrations, and therefore require long reaction times in conventional setups, the FIA approach effectively increases the contact between the reactants, thereby increasing the reaction rate, resulting in a detectable reaction within 40 s for an antigen concentration of 5 μg/mL. The analytical procedure, which as model system employed human serum albumin (HSA) and monoclonal anti-human serum albumin, was based on injecting the sample solution along with a known amount of tagged HSA, exploiting the fact that these two constituents are bound identically by the antibody but in a ratio depending on the concentration of albumin in the sample. Thus, determination of the amount of tagged material allowed calculation of the concentration of sample. The tag might be an enzyme, which via a coupled reaction can generate a species to be sensed—which approach thus allow amplification of the signal as a function of the exposure time between enzyme and ancillary reagent—or a radioactive element, or a fluorescent material. The determination might either be accomplished by optosensing directly on the site of the immobilized antibody, in which case the antibody–antigen binding subsequently is broken by a suitable solvent in order to regenerate it for the next samping cycle, or by first eluting the antigen and then pass it through a suitable detector.

As a link between chemistry and instruments FIA allows solutions to be handled more reproducibly than by hand and in a protected environment. Since solutions are moved and mixed in closed tubular conduits, any influence of light, oxygen, or other elements of the laboratory environment are eliminated. Unstable reagents can be generated *in situ*, and volatile, radioactive, or toxic materials that are handled in microvolumes are more easily contained. Such a link also facilitates multidetection, being an ideal vehicle for accommodation of several different detectors through which an injected sample material can be consecutively transported. Examples reported so far of such an approach comprise the combination of spectrophotometry with atomic absorption [579, 614] and potentiometric stripping analysis with atomic absorption [1263]. Compared to manual handling where the samples are measured sequentially by detectors operated individually and batchwise, the SPEC-FIA-AAS and PSA-FIA-AAS approaches reduce the extent of manual operations and thus the danger of cross contamination, and yield the well-known advantages of miniaturization. Yet, even more importantly, they offer the possibility of exploiting the concentration gradient formed from the injected sample material to trigger each detector to supply the readout from that element of the dispersed sample zone where the concentration lies precisely in the optimum operational range.

The term hyphenation was proposed by the late Thomas Hirschfeld [8.4] to emphasize a novel source of information obtained by a suitable combination of two mutually independent instrument responses with the purpose of enhancing tthe selectivity of response not attainable by a single instrument or even by the two individual instruments operated separately, but achieveable when these two instruments are appropriately linked—by a hyphen. Examples of well-known suitable combinations are GC-MS and GC-IR. The additional information is usually obtained at a high cost, since each additional dimension of data requires an additional expensive instrument. FIA may, however, be viewed as such a hyphen, since it allows exploitation of chemical reactions. In any FIA system the sample zone injected into a carrier stream of reagent yields a well-reproduced time-defined sample/reagent ratio profile, which in turn yields a range of concentration/kinetic-based information, since in the well-defined domain of the dispersing sample zone the components are competing for the reagent, the concentration of which may be varied from zero to a stoichiometric (i.e., equivalent) amount and then up to a large excess. Using a multiarray detector to obtain additional data dimension, multicomponent analysis is feasible, since each additional reagent used as carrier stream offers an inexpensive novel source of a rich data set for identification and quantification of analytes [8.5]. This was recently demonstrated in our laboratory, using a model system with Fe(II) as sample solution and *o*-phenanthroline as reagent and a Hewlett-Packard UV-VIS diode array instrument as detector. In contrast to a single-wavelength FIA readout (cf. Fig. 2.22) the multiwavelength readout offers an additional dimension. Since Fe(II) forms two complexes with *o*-phenanthroline, a 1:1 complex which is colorless and a 1:3 complex which absorbs at 510 nm, a trough is, for injection of samples of higher concentrations of Fe(II), obtained in the center of the dispersed Fe(II) sample zone, the shape of the composite readout being markedly affected by the pH and by the presence of Co, Ni, or Mg due to kinetic interactions. Further study of this system, with the aid of chemometrics [such as the partial least square (PLS) method] will undoubtedly lead to the discovery and evaluation of novel techniques for sample component quantification and identification by employing spectral, stability, and kinetic differences between species. Competition of functional groups for substoichiometric, equivalent, and superstoichiometric amount of a suitable reagent monitored by means of an FIA-FTIR hyphenated technique will one day be a powerful method of organic analysis.

8.3 CONTINUOUS MONITORING OF INDUSTRIAL PROCESSES BY FIA

For process control applications, FIA offers the benefit of real-time continuous monitoring by regularly aspirating sample solution (e.g., from a batch-type chemical reactor) into a carrier–wash solution, the detector of choice being operated in an impulse response mode. Thus a chemically modulated signal, continuously recorded as a series of peaks (injected material on a constant background of carrier–wash), offers a positive control of both the detector and of the process performance, detector fouling being prevented by means of a periodic wash sequence. Occasional alternate injections of standard solutions allows further control of detector performance thus ensuring that the magnitude of the recorded peaks truly reflects the concentration of the monitored species (and not deterioration of the response characteristics of the detector). During a typical 1 min measuring cycle FIA allows dilution of up to 20,000 times [8.6] (cf. Section 2.4.1), chemical derivatization, separation or other solution handling tasks so well established by FIA laboratory praxis. Chemical production [556, 680, 852, 1035, 8.7, 8.8], biological wastewater treatment [261, 1378, 8.9], and biotechnological processes [743, 826, 960, 961, 1225] were so far reported to be controlled by FIA techniques, and two FIA based process control chemical analyzers are now marketed (cf. Chapter 6).

There are three approaches for process control applications utilizing peak height (or vertical readout) as the source of information. They are based on sample injection, standard injection, and reagent injection.

Sample injection (Fig. 8.2) follows the scheme most frequently used in the laboratory applications of FIA: a solution (S) from a reactor, or a

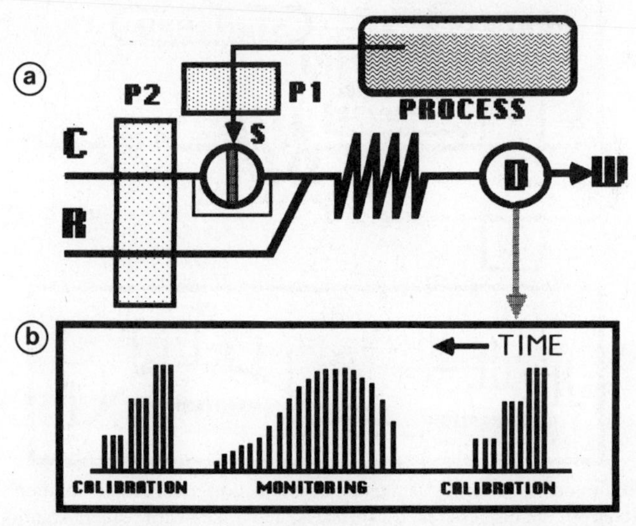

Figure 8.2. Continuous monitoring based on sample injection. (*a*) The sample, aspirated from a process reactor (process) by pump *P*1 is intermittently injected by means of valve *S* into carrier stream *C* which is merged downstream with reagent solution *R*, propelled by pump *P*2. (*b*) Prior to the monitoring sequence (monitoring) the system is calibrated by injecting a series of standards (calibration), which procedure is repeated after the monitoring period. Note that the baseline is reached between each injection, thereby allowing control of the performance of the analyzer itself.

stream to be monitored, is (after possible pretreatment) propelled by a pump (*P*1) into an injection valve, which, when turned, injects the analyte into the carrier stream (*C*), merging downstream with a suitable reagent (*R*). After passage through a reaction coil, within which the species to be monitored is produced, the stream enters the detector (*D*), which continuously records the composition of the stream. A sufficient amount of time is allowed to pass between individual injections, so that the baseline is reached, indicating that the system has been thoroughly washed and that the detector is functioning properly. Calibration can be performed before, after, or even during the monitoring period by means of suitable standard solutions (or by means of the products from a previous successfully processed batch). Computer control of the FIA system allows for suitable timing of events (injection, start–go pumping sequences, data collection, and recalibration).

The *standard injection* manifold (Fig. 8.3) does not require an injection valve and is therefore mechanically less complex. The analyte (*S*) is con-

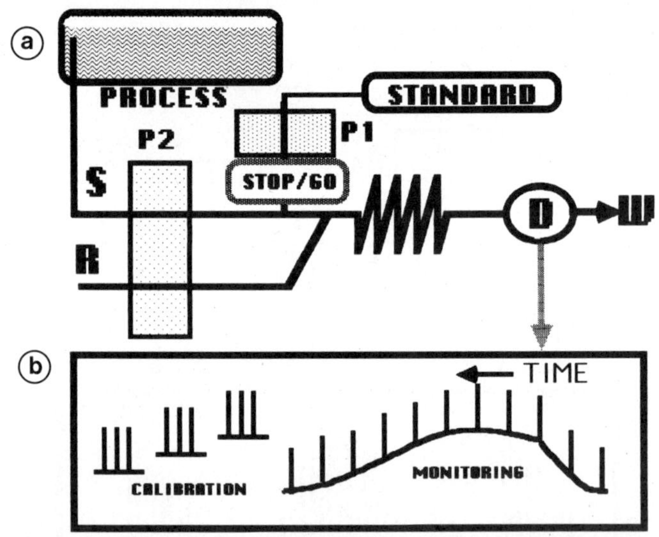

Figure 8.3. Continuous monitoring using standard injection. (*a*) Sample solution is continuously aspirated from the process reactor (process) and mixed with reagent solution *R*, both streams being propelled by pump *P*2, the resulting chemical reaction as monitored by detector *D* yielding a continuous readout (monitoring). At preselected intervals a precisely metered amount of standard solution is introduced by pump *P*1 by operating it in the GO mode for a selected period of time, which gives rise to a spike on top of the recorded signal (*b*), the amplitude of the spike permitting control of the response of the system via a prerecorded calibration run (calibration). Note that this approach does not require the use of an injection valve.

tinuously pumped (by *P*2) through the FIA channel, merged with reagent (*R*), and the subsequently produced species is monitored by detector (*D*). A vertical displacement of the continuously recorded signal reflects variations of the analyte concentration within the monitored process. An occasional injection of a standard (STD) by means of pump *P*1 produces a spike, which allows periodical control of the response of the system. (Depending on the selection of the injected volume of the standard solution and on its concentration, as well as on the overall dispersion in the FIA system, spikes of various amplitudes will be observed.) The drawback of the standard injection approach is that the FIA channel and the detector are not periodically washed by a cleansing carrier solution and might therefore become contaminated.

Reagent injection (Fig. 8.4), first designed to monitor the content of nutrients in seawater [279, 822], saves reagents by injecting small volume of these (by means of pump *P*1) only at the times when a readout is required, while the analyte solution (*S*) is pumped continuously through the FIA channel. The carrier stream (C) is used to dilute and/or precondition the sample solution, detector (*D*) monitoring the composition of the flowing stream. Since detectable species is formed only in the presence

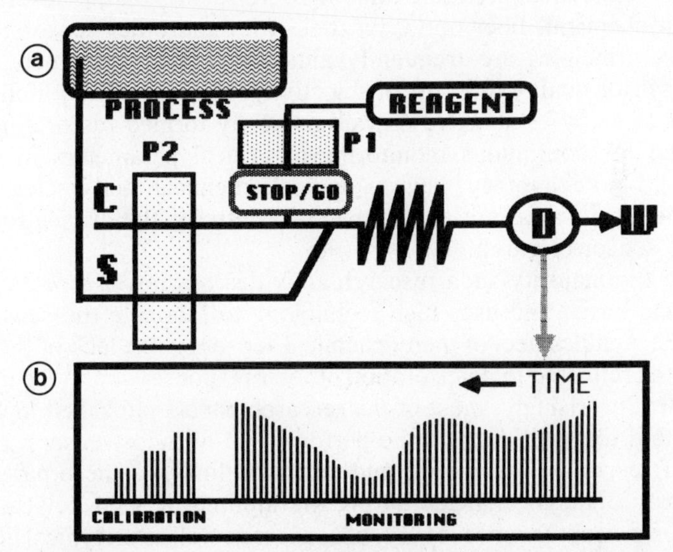

Figure 8.4. Continuous monitoring by reagent injection. (*a*) The monitored material (process) is by pump *P*2 continuously aspirated into the FIA system where reagent is periodically injected into carrier stream *C*, which upon merging with the sample stream forms a species measurable by detector *D*. (*b*) A series of peaks is obtained that reflects the change of analyte during the monitoring period. For calibration, standards with known levels of analyte are aspirated instead of sample solution (calibration).

of reagent, the readout has the form of a peak. Calibration is achieved by aspirating standard solutions and by repeated injections of reagent at different levels of aspirated standards.

While the above approaches to continuous-flow monitoring are derived from laboratory FIA systems, a novel approach to FIA process control is being developed at the Center for Process Analytical Chemistry (CPAC) at the University of Washington, Seattle, being based on the use of single-line FIA system, which replaces the mechanical complexity of the above schemes by a single-line manifold, yet results in a very complex, multiple peak readout, which is a "fingerprint" of the chemistries taking place. This simple system approach, using a combination of FIA titrations, and horizontal, vertical, and spectral readouts, relies, however, on a very sophisticated combination of detection and computational data handling exploiting the partial least square (PLS) method to decode the complex readout.

8.4 IMPULSE RESPONSE FIA FOR SENSOR AND MATERIAL TESTING

The recent analytical literature abounds with reports on the development of electrochemical, fiber optic, piezoelectric, and other sensors. Novel detection principles are frequently announced, and attainment of improved sensor quality in terms of selectivity, sensitivity, and lifetime are the goal of many established as well as newly formed research groups. The need for continuous monitoring of critical parameters in clinical chemistry, biotechnology, pharmaceutical, chemical, and nuclear industries, chemical warfare or environmental control is the driving force behind the sensor research.

While the majority such research activities has novel aspects, sound theoretical basis, and uses high technology to fabricate the sensing devices, the Achilles heel of many chemical sensors is the lack of their time stability as reflected in deterioration of their response—up to the ultimate lack of it. Surprisingly, most of the research papers published to date do not contain the vital data on the performance of these sensors, such as speed of response, sensitivity, and detection limit, as these parameters necessarily undergo changes during the lifetime of each device. Information on poisoning, or recovery, of sensors as well as on typical lifetimes at various operating conditions are even more scarce. Whether this is due to proprietory information or lack of data is difficult to judge.

Introduction of an automated, well reproducible rigorous testing procedure that would allow critical testing of chemical sensors should become a powerful tool for sensor development. There is increasing experimental evidence that FIA is suitable for sensor and material testing by means of the impulse-response technique.

Impulse-response FIA is based on repetitive action of a well defined zone of a selected chemical species (reagent) on a target (*T*) situated in a continuous unsegmented carrier stream (*C*) consisting of an inert fluid (Fig. 8.5). The target may be a sensor, the carrier stream may be an aqueous solution to which the sensor does not respond (or responds at the detection limit—thus establishing a continuous baseline). The injected species is the one to which the sensor is to respond. By continuously monitoring the sensor output, while repeatedly injecting, at a fixed frequency, the same species at the same concentration level in the form of a well-defined zone, a series of peaks will be obtained, which would reflect the response of the sensor. Thus the series of peaks will show the response to either remain unchanged (Fig. 8.5*b A*) or to be deteriorating (Fig. 8.5*b B*) over a selected test period. Since the procedure is automated, it is not prone to subjective judgment or manual operational errors. Indeed, instead of injecting repeatedly the same solution, a series of solutions, *R*1,

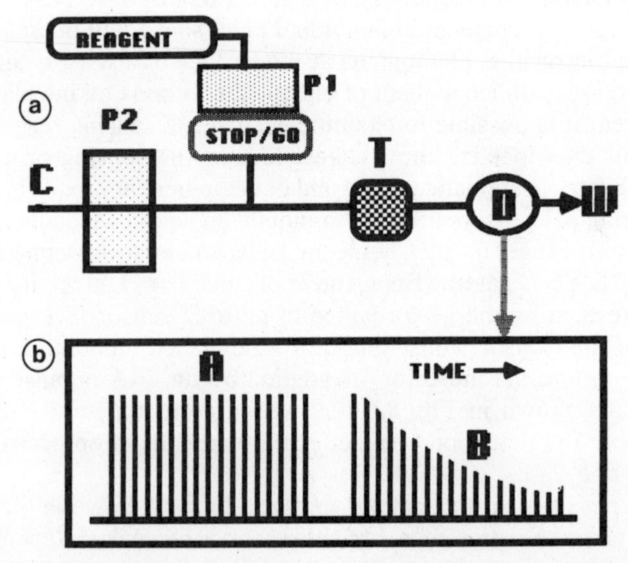

Figure 8.5. (*a*) Impulse-response FIA system comprising a target material *T* situated within carrier stream (*C*) propelled by pump *P*2. At preselected time intervals a metered volume of test liquid (reagent), aspirated by pump *P*1 by operating it for a fixed period of time (GO), is injected into the carrier stream, the effect of the exposure being monitored by detector *D*. In (*b*) A shows a run where the target is unaffected by the repeated bombardments of test solution, while B represents a situation where the target is gradually deteriorating due the repeated exposure.

*R*2, *R*3, of different concentrations, may be injected in preprogrammed sequences. Similarly, interferring or regenerative species, may be injected repeatedly or singly.

Target materials may range from inorganic materials to biologically active species. Thus, surface corrosion may be investigated by repeatedly bombarding the target with a suitable corrosive agent and by measuring the released material by AAS or ICP. The rate of dissolution of coatings of surfaces, of pharmaceutical products, the stability of dyes against bleaching, of the degree of immobilization of enzymes, the rate of dissolution of solids, all can be investigated by the FIA impulse-response technique with aid of suitable detectors. The lifetime and change of response characteristiques of sensors (fiber optic, electrochemical) can be tested and quantitatively evaluated as described above. Since target materials can be tested automatically with high reproducibility under prolonged periods, the FIA impulse response technique might become, once its potential has become generally recognized and exploited, the most important modification of the flow-injection method, perhaps even surpassing its applications as a tool for serial assays.

Besides examining a tendency of a series of response peaks, as illustrated in Fig. 8.5, a change in individual peak shape will be indicative of adsorption–desorption phenomena at the surface of a sensor. Since most chemical sensors utilize a chain of chemical reactions taking place in the sensing area, it is possible to examine each set of sensing reactions separately and thus localize the weakest link of the sensing process. Examples of such a systematic testing and development approach is the work on fiber optic pH sensors utilizing immobilized acid–base indicators, carried out by R. Singer at the Chemistry Department A in Denmark [8.10], and that of T. Yerian at tthe Department of Chemistry University of Washington investigating the performance of an urea sensor [8.11]. The time stability of this sensor, consisting of a sandwiched immobilized urease–covalently bound pH indicator, as studied by the FIA impulse response technique, is shown in Fig. 8.6 for two different types of enzyme immobilization, the detection being effected by means of optosensing (Sections 4.12 and 5.4).

Such types of recording can be analyzed to show how the lifetime and response characteristics of a pH sensor or a biosensor depend on the stability of the following:

1. The optical components (light source, detector and light losses in and the quality of the optical fibers), as well as on the physical changes occurring on the surface of the reflecting material in the flow cell.

2. The indicator stability in terms of its covalent attachment to the supporting material and as affected by photobleaching.

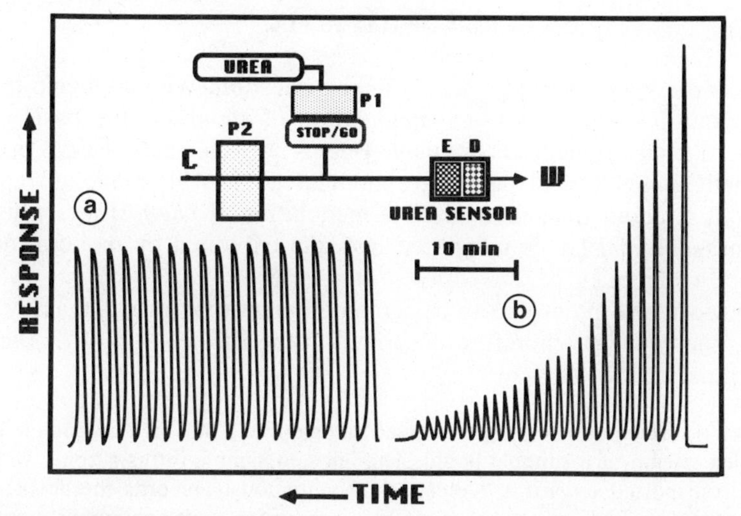

Figure 8.6. Impulse-response FIA system used for investigating the performance of a urea optosensor comprising a pad on which urease (*E*) and an acid–base indicator (*D*) are coimmobilized, the transmission of light to and from the flow cell being effected by means of optical fibers (not shown). First the indicator stability in terms of its covalent attachment to the supporting material was evaluated by preparing a pad with acid–base indicator alone and subjecting it repeatedly and alternately to acidic and alkaline standard solutions. A series of readouts similar to those shown in (*a*) confirmed that no loss of immobilized dye is encountered. Two types of urease immobilization were then tested by repetitive impulses of the same urea standard (10 m*M*), operating the FIA system in the stopped-flow mode (stop time 60 s). In (*b*) the indicator containing pad is sandwiched with an immunoaffinity membrane (Biodyne) to which urease is attached, yet the rapid decay in response as a function of time clearly indicates that the enzyme is gradually being washed out of the sensor. In (*a*) urease is coimmobilized by glutaraldehyde cross-linking on the cellulose indicator pad, the constant response signals demonstrating the stability of this sensor system [8.11].

3. The enzyme stability in terms of its physical or chemical attachment to the supporting material, and the activity loss as caused by thermal and chemical degradation during use (Fig. 8.6*A* and *B*).

4. The ability of the indicator–enzyme couple to respond to the analyte in a wide range of concentrations and to become regenerated/washed if a maximum level of analyte is repeatedly exceeded.

5. The speed of response and the reversibility of the enzyme-indicator couple used.

The use of FIA is not limited to optical sensors. It has been previously used to test ion selective electrodes [12, 68], and most recently to testing in the development of pneumatic gas voltammetric detectors [1050, 1051]. Also the surface treatment of glassy carbon electrodes has been evaluated by means of their voltammetric response observed in the FIA mode [853].

8.5 WHAT IS FIA?

This is the very last and truly intriguing question. We may begin by the observation that the physical modulation of *signals* is the basis of all modern measurement techniques. Beams are chopped, lasers pulsed, acoustic and electrical signals are modulated at a wide range of frequencies. It is about time that physical modulation of *chemistries*, which is the essence of FIA, is integrated into and exploited by modern instrumental techniques.

Our original intent was to design an efficient tool for serial assay, and flow injection was therefore described in the first edition of this monograph as:

> A method based on injection of a liquid sample into a moving unsegmented continuous stream of a suitable liquid. The injected sample forms a zone, which is then transported toward a detector that continuously records the absorbance, electrode potential, or any other physical parameter, as it continuously changes as a result of the passage of sample material through the flow cell.

Now we know that FIA has evolved into a general technique for solution handling and data gathering, applicable to many areas of chemical research and technology, and its scope is no longer limited to serial assays. Interestingly, however, even the new and yet unforseen modifications of flow injection conform with the original idea of:

> Information-gathering from a concentration gradient formed from an injected, well-defined zone of a fluid, dispersed into a continuous unsegmented stream of a carrier.

Discussions may ensue whether this is a definition of flow injection analysis or of a more general flow-injection technique. We believe, however, that definitions, if too rigorous for the sake of precision, might restrict the imagination lent to all of us by the uncontrollable spirits of inspiration.

Out of reality are our tales of imagination fashioned.

HANS CHRISTIAN ANDERSEN

REFERENCES

8.1 H. Bækmark Andersen, *Determination of Urea in Full Blood on the Basis of Flow Injection Analysis Using Optosensing* [in Danish]. M.Sc. Thesis, Tech. Univ. Denmark (1986).

8.2 A. J. Rasmussen, *Indirect Flow Injection Determination of Sulfate by Atomic Absorption Spectrophotometry* [in Danish]. Tech. Univ. Denmark (1986).

8.3 T. Buch-Rasmussen, *Application of FIA for Immunoassay* [in Danish]. M.Sc. Thesis, Tech. Univ. Denmark (1986).

8.4 T. Hirschfeld, *Anal. Chem.,* **52** (1980) 297A.

8.5 J. Růžička, *CHEMTECH*, **17** (1987) 96.

8.6 M. B. Garn, M. Gisin, H. Gross, W. Schmidt, and C. Thommen, *Anal. Chim. Acta* (in press).

8.7 D. K. Wolcott and D. G. Hunt, *Flow Injection Titration in Process Control.* Paper No. 353, 11. FACCS Meeting, Philadelphia, PA (1984).

8.8 D. Olson, Shell, Houston, TX (priv. comm.)

8.9 M. Gisin and Z. Jardas, *FIA-Based Single-Channel Sequential Monitoring of Phosphate and Sulphate in Industrial Effluents.* Paper No. 352, 11. FACCS Meeting, Philadelphia, PA (1984).

8.10 R. Singer, *Applications of Immobilized Indicators for Optosensing by Flow Injection Analysis* [in Danish]. M.Sc. Thesis, Tech. Univ. Denmark (1986).

8.11 T. Yerian, G. D. Christian, and J. Růžička, *Anal. Chem.* (in press).

LIST OF FIA REFERENCES

1. J. Růžička and E. H. Hansen, Flow Injection Analysis. Part I. A New Concept of Fast Continuous Flow Analysis. *Anal. Chim. Acta,* **78** (1975) 145. Danish Pat. Appl. No. 4846/74, September 1974; subsequent U.S. Pat. No. 4,022,575.
2. J. Růžička and J. W. B. Stewart, Flow Injection Analysis. Part II. Ultrafast Determination of Phosphorous in Plant Material by Continuous Flow Spectrophotometry. *Anal. Chim. Acta,* **79** (1975) 79.
3. J. W. B. Stewart, J. Růžička, H. Bergamin F$^{\underline{o}}$, and E. A. G. Zagatto, Flow Injection Analysis. Part III. Comparison of Continuous Flow Spectrophotometry and Potentiometry for the Rapid Determination of the Total Nitrogen Content in Plant Digests. *Anal. Chim. Acta,* **81** (1976) 371.
4. J. Růžička, J. W. B. Stewart, and E. A. G. Zagatto, Flow Injection Analysis. Part IV. Stream Sample Splitting and Its Application to the Continuous Spectrophotometric Determination of Chloride in Brackish Waters. *Anal. Chim. Acta,* **81** (1976) 387.
5. J. W. B. Stewart and J. Růžička, Flow Injection Analysis. Part V. Simultaneous Determination of Nitrogen and Phosphorous in Acid Digests of Plant Material with a Single Spectrophotometer. *Anal. Chim. Acta,* **82** (1976) 137.
6. J. Růžička and E. H. Hansen, Flow Injection Analysis. Part VI. The Determination of Phosphate and Chloride in Blood Serum by Dialysis and Sample Dilution. *Anal. Chim. Acta,* **87** (1976) 353.
7. D. Betteridge and J. Růžička, The Determination of Glycerol in Water by Flow Injection Analysis—A Novel Way of Measuring Viscosity. *Talanta,* **23** (1976) 409.
8. J. Růžička, E. H. Hansen, and E. A. G. Zagatto, Flow Injection Analysis. Part VII. Use of Ion-Selective Electrodes for Rapid Analysis of Soil Extracts and Blood Serum. Determination of Potassium, Sodium and Nitrate. *Anal. Chim. Acta,* **88** (1977) 1.
9. E. H. Hansen, J. Růžička, and B. Rietz, Flow Injection Analysis. Part VIII. Determination of Glucose in Blood Serum with Glucose Dehydrogenase. *Anal. Chim. Acta,* **89** (1977) 241.
10. J. Růžička, E. H. Hansen, and H. Mosbæk, Flow Injection Analysis. Part IX. A New Approach to Continuous Flow Titrations. *Anal. Chim. Acta,* **92** (1977) 235.

11. F. J. Krug, H. Bergamin F°, E. A. G. Zagatto, and S. S. Jørgensen, Rapid Determination of Sulphate in Natural Waters and Plant Digests by Continuous Flow Injection Turbidimetry. *Analyst*, **102** (1977) 503.
12. E. H. Hansen, A. K. Ghose, and J. Růžička, Flow Injection Analysis of Environmental Samples for Nitrate Using an Ion-Selective Electrode. *Analyst,* **102** (1977) 705.
13. E. H. Hansen, F. J. Krug, A. K. Ghose, and J. Růžička, Rapid Determination of Nitrogen, Phosphorous and Potassium in Fertilisers by Flow Injection Analysis. *Analyst*, **102** (1977) 714.
14. J. Růžička, E. H. Hansen, H. Mosbæk, and F. J. Krug, Pumping Pressure and Reagent Consumption in Flow Injection Analysis. *Anal. Chem.,* **49** (1977) 1858.
15. S. S. Jørgensen, H. Bergamin F°, E. A. G. Zagatto, F. J. Krug, and S. R. B. Bringel, Determinacao de Nitrito em Aguas Naturais Atraves do Sistema de Injecao em Fluxo Continuo [in Portuguese]. *Boletim Cientifico* (*CENA/ESALQ/USP*), **047** (1977) 1.
16. H. Bergamin F°, F. J. Krug, E. A. G. Zagatto, H. Fonseca, M. Graner, J. N. Nogueira, and A. V. K. O. Annicchino, Determinacao de Sulfito em Passa de Banana com Emprego do Electrodo com Separacao de Ar em Sistema de Fluxo Continuo [in Portuguese]. *Boletim Cientifico* (*CENA/ESALQ/USP*), **049** (1977) 1.
17. L. Sodek, J. Růžička, and J. W. B. Stewart, Rapid Determination of Proteins in Plant Material by Flow Injection Spectrophotometry with Trinitrobenzenesulfonic Acid. *Anal. Chim. Acta*, **97** (1978) 327.
18. H. Bergamin F°, F. Reis, and E. A. G. Zagatto, A New Device for Improving Sensitivity and Stabilization in Flow Injection Analysis. *Anal. Chim. Acta*, **97** (1978) 427.
19. B. Karlberg and S. Thelander, Extraction Based on the Flow Injection Principle. Part 1. Description of the Extraction System. *Anal. Chim. Acta,* **98** (1978) 1.
20. J. Růžička and E. H. Hansen, Flow Injection Analysis. Part X. Theory, Techniques and Trends. *Anal. Chim. Acta,* **99** (1978) 37.
21. E. H. Hansen, J. Růžička, and A. K. Ghose, Flow Injection Analysis for Calcium in Serum, Water and Waste Waters by Spectrophotometry and by Ion-Selective Electrode. *Anal. Chim. Acta*, **100** (1978) 151.
22. H. Bergamin F°, J. X. Medeiros, B. F. Reis, and E. A. G. Zagatto, Solvent Extraction in Continuous Flow Injection Analysis. Determination of Molybdenum in Plant Material. *Anal. Chim. Acta*, **101** (1978) 9.
23. H. Bergamin F°, E. A. G. Zagatto, F. J. Krug, and B. F. Reis, Merging Zones in Flow Injection Analysis. Part 1. Double Proportional Injector and Reagent Consumption. *Anal. Chim. Acta,* **101** (1978) 17.
24. W. E. van der Linden and R. Oostervink, Construction and Behaviour of

a Micro Flow-Through Copper(II)-Selective Electrode. *Anal. Chim. Acta,* **101** (1978) 419.

25. B. F. Reis, Determinacao Colorimetrica de Aluminio em Aguas Naturais, Plantas e Solos por Injecao em Fluxo Continuo [in Portuguese]. *ESALQ/ USP, Brazil,* (1978). (M. Sc. Thesis).
26. W. D. Basson and J. F. van Staden, Use of Non-Segmented High-Speed Continuous Flow Analysis for the Determination of Calcium in Animal Feeds. *Analyst,* **103** (1978) 296.
27. D. Betteridge, E. L. Dagless, B. Fields, and N. F. Graves, A Highly Sensitive Flow-Through Phototransducer for Unsegmented Continuous Flow Analysis Demonstrating High-Speed Spectrophotometry at the Parts per Billion Level and a New Method of Refractometric Determinations. *Analyst,* **103** (1978) 897.
28. W. D. Basson and J. F. van Staden, Low Level Determination of Hydrazine in Boiler Feed Water with an Unsegmented High-Speed Continuous Flow System. *Analyst,* **103** (1978) 998.
29. B. Karlberg and S. Thelander, Determination of Readily Oxidized Compounds by Flow Injection Analysis and Redox Potential Detection. *Analyst,* **103** (1978) 1154.
30. M. A. B. Regitano, Chromium in the Environment with Special Emphasis on its Behaviour in Soils. Royal Vet. Agr. Univ., Denmark, (1978). (B.Sc. Thesis).
31. K. Kina, K. Shirashi and N. Ishibashi, Ultramicro Solvent Extraction and Fluorimetry Based on the Flow Injection Method. *Talanta,* **25** (1978) 295.
32. D. Betteridge and B. Fields, Construction of pH Gradients in Flow Injection Analysis and Their Potential Use for Multielement Analysis in a Single Sample Bolus. *Anal. Chem.,* **50** (1978) 654.
33. D. Betteridge, Flow Injection Analysis. *Anal. Chem.,* **50** (1978) 832A.
34. J. Slanina, F. Bakker, A. G. M. Bruijn-Hes, and J. J. Möls, Fast Determination of Nitrate in Small Samples of Rain and Surface Waters by Means of UV Spectrophotometry and Flow Injection Analysis. *Fresenius Z. Anal. Chem.,* **189** (1978) 38.
35. B. Karlberg, Flow Injection Analysis—Teknik på Frammarsch [in Swedish]. *Svensk Kemisk Tidskrift,* **90** (1978) 26.
36. K. Kina and N. Ishibashi, Flow Injection Analysis—A Review [in Japanese]. *Dojin,* **10** (1978) 1.
37. E. Bylund, R. Andersson, and J. Å. Carlsson, Determination of Heavy Metals by Flow Injection Analysis (FIA) [in Swedish]. Pharmacia AB, Box 181, S-751 04, Uppsala, Sweden, (1978).
38. M. Koshino, Determination of Major Nutrients in Fertilizers [in Japanese]. *Bunseki,* **11** (1978) 803.
39. J. H. Dahl, Differentialkinetisk Analyse af Jordalkalimetallernes CDTA-

Komplexer under Anvendelse af Flow Injection Analysis Systemet [in Danish]. Royal Dan. School Pharm., Denmark, (1978). (Ph.D. Thesis).

40. A. U. Ramsing, Udvikling og Anvendelser af Voltammetrisk Måleudstyr i Flow Injection Analysis [in Danish]. Techn. Univ. Den., Denmark, (1978) (M.Sc. Thesis).
41. B. Karlberg, P. A. Johansson, and S. Thelander, Extraction Based on the Flow Injection Principle. Part II. Determination of Codeine as the Picrate Ion-Pair in Acetylsalicylic Acid Tablets. *Anal. Chim. Acta,* **104** (1979) 21.
42. J. Mindegaard, Flow Multi-Injection Analysis—A System for the Analysis of Highly Concentrated Samples Without Prior Dilution. *Anal. Chim. Acta,* **104** (1979) 185.
43. E. A. G. Zagatto, F. J. Krug, H. Bergamin F$^{\underline{o}}$, S. S. Jørgensen, and B. F. Reis, Merging Zones in Flow Injection Analysis. Part 2. Determination of Calcium, Magnesium and Potassium in Plant Material by Continuous Flow Injection Atomic Absorption and Flame Emission Spectrometry. *Anal. Chim. Acta,* **104** (1979) 279.
44. O. Åström, Single-Point Titrations. Part 4. Determination of Acids and Bases with Flow Injection Analysis. *Anal. Chim. Acta,* **105** (1979) 67.
45. J. H. Dahl, D. Espersen, and A. Jensen, Diffential Kinetic Analysis and Flow Injection Analysis. Part 1. The trans-1,2-Diaminocyclohexane-Tetraacetate Complexes of Magnesium and Strontium. *Anal. Chim. Acta,* **105** (1979) 327.
46. J. Růžička and E. H. Hansen, Stopped-Flow and Merging Zones—A New Approach to Enzymatic Assay by Flow Injection Analysis. *Anal. Chim. Acta,* **106** (1979) 207.
47. J. I. Braithwaite and J. N. Miller, Flow Injection Analysis with a Fluorimetric Detector for Determination of Glycine and Albumin. *Anal. Chim. Acta,* **106** (1979) 395.
48. B. F. Reis, H. Bergamin F$^{\underline{o}}$, E. A. G. Zagatto, and F. J. Krug, Merging Zones in Flow Injection Analysis. Part 3. Spectrophotometric Determination of Aluminium in Plant and Soil Materials with Sequential Addition of Pulsed Reagents. *Anal. Chim. Acta,* **107** (1979) 309.
49. D. Espersen and A. Jensen, Differential Kinetic Analysis and Flow Injection Analysis. Part II. The (2.2.1) Cryptates of Magnesium and Calcium. *Anal. Chim. Acta,* **108** (1979) 241.
50. E. A. G. Zagatto, B. F. Reis, H. Bergamin F$^{\underline{o}}$, and F. J. Krug, Isothermal Distillation in Flow Injection Analysis. Determination of Total Nitrogen in Plant Material. *Anal. Chim. Acta,* **109** (1979) 45.
51. L. Andersson, Simultaneous Spectrophotometric Determination of Nitrite and Nitrate by Flow Injection Analysis. *Anal. Chim. Acta,* **110** (1979) 123.
52. F. J. Krug, J.Růžička, and E. H. Hansen, Determination of Ammonia in Low Concentrations with Nessler's Reagent by Flow Injection Analysis. *Analyst,* **104** (1979) 47.

53. M. F. Giné, E. A. G. Zagatto, and H. Bergamin F$^{\underline{o}}$, Semiautomatic Determination of Manganese in Natural Waters and Plant Digests by Flow Injection Analysis. *Analyst,* **104** (1979) 371.

54. W. D. Basson and J. F. van Staden, Direct Determination of Calcium in Milk on a Non-Segmented Continuous Flow System. *Analyst,* **104** (1979) 419.

55. J. X. Medeiros, Determinacao de Molibdenio em Material de Plantas por Extracao com Solventes em Fluxo Continuo [in Portuguese]. *CENA/ESALQ/USP, Brazil,* (1979). (M.Sc. Thesis).

56. D. Espersen, Differentialkinetisk Analyse under Anvendelse af Flow Injection Analysis Systemet. Samtidig Bestemmelse af Magnesium og Calcium ved hjælp af (2.2.1) Cryptaterne [in Danish]. Royal Dan. School Pharm., Denmark, (1979). (Ph.D. Thesis).

57. H. Baadenhuijsen and H. E. H. Seuren-Jacobs, Determination of Total CO_2 in Plasma by Automated Flow Injection Analysis. *Clin. Chem.,* **25** (1979) 443.

58. G. Rule and W. R. Seitz, Flow Injection Analysis with Chemiluminescence Detection. *Clin. Chem.,* **25** (1979) 1635.

59. E. H. Hansen and J. Růžička, The Principles of Flow Injection Analysis as Demonstrated by Three Lab Exercises. *J. Chem. Educ.,* **56** (1979) 677.

60. J. Růžička, E. H. Hansen, A. K. Ghose, and H. A. Mottola, Enzymatic Determination of Urea in Serum Based on pH Measurement with the Flow Injection Method. *Anal. Chem.,* **51** (1979) 199.

61. A. N. Strohl and D. J. Curran, Flow Injection Analysis with Reticulated Vitreous Carbon Flow-Through Electrodes. *Anal. Chem.,* **51** (1979) 1045.

62. W. R. Wolf and K. K. Stewart, Automated Multiple Flow Injection Analysis for Flame Atomic Absorption Spectrometry. *Anal. Chem.,* **51** (1979) 1201.

63. J. Kawase, A. Nakae, and M. Yamenaka, Determination of Anionic Surfactants by Flow Injection Analysis Based on Ion-Pair Extraction. *Anal. Chem.,* **51** (1979) 1640.

64. J. Růžička and E. H. Hansen, Flow Injection Analysis. *CHEMTECH,* **9** (1979) 756.

65. J. Růžička and E. H. Hansen, Flow Injection Analysis—A New Approach to Quantitative Measurements. *NBS Spec. Publ.,* **519** (1979) 501.

66. N. Yoza, Y. Aoyagi, S. Ohashi, and A. Tateda, Flow Injection System for Atomic Absorption Spectrometry. *Anal. Chim. Acta,* **111** (1979) 163.

67. H. K. Chan and A. G. Fogg, Flow Injection Determination of Meptazinol with Electrochemical Detection. *Anal. Chim. Acta,* **111** (1979) 281.

68. E. H. Hansen, J. Růžička, and A. K. Ghose, Rapid Determination of Nitrogen-Containing Compounds by Flow Injection Potentiometry. STI/PUB/535 (Int. Atom. Energ. Agen., Vienna) (1980) 77.

69. S. S. Jørgensen and M. A. B. Regitano, Rapid Determination of Chromium(VI) by Flow Injection Analysis. *Analyst,* **105** (1980) 292.
70. L. R. Snyder, Continuous Flow Analysis: Present and Future. *Anal. Chim. Acta,* **114** (1980) 3.
71. J. Růžička, and E. H. Hansen, Flow Injection Analysis. Principles, Applications and Trends. *Anal. Chim. Acta,* **114** (1980) 19.
72. H. Poppe, Characterization and Design of Liquid Phase Flow-Through Detector Systems. *Anal. Chim. Acta,* **114** (1980) 59.
73. R. Tijssen, Axial Dispersion and Flow Phenomena in Helically Coiled Tubular Reactors for Flow Analysis and Chromatography. *Anal. Chim. Acta,* **114** (1980) 71.
74. J. H. M. van den Berg, R. S. Deelder, and H. G. M. Egberink, Dispersion Phenomena in Reactors for Flow Analysis. *Anal. Chim. Acta,* **114** (1980) 91.
75. J. M. Reijn, W. E. van der Linden, and H. Poppe, Some Theoretical Aspects of Flow Injection Analysis. *Anal. Chim. Acta,* **114** (1980) 105.
76. K. K. Stewart, J. F. Brown, and B. M. Golden, A Microprocessor Control System for Automated Multiple Flow Injection Analysis. *Anal. Chim. Acta,* **114** (1980) 119.
77. B. Karlberg and S. Thelander, Extraction Based on the Flow Injection Principle. Part 3. Fluorimetric Determination of Vitamin B1 (Thiamine) by the Thiochrome Method. *Anal. Chim. Acta,* **114** (1980) 129.
78. J. B. Landis, Rapid Determination of Corticosteroids in Pharmaceuticals by Flow Injection Analysis. *Anal. Chim. Acta,* **114** (1980) 155.
79. A. Ramsing, J. Růžička, and E. H. Hansen, A New Approach to Enzymatic Assay Based on Flow Injection Spectrophotometry with Acid-Base Indicators. *Anal. Chim. Acta,* **114** (1980) 165.
80. C. S. Lim, J. N. Miller, and J. W. Bridges, Automation of an Energy-Transfer Immunoassay by Using Stopped-Flow Injection Analysis with Merging Zones. *Anal. Chim. Acta,* **114** (1980) 183.
81. M. F. Giné, H. Bergamin F$^{\underline{o}}$, E. A. G. Zagatto, and B. F. Reis, Simultaneous Determination of Nitrate and Nitrite by Flow Injection Analysis. *Anal. Chim. Acta,* **114** (1980) 191.
82. I. Kågevall, O. Åström, and A. Cedergren, Determination of Water by Flow Injection Analysis with the Karl Fischer Reagent. *Anal. Chim. Acta,* **114** (1980) 199.
83. J. L. Burguera, A. Townshend, and S. Greenfield, Flow Injection Analysis for Monitoring Chemiluminescent Reactions. *Anal. Chim. Acta,* **114** (1980) 209.
84. P. A. Johansson, B. Karlberg, and S. Thelander, Extractions Based on the Flow Injection Principle. Part 4. Determination of Extraction Constants. *Anal. Chim. Acta,* **114** (1980) 215.

85. H. Kagenow and A. Jensen, Differential Kinetic Analysis and Flow Injection Analysis. Part 3. The (2.2.2.) Cryptates of Magnesium, Calcium and Strontium. *Anal. Chim. Acta,* **114** (1980) 227.

86. J. W. Dieker and W. E. van der Linden, Determination of Iron(II) and Iron(III) by Flow Injection and Amperometric Detection with a Glassy Carbon Electrode. *Anal. Chim. Acta,* **114** (1980) 267.

87. S. Baban, D. Beetlestone, D. Betteridge, and P. Sweet, The Determination of Sulphate by Flow Injection Analysis with Exploitation of pH Gradients and EDTA. *Anal. Chim. Acta,* **114** (1980) 319.

88. N. Ishibashi, K. Kina, and Y. Goto, Selective Determination of Gallium by Flow Injection Fluorometry. *Anal. Chim. Acta,* **114** (1980) 325.

89. Y. Hirai, N. Yoza, and S. Ohashi, Flow Injection Analysis of Inorganic Polyphosphates. *Anal. Chim. Acta,* **115** (1980) 269.

90. O. Klinghoffer, J. Růžička, and E. H. Hansen, Flow Injection Analysis of Traces of Lead and Cadmium by Solvent Extraction with Dithizone. *Talanta,* **27** (1980) 169.

91. J. Slanina, F. Bakker, A. Bruyn-Hes, and J. J. Möls, A Computer Controlled Multichannel Continuous Flow Analysis System Applied to the Measurement of Nitrate, Chloride and Ammonium Ions in Small Samples of Rain Water. *Anal. Chim. Acta,* **113** (1980) 331.

92. Z. Fang, On the Modernization of Soil and Plant Analysis [in Chinese]. *Torang Tongbao,* **2** (1979) 34.

93. Z. Fang, Flow Injection Analysis. A Review [in Chinese]. *Fenxi Huaxue,* **9** (1981) 369.

94. L. Sun, Z. Gao, L. Li, X. Yu, and Z. Fang, The Flow Injection Analysis of Available Phosphorous in Soils [in Chinese]. *Fenxi Huaxue,* **9** (1981) 586.

95. B. W. Renoe, K. K. Stewart, G. R. Beecher, M. R. Wills, and J. Savory, Automated Multiple Flow Injection Analysis in Clinical Chemistry: Determination of Albumin with Bromcresol Green. *Clin. Chem.,* **26** (1980) 331.

96. D. Espersen, H. Kagenow, and A. Jensen, Flow Injection Analysis and Differential Kinetic Analysis. *Arch. Pharm. Chem. Sci. Ed.,* **8** (1980) 53.

97. K. Kina and N. Ishibashi, Rapid Automated Analysis by Flow Injection Analysis [in Japanese]. *Kagaku (Kyoto),* **33** (1978) 1001.

98. B. Fields, Studies in Flow Injection Analysis. *Anal. Proc.,* **16** (1979) 4.

99. J. L. Burguera and A. Townshend, Monitoring Chemiluminescent Reactions by Flow Injection Analysis. *Anal. Proc.,* **16** (1979) 263.

100. K. Kina and N. Ishibashi, Flow Injection Analysis. Its Principles and Uses [in Japanese]. *A & R,* **17** (1979) 1.

101. K. Kina, Determination of Anionic Surfactants by Flow Injection Analysis Based on Ion-Pair Extraction [in Japanese]. *Dojin,* **14** (1979) 9.

102. J. A. Lown, R. Koile, and D. C. Johnson, Amperometric Flow-Through Wire Detector: A Practical Design with High Sensitivity. *Anal. Chim. Acta,* **116** (1980) 33.

103. J. A. Lown and D. C. Johnson, Anodic Detection of Arsenic(III) in a Flow-Through Platinum Electrode for Flow Injection Analysis. *Anal. Chim. Acta,* **116** (1980) 41.

104. T. Korenaga, Apparatus for Measuring Chemical Oxygen Demand Based on Flow Injection Analysis [in Japanese]. *Bunseki Kagaku,* **29** (1980) 222.

105. T. Ito, H. Kawaguchi, and A. Mizuike, Inductively Coupled Plasma Emission Spectrometry of Microlitre Samples by a Flow Injection Technique [in Japanese]. *Bunseki Kagaku,* **29** (1980) 332.

106. Y. Hirai, N. Yoza, and S. Ohashi, Flow Injection Analysis of Inorganic Ortho- and Polyphosphates Using Ascorbic Acid as Reductant of Molybdophosphate. *Chem. Lett.,* **5** (1980) 499.

107. J. Růžička and E. H. Hansen, Write On: Flow Injection Analysis. *CHEMTECH,* **10** (1980) 202.

108. B. W. Renoe and A. Obrien, Calcium by Flow Injection Atomic Absorption. *Clin Chem.,* **26** (1980) 1021.

109. N. W. Holt, Flow Injection Analysis—Adaption to a Small Laboratory. *Can. J. Plant Sci.,* **60** (1980) 767.

110. H. Bergamin F$^{\underline{o}}$, B. F. Reis, A. O. Jacintho, and E. A. G. Zagatto, Ion Exchange in Flow Injection Analysis. Determination of Ammonium Ions at the Microgram per Litre Level in Natural Waters with Pulsed Nessler Reagent. *Anal. Chim. Acta,* **117** (1980) 81.

111. J. Slanina, W. A. Lingerak, and F. Bakker, The Use of Ion-Selective Electrodes in Manual and Computer Controlled Flow Injection Systems. *Anal. Chim. Acta,* **117** (1980) 91.

112. A. U. Ramsing, J. Janata, J. Růžička, and M. Levy, Miniaturization in Analytical Chemistry—A Combination of Flow Injection Analysis and Ion-Sensitive Field-Effect Transistors for Determination of pH, and Potassium and Calcium Ions. *Anal. Chim. Acta,* **118** (1980) 45.

113. T. Fujinaga, S. Okazaki, and T. Hori, A Flow-Coulometric Method for the Rapid Determination of Orthophosphate Ion [in Japanese]. *Bunseki Kagaku,* **29** (1980) 367.

114. W. D. Basson and J. F. van Staden, Simultaneous Determination of Sodium, Potassium, Magnesium and Calcium in Surface, Ground and Domestic Water by Flow Injection Analysis. *Fresenius Z. Anal. Chem.,* **302** (1980) 370.

115. P. Maitoza and D. C. Johnson, Detection of Metal Ions Without Interference from Dissolved Oxygen by Reverse Pulse Amperometry in Flow Injection Systems and Liquid Chromatography. *Anal. Chim. Acta,* **118** (1980) 233.

116. L. Nord and B. Karlberg, Extraction Based on the Flow Injection Principle. Part 5. Assessment with a Membrane Phase Separator for Different Organic Solvents. *Anal. Chim. Acta,* **118** (1980) 285.

117. T. Korenaga, A Flow Injection Analyzer for Chemical Oxygen-Demand Using Potassium Permanganate. *Chem. Biomed. Environ. Instrum.,* **10** (1980) 273.

118. C. E. Shideler, B. W. Renoe, J. Crump, M. R. Wills, J. Savory, and K. K. Stewart, Automated Multiple Flow-Injection Analysis in Clinical Chemistry: Determination of Total Protein with Biuret Reagent. *Clin. Chem.,* **26** (1980) 1454.

119. B. J. Compton, J. R. Weber, and W. C. Purdy, Stop-Flow Analysis: An Aid in the Diagnosis and Optimization of Continuous-Flow Systems. *Anal. Lett. B,* **13** (1980) 861.

120. T. Korenaga, Flow Injection Analysis Using Potassium Permanganate: An Approach for Measuring Chemical Oxygen Demand in Organic Wastes and Waters. *Anal. Lett. A,* **13** (1980) 1001.

121. N. Yoza, Flow Injection Analysis [in Japanese]. *Bunseki,* **8** (1980) 555.

122. R. Kuroda, T. Mochizuki, and K. Oguma, Rapid Determination of Copper in Various Copper Based Alloys by Flow Injection Analysis [in Japanese]. *Bunseki Kagaku,* **29** (1980) T73.

123. T. Korenaga and H. Ikatsu, Flow Injection Analysis of Chemical Oxygen Demand in Waste Waters from Laboratories [in Japanese]. *Bunseki Kagaku,* **29** (1980) 497.

124. K. Kina and K. Ueno, Flow Injection Analysis Opens New Aspects in Analytical Chemistry [in Japanese]. *Kagaku (Kyoto),* **36** (1980) 662.

125. T. Korenaga, H. Ikatsu, T. Moriwake, and T. Takahashi, Semiautomated Determination of COD in Wastewater Samples [in Japanese]. *Mem. Sch. Eng., Okayama Univ.,* **14** (1980) 119.

126. B. F. Reis, E. A. G. Zagatto, A. O. Jacintho, F. J. Krug, and H. Bergamin F$^{\underline{o}}$, Merging Zones in Flow Injection Analysis. Part 4. Simultaneous Spectrophotometric Determination of Total Nitrogen and Phosphorous in Plant Material. *Anal. Chim. Acta,* **119** (1980) 305.

127. F. Fukamachi and N. Ishibashi, Flow Injection Atomic Absorption Spectrometry with Organic Solvents. *Anal. Chim. Acta,* **119** (1980) 383.

128. T. Yamane and T. Fukasawa, Flow Injection Determination of Trace Vanadium with Catalytic Photometric Detection. *Anal. Chim. Acta,* **119** (1980) 389.

129. S. M. Ramasamy, M. S. A. Jabbar, and H. A. Mottola, Flow Injection Analysis Based on Two Consecutive Reactions at a Gas-Solid Interface for Determination of Bromine and Chlorine. *Anal. Chem.,* **52** (1980) 2062.

130. D. Betteridge, Analytical Chemistry—The Numbers Game. *Chem. Brit.,* **16** (1980) 646.

131. T. N. Morrison, K. G. Schick, and C. O. Huber, Determination of Ethanol by Air-Stream Separation with Flow Injection and Electrochemical Detection at a Nickel Oxide Electrode. *Anal. Chim. Acta,* **120** (1980) 75.

132. E. A. G. Zagatto, A. O. Jacintho, J. Mortatti, and H. Bergamin F$^{\underline{o}}$, An Improved Flow Injection Determination of Nitrite in Waters by Using Intermittent Flows. *Anal. Chim. Acta,* **120** (1980) 399.

133. N. Yoza, Y. Kurokawa, Y. Hirai, and S. Ohashi, Flow Injection Determinations of Polyphosphates Based on Coloured Metal Complexes of Xylenol Orange and Methylthymol Blue. *Anal. Chim. Acta,* **121** (1980) 281.

134. E. J. Duffield, G. J. Moody, and J. D. R. Thomas, Development of Ion-Selective Electrodes and Flow Injection Analysis for Sulphides and Thiols. *Anal. Proc.,* **17** (1980) 533.

135. S. Baban, Recent Developments in Flow Injection Analysis: Determination of Bismuth, Thorium and Copper with Pyrocatechol Violet by Exploitation of pH. *Anal. Proc.,* **17** (1980) 535.

136. J. Kawase, Automated Determination of Cationic Surfactants by Flow Injection Analysis Based on Ion-Pair Extraction. *Anal. Chem.,* **52** (1980) 2124.

137. K. Kina, Flow Injection Analysis [in Japanese]. *Dojin,* **15** (1980) 8.

138. K. Kina, Automated Multiple Flow-Injection Analysis in Clinical Chemistry: Determination of Albumin with Bromcresol Green [in Japanese]. *Dojin,* **16** (1980) 6.

139. K. Kina, Introduction to Flow Injection Analysis Systems [in Japanese]. *Dojin,* **16** (1980) 8.

140. J. L. Burguera, M. Burguera, F. Millan, and A. Townshend, Flow Injection Analysis Using Chemiluminescent Reactions [in Spanish]. *Acta Cient. Venez.,* **3** (1980) 221.

141. C. B. Ranger, Flow Injection Analysis. Principles, Techniques, Applications and Design. *Anal. Chem.,* **53** (1981) 20A.

142. J. T. Vanderslice, K. K. Stewart, A. G. Rosenfeld, and D. Higgs, Laminar Dispersion in Flow-Injection Analysis. *Talanta,* **28** (1981) 11.

143. U. Baltensperger and R. Eggli, Characterization of an Amperometric Flow-Through Detector with a Renewable Stationary Mercury Electrode. *Anal. Chim. Acta,* **123** (1981) 107.

144. B. Persson and L. Rosén, Flow Injection Determination of Isosorbide Dinitrate with Polarographih Detection. *Anal. Chim. Acta,* **123** (1981) 115.

145. B. F. Reis, A. O. Jacintho, J. Mortatti, F. J. Krug, E. A. G. Zagatto, H. Bergamin F$^{\underline{o}}$, and L. C. R. Pessenda, Zone-Sampling Processes in Flow Injection Analysis. *Anal. Chim. Acta,* **123** (1981) 221.

146. J. M. Reijn, W. E. van der Linden, and H. Poppe, Dispersion in Open Tubes and Tubes Packed with Large Glass Beads. The Single Bead String Reactor. *Anal. Chim. Acta,* **123** (1981) 229.

147. T. R. Williams, S. W. McElvany, and E. C. Ighodalo, Determination of

Sulfur Dioxide in Solutions by Pyridinium Bromide Perbromide and Titrimetric and Flow Injection Procedures. *Anal. Chim. Acta,* **123** (1981) 351.

148. J.-H. B. Hansen, Flow Injection Analysis. Spektrofotometrisk Bestemmelse af Chlorid, Nitrit og Nitrat [in Danish]. Dan. Eng. Acad., Denmark, (1981). (M. Sc. Thesis).
149. W. J. Blaedel and J. Wang, Symmetrical Two-Electrode Pulsed-Flow Detector for Liquid Chromatography. *Anal. Chem.,* **53** (1981) 78.
150. H. L. Pardue and B. Fields, Kinetic Treatment of Unsegmented Flow Systems. Part 1. Subjective and Semiquantitative Evaluations of Flow Injection Systems with Gradient Chamber. *Anal. Chim. Acta,* **124** (1981) 39.
151. H. L. Pardue and B. Fields, Kinetic Treatment of Unsegmented Flow Systems. Part 2. Detailed Treatment of Flow Injection Systems with Gradient Chamber. *Anal. Chim. Acta,* **124** (1981) 65.
152. A. U. Ramsing, Automatisering af Kemiske Analyser Baseret på Flow Injection Analysis Princippet under Anvendelse af Spektrofotometrisk og Potentiometrisk Detektion [in Danish]. Tech. Univ. Denm., Denmark, (1981). (Ph.D. Thesis).
153. J. Růžička and E. H. Hansen, *Flow Injection Analysis,* Wiley, New York (1981).
154. S. S. Olsen, Enzymkinetiske Målinger ved hjælp af Flow Injection Analysis med Specielt Henblik på Klinisk Kemiske Anvendelser [in Danish]. Tech. Univ. Denm., Denmark, (1981). (M.Sc. Thesis).
155. P. L. Meschi and D. C. Johnson, The Amperometric Response of Tubular Electrodes Applied to Flow Injection Determinations. *Anal. Chim. Acta,* **124** (1981) 303.
156. P. L. Meschi, D. C. Johnson, and G. R. Luecke, The Coulometric Response of Tubular Electrodes Applied to Flow Injection Determinations. *Anal. Chim. Acta,* **124** (1981) 315.
157. B. C. Madsen, Utilization of Flow Injection with Hydrazine Reduction and Photometric Detection for the Determination of Nitrate in Rain-Water. *Anal. Chim. Acta,* **124** (1981) 437.
158. K. K. Stewart and A. G. Rosenfeld, Automated Titrations: The Use of Automated Multiple Flow Injection Analysis for the Titration of Discrete Samples. *J. Autom. Chem.,* **3** (1981) 30.
159. S. J. Lyle and M. I. Saleh, Observations on a Dropping-Mercury Electrochemical Detector for Flow Injection Analysis and HPLC. *Talanta,* **28** (1981) 251.
160. J. F. van Staden and W. D. Basson, Automated Flow-Injection Analysis of Urinary Inorganic Sulphates. *Lab. Pract.,* **29** (1980) 1279.
161. K. Hirayama and N. Unohara, Studies on Spectrophotometric-Catalytic Determination of Trace Amounts of Metals. Part 5. Spectrophotometric Determinations of Trace Vanadium by Means of F.I.A. Based on the Cat-

alytic Oxidation of Bindschedler's Green Leuco Base by Potassium Bromate [in Japanese]. *Nippon Kagaku Kaishi,* **1** (1981) 98.

162. D. Betteridge, E. L. Dagless, B. Fields, P. Sweet, and D. R. Deans, Analytical Chemistry at the Interface. *Anal. Proc.,* **18** (1981) 26.

163. Y. Hirai, N. Yoza, and S. Ohashi, Flow Injection System as a Post-Column Reaction Detector for High-Performance Liquid Chromatography of Phosphinate, Phosphonate and Orthophosphate. *J. Chromatogr.,* **206** (1981) 501.

164. G. Ham, Refractive Index Effect in Flow Injection Analysis. *Anal. Proc.,* **18** (1981) 69.

165. T. R. Lindstrom, The Determination of Mercury at Trace Levels by Flow Injection Analysis with Electrochemical Detection. *Diss. Abstr. Int. B,* **41** (1981) 3021.

166. E. A. G. Zagatto, A. O. Jacintho, L. C. R. Pessenda, F. J. Krug, B. F. Reis, and H. Bergamin F$^{\underline{o}}$, Merging Zones in Flow Injection Analysis. Part 5. Simultaneous Determination of Aluminium and Iron in Plant Digests by a Zone-Sampling Approach. *Anal. Chim. Acta,* **125** (1981) 37.

167. K. K. Stewart, G. R. Beecher, and P. E. Hare, Rapid Analysis of Discrete Samples. The Use of Nonsegmented Continuous Flow. *Anal. Biochem.,* **70** (1976) 167.

168. R. E. Shoup, C. S. Brunlett, P. T. Kissinger, and A. W. Jacobs, Thin Transducer as Detector for Trace Organics. *Ind. Res. Dev.,* **23** (1981) 148.

169. F. J. Krug, J. Mortatti, L. C. R. Pessenda, E. A. G. Zagatto, and H. Bergamin F$^{\underline{o}}$, Flow Injection Spectrophotometric Determination of Boron in Plant Material with Azomethine-H. *Anal. Chim. Acta,* **125** (1981) 29.

170. E. B. Schalscha, T. Schirado, and I. Vergara, Flow Injection Analysis of Nitrate in Soil Extracts—Evaluation of a Nitrate-Selective Flow Electrode Method. *J. Soil Sci. Soc. Amer.,* **45** (1981) 446.

171. W. D. Basson and J. F. van Staden, Simultaneous Determination of Chloride and Sulphate in Natural Waters by Flow-Injection Analysis. *Water Res.,* **15** (1981) 333.

172. J. L. Burguera, M. Burguera, and A. Townshend, Determination of Zinc and Cadmium by Flow Injection Analysis and Chemiluminescence. *Anal. Chim. Acta,* **127** (1981) 199.

173. M. E. Meyerhoff and Y. M. Fraticelli, Flow Injection Determination of Ammonia-N Using Polymer Membrane Electrode-Based Gas Sensing System. *Anal. Lett.,* **14** (1981) 415.

174. S. M. Ramasamy and H. A. Mottola, Flow Injection (Closed-Loop Configuration) Catalytic Determination of Copper in Human Serum. *Anal. Chim. Acta,* **127** (1981) 39.

175. T. Korenaga and H. Ikatsu, Effect of Silver Salt on the Determination of Chemical Oxygen Demand by Flow Injection Analysis [in Japanese]. *Nippon Kagaku Kaishi,* **4** (1981) 618.

176. J. M. Reijn, W. E. van der Linden, and H. Poppe, Transport Phenomena

in Flow Injection Analysis Without Chemical Reaction. *Anal. Chim. Acta,* **126** (1981) 1.

177. Y. Hirai, T. Yokoyama, N. Yoza, T. Tarutani, and S. Ohashi, Flow Injection Analysis of Silicic Acid [in Japanese]. *Bunseki Kagaku,* **30** (1981) 350.
178. M. L. Grayeski, J. Mullin, W. R. Seitz, and E. Zygowicz, Flow Injection Analysis with Chemiluminescence Detection: Recent Advances and Clinical Applications. *Biolumin. Chemilumin.* (2nd Int. Symp. Anal. Appl. Biolumin. Chemilumin.) (1981) 623.
179. D. J. Leggett, N. H. Chen, and D. S. Mahadevappa, Flow Injection Method for Sulfide Determination by the Methylene Blue Method. *Anal. Chim. Acta,* **128** (1981) 163.
180. T. Korenaga and H. Ikatsu, Continuous Flow InjectionAnalysis of Aqueous Environmental Samples for Chemical Oxygen Demand. *Analyst,* **106** (1981) 653.
181. C. C. Painton and H. A. Mottola, Chemical Kinetic Contributions to Practical Dispersion in Unsegmented Continuous-Flow Determinations (Flow Injection Analysis). *Anal. Chem.,* **53** (1981) 1713.
182. Y. Hirai, N. Yoza, and S. Ohashi, Flow Injection Analysis of Phosphates in Environmental Waters [in Japanese]. *Bunseki Kagaku,* **30** (1981) 465.
183. A. U. Ramsing, J. Růžička, and E. H. Hansen, The Principles and Theory of High-Speed Titrations by Flow Injection Analysis. *Anal. Chim. Acta,* **129** (1981) 1.
184. B. D. Mindel and B. Karlberg, A Sample Pretreatment System for Atomic Absorption Using Flow Injection Analysis. *Lab. Pract.,* **30** (1981) 719.
185. T. Imasaka, T. Harada, and N. Ishibashi, Fluorimetric Determination of Gallium with Lumogallion by Flow Injection Analysis Based on Solvent Extraction. *Anal. Chim. Acta,* **129** (1981) 195.
186. B. C. Madsen and J. R. Murphy, Flow Injection and Photometric Determination of Sulfate in Rainwater with Methylthymol Blue. *Anal. Chem.,* **53** (1981) 1924.
187. R. T. Lindstrom and D. C. Johnson, Evaluations of n_{app} for the Underpotential Deposition of Mercury on Gold by Flow Injection Coulometry. *Anal. Chem.,* **53** (1981) 1855.
188. S. Greenfield, FIA Weds ICP—A Marriage of Convenience. *Ind. Res. Dev.,* **23** (1981) 140.
189. A. Deratani and B. Sebille, Metal Extraction with a Thiol Hydrophilic Resin. *Anal. Chem.,* **53** (1981) 1742.
190. A. D. Basson, J. F. van Staden, and P. M. Catttin, Determination of Phosphorous as Molybdovanadophosphoric Acid in Phosphate Rock with a Flow-Injection Procedure. *Fresenius Z. Anal. Chem.,* **307** (1981) 373.
191. T. Yamane, Flow Injection Determination of Traces of Cobalt by Catalysis

of the SPADNS-Hydrogen Peroxide Reaction with Spectrophotometric Detection. *Anal. Chim. Acta,* **130** (1981) 65.

192. H. W. van Rooijen and H. Poppe, An Electrochemical Reactivation Method for Solid Electrodes Used in Electrochemical Detectors for High-Performance Liquid Chromatography and Flow Injection Analysis. *Anal. Chim. Acta,* **130** (1981) 9.
193. J. N. Miller, Luminescence Detection in Flow Injection Analysis. *Anal. Proc.,* **18** (1981) 264.
194. T. A. Kelly and G. D. Christian, Fluorometer for Flow Injection Analysis with Application to Oxidase Enzyme Dependent Reactions. *Anal. Chem.,* **53** (1981) 2110.
195. B. W. Renoe, C. E. Shideler, and J. Savory, Use of a Flow-Injection Sample Manipulator as an Interface Between a "High-Performance" Liquid Chromatograph and an Atomic Absorption Spectrophotometer. *Clin. Chem.,* **27** (1981) 1546.
196. E. H. Hansen, Flow Injection Analysis: New Analytical Methods Based on the Use of Potentiometric and Spectrophotometric Flow-Through Detectors. *Anal. Proc.,* **18** (1981) 261.
197. J. Růžička, Theory and Principles of Flow Injection Analysis. *Anal. Proc.,* **18** (1981) 267.
198. Y. Kato, Liquid-Flow Electroanalytical Techniques [in Japanese]. *Bunseki,* **8** (1981) 573.
199. L. Gorton and L. Ögren, Flow Injection Analysis for Glucose and Urea with Enzyme Reactors and On-Line Dialysis. *Anal. Chim. Acta,* **130** (1981) 45.
200. A. G. Fogg and D. Bhanot, Further Voltammetric Studies of Synthetic Food Colouring Matters at Glassy Carbon and Carbon Paste Electrodes Using Static and Flowing Systems. *Analyst,* **106** (1981) 883.
201. J. Růžička, Flow Injection Analysis and Its Future Development. *Proc. Int. Microchem. Symp.,* **8** (1981) 288.
202. A. O. Jacintho, E. A. G. Zagatto, B. F. Reis, L. C. R. Pessenda, and F. J. Krug, Merging Zones in Flow Injection Analysis. Part 6. Determination of Calcium in Natural Waters, Soil and Plant Materials with Glyoxalbis(2-Hydroxylanil). *Anal. Chim. Acta,* **130** (1981) 361.
203. R. Kuroda and T. Mochizuki, Continuous Spectrophotometric Determination of Copper, Nickel and Zinc in Copper-Base Alloys by Flow Injection Analysis. *Talanta,* **28** (1981) 389.
204. Z. Gao and M. Lu, Flow Injection Analysis by Ion-Selective Electrodes. Use of a Solid Membrane Chloride Ion-Selective Electrode for Determination of Chloride in Soil-Water Extracts [in Chinese]. *Huanjing Kexue,* **2** (1981) 376.
205. H. Cui and Z. Fang, Flow Injection Analysis with Ion Selective Electrodes and its Application in Environmental Analysis. A Review [in Chinese]. *Zhongguo Huanjing Jiance,* **8** (1982) 1.

206. A. O. Jacintho, E. A. G. Zagatto, F. Bergamin F$^{\underline{o}}$, F. J. Krug, B. F. Reis, R. E. Bruns, and B. R. Kowalski, Flow Injection Systems with Inductively-Coupled Argon Plasma Atomic Emission Spectrometry. Part 1. Fundamental Considerations. *Anal. Chim. Acta,* **130** (1981) 243.

207. F. J. Krug, L. C. R. Pessenda, E. A. G. Zagatto, A. O. Jacintho, and B. F. Reis, Spectrophotometric Flow Injection Determination of Chloride in Ethanol. *Anal. Chim. Acta,* **130** (1981) 409.

208. O. Åström, Flow Injection Analysis for the Determination of Bismuth by Atomic Absorption Spectrometry with Hydride Generation. *Anal. Chem.,* **54** (1982) 90.

209. P. L. Meschi, Tubular Electrodes in Flow Injection Analysis. *Diss. Abstr. Int. B,* **42** (1981) 1444.

210. H. B. Hanekamp, P. Bos, and O. Vittori, The Applicability of Phase-Sensitive Alternating Current Measurements in Flow-Through Detection. *Anal. Chim. Acta,* **131** (1981) 149.

211. P. C. A. Ooms, G. P. Leendertse, H. A. Das, and U. A. Th. Brinkman, Multielement Isotope Dilution Analysis by Means of Radiometric Titration. *J. Radioanal. Chem.,* **67** (1981) 5.

212. S. M. Ramasamy and H. A. Mottola, Repetitive Determination of Sulfur Dioxide Gas in Air Samples by Flow Injection and Chemical Reaction at a Gas-Liquid Interface. *Anal. Chem.,* **54** (1982) 283.

213. S. F. Simpson and F. J. Holler, Design and Evaluation of a Potentiometric Detection System for Flow Injection Titrimetry. *Anal. Chem.,* **54** (1982) 43.

214. D. C. Shelly, T. M. Rossi, and I. M. Warner, Multiple Solvent Extraction System with Flow Injection Technology. *Anal. Chem.,* **54** (1982) 87.

215. T. E. Hu, Potentiometric Stripping as a Detector for Flow Injection Analysis of Metal Ions. *Diss. Abstr. Int. B,* **42** (1981) 1874.

216. H. A. Mottola, Continuous Flow Analysis Revisited. *Anal. Chem.,* **53** (1981) 1312A.

217. A. Ivaska and T. H. Ryan, Application of a Voltammetric Flow-Through Cell to Flow Injection Analysis. *Coll. Czech. Chem. Comm.,* **46** (1981) 2865.

218. S. Hughes, P. L. Meschi, and D. C. Johnson, Amperometric Detection of Simple Alcohols in Aqueous Solutions by Application of a Triple-Pulse Potential Waveform at Platinum Electrodes. *Anal. Chim. Acta,* **132** (1981) 1.

219. S. Hughes and D. C. Johnson, Amperometric Detection of Simple Carbohydrates at Platinum Electrodes in Alkaline Solutions by Application of a Triple-Pulse Potential Waveform. *Anal. Chim. Acta,* **132** (1981) 11.

220. D. Betteridge and B. Fields, The Application of pH Gradients in Flow-Injection Analysis. A Method for Simultaneous Determination of Binary Mixtures of Metal Ions in Solution. *Anal. Chim. Acta,* **132** (1981) 139.

221. I. Kågevall, O. Åström, and A. Cedergren, Minimization of Interference Effects from Iodine-Consuming Samples in the Determination of Water

with the Karl Fischer Reagent in a Flow Injection System. *Anal. Chim. Acta,* **132** (1981) 215.

222. K. K. Stewart, Flow Injection Analysis. A Review of Its Early History. *Talanta,* **28** (1981) 789.
223. J. F. Tyson and A. B. Idris, Flow Injection Sample Introduction for Atomic-Absorption Spectrometry. Applications of a Simplified Model for Dispersion. *Analyst,* **106** (1981) 1125.
224. T. B. Goad, Microprocessor Controlled Flow Injection Analysis. Univ. Coll. Swansea, UK (1982). (Ph.D. Thesis).
225. A. G. Fogg and N. K. Bsebsu, Differential-Pulse Voltammetric Determination of Phosphate as Molybdovanadophosphate at a Glassy Carbon Electrode and Assessment of Eluents for the Flow Injection Voltammetric Determination of Phosphate, Silicate, Arsenate and Germanate. *Analyst,* **106** (1981) 1288.
226. P. J. Worsfold, J. Růžička, and E. H. Hansen, Rapid Automated Enzymatic Method for the Determination of Alcohol in Blood and Beverages Using Flow Injection Analysis. *Analyst,* **106** (1981) 1309.
227. J. F. Tyson, Low Cost Continuous Flow Analysis. Flow Injection Techniques in Atomic Absorption Spectrometry. *Anal. Proc.,* **18** (1981) 542.
228. L. J. Sun, Z. Gao, L. Li, and Z. L. Fang, The Determination of Total Nitrogen in Soil Digests by Flow Injection Analysis [in Chinese]. *Turang Tongbao,* **5** (1981) 38.
229. Z. Gao and Z. L. Fang, Ion Selective Electrodes in Flow Injection Analysis. A Review [in Chinese]. *Lizi Xuanze Dianji Tongxun,* **1** (1981) 44.
230. W. Coakley, *Handbook of Automated Analysis. Continuous Flow Techniques.* Marcel Dekker, Inc., New York (1981).
231. T. A. Kelly, Electrochemical, Fluorimetric and Flow Injection Analysis of Enzyme and Immunochemical Systems. *Diss. Abstr. Int. B,* **42** (1981) 2356.
232. B. Fields, Studies in Flow Injection Analysis. Univ. Coll. Swansea, U.K., 1981. (Ph.D. Thesis).
233. L. Gorton, A Study of Modified Electrodes and Enzyme Reactors. Univ. of Lund, Sweden (1981). (Ph.D. Thesis).
234. L. Ögren, Enzyme Reactors in Analytical Detection Systems. Theory and Applications. Univ. of Lund, Sweden (1981). (Ph.D. Thesis).
235. S. M. Ramasamy, Studies in Unsegmented Continuous Flow Analysis. Part I. Repetitive Determinations of Copper(II) Catalyst. Part II. Repetitive Determinations of Bromine and Chlorine in Gaseous Samples at a Gas-Solid Interface. *Diss. Abstr. Int. B.,* **42** (1981) 1875.
236. H. Salvesen, Flow Injection Analyse (FIA). Bestemmelse av Summen av Kationer og Anioner i Vann med FIA Sammenliknet med lonekromatografi [in Norwegian]. Univ. of Oslo, Norway (1981). (M.Sc. Thesis).
237. L. C. R. Pessenda, Determinacao de Baixas Concentracoes de Ortofosfato em Aguas Naturais com Emprego de Resina de Troca Ionica em Sistema

de Injecao em Fluxa [in Portuguese]. *CENA/ESALQ/USP, Brazil* (1981). (M.Sc. Thesis).

238. K. Kina, Flow Injection Analysis (FIA) of Ammonia by Isothermal Distillation [in Japanese]. *Dojin,* **19** (1981) 7.

239. K. Kina, Rapid Determination of Corticosteroids in Medicals by Means of the Flow Injection Method [in Japanese]. *Dojin,* **20** (1981) 13.

240. K. Kina, Introduction to the Principles of Flow Injection Analysis. VI. Experiments and Practical Applications [in Japanese]. *Dojin,* **21** (1981) 10.

241. T. Mochizuki and R. Kuroda, Determination of Beryllium in Copper-Beryllium Alloys by Flow Injection Spectrophotometry. *Fresenius Z. Anal. Chem.,* **309** (1981) 363.

242. E. Zminkowska-Halliop, E. Soczewinski, and J. Matysik, Amperometric Flow-Through Detector of Very Small Detection Volume. *Chem. Anal. (Warzaw),* **26** (1981) 161.

243. D. J. Leggett, N. H. Chen, and D. S. Mahadevappa, Flow Injection Methods for Determination of Iron(III). *Ind. J. Chem.,* **20A** (1981) 1051.

244. A. U. Ramsing and J. Růžička, Simultaneous Detection of Ca^{2+} and pH in Flow Injection Analysis with Ion-Sensitive Field Effect Transistors. A Model System for in Vivo Measurements. *Proc. Int. Conf. Nijmegen (Holland),* (1981) 134.

245. E. A. G. Zagatto, A. O. Jacintho, B. F. Reis, F. J. Krug, H. Bergamin F$^{\underline{o}}$, L. C. R. Pessenda, J. Mortatti, and M. F. Giné, Manual de Analises de Plantas e Aguas Empregando Sistemas de Injecao em Fluxo [in Portuguese]. Univ. de Sao Paulo, CENA, Brazil, (1981).

246. F. Oshima, Flow Injection Analysis of Ammonium Nitrogen in Environmental Waters [in Japanese]. *Fukuoka Kyoiku Daigaku Kiyo,* **31** (1981) 57.

247. E. L. Guldberg, A. S. Attiyat, and G. D. Christian, Amperometric Systems for the Determination of Oxidase Enzyme Dependent Reactions by Continuous Flow and Flow Injection Analysis. *J. Autom. Chem.,* **2** (1980) 189.

248. J. Harrow, J. Janata, R. L. Stephen, and W. J. Kolff, Portable System for Simultaneous Measurements of Blood Elecctrolytes. *Proc. EDTA,* **17** (1980) 179.

249. J. Matysik, E. Soczewinski, E. Zminkowska-Halliop, and M. Przegalinski, Determination of *o*-Diphenols by Flow Injection Analysis with an Amperometric Detector. *Chem. Anal. (Warzaw),* **26** (1981) 463.

250. R. Virtanen, A Flow Injection Analyzer with Multiple ISE-Detector. *Anal. Chem. Symp. Ser.,* **8** (1981) 375.

251. H. Müller, Chloride and Cyanide Determination by Use of the Flow Injection Method Using Ion-Selective Flow-Type Electrodes. *Anal. Chem. Symp. Ser.,* **8** (1981) 279.

252. T. Yamane, Flow Injection Determination of Trace Cobalt(II) [in Japanese]. *Nippon Kagaku Kaishi,* **1** (1982) 93.

253. J. Růžička, E. H. Hansen, and A. U. Ramsing, Flow Injection Analyzer

for Students, Teaching and Research. Spectrophotometric Methods. *Anal. Chim. Acta,* **134** (1982) 55.

254. J. F. van Staden, Automated Turbidimetric Determination of Sulphate in Surface, Ground and Domestic Water by Flow Injection Analysis. *Fresenius Z. Anal. Chem.,* **310** (1982) 239.
255. B. S. Hui and C. O. Huber, Amperometric Detection of Amines and Amino Acids in Flow Injection Systems with a Nickel Oxide Electrode. *Anal. Chim. Acta,* **134** (1982) 211.
256. O. Kondo, H. Miyata, and K. Toei, Determination of Sulfate in River Water by Flow Injection Analysis. *Anal. Chim. Acta,* **134** (1982) 353.
257. J. Růžička and E. H. Hansen, Flow Injection Analysis and Its Early History (Letter to the Editor). *Talanta,* **29** (1982) 157.
258. W. D. Basson, Consecutive Determination of Nitrogen, Phosphorous and Calcium, in Animal Feeds on a Single Channel Flow Injection Analyzer with a Common Analytical Manifold. *Fresenius Z. Anal. Chem.,* **311** (1982) 23.
259. T. Mochizuki and R. Kuroda, Rapid Continuous Determination of Aluminium in Copper-Base Alloys by Flow Injection Spectrophotometry. *Fresenius Z. Anal. Chem.,* **311** (1982) 11.
260. K. Mori, T. Imasaka, N. Ishibashi, and C. Jin, Application of Microcomputer for Data Processing in Flow Injection Analysis [in Japanese]. *Bunseki Kagaku,* **31** (1982) 103.
261. T. Korenaga and H. Ikatsu, Fully Automated System for Continuous Monitoring of Chemical Oxygen Demand by Flow Injection Analysis [in Japanese]. *Bunseki Kagaku,* **31** (1982) 135.
262. B. F. Rocks and C. Riley, Flow Injection Analysis. A New Approach to Quantitative Measurements in Clinical Chemistry. *Clin. Chem.,* **28** (1982) 409.
263. B. F. Rocks, R. A. Sherwood, and C. Riley, Direct Determination of Therapeutic Concentrations of Lithium in Serum by Flow Injection Analysis with Atomic Absorption Spectroscopy Detection. *Clin. Chem.,* **28** (1982) 440.
264. S. Olsen, J. Růžička, and E. H. Hansen, Gradient Techniques in Flow Injection Analysis. Stopped-Flow Measurements of the Activity of Lactate Dehydrogenase with Electronic Dilution. *Anal. Chim. Acta,* **136** (1982) 101.
265. B. Olsson, Determination of Hydrogen Peroxide in a Flow System with Microperoxidase as Catalyst for the Luminol Chemiluminescence Reaction. *Anal. Chim. Acta,* **136** (1982) 113.
266. J. Möller, FIA—A New Analytical Method [in German]. *Labor Praxis,* **6** (1982) 278.
267. Z. Fang, L. Sun, Z. Gao, Y. Zhu, X. Wang, and L. Li, Determination of Total Phosphorous in Soil Digests by Flow Injection Analysis [in Chinese]. *Turang Tongbao,* **4** (1982) 40.

268. Z. Fang, Development of Flow Injection Analysis. A Review [in Chinese]. *Turangzue Jinzhan,* **10** (1982) 48.

269. K. Kina, Flow Injection Analysis. Determination of Glucose [in Japanese]. *Dojin,* **22** (1982) 8.

270. J. Růžička and A. U. Ramsing, Flow Injection Analysis Using Ion-Sensitive Field Effect Transistors. A Model System for Discrete Assays and Continuous in Vitro Monitoring of pH and pCa. *Scand. J. Clin. Lab. Invest.,* **42** (1982) 35.

271. M. Trojanowicz, W. Matuszewski, and A. Hulanicki, Flow Injection Potentiometric Determination of Residual Chlorine in Water. *Anal. Chim. Acta,* **136** (1982) 85.

272. D. J. Leggett, N. H. Chen, and D. S. Mahadevappa, Rapid Determination of Residual Chlorine by Flow Injection Analysis. *Analyst,* **107** 1982) 433.

273. G. Svensson and T. Anfält, Rapid Determination of Ammonia in Whole Blood and Plasma Using Flow Injection Analysis. *Clin. Chem. Acta,* **119** (1982) 7.

274. T. Korenaga, The Continuous Determination of Filtered Chemical Oxygen Demand with Potassium Dichromate by Means of Flow Injection Analysis. *Bull. Chem. Soc. Jpn.,* **55** (1982) 1033.

275. J. Růžička, Flow Injection Methods. A New Tool for Instrumental Analysis. *Phil. Trans. R. Soc. Lond.,* **A305** (1982) 645.

276. A. G. Fogg and N. K. Bsebsu, Flow Injection Voltammetric Determination of Phosphate. Direct Injection of Phosphate into Molybdate Reagent. *Analyst,* **107** (1982) 566.

277. D. J. Leggett, N. H. Chen, and D. S. Mahadevappa, Flow Injection Analysis of Aromatic Sulphonyl Haloamines. *Fresenius Z. Anal. Chem.,* **311** (1982) 687.

278. U. Rydevik, L. Nord, and F. Ingman, Automatic Lactate Determination by Reagent Injection Analysis. *Int. J. Sports Med.,* **3** (1982) 47.

279. K. S. Johnson and R. L. Petty, Determination of Phosphate in Seawater by Flow Injection Analysis with Injection of Reagent. *Anal. Chem.,* **54** (1982) 1185.

280. T. Hara, M. Toriyama, and M. Imaki, The Flow Injection Analysis of D-Glucose Using a Flow Cell with Immobilized Peroxidase and Its Application to Serum. *Bull. Chem. Soc. Jpn.,* **55** (1982) 1854.

281. J. Mortatti, F. J. Krug, L. C. R. Pessenda, E. A. G. Zagatto, and S. S. Jørgensen, Determination of Iron in Natural Waters and Plant Material with 1,10-Phenanthroline by Flow Injection Analysis. *Analyst,* **107** (1982) 659.

282. M. Trojanowicz and W. Matuszewski, Limitation of Linear Response in Flow Injection Systems with Ion-Selective Electrodes. *Anal. Chim. Acta,* **138** (1982) 71.

283. T. Yao, Y. Kobayashi, and S. Musha, Flow Injection Analysis for L-Lactate

with Immobilized Lactate Dehydrogenase. *Anal. Chim. Acta,* **138** (1982) 81.

284. J. F. van Staden, Automated Simultaneous Determination of Nitrate and Nitrite by Pre-Valve Reduction of Nitrate in a Flow Injection System. *Anal. Chim. Acta,* **138** (1982) 403.
285. J. Wang and H. Dewald, A Porous-Jet Flow-Through Electrode. *Talanta,* **29** (1982) 453.
286. C. B. Ranger, Rapid Sample Pretreatment and Analysis Using Automated FIA. *Am. Lab.,* **14** (1982) 56.
287. T. A. Kelly and G. D. Christian, Capillary Flow Injection Analysis for Enzyme Assay with Fluorescence Detection. *Anal. Chem.,* **54** (1982) 1444.
288. J. Janata and J. Růžička, Combination of Flow Injection Analysis and Voltammetry. *Anal. Chim. Acta,* **139** (1982) 105.
289. U. Forsman and A. Karlsson, Polarographic Determination of Penicilloic Acid in Penicillin Preparations with a Flow Injection System. *Anal. Chim. Acta,* **139** (1982) 133.
290. J. B. Kafil and C. O. Huber, Flow Injection Sample Processing with Nickel Oxide Electrode Amperometric Detection of Amino Acids Separated by Ion-Exchange Chromatography. *Anal. Chim. Acta,* **139** (1982) 347.
291. C. W. Holy, Commentary: Flow Injection Analysis—An Idea Incomplete? *J. Autom. Chem.,* **4** (1982) 111.
292. B. Karlberg, Flow Injection Analysis. *Chem. Deriv. Anal. Chem.,* **2** (1982) 1.
293. T. Mochizuki, Y. Toda, and R. Kuroda, Flow Injection Analysis of Silicate Rocks for Total Iron and Aluminium. *Talanta,* **29** (1982) 659.
294. T. Korenaga, H. Ikatsu, and T. Moriwake, The Oxidation Behaviour of Various Organic Compounds on the Determination of Chemical-Oxygen Demand by Means of Flow Injection Analysis with Acidic Permanganate. *Bull. Chem. Soc. Jpn.,* **55** (1982) 2622.
295. C. Wyganowski, S. Motomizu, and K. Toei, Spectrophotometric Determination of Aluminium in River Water with Bromopyrogallol Red and *n*-Tetradecyltrimethyl Ammonium Bromide by Flow Injection Analysis. *Anal. Chim. Acta,* **140** (1982) 313.
296. W. E. van den Linden, Flow Injection Analysis: The Manipulation of Dispersion. *Trends Anal. Chem.,* **1** (1982) 188.
297. J. Růžička, The Flow Injection Method—A New Tool for Instrumental Analysis. *Rec. Adv. Anal. Spectrosc.* (Pergamon Press), (1982) 285.
298. J. F. van Staden, Automated Prevalve Sample Filtration in Flow Injection Analysis. Determination of Sulphate in Water Removing Suspended Solids and Colour Before Sampling. *Fresenius Z. Anal. Chem.,* **312** (1982) 438.
299. D. Betteridge, Flow Injection Analysis in the Teaching Laboratory. *Fresenius Z. Anal. Chem.,* **312** (1982) 441.
300. A. G. Fogg, N. K. Bsebsu, and M. A. Abdalla, Flow Injection Voltam-

metric Determination of Nitrite by Reduction at a Glassy Carbon Electrode in Acidic Bromide or Chloride Media. *Analyst,* **107** (1982) 1040.

301. B. F. Rocks and C. Riley, Flow Injection Analysis. *TIBS,* **7** (1982) 315.
302. S. Nakahara, M. Yamada, and S. Suzuki, Chemiluminescence for the Determination of Traces of Cobalt(II) by Continuous and Flow Injection Methods. *Anal. Chim. Acta,* **141** (1982) 255.
303. T. Korenaga and H. Ikatsu, The Determination of Chemical Oxygen Demand in Waste Waters with Dichromate by Flow Injection Analysis. *Anal. Chim. Acta,* **141** (1982) 301.
304. L. Fossey and F. F. Cantwell, Characterization of Solvent Extraction/Flow Injection Analysis with Constant Pressure Pumping and Determination of Procyclidine Hydrochloride in Tablets. *Anal. Chem.,* **54** (1982) 1693.
305. T. Korenaga and H. Ikatsu, Flow Injection Analysis System with Personal Computer Application to a Fully Automated Method for the Measurement of Dissolved Chemical Oxygen Demand in River Waters [in Japanese]. *Bunseki Kagaku,* **31** (1982) 517.
306. T. M. Rossi, D. C. Shelly, and I. M. Warner, Optimization of a Flow Injection Analysis System for Multiple Solvent Extraction. *Anal. Chem.,* **54** (1982) 2056.
307. K. Ogata, K. Taguchi, and T. Imanari, Phase Separator for Flow Injection Analysis. *Anal. Chem.,* **54** (1982) 2127.
308. W. S. Gardner and H. A. Vanderploeg, Microsample-Filtering Device for Liquid Chromatography or Flow Injection Analysis. *Anal. Chem.,* **54** (1982) 2129.
309. M. Thompson and U. J. Krull, Bilayer Lipid Membrane Electrochemistry in a Flow Injection System. *Anal. Chim. Acta,* **142** (1982) 207.
310. R. L. Petty, W. C. Michel, J. P. Snow, and K. S. Johnson, Determination of Total Primary Amines in Seawater and Plant Nectar with Flow Injection Sample Processing and Fluorescence Detection. *Anal. Chim. Acta,* **142** (1982) 299.
311. T. Mochizuki and R. Kuroda, Flow Injection Analysis of Silicate Rocks for Titanium. *Analyst,* **107** (1982) 1255.
312. J. G. Williams, M. Holmes, and D. G. Porter, Titration of Spoilt Beer Samples by Flow Injection Analysis. *J. Autom. Chem.,* **4** (1982) 176.
313. E. H. Hansen and J. Růžička, Correspondence. Flow Injection Analysis: An Idea Complete—But Yet Far from Fully Exploited. *J. Autom. Chem.,* **4** (1982) 193.
314. K. K. Stewart, Correspondence. A Reply to H. W. Holy. *J. Autom. Chem.,* **4** (1982) 193.
315. M. Yamada and S. Suzuki, Chemiluminescent Determination of Traces of Copper(II) by the Flow Injection Method. *Chem. Lett.,* **11** (1982) 1747.
316. J. Wang and H. D. Dewald, Flow Injection Analysis of Oxidizable Species with Reverse-Pulse Amperometric Detection. *Talanta,* **29** (1982) 901.

317. J. M. Harris, Flow Injection of Ultra-Trace Level Samples into Laser-Based Detectors. *Anal. Chem.,* **54** (1982) 2337.

318. K. K. Stewart and A. G. Rosenfeld, Exponential Dilution Chambers for Scale Expansion in Flow Injection Analysis. *Anal. Chem.,* **54** (1982) 2368.

319. T. Yokoyama, Y. Hirai, N. Yoza, T. Tarutani, and S. Ohashi, Spectrophotometric Determination of Silicic Acid by Flow Injection Analysis. *Bull. Chem. Soc. Jpn.,* **55** (1982) 3477.

320. K. Ogata, K. Taguchi, and T. Imanari, Determination of Orthophosphate by Flow Injection Analysis Based on Solvent Extraction [in Japanese]. *Bunseki Kagaku,* **31** (1982) 641.

321. Y. Hirai, N. Yoza, and S. Ohashi, Flow Injection Determination of Sulfite Based on the Oxidation of Phosphonic Acid with Sulfite [in Japanese]. *Bunseki Kagaku,* **31** (1982) 681.

322. T. A. Kelly and G. D. Christian, Homogeneous Enzymatic Fluorescence Immunoassay of Serum IgG by Continuous Flow Injection Analysis. *Talanta,* **29** (1982) 1109.

323. A. Haemmerli and J. Janata, A Flow Injection System for Measurement of Chemical Response Time of Microelectrodes. *Anal. Chim. Acta,* **144** (1982) 115.

324. A. G. Fogg, N. K. Bsebsu, and M. A. Abdalla, Indirect Flow Injection Voltammetric Determination of Aromatic Amines by Monitoring at a Glassy Carbon Electrode the Excess of Nitrite Remaining after Their Diazotization. *Analyst,* **107** (1982) 1462.

325. G. Gerhardt and R. N. Adams, Determination of Diffusion Coefficients by Flow Injection Analysis. *Anal. Chem.,* **54** (1982) 2618.

326. S. Nakashima, M. Yasi, M. Zenki, A. Takahashi, and K. Toei, Determination of Trace Amounts of Nitrite by Flow Injection Spectrophotometry [in Japanese]. *Bunseki Kagaku,* **31** (1982) 732.

327. J. L. Burguera and M. Burguera, Determination of Iodine by a Chemiluminescent Reaction Using Flow Injection Analysis. *An. Quim.,* **78.B.** (1982) 307.

328. R. Karliček, Flow Injection Analysis and Its Use in Drug Analysis [in Czech]. *Česk. Farm.,* **31** (1982) 190.

329. J. L. Burguera, M. Burguera, and M. Gallignani, Determination of Terbium by Flow Injection Analysis and Fluorescence. *Acta Cient. Venez.,* **33** (1982) 99.

330. C. Ridder, E. H. Hansen, and J. Růžička, Flow Injection Analysis of Glucose in Human Serum by Chemiluminescence. *Anal. Lett.,* **15** (1982) 1751.

331. B. F. Rocks, R. A. Sherwood, L. M. Bayford, and C. Riley, Zinc and Copper Determination in Microsamples of Serum by Flow Injection and Atomic Absorption Spectroscopy. *Ann. Clin. Biochem.,* **19** (1982) 338.

332. K. Kina, Potentiometric Determinations by Means of Flow Injection Analysis [in Japanese]. *Dojin,* **25** (1982) 7.

333. K. Kina, Flow Injection Analysis for Students, Teaching and Research [in Japanese]. *Dojin,* **23** (1982) 9.

334. T. J. Sly, D. Betteridge, D. Wibberley, and D. G. Porter, An Improved Flow-Through Phototransducer. *J. Autom. Chem.,* **4** (1982) 186.

335. F. Orak, Flow Injection Techniques in Automated Analysis [in Turkish]. *Doga Ser. C,* **6** (1982) 109.

336. S. A. McClintock and W. C. Purdy, A Microprocessor Controlled Potentiometric Detection System. *Anal. Lett.,* **15** (1982) 1001.

337. J. L. Burguera, M. Burguera, and A. Townshend, The Principles, Applications and Trends of Flow Injection Analysis for Monitoring Chemiluminescent Reactions. *Rev. Roum. Chim.,* **27** (1982) 879.

338. J. Růžička and E. H. Hansen, Recent Developments in Flow Injection Analysis. Gradient Techniques and Hydrodynamic Injection. *Anal. Chim. Acta,* **145** (1983) 1.

339. H. Poppe, The Performance of Some Liquid Phase Flow-Through Detectors. *Anal. Chim. Acta,* **145** (1983) 17.

340. H. A. Mottola, Enzymatic Preparations in Analytical Continuous-Flow Systems. *Anal. Chim. Acta,* **145** (1983) 27.

341. Z. Feher, G. Horvai, G. Nagy, Z. Niegreisz, K. Toth, and E. Pungor, A Polarographic and Spectrophotometric Routine Analyzer for Assaying Content Uniformity in Pharmaceutical Quality Control. *Anal. Chim. Acta,* **145** (1983) 41.

342. S. Angelova, Optimal Speed as a Function of System Performance for Continuous Flow Analyzers. *Anal. Chim. Acta,* **145** (1983) 51.

343. J. M. Reijn, H. Poppe, and W. E. van der Linden, A Possible Approach to the Optimization of Flow Injection Analysis. *Anal. Chim. Acta,* **145** (1983) 59.

344. G. Johansson, L. Ögren, and B. Olsson, Enzyme Reactors in Unsegmented Flow Injection Analysis. *Anal. Chim. Acta,* **145** (1983) 71.

345. B. Olsson and L. Ögren, Optimization of Peroxidase Immobilization and of the Design of Packed-Bed Enzyme Reactors for Flow Injection Analysis. *Anal. Chim. Acta,* **145** (1983) 87.

346. B. Olsson, L. Ögren, and G. Johansson, An Enzymatic Flow Injection Method for the Determination of Oxygen. *Anal. Chim. Acta,* **145** (1983) 101.

347. G. L. Abdullahi, J. N. Miller, H. N. Sturley, and J. W. Bridges, Studies of Drug-Protein Binding Interactions by Flow Injection Analysis with Fluorimetric Detection. *Anal. Chim. Acta,* **145** (1983) 109.

348. P. J. Worsfold, The Bio-Analytical Potential of Flow Injection Analysis. *Anal. Chim. Acta,* **145** (1983) 117.

349. H. Kagenow and A. Jensen, Kinetic Determination of Magnesium and Calcium by Stopped-Flow Injection Analysis. *Anal. Chim. Acta,* **145** (1982) 125.

350. G. Nakagawa, H. Wada, and C. Wei, Spectrophotometric Determination of Calcium with a Flow Injection System. *Anal. Chim. Acta,* **145** (1982) 135.

351. Z. Fang and S. Xu, Determination of Molybdenum at Microgram per Litre Levels by Catalytic Spectrophotometric Flow Injection Analysis. *Anal. Chim. Acta,* **145** (1983) 143.

352. L. Nord and B. Karlberg, Sample Preconcentration by Continuous Flow Extraction with a Flow Injection Atomic Absorption Detection System. *Anal. Chim. Acta,* **145** (1983) 151.

353. J. F. Tyson, J. M. H. Appleton, and A. B. Idris, Flow Injection Calibration Methods for Atomic Absorption Spectrometry. *Anal. Chim. Acta,* **145** (1983) 159.

354. E. A. G. Zagatto, A. O. Jacintho, F. J. Krug, B. F. Reis, R. E. Bruns, and M. C. U. Araujo, Flow Injection Systems with Inductively-Coupled Argon Plasma Atomic Emission Spectrometry. Part 2. The Generalized Standard Addition Method. *Anal. Chim. Acta,* **145** (1983) 169.

355. F. J. Krug, E. A. G. Zagatto, B. F. Reis, O. Bahia F°, A. O. Jacintho, and S. S. Jørgensen, Turbidimetric Determination of Sulphate in Plant Digests and Natural Waters by Flow Injection Analysis with Alternating Streams. *Anal. Chim. Acta,* **145** (1983) 179.

356. H. Lundbäck, Amperometric Determination of Hydrogen Peroxide in Pickling Baths for Copper and Copper Alloys by Flow Injection Analysis. *Anal. Chim. Acta,* **145** (1983) 189.

357. R. C. Schothorst, J. M. Reijn, H. Poppe, and G. den Boef, The Application of Strongly Reducing Agents in Flow Injection Analysis. Part 1. Chromium (II) and Vanadium(II). *Anal. Chim. Acta,* **145** (1983) 197.

358. V. K. Mahant, J. N. Miller, and H. Thakrar, Flow Injection Analysis with Chemiluminescence Detection in the Determination of Fluorescein- and Fluorescamine-Labelled Species. *Anal. Chim. Acta,* **145** (1983) 203.

359. P. van den Winkel, G. de Backer, M. Vandeputte, N. Mertens, L. Dryon, and D. L. Massart, Performance and Characteristics of the Fluoride-Selective Electrode in a Flow Injection System. *Anal. Chim. Acta,* **145** (1983) 207.

360. M. Mascini and G. Palleschi, A Flow-Through Detector for Simultaneous Determinations of Glucose and Urea in Serum Samples. *Anal. Chim. Acta,* **145** (1983) 213.

361. M. Strandberg and S. Thelander, A Microprocessor-Controlled Flow Injection Analyser for the Determination of Terbutaline Sulphate. *Anal. Chim. Acta,* **145** (1983) 219.

362. D. Betteridge, W. C. Cheng, E. L. Dagless, P. David, T. B. Goad, D. R. Deans, D. A. Newton, and T. B. Pierce, An Automated Viscometer Based on High-Precision Flow Injection Analysis. Part I. Apparatus for High-Precision Flow Injection Analysis. *Analyst,* **108** (1983) 1.

363. D. Betteridge, W. C. Cheng, E. L. Dagless, P. David, T. B. Goad, D. R. Deans, D. A. Newton, and T. B. Pierce, An Automated Viscometer Based on High-Precision Flow Injection Analysis. Part II. Measurement of Viscosity and Diffusion Coefficients. *Analyst,* **108** (1983) 17.

364. T. Deguchi, A. Tanaka, I. Sanemasa, and H. Nagai, Flow Injection Analysis of Trace Amounts of Iodide Ion Using a Catalytic Method [in Japanese]. *Bunseki Kagaku,* **32** (1983) 23.

365. D. J. Hooley and R. E. Dessy, Continuous Flow Kinetic Techniques in Flow Injection Analysis. *Anal. Chem.,* **55** (1983) 313.

366. A. Hu, R. E. Dessy, and A. Graneli, Potentiometric Stripping with Matrix Exchange Techniques in Flow Injection Analysis of Heavy Metals in Groundwaters. *Anal. Chem.,* **55** (1983) 320.

367. C. Riley and B. F. Rocks, Comments: Flow Injection Analysis—The End of the Beginning? Segmented-Flow Analysis—The Beginning of the End? *J. Autom. Chem.,* **5** (1983) 1.

368. H. I. Tarlin, Comments on "Flow Injection Analysis—An Idea Incomplete." *J. Autom. Chem.,* **5** (1983) 2.

369. C. Riley, B. F. Rocks, R. A. Sherwood, L. H. Aslett, and P. R. Oldfield, A Stopped-Flow/Flow Injection System for Automation of α_2 Macroglobulin Kinetic Studies. *J. Autom. Chem.,* **5** (1983) 32.

370. J. Wang, H. D. Dewald, and B. Greene, Anodic Stripping Voltammetry of Heavy Metals with a Flow Injection System. *Anal. Chim. Acta,* **146** (1983) 45.

371. D. Betteridge and B. Fields, Two-Point Kinetic Simultaneous Determination of Cobalt(II) and Nickel(II) in Aqueous Solution Using Flow Injection Analysis (FIA). *Fresenius Z. Anal. Chem.,* **314** (1983) 386.

372. J. F. Tyson, J. M. H. Appleton, and A. B. Idris, Flow Injection Sample Introduction Methods for Atomic-Absorption Spectrometry. *Analyst,* **108** (1983) 153.

373. J. Möller, FIA—New Techniques and Applications [in German]. *Labor Praxis,* **7** (1983) 1962.

374. M. Yamada, T. Nakada, and S. Suzuki, The Determination of Sulfite in a Flow Injection System with Chemiluminescence Detection. *Anal. Chim. Acta,* **147** (1983) 401.

375. S. Motomizu, T. Wakimoto, and K. Toei, Determination of Trace Amounts of Phosphate in River Water by Flow Injection Analysis. *Talanta,* **30** (1983) 333.

376. K. W. Pratt and D. C. Johnson, The Vibrating Wire Electrode as am Amperometric Detector for Flow Injection Systems. *Anal. Chim. Acta,* **148** (1983) 87.

377. P. W. Alexander and U. Akapongkul, Amperometric Determination of Metal Ions in a Flow Injection System with a Copper-Amalgam Electrode. *Anal. Chim. Acta,* **148** (1983) 103.

378. E. H. Hansen, J. Růžička, F. J. Krug, and E. A. G. Zagatto, Selectivity in Flow Injection Analysis. *Anal. Chim. Acta,* **148** (1983) 111.

379. G. R. Beecher, K. K. Stewart, and P. E. Hare, Automated High Speed Analysis of Discrete Samples. The Use of Nonsegmented, Continuous Flow. In *Protein Nutritional Quality of Foods and Feeds* (M. Friedman, Ed.). Marcel Dekker, New York, Part I (1975) 411.

380. C. B. Ranger, Flow Injection Analysis Is a Modern Technique for Using Unsegmented Flowing Streams to Perform Multiparameter Analysis. *Ind. Res. Dev.,* **21** (1979) 134.

381. H. Ma and H. Yan, The Development of an Amperometric Detector for Flow Injection Analysis [in Chinese]. *Kexue Tongbao,* **27** (1982) 959.

382. K. Toei and T. Korenaga, Analytical Chemistry and Microcomputers. III. Uses: Flow Injection Analysis [in Japanese]. *Bunseki,* **10** (1982) 746.

383. T. Uchida, C. S. Wei, C. Iida, and H. Wada, Simultaneous Determination of Calcium and Magnesium in Serum with Flow Injection-Atomic Absorption System [in Japanese]. *Nagoya Kogyo Daisako Gakuho,* **33** (1982) 97.

384. H. Ma and H. Yan, Application of Flow Injection Analysis to Environmental Monitoring [in Chinese]. *Huanjing Kexue,* **4** (1983) 59.

385. S. Greenfield, Inductively Coupled Plasma-Atomic Emission Spectroscopy (ICP-AES) with Flow Injection Analysis (FIA). *Spectrochim. Acta,* **38 B** (1983) 93.

386. T. Yao and Y. Kobayashi, Amperometric Determination of Glucose, Galactose, Free Cholesterol and Choline in Serums by Flow Injection Analysis with Immobilized Enzyme Reactors [in Japanese]. *Bunseki Kagaku,* **32** (1983) 253.

387. T. Kojima, Y. Hara, and F. Morishita, Flow Injection Analysis Using Immobilized Enzyme Reagent. *Bunseki Kagaku,* **32** (1983) E101.

388. T. Yamane and H. A. Mottola, The Transient Oxidation of Brucine in Solution as a Tool for the Determination of Chromium(VI) and Brucine. *Anal. Chim. Acta,* **146** (1983) 181.

389. A. G. Fogg, A. Y. Chamsi, and M. A. Abdalla, Flow Injection Voltammetric Determination of Nitrate after Reduction to Nitrite. *Analyst,* **108** (1983) 464.

390. T. P. Lynch, A. F. Taylor, and J. N. Wilson, Fully Automatic Flow Injection System for the Determination of Uranium at Trace Levels in Ore Leachates. *Analyst,* **108** (1983) 470.

391. J. C. de Andrade, J. C. Rocha, C. Pasquini, and N. Baccan, Effect of On-Line Complex Formation Kinetics on the Flow Injection Analysis Signal. The Spectrophotometric Determination of Chromium(VI). *Analyst,* **108** (1983) 621.

392. C. Pasquini, W. A. de Oliveira, and C. Pasquini, Direct Reading of Signals Obtained in Flow Injection Analysis. *Quim. Nova,* **5** (1982) 51.

393. C. B. Ranger, Flow Injection Analysis. A New Approach to Near-Real-

Time Process Monitoring. *Autom. Stream Anal. for Process Control,* **1** (1982) 39.

394. C. Riley, L. H. Aslett, B. F. Rocks, R. A. Sherwood, J. D. M. Watson, and J. Morgon, Controlled Dispersion Analysis: Flow Injection Analysis Without Injection. *Clin. Chem.,* **29** (1983) 332.
395. B. F. Rocks, R. A. Sherwood, Z. J. Turner, and C. Riley, Serum Iron and Total Iron-Binding Capacity Determination by Flow Injection Analysis with Atomic Absorption Detection. *Ann. Clin. Biochem.,* **20** (1983) 72.
396. B. F. Rocks, R. A. Sherwood, and C. Riley, Comment: More on Flow Injection/Atomic Absorption Analysis for Electrolytes. *Clin. Chem.,* **29** (1983) 569.
397. J. L. Burguera, M. Burguera, M. Gallignani, and O. M. Alarcon, More on Flow Injection//Atomic Absorption Analysis for Electrolytes. *Clin. Chem.,* **29** (1983) 568.
398. R. Karliček, Device for Flow Injection Analysis [in Czech]. *Chem. Listy,* **77** (1983) 100.
399. K. Honda, J. Sekino, and K. Imai, Bis(2,4-Dinitrophenyl) Oxalate as a Chemiluminescence Reagent in Determination of Fluorescent Compounds by Flow Injection Analysis. *Anal. Chem.,* **55** (1983) 940.
400. D. J. Leggett, N. H. Chen, and D. S. Mahadevappa, A Flow Injection Method for Analysis of Residual Chlorine by the DPD Procedure. *Fresenius Z. Anal. Chem.,* **315** (1983) 47.
401. L. Sun, Z. Sun, L. Li, and Z. Fang, The Determination of Soil Available Boron with Azomethine-H by a Flow Injection Spectrophotometric Method [in Chinese]. *Turang Tongbao,* **5** (1983) 41.
402. P. Marstorp, T. Anfält, and L. Andersson, Determination of Oxidized Ketone Bodies in Milk by Flow Injection Analysis. *Anal. Chim. Acta,* **149** (1983) 281.
403. H. Wada, A. Yuchi, and G. Nakagawa, Spectrophotometric Determination of Magnesium by Flow Injection Analysis with a Ligand Buffer for Masking Calcium. *Anal. Chim. Acta,* **149** (1983) 291.
404. U. J. Krull and M. Thompson, Flow Injection Analysis with a Bilayer Lipid Membrane Detector. *Trends Anal. Chem.,* **2** (1983) 6.
405. A. K. Covington and A. Sibbald, Offset-Gate Chemical-Sensitive Field Effect Transistor with Electrolytically Programmable Selectivity. Eur. Pat. Appl. No. 821124, (1982).
406. T. Hara, M. Toriyama, and K. Tsukagoshi, Determination of a small Amount of Biological Constituent by Use of Chemiluminescence. 1. The Flow Injection Analysis of Protein. *Bull. Chem. Soc. Jpn.,* **56** (1983) 1382.
407. G. B. Marshall and D. Midgley, Potentiometric Determination of Sulphite by Use of Mercury(I) Chloride-Mercury(II) Sulphide Electrodes in Flow Injection Analysis and in Air-Gap Electrodes. *Analyst,* **108** (1983) 701.
408. J. F. van Staden, Determination of Creatinine in Urine and Serum by Flow

Injection Analysis Using the Jaffe Reaction. *Fresenius Z. Anal. Chem.,* **315** (1983) 141.

409. B. Rössner and G. Schwedt, Methods for the Determination of Inorganic Anions. 1. Photometric Trace Analysis of Chloride in Air, Water and Technical Products in the Fe(II)/Hg-TPTZ-System. Manual Continuous-Flow and Flow Injection Techniques. *Fresenius Z. Anal. Chem.,* **315** (1983) 197.
410. A. G. Fogg, M. A. Ali, and M. A. Abdalla, On-Line Bromimetric Determination of Phenol, Aniline, Aspirin and Isoniazid Using Flow Injection Voltammetry. *Analyst,* **108** (1983) 840.
411. M. Tamano and J. Keketsu, Flow Injection Determination of ppt Levels of Cobalt(II) in Water by the Use of Contact Catalytic Reaction [in Japanese]. *Nippon Kagaku Kaishi,* **7** (1983) 1023.
412. J. Wang and B. A. Freiha, Selective Voltammetric Detection Based on Adsorptive Preconcentration for Flow Injection Analysis. *Anal. Chem.,* **55** (1983) 1285.
413. D. Betteridge, T. J. Sly, A. P. Wade, and J. E. W. Tillman, Computer-Assisted Optimization for Flow Injection Analysis of Isoprenaline. *Anal. Chem.,* **55** (1983) 1292.
414. O. Åström, New Approaches to Analytical Methods for Bismuth, Water, Acids and Bases Using Flow Injection Analysis. Univ. of Umeå, Sweden (1983). (Ph.D. Thesis).
415. F. J. Krug, B. F. Reis, M. F. Giné, E. A. G. Zagatto, A. O. Jacintho, and J. R. Ferreira, Zone Trapping in Flow Injection Analysis. Spectrophotometric Determination of Low Levels of Ammonium Ion in Natural Waters. *Anal. Chim. Acta,* **151** (1983) 39.
416. M. Trojanowicz and W. Matuszewski, Potentiometric Flow Injection Determination of Chloride. *Anal. Chim. Acta,* **151** (1983) 77.
417. H. Kimura, K. Oguma, and R. Kuroda, Atomic Absorption Spectrophotometric Determination of Calcium in Silicate Rocks by a Flow Injection Method [in Japanese]. *Bunseki Kagaku,* **32** (1983) 179.
418. B. Karlberg, Automation of Wet Chemical Procedures Using FIA. *Am. Lab.,* **15** (1983) 73.
419. H. Ma and H. Yan, Study on Application of Flow Injection Analysis (FIA) with an Amperometric Detector—Determination of Silver Ion in Wastewaters by FIA [in Chinese]. *Huanjing Huaxue,* **1** (1982) 422.
420. T. Yamane, Determination of GOT and GPT in Blood Serum by Flow Injection Analysis Using Pyrovate Oxidase [in Japanese]. *Mem. Fac. Liberal Arts Educ. (Univ. of Yamanashi),* **32** (1981) 52.
421. O. Åström, Analysis of Substances which Form Volatile Hydrides. Brit. UK Pat. Appl. No. 821208, (1982). Pat. No. 2,099,579.
422. R. Q. Thomson, The Characterization of Nylon Open-Tubular Immobilized Enzyme Reactors Incorporated in Stopped-Flow and Continuous Flow Systems. *Diss. Abstr. Int. B,* **43** (1983) 2893.

423. S. Motomizu, Fundamental Study of the Determination of Iron and Cobalt Using 2-Nitroso-5-Dimethylaminophenol as Colour Reagent by a Continuous Flow Stream Propelled by Gas Pressure [in Japanese]. *Bunseki Kagaku,* **32** (1983) 191.

424. S. Xu and Z. Fang, Sinultaneous Spectrophotometric Determination of Nitrate and Nitrite in Water and Soil Extracts by Flow Injection Analysis [in Chinese]. *Fenxi Huaxue,* **11** (1983) 93.

425. J. F. van Staden, Simultaneous Determination of Protein (Nitrogen), Phosphorous and Calcium in Animal Feed Stuffs by Multichannel Flow Injection Analysis. *J. Assoc. Anal. Chem.,* **66** (1983) 718.

426. J. Karlsson, I. Jacobs, B. Sjödin, P. Tesch, P. Kaiser, O. Sahl, and B. Karlberg, Semi-Automatic Blood Lactate Assay. Experiences from an Exercise Laboratory. *Int. J. Sports Med.,* **4** (1983) 52.

427. K. K. Stewart, Flow Injection Analysis—New Tool for Old Assays—New Approach to Analytical Measurements. *Anal. Chem.,* **55** (1983) 931A.

428. S. Olsen, L. C. R. Pessenda, J. Růžička, and E. H. Hansen, Combination of Flow Injection Analysis with Flame Atomic Absorption Spectrophotometry. Determination of Trace Amounts of Heavy Metals in Polluted Seawater. *Analyst,* **108** (1983) 905.

429. T. E. Edmonds and G. Coutts, Flow Injection Analysis System for Determining Soil pH. *Analyst,* **108** (1983) 1013.

430. W. E. van der Linden, Membrane Separation in Flow Injection Analysis. Gas Diffusion. *Anal. Chim. Acta,* **151** (1983) 359.

431. M. Valcarcel and M. D. Luque de Castro, Flow Injection Analysis. An Important Methodogic Invention in Analytical Chemistry [in Spanish]. *Quim. Anal.,* **1** (1983) 201.

432. N. Yoza, H. Hirano, M. Okamura, S. Ohashi, Y. Hirai, and K. Tomokuni, Measurement of Enzymatic Activity of Inorganic Pyrophosphatase for Pyrophosphate by Flow Injection Analysis. *Chem. Lett.,* **9** (1983) 1433.

433. T. Odashima, New Analytical Methods for Phosphorous [in Japanese]. *Mizu Shori Gijutsu,* **23** (1982) 1063.

434. H. Baadenhuijsen and T. Zelders, The Use of a Microcomputer System for Peak Recognition, Data Processing and Representation in Continuous Flow Analysis. *J. Autom. Chem.,* **5** (1983) 18.

435. E. H. Hansen and J. Růžička, Apparatus for Flow Injection Analysis. Brit. UK Pat. Appl. No. 830309, (1983). Pat. No. 2,104,657.

436. N. D. Byington, Flow Injection Atomic Absorption Assay of Copper and Zinc in the Plasma of Age Dependent Audiogenic Seizure Susceptible Mice. *Diss. Abstr. Int. B,* **43** (1983) 3228.

437. T. Mise, Determination of Anionic Surfactants by Flow Injection Analysis Based on Solvent Extraction [in Japanese]. *Miyakojo Kogyo Koto Semmon Gakko Kenkyo Hokoku,* **17** (1983) 7.

438. M. Hallas, Flow Injection Analysis—A New Environmentally Safe Ana-

lytical System for Future Laboratories [in Danish]. *Dan. Kemi,* **64** (1983) 4.

439. S. Olsen, Trace Analysis of Heavy Metals in Seawater. Flow Injection Analysis and Atomic Absorption Spectrophotometry [in Danish]. *Dan. Kemi,* **64** (1983) 68.
440. E. H. Hansen and J. Ruzicka, FIA is Already a Routine Tool in Brazil. *Trends Anal. Chem.,* **2** (1983) 5.
441. J. Růžička, Flow Injection Analysis—From Test Tube to Integrated Microconduits. *Anal. Chem.,* **55** (1983) 1040A.
442. R. A. Leach, J. Růžička, and J. M. Harris, Spectrophotometric Determination of Metals at Trace Levels by Flow Injection and Series Differential Detection. *Anal. Chem.,* **55** (1983) 1669.
443. A. P. Wade, Computer Assisted Optimisation of Chemical Systems, in Particular Flow Injection Analysis. *Anal. Proc.,* **20** (1983) 108.
444. C. B. Elliott, Application of Flow-Injection Analysis to Enzymatic Fluorescence Kinetic Methods. *Diss. Abstr. Int. B,* **43** (1983) 3966.
445. K. Ogata, S. Tanabe, and T. Imanari, Flame Atomic Absorption Spectrophotometry Coupled with Solvent Extraction/Flow Injection Analysis. *Chem. Pharm. Bull.,* **31** (1983) 1419.
446. M. E. Meyerhoff and P. M. Kovach, An Ion Selective Electrode/Flow Injection Analysis Experiment. Determination of Potassium in Serum. *J. Chem. Educ.,* **60** (1983) 766.
447. L. Fossey and F. F. Cantwell, Simultaneous Monitoring of Both Phases in the Solvent Extraction/Flow Injection Analysis of Dramamine Tablets. *Anal. Chem.,* **55** (1983) 1882.
448. J. J. Donkerbroek, A. C. Veltkamp, C. Gooijer, N. H. Velthorst, and R. W. Frei, Quenched Room Temperature Phosphorescence Detection for Flow Injection Analysis and Liquid Chromatography. *Anal. Chem.,* **55** (1983) 1886.
449. N. Thøgersen, J. Janata, and J. Růžička, Flow Injection Analysis and Cyclic Voltammetry. *Anal. Chem.,* **55** (1983) 1986.
450. E. H. Hansen, The Early History of Flow Injection Analysis. *In Focus,* **6** (1983) 12.
451. B. Karlberg and S. Twengström, Applications Based on Gas Diffusion and Flow Injection Analysis. *In Focus,* **6** (1983) 14.
452. H. Ma, L. Jin, and H. Yan, Flow Injection Analysis of Traces of Free Cyanide in Surface and Ground Waters with an Amperometric Flow-Through Detector [in Chinese]. *Kexue Tongbao,* **28** (1983) 1145.
453. H. Ma and H. Yan, An Amperometric Detector for Flow Injection Analysis [in Chinese]. *Yigi Yibiao Xuebao,* **4** (1983) 44.
454. L. Wang, S. Zhang, Y. Zhu, and Q. Wang, Rapid Determination of Sulphide in Waters by Flow Injection Analysis [in Chinese]. *Huanjing Huaxue,* **2** (1983) 64.

455. H. Ma, L. Jin, and H. Yan, Studies on the Application of an Amperometric Gold Tube Electrode Flow Through Detector. II. The Determination of Micro Quantities of Cyanide in Surface Waters [in Chinese]. *Huanjing Huaxue,* **2** (1983) 58.

456. K. Kina, Flow Injection Analysis. A Review [in Japanese]. *Dojin,* **25** (1982) 9.

457. J. Mortatti, F. J. Krug, and H. Bergamin F$^{\underline{o}}$, Spectrophotometric Determination of Lead in Natural Waters and Plant Material by Flow Injection Analysis [in Portuguese]. *Energ. Nucl. Agric. (Piracicaba),* **4** (1982) 82.

458. N. Zhou, W. Frech, and E. Lundberg, Rapid Determination of Lead, Bismuth, Antimony and Silver in Steels by Flame Atomic Absorption Spectrometry Combined with Flow Injection Analysis. *Anal. Chim. Acta,* **153** (1983) 23.

459. R. C. Schothorst and G. den Boef, The Application of Strongly Reducing Agents in Flow Injection Analysis. Part 2. Chromium(II). *Anal. Chim. Acta,* **153** (1983) 133.

460. T. Yao, Flow Injection Analysis for Cholinesterase in Blood Serum by Use of a Choline-Sensitive Electrode as an Amperometric Detector. *Anal. Chim. Acta,* **153** (1983) 169.

461. H. Wade, G. Nakagawa, and K. Ohshita, Spectrophotometric Determination of Traces of Iron with 2-(3,5-Dibromo-2-Pyridylazo)-5-[*n*-Ethyl-*n*-(3-Sulfopropyl)Amino]Phenol and its Application in Flow Injection Analysis. *Anal. Chim. Acta,* **153** (1983) 199.

462. J. L. Burguera and M. Burguera, Determination of Cadmium in Human Urine by Extraction with Dithizone in a Flow Injection System. *Anal. Chim. Acta,* **153** (1983) 207.

463. G. C. M. Bourke, G. Stedman, and A. P. Wade, The Spectrophotometric Determination of Hydroxylamine Alone and in the Presence of Hydrazine by Flow Injection Analysis. *Anal. Chim. Acta,* **153** (1983) 277.

464. J. Wang and H. D. Dewald, Potential Scanning Voltametric Detection for Flow Injection Systems. *Anal. Chim. Acta,* **153** (1983) 325.

465. T. Yao, Y. Kobayashi, and M. Sato, Amperometric Determination of Phospholipids in Blood Serum with a Lecithin-Sensitive Electrode in a Flow Injection System. *Anal. Chim. Acta,* **153** (1983) 337.

466. A. G. Fogg, G. C. Cripps, and B. J. Birch, Static and Flow Injection Voltammetric Determination of Total Phosphate and Soluble Silicate in Commercial Washing Powders at a Glassy Carbon Electrode. *Analyst,* **108** (1983) 1485.

467. M. O. Babiker and J. A. W. Dalziel, Studies on the Determination of Sulphide Using *n,n*-Diethyl-*p*-Phenylenediamine. *Anal. Proc.,* **20** (1983) 609.

468. S. Motomizu, M. Oshima, and K. Toei, Fluorimetric Determination of Boron with Chromotropic Acid by Continuous Flow System [in Japanese]. *Bunseki Kagaku,* **32** (1983) 458.

469. C. J. Patton, Design, Characterization and Applications of a Miniature Continuous Flow Analysis System. *Diss. Abstr. Int. B,* **44** (1983) 788.

470. H. Wada, T. Ishizuki, and G. Nakagawa, Synthesis of 2-(2-Thiazolylazo)-4-Methyl-5-(Sulfopropylamino)Benzoic Acid and the Application to the Flow Injection Analysis of Copper(II). *Mikrochim. Acta,* **3** (1983) 235.

471. J. C. Andrade, C. Pasquini, N. Baccan, and J. C. van Loon, Cold Vapour Atomic Absorption Determination of Mercury by Flow Injection Analysis Using a Teflon Membrane Phase Separator Coupled to the Absorption Cell. *Spectrochim. Acta,* **38** (1983) 1329.

472. J. L. Burguera, M. Burguera, and M. Gallignani, Direct Determination of Sodium and Potassium in Blood Serum by Flow Injection and Atomic Absorption Spectrophotometry. *An. Acad. Bras. Cienc.,* **55** (1983) 209.

473. A. J. Frend, G. J. Moody, J. D. R. Thomas, and B. J. Birch, Flow Injection Analysis with Tubular Membrane Ion-Selective Electrodes in the Presence of Anionic Surfactants. *Analyst,* **108** (1983) 1357.

474. B. P. Bubnis, M. R. Straka, and G. E. Pacey, Metal Speciation by Flow Injection Analysis. *Talanta,* **30** (1983) 841.

475. H. Morita, T. Kimoto, and S. Shimomura, Flow Injection Analysis of Mercury/Cold Vapour Atomic Fluorescence Spectrophotometry. *Anal. Lett.,* **16** (1983) 1187.

476. M. H. Shah and J. T. Stewart, Amperometric Determination of Isoniazid in a Flowing Stream at the Glassy Carbon Electrode. *Anal. Lett.,* **16** (1983) 913.

477. J. Thomsen, K. S. Johnson, and R. L. Petty, Determination of Reactive Silicate in Sea Water by Flow Injection Analysis. *Anal. Chem.,* **55** (1983) 2378.

478. C. W. Bradberry and R. N. Adams, Flow Injection Analysis with an Ezyme Reactor Bed for Determination of Ascorbic Acid in Brain Tissue. *Anal. Chem.,* **55** (1983) 2439.

479. J. Harrow and J. Janata, Comparison of Sample Injection Systems for Flow Injection Analysis. *Anal. Chem.,* **55** (1983) 2461.

480. H. Furuya and K. Nakayama, Flow Injection Analysis of Chloride Ion in Water Using Ion Selective Electrode [in Japanese]. *J. Japan Water Works Assoc.,* **52** (1983) 51.

481. K. S. Jonson and R. L. Petty, Determination of Nitrate and Nitrite in Seawater by Flow Injection Analysis. *Limnol. Oceanogr.,* **28** (1983) 1260.

482. C. D. C. Painton, Chemical Contributions to Dispersion. Their Analytical Impact in Flow Injection Sample Processing Systems. *Diss. Abstr. Int. B,* **44** (1983) 788.

483. D. E. Weisshaar, Application of Kelgraph to Electrochemical Detectors for Flow Injection Analysis and High Performance Liquid Chromatography. *Gov. Rep. Announce. Index (U.S.),* **83**(20) (1983) 4961.

484. C. Macckoya, F. Mizuniwa, K. Usami, and K. Osumi, Flow Injection De-

termination of Parts per Trillion Levels of Cobalt(II) in Water by the Use of Contact Catalytic Reaction [in Japanese]. *Nippon Kagaku Kaishi,* **7** (1983) 1023.

485. M. Tachibana, T. Imamura, M. Saito, and K. Kina, Rapid Determination of Copper in Serum by Flow Injection Analysis with Water Soluble Azo Dye [in Japanese]. *Bunseki Kagaku,* **32** (1983) 776.
486. N. H. Chen, Applications of Flow Injection Methods of Analysis. *Diss. Abstr. Int. B,* **43** (1983) 3581.
487. T. P. Tougas, An Electrochemical Detector for High Performance Liquid Chromatography and Flow Injection Analysis Based on a Reticulated Vitreous Carbon Working Electrode. *Diss. Abstr. Int. B,* **43** (1983) 3971.
488. P. J. Worsfold, Introduction to Flow Injection Analysis. Collective Summary of Five Papers Presented at a Meeting of the North East Region of SAC, December 8, 1982, in Sheffield. *Anal. Proc.,* **20** (1983) 486.
489. J. F. Tyson, Flow Injection Methods and Atomic Absorption Spectrophotometry. *Anal. Proc.,* **20** (1983) 488.
490. R. Eggli and R. Asper, Electrochemical Flow-Through Detector for the Determination of Cysteine and Related Compounds. *Anal. Chim. Acta,* **101** (1978) 253.
491. J. Wang and B. A. Freiha, Flow Electrolysis at a Porous Tubular Electrode with Internal Stirring. *Anal. Chim. Acta,* **151** (1983) 109.
492. W. L. Caudill, A. G. Ewing, S. Jones, and R. M. Wightman, Liquid Chromatography with Rapid Scanning Electrochemical Detection at Carbon Electrodes. *Anal. Chem.,* **55** (1983) 1877.
493. P. Roehrig, C. M. Wolff, and J. P. Schwing, Repetitive Enzymatic Determination of Glucose with Regeneration and Recycling of Coenzyme and Enzymes. *Anal. Chim. Acta,* **153** (1983) 181.
494. M. van Son, R. C. Schothorst, and G. den Boef, Determination of Total Ammoniacal Nitrogen in Water by Flow Injection Analysis and a Gas Diffusion Membrane. *Anal. Chim. Acta,* **153** (1983) 271.
495. T. A. H. M. Janse, P. F. A. van der Wiel, and G. Kateman, Experimental Optimization Procedures in the Determination of Phosphate by Flow Injection Analysis. *Anal. Chim. Acta,* **155** (1983) 89.
496. M. F. Giné, B. F. Reis, E. A. G. Zagatto, F. J. Krug, and A. O. Jacintho, A Simple Procedure for Standard Additions in Flow Injection Analysis. Spectrophotometric Determination of Nitrate in Plant Extracts. *Anal. Chim. Acta,* **155** (1983) 131.
497. O. F. Kamson and A. Townshend, Ion Exchange Removal of Some Interferences on the Determination of Calcium by Flow Injection Analysis and Atomic Absorption Spectrometry. *Anal. Chim. Acta,* **155** (1983) 253.
498. M. Yamada, T. Komatsu, S. Nakahara, and S. Suzuki, Improved Chemiluminescence Determination of Traces of Cobalt(II) by Continuous Flow and Flow Injection Methods. *Anal. Chim. Acta,* **155** (1983) 259.

499. S. Nakashima, M. Yagi, M. Zenki, A. Takahashi, and K. Toei, Spectrophotometric Determination of Nitrite in Natural Waters by Flow Injection Analysis. *Anal. Chim. Acta,* **155** (1983) 263.

500. W. E. van der Linden, The Optimum Composition of pH Sensitive Acceptor Solutions for Membrane Separation in Flow Injection Analysis. *Anal. Chim. Acta,* **155** (1983) 273.

501. J. M. Reijn, Flow Injection Analysis. Univ. of Amsterdam, Holland (1981). (Ph.D. Thesis).

502. S. M. Wolfrum, P. F. A. van der Wiel, and P. C. Thijssen, A Computer Controlled System for Automated Flow Injection Analysis. *Lab. Microcomputer,* **2** (1983) 4.

503. J. Wang and B. A. Freiha, Preconcentration and Differential Pulse Voltammetry of Butylated Hydroxyanisole at a Carbon Paste Electrode. *Anal. Chim. Acta,* **154** (1983) 87.

504. H. J. Wieck, Characterization of Immobilized Enzyme Chemically Modified Electrodes and Their Application in Flow Injection Analysis. *Diss. Abstr. Int. B.,* **44** (1983) 1449.

505. J. N. Miller, Flow Injection Analysis–Flexible and Convenient Automation. *Lab. Pract.,* Nov. (1983) 13.

506. A. P. Wade, Optimisation of Flow Injection Analysis and Polarography by the Modified Simplex Method. *Anal. Proc.,* **20** (1983) 523.

507. H. W. van Rooijen and H. Poppe, Noise and Drift Phenomena in Amperometric and Coulometric Detectors for HPLC and FIA. *J. Liq. Chromatogr.,* **6** (1983) 2231.

508. E. A. Jones, The Determination by Flow Injection Analysis of Iron, Sulphate, Silver and Cadmium. *Techn. Rep. Mintek.,* **M111** (1983) 32.

509. K. Kusube, K. Abe, O. Hiroshima, Y. Ishiguro, S. Ishikawa, and H. Hoshida, Electrochemical Derivatization of Thiamine in a Flow Injection System—Application to Thiamine Analysis. *Chem. Pharm. Bull.,* **31** (1983) 3589.

510. A. Schelter-Graf, H. Huck, and H. L. Schmidt, A Rapid and Accurate Determination of Ethanol Using an Oxidase-Electrode in a Flow Injection System. *Z. Lebensm. Unters. Forsch.,* **177** (1983) 356.

511. S. Nakashima, M. Yagi, M. Zenki, M. Doi, and K. Toei, Determination of Sulphate in Natural Water by Flow Injection Analysis. *Fresenius Z. Anal. Chem.,* **317** (1984) 29.

512. A. G. Fogg and N. K. Bsebsu, Sequential Flow Injection Voltammetric Determination of Phosphate and Nitrite by Injection of Reagents into a Sample Stream. *Analyst,* **109** (1984) 19.

513. J. F. Tyson and A. B. Idris, Determination of Chromium in Steel by Flame Atomic Absorption Spectrometry Using a Flow Injection Standard Additions Method. *Analyst,* **109** (1984) 23.

514. D. Betteridge, N. G. Courtney, T. J. Sly, and D. G. Porter, Development

of a Flow Injection Analyser for the Post-Column Detection of Sugars Separated by High-Performance Liquid Chromatography. *Analyst,* **109** (1984) 91.

515. J. Wang and H. D. Dewald, Subtractive Anodic Stripping Voltammetry with Flow Injection Analysis. *Anal. Chem.,* **56** (1984) 156.
516. G. Decristoforo and B. Danielsson, Flow Injection Analysis with Enzyme Thermistor Detector for Automated Determination of β-Lactams. *Anal. Chem.,* **56** (1984) 263.
517. J. T. Vanderslice, G. R. Beecher, and A. G. Rosenfeld, Determination of 1st Order Reaction Rate Constants by Flow Injection Analysis. *Anal. Chem.,* **56** (1984) 268.
518. K. E. Lawrence, G. W. Rice, and V. A. Fassel, Direct Liquid Sample Introduction for Flow Injection Analysis and Liquid Chromatography with Inductively Coupled Argon Plama Spectrometric Detection. *Anal. Chem.,* **56** (1984) 289.
519. J. T. Vanderslice, G. R. Beecher, and A. G. Rosenfeld, Dispersion and Diffusion Coefficients in Flow Injection Analysis. *Anal. Chem.,* **56** (1984) 292.
520. D. Zöltzer and G. Schwedt, Comparison of Continuous Flow (CFA) and Flow Injection (FIA) Techniques for the Photometric Determination of Traces of Aluminium in Water and Soil Samples. *Fresenius Z. Anal. Chem.,* **317** (1984) 422.
521. G. Schulze, M. Husch, and W. Frenzel, Flow Injection Potentiometric Stripping Analysis and Potentiometric Stripping Coulometry. *Mikrochim. Acta,* **1** (1984) 191.
522. N. Yoza, T. Miyaji, Y. Hirai, and S. Ohashi, Determination of Complexing Abilities of Ligands for Metal Ions by Flow Injection Analysis and High Performance Liquid Chromatography. I. Principles of the Substitution Method. *J. Chromatogr.,* **283** (1984) 89.
523. H. Imai, H. Yoshida, T. Masujima, and T. Owa, Determination of Indole Derivatives by Flow Injection Method with Chemiluminescence Detection [in Japanese]. *Bunseki Kagaku,* **33** (1984) 110.
524. T. Yamane and M. Kamijo, Determination of Water Hardness by Flow Injection Spectrophotometry [in Japanese]. *Bunseki Kagaku,* **33** (1984) 110.
525. T. Zaitsu, M. Maehara, and K. Toei, Flow Injection Analysis by Using Turbidimetry for Chloride in River Water [in Japanese]. *Bunseki Kagaku,* **33** (1984) 149.
526. J. F. Tyson, Extended Calibration of Flame Atomic Absorption Instruments by a Flow Injection Peak Width Method. *Analyst,* **109** (1984) 319.
527. K. Bäckström, L. G. Danielsson, and L. Nord, Sample Work-Up for Graphite Furnace Atomic Absorption Spectrometry Using Continuous Flow Extraction. *Analyst,* **109** (1984) 323.
528. C. W. McLeod, P. J. Worsfold, and A. G. Cox, Simultaneous Multi-Ele-

ment Analysis of Blood Serum by Flow Injection/Inductively Coupled Plasma Atomic Emission Spectrometry. *Analyst,* **109** (1984) 327.

529. F. L. Boza, M. D. Luque de Castro, and M. Valcarcel Cases, Catalytic-Fluorimetric Determination of Copper at the Nanograms per Millilitre Level by Flow Injection Analysis. *Analyst,* **109** (1984) 333.
530. P. J. Worsfold and A. Hughes, A Model Immunosassay Using Automated Flow Injection Analysis. *Analyst,* **109** (1984) 339.
531. M. Bos, J. H. H. G. van Willigen, and W. E. van der Linden, Flow Injection Analysis with Tensammetric Detection for the Determination of Detergents. *Anal. Chim. Acta,* **156** (1984) 71.
532. C. Pasquini and W. A. de Oliveira, Comparison of Merging Zones, Injection of Reagent and Single-Line Manifolds for Enthalpimetric Flow Injection Analysis. *Anal. Chim. Acta,* **156** (1984) 307.
533. E. A. Jones, Spectrophotometric Determination of Sulphate in Sodium Hydroxide Solutions by Flow Injection Analysis. *Anal. Chim. Acta,* **156** (1984) 313.
534. D. MacKoul, D. C. Johnson, and K. G. Schick, Effect of Variation in Flow Rate on Amperometric Detection in Flow Injection Analysis. *Anal. Chem.,* **56** (1984) 436.
535. A. S. Attiyat and G. D. Christian, Nonaqueous Solvents as Carrier or Sample Solvent in Flow Injection Analysis/Atomic Absorption Spectrometry. *Anal. Chem.,* **56** (1984) 439.
536. D. J. Curran and T. P. Tougas, Electrochemical Detector Based on Reticulated Vitreous Carbon Working Electrode for Liquid Chromatography and Flow Injection Analysis. *Anal. Chem.,* **56** (1984) 672.
537. H. Müller and G. Wallaschek, Determination of Water in Organic Solvents by Means of the Karl Fischer Reagent with Flow Injection Analysis [in German]. *Z. Chem.,* **24** (1984) 75.
538. J. L. Burguera and M. Burguera, Determination of Sulphur Anions by Flow Injection with a Molecular Emission Cavity Detector. *Anal. Chim. Acta,* **157** (1984) 177.
539. M. Thompson, U. J. Krull, and L. Bendell-Young, Biosensors and Their Uses in Flow Injection Systems. *Anal. Proc.,* **20** (1984) 568.
540. M. D. Luque de Castro and M. Valcarcel Cases, Simultaneous Determinations in Flow Injection Analysis. A Review. *Analyst,* **109** (1984) 413.
541. C. C. Painton and H. A. Mottola, Kinetics in Continuous Flow Sample Processing. Chemical Contributions to Dispersion in Flow Injection Techniques. *Anal. Chim. Acta,* **158** (1984) 67.
542. H. J. Wieck, G. H. Heider, Jr., and A. M. Yacynych, Chemically Modified Reticulated Vitreous Carbon Electrode with Immobilized Enzyme as a Detector in Flow Injection Determination of Glucose. *Anal. Chim. Acta,* **158** (1984) 137.
543. M. Gallignani, J. L. Burguera, and M. Burguera, Determination of Calcium

and Magnesium in Blood Sera by Flow Injection Analysis and Atomic Absorption Spectrometry [in Spanish]. *Acta Cient. Venez.*, **33** (1982) 371.

544. S. Kato, M. Toyoshima, M. Washida, and K. Sagisaka, Flow Injection Determination of Phosphorous in Phosphorous Deoxidized Copper [in Japanese]. *Sumitomo Keikinzoku Giho,* **24** (1983) 108.
545. J. L. Burguera and M. Burguera, New Applications of Flow Injection Analysis in Analytical Chemistry [in Spanish]. *Acta Cient. Venez.*, **33** (1982) 375.
546. H. Lundbäck, G. Johansson, and O. Holst, Determination of Hydrogen Peroxide for Application in Aerobic Cell Systems Oxygenated via Hydrogen Peroxide. *Anal. Chim. Acta,* **155** (1983) 47.
547. T. Yao, N. Nakanishi, and T. Wasa, Flow Injection Analysis for Glucose with a Chemically Modified Enzyme Membrane Electrode [in Japanese]. *Bunseki Kagaku,* **33** (1984) 213.
548. R. Kuroda, I. Ida, and K. Oguma, Determination of Phosphorous in Silicate Rocks by Flow Injection Method of Analysis. *Mikrochim. Acta,* **1** (1984) 377.
549. T. Yamane, Catalytic Determination of Traces of Cobalt by the Protocatechuic Acid-Hydrogen Peroxide Reaction in a Flow Injection System. *Mikrochim. Acta,* **1** (1984) 425.
550. B. F. Rocks, R. A. Sherwood, and C. Riley, Direct Determination of Calcium and Magnesium in Serum Using Flow Injection Analysis and Atomic Absorption Spectroscopy. *Ann. Clin. Biochem.*, **21** (1984) 51.
551. B. P. Bubnis, Flow Injection Analysis and Functionalized Crown Ethers for Trace Metal Analysis. *Diss. Abstr. Int. B,* **44** (1984) 2413.
552. R. Smith, R. J. Huber, and J. Janata, Electrostatically Protected Ion-Sensitive Field Effect Transistors. *Sens. Actuators.*, **5** (1984) 127.
553. T. Hara, M. Toriyama, and K. Tsukagoshi, Determination of a Small Amount of a Biological Constituent by the Use of Chemiluminescence. II. Determination of Albumin as a Model Protein by Means of the Flow Injection Analysis Using a Cobalt(III) Complex Compound as a Catalyst. *Bull. Chem. Soc. Jpn.*, **57** (1984) 289.
554. J. M. Reijn, H. Poppe, and W. E. van der Linden, Kinetics in a Single Bead String Reactor for Flow Injection Analysis. *Anal. Chem.*, **56** (1984) 943.
555. J. C. de Andrade, J. C. Rocha, and N. Baccan, On-Line Oxidation of Cr(III) to Cr(VI) for Use with the Flow Injection Analysis Technique. *Analyst,* **109** (1984) 645.
556. C. C. Lee and B. D. Pollard, Determination of the Iodine Value of Fatty Acids by a Flow Injection Method. *Anal. Chim. Acta,* **158** (1984) 157.
557. P. W. Hansen, Determination of Fungal α-Amylase by Flow Injection Analysis. *Anal. Chim. Acta,* **158** (1984) 375.

558. J. Wang and H. D. Dewald, Background-Current Subtraction in Voltammetric Detection for Flow Injection Analysis. *Talanta,* **31** (1984) 387.

559. H. Müller and V. Müller, Principles and Applications of Flow Injection Analysis [in German]. *Z. Chem.,* **24** (1984) 81.

560. R. S. Brazell, R. W. Holmberg, and J. H. Moneyhun, Application of High-Performance Liquid Chromatography Flow Injection Analysis for the Determination of Polyphosphoric Acids in Phosphorous Smokes. *J. Chromatogr.,* **290** (1984) 163.

561. H. Müller, Determination of Iodide, Cyanide, Molybdenum(VI) and Tungsten(VI) with the Flow-Through Iodide-Selective Electrode Using the Flow Injection Method. *Anal. Chem. Symp. Ser.,* **18** (1984) 353.

562. A. S. Attiyat and G. D. Christian, Flow Injection Analysis—Atomic Absorption Determination of Serum Zinc. *Clin. Chim. Acta,* **137** (1984) 151.

563. S. Ikeda, H. Satake, and Y. Kohri, Flow Injection Analysis with an Amperometric Detector Utilizing the Redox Reaction of Iodate Ion. *Chem. Lett.,* **6** (1984) 873.

564. A. Fernandez, M. D. Luque de Castro and M. Valcarcel, Comparison of Flow Injection Analysis Configurations for Differential Kinetic Determination of Cobalt and Nickel. *Anal. Chem.,* **56** (1984) 1146.

565. Y. Israel and R. M. Barnes, Standard Addition Method in Flow Injection Analysis with Inductively Coupled Plasma Atomic Emission Spectrometry. *Anal. Chem.,* **56** (1984) 1188.

566. L. Nondek, Band Broadening in Solid-Phase Derivatization Reactions for Irreversible First-Order Reactions. *Anal. Chem.,* **56** (1984) 1192.

567. H. Engelhardt and R. Klinkner, Phosphate Determination by Flow Injection Analysis with Geometrically Deformed Open Tubes [in German]. *Fresenius Z. Anal. Chem.,* **317** (1984) 671.

568. H. Engelhardt and R. Klinkner, Phosphate Determination by Flow Injection Analysis Using Geometrically Deformed Open Tubes [in German]. *Fresenius Z. Anal. Chem.,* **319** (1984) 277.

569. A. Jensen and M. Hallas, Flow Injection Analysis. A New System of Analysis with Extensive Potential Applications for Pharmaceutically Relevant Analyses [in Danish]. *Farmaceut. Tidende.,* **26** (1984) 609.

570. H. Wada, G. Nakagawa, and K. Ohshita, Synthesis of *o,o'*-Dithydroxyazo Compounds and Their Application to the Determination of Magnesium and Calcium by Flow Injection Analysis. *Anal. Chim. Acta,* **159** (1984) 289.

571. M. Yamada and S. Suzuki, Micellar Enhanced Chemiluminescence of 1,10-Phenanthroline for the Determination of Ultratraces of Copper(II) by the Flow Injection Method. *Anal. Lett. A,* **17** (1984) 251.

572. P. W. Alexander, M. Trojanowicz, and P. R. Haddad, Indirect Potentiometric Determination of Metal Ions by Flow Injection Analysis with a Copper Electrode. *Anal. Lett. A,* **17** (1984) 309.

573. M. J. Medina, J. Bartroli, J. Alonso, M. Blanco, and J. Fuentes, Direct

Determination of Glucose in Blood Serum Using Trinder's Reaction. *Anal. Lett. B,* **17** (1984) 385.

574. K. M. Korfhage, K. Ravichandran, and R. P. Baldwin, Phthalocyanine Containing Chemically Modified Electrodes for Electrochemical Detection in Liquid Chromatography/Flow Injection Systems. *Anal. Chem.,* **56** (1984) 1514.
575. M. R. Straka, G. E. Pacey, and G. Gordon, Residual Ozone Determination by Flow Injection Analysis. *Anal. Chem.,* **56** (1984) 1973.
576. F. Malamas, M. Bengtsson, and G. Johansson, On-Line Trace Metal Enrichment and Matrix Isolation in Atomic Absorption Spectrometry by a Column Containing Immobilized 8-Quinolinol in a Flow Injection System. *Anal. Chim. Acta,* **160** (1984) 1.
577. C. Silfwerbrand-Lindh, L. Nord, L.-G. Danielsson, and F. Ingman, The Analysis of Aqueous Solutions with Ethanol Soluble Reagents in a Flow Injection System. Spectrophotometric Determination of Uranium. *Anal. Chim. Acta,* **160** (1984) 11.
578. T. P. Lynch, Determination of Free Cyanide in Mineral Leachates. *Analyst,* **109** (1984) 421.
579. T. P. Lynch, N. J. Kernoghan, and J. N. Wilson, Speciation of Metals in Solution by Flow Injection Analysis. Part 1. Sequential Spectrophotometric and Atomic Absorption Detectors. *Analyst,* **109** (1984) 839.
580. A. G. Fogg, A. Y. Chamsi, A. A. Barros, and J. O. Cabral, Flow Injection Voltammetric Determination of Hypochlorite and Hypobromite as Bromine by Injection into an Acidic Bromide Eluent and the Indirect Determination of Ammonia and Hydrazine by Reaction with an Excess of Hypobromite. *Analyst,* **109** (1984) 901.
581. T. P. Lynch, N. J. Kernoghan, and J. N. Wilson, Speciation of Metals in Solution by Flow Injection Analysis. Part 2. Determination of Iron(III) and Iron (II) in Mineral Process Liquors by Simultaneous Injection into Parallel Streams. *Analyst,* **109** (1984) 843.
582. A. G. Fogg and A. M. Summan, Simple Wall-Jet Detector Cell Holding Either a Solid Electrode or a Sessile Mercury-Drop Electrode and an Illustration of Its Use in the Oxidative and Reductive Flow Injection Voltammetric Determination of Food Colouring Matters. *Analyst,* **109** (1984) 1029.
583. M. W. Brown and J. Růžička, Parameters Affecting Sensitivity and Precision in the Combination of Flow Injection Analysis with Flame Atomic Absorption Spectrophotometry. *Analyst,* **109** (1984) 1091.
584. T. M. Rossi, D. C. Shelly, and I. M. Warner, Optimization of a Flow Injection Analysis System for Multiple Solvent Extraction. *Energy. Res. Abstr.,* **9** (1984) No. 8101.
585. H. Weicker, FIA (Flow Injection Analysis) Is Also for Lactate Determination. *Labor Praxis,* **8** (1984) 300.

586. C. Okumoto, M. Nagashima, S. Mizoiri, M. Kazama, and K. Akiyama, Flow Injection Analysis of Cyanide in Wastewater from Metal Plating Processes [in Japanese]. *Eisei Kagaku,* **30** (1984) 7.

587. T. Hara, M. Toriyama, and K. Tsukagoshi, Determination of a Small Amount of Biological Constituent by Use of Chemiluminescence. III. The Flow Injection Analysis of Protein by Direct Injection. *Bull. Chem. Soc. Jpn.,* **57** (1984) 1551.

588. W. R. Seitz and M. L. Grayeski, Flow Injection Analysis: A New Approach to Laboratory Automation. *J. Clin. Lab. Autom.,* **4** (1984) 169.

589. T. Chow, S. Yoshida, M. Itoh, S. Hirose, and T. Takeda, Determination of Reduced Type Nicotinamide Adenine Dinucleotide by Flow Injection Analysis Using an Immobilized Enzyme Voltammetry System [in Japanese]. *Bunseki Kagaku,* **33** (1984) 310.

590. N. Yoza, Flow Injection Analysis [in Japanese]. *Bunseki,* **7** (1984) 513.

591. Hitachi Chemical Co., Ltd, Reagent for Spectrophotometric Determination of Total Proteins. (In Japanese). Kokai Tokkyo Koho, Pat. Appl. No. 8483059, (1984).

592. H. A. Mottala, C. M. Wolf, A. Iob, and R. Gnanasekaran, Potentiometric and Amperometric Detection in Flow Injection Enzymatic Determination. *Anal. Chem. Symp. Ser.,* **18** (1984) 49.

593. W. Kemula, J. Debrowski, and W. Kutner, Flow Polarographic Detector. Pol. Pat. Appl. No. 831025, (1983). Pat. No. 119995.

594. F. Kikui and T. Hayakawa, Semiautomatic Method for Determination of Iron(II) and Iron(III) by Using Flow Injection and a Coulometric Monitor [in Japanese]. *Sumito Tokushu Kinzoku Giho,* **7** (1984) 33.

595. J. Wang and B. A. Freiha, Thin-Layer Flow Cell with a Rotating Disk Electrode. *J. Electroanal. Chem. Interfacial Electrochem.,* **164** (1984) 79.

596. H. Weicker, H. Hägele, B. Kornes, and A. Werner, Determination of Alanine, Lactate, Pyruvate, β-Hydroxybutyrate and Acetoacetate by Flow Injection Analysis (FIA). *Int. J. Sports Med.,* **5** (1984) 47.

597. M. H. Ho, Microprocessor-Controlled Flow Injection Analyzer for Biochemical Applications. *Biomed. Sci. Instrum.,* **20** (1984) 93.

598. R. A. Mowery, Jr., Flow Injection Analysis: Potential for Process Composition Measurement. *InTech.,* **31** (1984) 51.

599. R. Kuroda, Flow Injection Analysis [in Japanese]. *Gendai Kagaku,* **158** (1984) 48.

600. M. Tojanowicz, W. Augustyniak, and A. Hulanicki, Photometric Flow Injection Measurements with Flow Cell Employing Light Emitting Diodes. *Mikrochim. Acta,* **2** (1984) 17.

601. Y. Baba, N. Yoza, and S. Ohashi, Simultaneous Determination of Phosphate and Phosphonate by Flow Injection Analysis with Parallel Detection System. *J. Chromatogr.,* **295** (1984) 153.

602. T. Yao and T. Wasa, Rapid Determination of Cholinesterase Activity in Blood Serum [in Japanese]. *Bunseki Kagaku,* **33** (1984) 342.

603. T. Yamane, Flow Injection Determination of Hydrogen Peroxide by Means of the Manganese-Catalyzed Oxidation of Hydroxynaphthol Blue. *Bunseki Kagaku,* **33** (1984) E203.

604. N. Yoza, T. Shuto, Y. Baba, A. Tanaka, and S. Ohashi, Determination of Complexing Abilities of Ligands for Metal Ions by Flow Injection Analysis and High-Performance Liquid Chromatography. II. Copper(II) Complexes of Aminopolycarboxylic Acids. *J. Chromatogr.,* **298** (1984) 419.

605. K. Matsumoto, K. Ishida, T. Nomura, and Y. Osajima, Conductometric Flow Injection Analysis of the Organic Acid Content in Citrus Fruits. *Agric. Biol. Chem.,* **48** (1984) 2211.

606. S. M. Harden and W. K. Nonidez, Determination of Orthophosphate by Flow Injection Analysis with Amperometric Detection. *Anal. Chem.,* **56** (1984) 2218.

607. T. Braun and W. S. Lyon, The Epidemiology of Research on Flow Injection Analysis. Un Unconventional Approach. *Fresenius Z. Anal. Chem.,* **319** (1984) 74.

608. J. Růžička and E. H. Hansen, Integrated Microconduits for Flow Injection Analysis. *Anal. Chim. Acta,* **161** (1984) 1.

609. R. C. Schothorst, J. J. F. van Veen, and G. den Boef, The Application of Strongly Reducing Agents in Flow Injection Analysis. Part 3. Vanadium(II). *Anal. Chim. Acta,* **161** (1984) 27.

610. F. J. Krug, O. Bahia Fº, and E. A. G. Zagatto, Determination of Molybdenum in Steels by Flow Injection Spectrophotometry. *Anal. Chim. Acta,* **161** (1984) 245.

611. P. Linares, M. D. Luque de Castro, and M. Valcarcel, Spectrofluorimetric Flow Injection Determination of Cyanide. *Anal. Chim. Acta,* **161** (1984) 257.

612. R. R. Liversage, J. C. van Loon, and J. C. de Andrade, A Flow Injection/Hydride Generation System for the Determination of Arsenic by Inductively-Coupled Plasma Atomic Emission Spectrometry. *Anal. Chim. Acta,* **161** (1984) 275.

613. T. P. Tougas and D. J. Curran, Stopped-Flow Linear Sweep Voltammetry at the Reticulated Vitreous Carbon Electrode in a Flow Injection System. Determination of Dopamine in the Presence of Ascorbic Acid. *Anal. Chim. Acta,* **161** (1984) 325.

614. J. L. Burguera and M. Burguera, Flow Injection Spectrophotometry Followed by Atomic Absorption Spectrometry for the Determination of Iron(II) and Total Iron. *Anal. Chim. Acta,* **161** (1984) 375.

615. J. W. Keller, Enzyme Assay by Repetitive Flow Injection Analysis. Application to the Assay of Hog Kidney Aminoacylase. *Anal. Lett.,* **17** (1984) 589.

616. P. W. Alexander, P. R. Haddad, and M. Trojanowicz, Potentiometric Flow Injection Determination of Copper Complexing Inorganic Anions with a Copper Wire Indicator Electrode. *Anal. Chem.,* **56** (1984) 2417.

617. H. Narasaki and M. Ikeda, Automated Determination of Arsenic and Selenium by Atomic Absorption Spectrometry with Hydride Generation. *Anal. Chem.,* **56** (1984) 2059.

618. D. S. Austin, J. A. Polta, T. Z. Polta, A. P. C. Tang, T. D. Cabelka, and D. C. Johnson, Electrocatalysis at Platinum Electrodes for Anodic Electroanalysis. *J. Electroanal. Chem. Interfacial Electrochem.,* **168** (1984) 227.

619. J. F. van Staden and H. R. van Vliet, Flow Injection Analysis for Determining Total Alkalinity in Surface, Ground and Domestic Water Using the Automated Bromocresol Green Method [in Afrikaans]. *Water SA,* **10** (1984) 168.

620. J. A. Wise, Flow Injection Anodic Stripping Voltammetry. *Diss. Abstr. Int. B,* **45** (1984) 176.

621. H. Cui, Z. Zhu, and Z. Fang, Determination of Cyanide in Soil and Water by Flow Injection Analysis [in Chinese]. *Huanjing Huaxue,* **3** (1984) 48.

622. S. Zhang, L. Sun, H. Jiang, and Z. Fang, Determination of Copper, Zinc, Iron, Manganese, Sodium, Potassium and Magnesium in Plants and Soil by Flow Injection Analysis [in Chinese]. *Guangpruxe Yu Guangpu Fenxi,* **4** (1984) 42.

623. Hitachi Chemicals Co., Ltd, Reagent for Bilirubin Determination in Body Fluids [in Japanese]. Jpn. Kokai Tokkyo Koho, Pat. No. 84109863 (1984).

624. Hitachi Chemicals Co., Ltd, Reagent for Determination of Neutral Lipids [in Japanese]. Jpn. Kokai Tokkyo Koho, Pat. No. 84109196 (1984).

625. Hitachi Chemicals Co., Ltd, Reagent for Cholesterol Determination [in Japanese]. Jpn. Kokai Tokkyo Koho, Pat. No. 84109100 (1984).

626. Hitachi Chemicals Co., Ltd, Reagent for Glucose Determination [in Japanese]. Jpn. Kokai Tokkyo Koho, Pat. No. 84109197 (1984).

627. H. Cui and Z. Fang, Flow Injection Analysis of Iron in Soil Extracts [in Chinese]. *Fenxi Huaxue,* **12** (1984) 759.

628. Z. Fang, S. Xu, and S. Zhang, The Determination of Trace Amounts of Nickel by On-Line FIA Ion Exchange Preconcentration Atomic Absorption Spectrometry [in Chinese]. *Fenxi Huaxue,* **12** (1984) 997.

629. T. Aoki, S. Uemura, and M. Munemori, Flow Injection Analysis with Membrane Separation. Determination of Ammonia in Blood and Urine [in Japanese]. *Bunseki Kagaku,* **33** (1984) 505.

630. K. Matsumoto, H. Ukeda, and Y. Osajima, Flow Injection Analysis of Reduced Nicotinamide Adenine Dinucleotide Using β-Naphthoquinone-4-Sulfonate as a Mediator. *Agric. Biol. Chem.,* **48** (1984) 1879.

631. K. L. Lu and Y. M. Chen, Catalytic Photometric Method for the Determination of Molybdenum in Plants with Flow Injection Analysis [in Chinese]. *Proc. Natl. Sci. Counc. Repub. China, Part A,* **8** (1984) 85.

632. L. C. Davis and G. A. Radke, Chemically Coupled Spectrophotometric Assays Based on Flow Injection Analysis. Determination of Nitrogenase by Assays for Creatinine, Ammonia, Hydrazine, Phosphate and Dithionite. *Anal. Biochem.,* **140** (1984) 434.

633. Hitachi Chemicals Co., Ltd, Reagent for Uric Acid Determination [in Japanese]. Jpn. Kokai Tokkyo Koho, Pat. No. 84109200 (1984).

634. Q. Wei, Spectrophotometric Determination of Calcium by Flow Injection Analysis [in Chinese]. *Jilin Daxue Kexue Xuebao,* **3** (1984) 113.

635. M. Koupparis, P. Macheras, and C. Reppas, Application of Automated Flow Injection Analysis (FIA) to Dissolution Studies. *Int. J. Pharm.,* **20** (1984) 325.

636. G. E. Pacey and B. P. Bubnis, Flow Injection Analysis as a Tool for Metal Speciation. *Am. Lab.,* **16** (1984) 17.

637. R. C. Schothorst, M. van Son, and G. den Boef, The Application of Strongly Reducing Agents in Flow Injection Analysis. Part 4. Uranium(III). *Anal. Chim. Acta,* **162** (1984) 1

638. I. Nordin-Andersson, O. Åström, and A. Cedergren, Determination of Water by Flow Injection Analysis with the Karl Fischer Reagent. Minimization of Effects Caused by Differences in Physical Properties of the Samples. *Anal. Chim. Acta,* **162** (1984) 9.

639. J. Wang and H. D. Dewald, Theoretical and Experimental Aspects of the Response of Stripping Voltammetry in Flow Injection Systems. *Anal. Chim. Acta,* **162** (1984) 189.

640. A. Rios, M. D. Luque de Castro, and M. Valcarcel, New Approach to the Simultaneous Determination of Pollutants in Waste Waters by Flow Injection Analysis. Part I. Anionic Pollutants. *Analyst,* **109** (1984) 1487.

641. S. Nakashima, M. Yagi, M. Zenki, A. Takahashi, and K. Toei, Determination of Nitrate in Natural Waters by Flow Injection Analysis. *Fresenius Z. Anal. Chem.,* **319** (1984) 506.

642. R. A. Leach and J. M. Harris, Thermal Lens Absorption Measurements by Flow Injection into Supercritical Fluid Solvents. *Anal. Chem.,* **56** (1984) 2801.

643. B. C. Madsen and M. S. Kromis, Flow Injection and Photometric Determination of Hydrogen Peroxide in Rain Water with *n*-Ethyl-*n*-(Sulfopropyl)Aniline Sodium Salt. *Anal. Chem.,* **56** (1984) 2849.

644. Y. Yang and R. E. Hairrell, Single Laser Crossed Beam Thermal Lens Detection for Short Path Length Samples and Flow Injection Analysis. *Anal. Chem.,* **56** (1984) 3002.

645. C. Riley, B. F. Rocks, and R. A. Sherwood, Flow Injection Analysis in Clinical Chemistry. *Talanta,* **31** (1984) 879.

646. Z. Feher, G. Nagy, L. Bezur, J. Szovik, K. Toth, and E. Pungor, Use of a Novel Apparatus Based on the Injection Principle for Automatic Voltammetric Analysis. *Hung. Sci. Instrum.,* **45** (1979) 1.

647. H. F. R. Reijnders, J. F. van Staden, and B. Griepink, Flow Through Determination of Sulphate in Water Using Different Methods: A Comparison. *Fresenius Z. Anal. Chem.*, **300** (1980) 273.

648. J. F. Brown, K. K. Stewart, and D. Higgs, Microcomputer Control and Data System for Automated Multiple Flow Injection Analysis. *J. Autom. Chem.*, **3** (1981) 182.

649. L.-G. Ekström, An Automated Method for Determination of Free Fatty Acids. *J. Amer. Oil Chem. Soc.*, **58** (1981) 935.

650. W. Huber, Mechanization of Photometric Analysis with the Component Parts of Liquid Chromatography. Fast Determination of Formaldehyde [in German]. *Fresenius Z. Anal. Chem.*, **309** (1981) 386.

651. L. Nord and B. Karlberg, An Automated Extraction System for Flame Atomic Absorption Spectrometry. *Anal. Chim. Acta,* **125** (1981) 199.

652. M. Vandeputte, Fluor: Analyse en Voorkomen in Pollutieindikatoren [in Dutch]. Vrije Univ., Brussels, Belgium (1981/82). (Ph.D. Thesis).

653. N. Mertens, Fluor: Bepaling in Commerciele Dranken. Een Preliminaire Studie van de Fluoride Elektrode in een F.I.A. System [in Dutch]. Vrije Univ., Brussels, Belgium (1981/82). (M.Sc. Thesis).

654. B. Karlberg and J. Möller, Flow Injection Analysis—A New Technique for the Automation of Wet-Chemical Procedures. *In Focus,* **5** (1982) 5.

655. M. D. Lerique, F.I.A.—Or Injection into a Continuously Moving Stream [in French]. *Chemie Magazine,* **6** (1982) 49.

656. P. W. Alexander, R. J. Finlayson, L. E. Smythe, and A. Thalib, Rapid Flow Analysis with Inductively Coupled Plasma Atomic-Emission Spectroscopy Using a Micro-Injection Technique. *Analyst,* **107** (1982) 1335.

657. W. L. Caudill, J. O. Howell, and R. M. Wightman, Flow Rate Independent Amperometric Cell. *Anal. Chem.,* **54** (1982) 2532.

658. C. H. P. Bruins, D. A. Doornbos, and K. Brunt, The Hydrodynamics of the Amperometric Detector Flow Cell with a Rotating Disk Electrode. *Anal. Chim. Acta,* **140** (1982) 39.

659. K. Brunt, Comparison Between the Performances of an Electrochemical Detector Flow Cell in a Potentiometric and an Amperometric Measuring System Using Glucose as a Test Compound. *Analyst,* **107** (1982) 1261.

660. D. Pilosof and T. A. Nieman, Microporous Membrane Flow Cell with Non-immobilized Enzyme for Chemiluminescent Determination of Glucose. *Anal. Chem.,* **54** (1982) 1698.

661. T. Ito, E. Nakagawa, H. Kawaguchi, and A. Mizuike, Semi-Automatic Microlitre Sample Injection into an Inductively Coupled Plasma for Simultaneous Multielement Analysis. *Mikrochim. Acta,* **I** (1982) 423.

662. P. W. Alexander and A. Thalib, Nonsegmented Rapid-Flow Analysis with Ultraviolet/Visible Spectrophotometric Determination for Short Sampling Times. *Anal. Chem.,* **55** (1983) 497.

663. J. Růžička and E. H. Hansen, *Flow Injection Analysis* [in Japanese. Translated by N. Ishibashi and N. Yoza]. Kagakudonin, Kyoto, Japan (1983).

664. K. Ueno and K. Kina, *Introduction to Flow Injection Analysis. Experiments and Applications* [in Japanese]. Kodansha Scientific, Tokyo, Japan (1983).

665. M. Valcarcel Cases and M. D. Luque de Castro, *Análisis por Inyección en Flugo* [in Spanish]. Imprenta San Pablo, Cordoba, Spain (1984).

666. S. Motomizu, H. Mikasa, M. Oshima, and K. Toei, Continuous Flow Method for the Determination of Phosphorous Using the Fluorescence Quenching of Rhodamine 6G with Molybdophosphate [in Japanese]. *Bunseki Kagaku,* **33** (1984) 116.

667. D. Betteridge, Simulation and Modelling in Chemical Analysis. *Anal. Proc.,* **21** (1984) 139.

668. J. A. Wise, Flow Injection Anodic Stripping Voltammetry. *Diss. Abstr. Int. B,* **45** (1984) 176.

669. A. Rios, M. D. Luque de Castro, and M. Valcarcel, Spectrophotometric Determination of Cyanide by Unsegmented Flow Methods. *Talanta,* **31** (1984) 673.

670. K. Li and P. Hua, Flow Injection Catalytic Photometric Determination of Trace Iodine in Rock and Ore Samples [in Chinese]. *Yanshi Kuangwu Ji Ceshi,* **3** (1984) 314.

671. K. G. Schick, High-Speed, On-Stream Acid-Base Titration Utilizing Flow Injection Analysis. *Adv. Instrum.,* **39** (1984) 279.

672. T. Odashima and K. Satoh, A New Analysis for Sulphur Compounds [in Japanese]. *Mizu Shori Gijutsu,* **25** (1984) 647.

673. S. Xu and Z. Fang, Recent Developments in Flow Injection Analysis [in Chinese]. *Huaxue Tongbao,* **12** (1984) 22.

674. H. Ishida, M. Maeda, and A. Tsuji, Enzyme Immunoassay of α-Fetoprotein Based on Chemiluminescence Reaction Using a Flow Injection Analysis System [in Japanese]. *Rinsho Kagaku,* **13** (1984) 129.

675. T. Owa, T. Masujima, H. Yoshida, and H. Imai, Determination of Tetracycline in Plasma by Flow Injection Method with Chemiluminescence Detection [in Japanese]. *Bunseki Kagaku,* **33** (1984) 568.

676. H. Ma and H. Yan, Application of Amperometric Flow-Through Gold Tubular Detector. Determination of Trace Sulphide Ion by Flow Injection Analysis [in Chinese]. *Huanjing Huaxue,* **3** (1984) 75.

677. T. Takeda, S. Yoshida, K. Oda, and S. Hirose, Continuous Flow Injection Analysis of Total Bile Acid in Serum [in Japanese]. *Rinsho Kagaku,* **13** (1984) 134.

678. M. Tabata, C. Fukunaga, M. Ohyabu, and T. Murachi, Highly Sensitive Flow Injection Analysis of Glucose and Uric Acid in Serum Using an Immobilized Enzyme Column and Chemiluminescence. *J. Appl. Biochem.,* **6** (1984) 251.

679. G. Decristoforo and F. Knauseder, Rapid Determination of Cephalosporins with an Immobilized Enzyme Reactor and Sequential Subtractive Spectrophotometric Detection in an Automated Flow-Injection System. *Anal. Chim. Acta,* **163** (1984) 73.

680. A. Schelter-Graf, H. L. Schmidt, and H. Huck, Determination of the Substrates of Dehydrogenases in Biological Material in Flow-Injection Systems with Electrocatalytic NADH Oxidation. *Anal. Chim. Acta,* **163** (1984) 299.

681. B. F. Rocks, R. A. Sherwood, and C. Riley, Controlled-Dispersion Flow Analysis in Clinical Chemistry: Determination of Albumin, Triglycerides and Theophylline. *Analyst,* **109** (1984) 847.

682. T. Sakai and N. Ohno, Flow Injection Analysis of Trace Amounts of Iron with 2-Nitroso-5-(*n*-Ethyl-*n*-Sulfopropylamino)Phenol [in Japanese]. *Bunseki Kagaku,* **33** (1984) 331.

683. Z. Fang, S. Xu, and S. Zhang, The Determination of Trace Amounts of Heavy Metals in Waters by a Flow-Injection System Including Ion-Exchange Preconcentration and Flame Atomic Absorption Spectrometric Detection. *Anal. Chim. Acta,* **164** (1984) 41.

684. Z. Fang, J. Růžička, and E. H. Hansen, An Efficient Flow-Injection System with On-Line Ion-Exchange Preconcentration for the Determination of Trace Amounts of Heavy Metals by Atomic Absorption Spectrometry. *Anal. Chim. Acta,* **164** (1984) 23.

685. R. A. Leach and J. M. Harris, Real-Time Thermal Lens Absorption Measurements with Application to Flow-Injection Systems. *Anal. Chim. Acta,* **164** (1984) 91.

686. P. J. Worsfold, J. Farrelly, and M. S. Matharu, A Comparison of Spectrophotometric and Chemiluminescence Methods for the Determination of Blood Glucose by Flow Injection Analysis. *Anal. Chim. Acta,* **164** (1984) 103.

687. S. Alegret, J. Alonso, J. Bartroli, J. M. Paulis, J. L. F. C. Lima, and A. A. S. C. Machado, Flow Through Tubular PVC Matrix Membrane Electrode Without Inner Reference Solution for Flow Injection Analysis. *Anal. Chim. Acta,* **164** (1984) 147.

688. L. Nord and B. Karlberg, Extraction Based on the Flow-Injection Principle. Part 6. Film Formation and Dispersion in Liquid-Liquid Segmented Flow Extraction Systems. *Anal. Chim. Acta,* **164** (1984) 233.

689. J. N. Miller, Flow Injcction Analysis: Fundamentals and Recent Developments. *Anal. Proc.,* **21** (1984) 372.

690. D. Betteridge, A. F. Taylor, and A. P. Wade, Optimization of Conditions for Flow Injection Analysis. *Anal. Proc.,* **21** (1984) 373.

691. A. Shaw, Practical Aspects of Flow Injection Analysis. *Anal. Proc.,* **21** (1984) 375.

692. P. J. Worsfold, Clinical Applications of Flow Injection Analysis. *Anal. Proc.,* **21** (1984) 376.

693. J. F. Tyson, Flow Injection Analysis Combined with Atomic Absorption Spectrometry. *Anal. Proc.,* **21** (1984) 377.

694. F. Morishita, Y. Hara, and T. Kojima, Flow Injection Analysis of L-Lactate with Lactate Dehydrogenase Immobilized Open Tubalar Reactor [in Japanese]. *Bunseki Kagaku,* **33** (1984) 642.

695. T. Yamane and Y. Nozawa, Catalytic Determination of Trace Amounts of Manganese with a Flow Injection System [in Japanese]. *Bunseki Kagaku,* **33** (1984) 652.

696. T. Korenaga and K. Okada, Automated System for Total Phosphorous in Waste Waters by Flow Injection Analysis [in Japanese]. *Bunseki Kagaku,* **33** (1984) 683.

697. K. Ogata, S. Soma, I. Koshiishi, S. Tanabe, and T. Imanari, Determination of Silicate by Flow Injection Analysis Coupled with Suppression Column and Solvent Extraction System. *Bunseki Kagaku,* **33** (1984) E535.

698. C. Gooijer, N. H. Velthorst, and R. W. Frei, Phosphorescence in Liquid Solutions. A Promising Detection Principle in Liquid Chromatography and Flow Injection Analysis. *Trends Anal. Chem.,* **3** (1984) 259.

699. L. Nord, Extraction in Liquid-Liquid Segmented Flow Systems Applied to the Mechanization of Sample Pretreatment. Royal Inst. Techn., Stockholm, Sweden, (1984). (Ph.D. Thesis).

700. F. Lázaro, M. D. Luque de Castro, and M. Valcarcel, Stopped-Flow Injection Determination of Copper(II) at the ng per ml Level. *Anal. Chim. Acta,* **165** (1984) 177.

701. A. Fernández, M. A. Gómez-Nieto, M. D. Luque de Castro, and M. Valcárcel, A Flow-Injection Manifold Based on Splitting the Sample Zone and a Confluence Point Before a Single Detector Unit. *Anal. Chim. Acta,* **165** (1984) 217.

702. D. Betteridge, C. Z. Marczewski, and A. P. Wade, A Random Walk Simulation of Flow Injection Analysis. *Anal. Chim. Acta,* **165** (1984) 227.

703. T. Yao, M. Sato, Y. Kobayashi, and T. Wasa, Flow Injection Analysis for Glucose by the Combined Use of an Immobilized Glucose Oxidase Reactor and a Peroxidase Electrode. *Anal. Chim. Acta,* **165** (1984) 291.

704. M. Masoom and A. Townshend, Applications of Immobilized Enzymes in Flow Injection Analysis. *Anal. Proc.,* **22** (1985) 6.

705. A. T. Faizullah and A. Townshend, Flow Injection Analysis with Chemiluminescence Detection: Determination of Hydrazine. *Anal. Proc.,* **22** (1985) 15.

706. A. Hughes and P. J. Worsfold, Monitoring of Immunoprecipitin Reactions Using Flow Injection Analysis. *Anal. Proc.,* **22** (1985) 16.

707. J. F. Tyson and J. M. H. Appleton, Concentration Gradients for Calibration Purposes. *Anal. Proc.,* **22** (1985) 17.

708. J. E. Newbery and M. P. Lopez Hadded, Amperometric Determination of Nitrite by Oxidation at A Glassy Carbon Electrode. *Analyst,* **110** (1985) 81.

709. M. G. Glaister, G. J. Moody, and J. D. R. Thomas, Studies on Flow Injection Analysis with Sulphide Ion-Selective Electrodes. *Analyst,* **110** (1985) 113.

710. O. K. Borggaard and S. S. Jørgensen, Determination of Silicon in Soil Extracts by Flow Injection Analysis. *Analyst,* **110** (1985) 177.

711. J. Carlos de Andrade, J. C. Rocha, and N. Baccan, Sequential Spectrophotometric Determination of Chromium(III) and Chromium(VI) Using Flow Injection Analysis. *Analyst,* **110** (1985) 197.

712. S. D. Hartenstein, J. Růžička, and G. D. Christian, Sensitivity Enhancements for Flow Injection Analysis—Inductively Coupled Plasma Atomic Emission Spetrometry Using an On-Line Preconcentrating Ion-Exchange Column. *Anal. Chem.,* **57** (1985) 21.

713. C. J. Yuan and C. O. Huber, Determination of Proteins and Denaturation Studies by Flow Injection with a Nickel Oxide Electrode. *Anal. Chem.,* **57** (1985) 180.

714. F. F. Cantwell and J. A. Sweileh, Hydrodynamic and Interfacial Origin of Phase Segmentation in Solvent Extraction/Flow Injection Analysis. *Anal. Chem.,* **57** (1985) 329.

715. J. A. Sweileh and F. F. Cantwell, Sample Introduction by Solvent Extraction/Flow Injection to Eliminate Interferences in Atomic Absorption Spectroscopy. *Anal. Chem.,* **57** (1985) 420.

716. J. A. Polta, I.-H. Yeo, and D. C. Johnson, Flow-Injection System for the Rapid and Sensitive Assay of Concentrated Aqueous Solutions of Strong Acids and Bases. *Anal. Chem.,* **57** (1985) 563.

717. P. Petak, Possible Applications of Electrochemical Methods for Detection of Substances in Flowing Liquids [in Czech]. *Rudy,* **32** (1984) 173.

718. R. A. Mowery, Jr., The Potential of Flow Injection Techniques for On-Stream Process Applications. *Proc. Jt. Symp. Instrum. Control 80's* (1984) 51.

719. M. Masoom and A. Townshend, Determination of Glucose in Blood by Flow Injection Analysis and an Immobilized Glucose Oxidase Column. *Anal. Chim. Acta,* **166** (1984) 111.

720. P. W. Alexander and U. Akapongkul, Differential Pulse Voltammetry with Fast Pulse Repetition Times in a Flow-Injection System with a Copper-Amalgam Electrode. *Anal. Chim. Acta,* **166** (1984) 119.

721. H. D. Dewald and J. Wang, Spectroelectrochemical Detector for Flow-Injection Systems and Liquid Chromatography. *Anal. Chim. Acta,* **166** (1984) 163.

722. R. M. Smith and T. G. Hurdley, Spectrophotometric Determination of Copper as a Dithiocarbamate by Flow Injection Analysis. *Anal. Chim. Acta,* **166** (1984) 271.

723. I. Schneider, Determination of Penicillin V in Fermentation Samples by Flow Injection Analysis. *Anal. Chim. Acta,* **166** (1984) 293.

724. P. Linares, M. D. Luque de Castro, and M. Valcárcel, Fluorimetric Determination of Pyridoxal and Pyridoxal-5-Phosphate by Flow Injection Analysis. *Anal. Lett. B,* **18** (1985) 67.

725. K. Toda, I. Sanemasa and T. Deguchi, Flow Injection Analysis of Fluoride Ions Using Catalytic Reactions [in Japanese]. *Bunseki Kagaku,* **34** (1985) 31.

726. Y. Suzuki and T. Inoue, Determination of Cyanide Ion as Its Fluorescent Derivative by Flow Injection Analysis [in Japanese]. *Bunseki Kagaku,* **34** (1985) 53.

727. S. A. McClintock, J. R. Weber, and W. C. Purdy, The Design of a Computer-Controlled Flow-Injection Analyzer: An Undergraduate Experiment. *J. Chem. Educ.,* **62** (1985) 65.

728. N. Yoza, Flow Injection Analysis and Its Application to Environmental Analysis [in Japanese]. *Kankyo-to-Sokuteigijutsu,* **10** (1982) 22.

729. P. Hemmings, Automated Flow Injection Analysis for Anions in Waters. Univ. Birmingham, UK (1983). (Ph.D. Thesis).

730. P. Langer, Investigations for Sulphate Determination in Natural Waters by Means of Flow Injection Analysis [in German]. Tech. Univ. Berlin, Germany (1983). (M.Sc. Thesis).

731. J. L. Burguera and M. Burguera, The Principles, Applications and Trends of Flow Injection Analysis for Monitoring Chemiluminescent Reactions. *Acta Cient. Venez.,* **34** (1983) 79.

732. J. A. Polta and D. C. Johnson, The Direct Electrochemical Detection of Amino Acids at a Platinum Electrode in an Alkaline Chromatographic Effluent. *J. Liq. Chromatogr.,* **6** (1983) 1729.

733. T. S. C. Wang, Analytical Chemistry of Bilirubins: New Methods, Caffeine Complexation, Stability, and Biosynthesis of Conjugates. *Diss. Abstr. Int. B,* **44** (1983) 1109.

734. M. Gallignani, J. L. Burguera, and M. Burguera, Development and Comparison of Methods for the Determination of Selenium in Biological Tissues and/or Fluids [in Spanish]. *Acta Scient. Venez.,* **34** (1983) 449.

735. A. M. Gunn, An Automated Hydride Generation Atomic Absorption Spectrometric Method for the Determination of Total Arsenic in Raw and Potable Waters. *Tech. Rep. TR-Water Res. Cent.,* No. TR191 (1983) 31.

736. R. F. Browner, Sample Introduction for Inductively Coupled Plasmas and Flames. *Trends Anal. Chem.,* **2** (1983) 121.

737. J. Michel, Investigations for the Determination of the Content of Ionized Materials in Water by Means of Flow Injection Analysis [in German]. Tech. Univ. Berlin, Germany (1984). (M.Sc. Thesis).

738. R. F. Browner and A. W. Boorn, Sample Introduction: The Achilles' Heel of Atomic Spectroscopy. *Anal. Chem.,* **56** (1984) 786A.

739. R. F. Browner and A. W. Boorn, Sample Introduction Techniques for Atomic Spectroscopy. *Anal. Chem.,* **56** (1984) 875A.

740. S. Honda, T. Konishi, and H. Chiba, Evaluation of Dual-Wavelength Spectrophotometry for Drug Level Monitoring. *Anal. Chem.*, **56** (1984) 2352.

741. E. Anders, R. Voigtländer, H. W. Rüttinger, and H. Matschiner, Determination of Ammonium in Water and in Blood Sera from Fish by Flow Injection Analysis [in German]. *Z. Binnenfischerei DDR,* **31** (1984) 331.

742. G. Warren, An Introduction to Flow Injection Analysis. *Lab. News.*, Nov. (1984) 33.

743. C. F. Mandenius, B. Danielsson, and B. Mattiasson, Evaluation of a Dialysis Probe for Continuous Sampling in Fermentors and in Complex Media. *Anal. Chim. Acta,* **163** (1984) 135.

744. F. Winquist, A. Spetz, and I. Lundström, Determination of Urea with an Ammonia Gas-Sensitive Semiconductor Device in Combination with Urease. *Anal. Chim. Acta,* **163** (1984) 143.

745. F. Winquist, A. Spetz, and I. Lundström, Determination of Ammonia in Air and Aqueous Samples with a Gas-Sensitive Semiconductor Capacitor. *Anal. Chim. Acta,* **164** (1984) 127.

746. E. Watanabe, S. Tokimatsu, and K. Toyama, Simultaneous Determination of Hypoxanthine, Inosine, Inosine-5′-Phosphate and Adenosine-5′-Phosphate with a Multielectrode Enzyme Sensor. *Anal. Chim. Acta,* **164** (1984) 139.

747. H. Lundbäck, A Study of Some Methods Involving the Determination of Hydrogen Peroxide, in Particular in Flow Injection Analysis. Univ. Lund, Sweden (1984). (Ph.D. Thesis).

748. F. Lázaro, M. D. Luque de Castro, and M. Valcárcel, Simultaneous Catalytic-Fluorimetric Determination of Copper and Mercury by Flow Injection Analysis. *Fresenius Z. Anal. Chem.*, **320** (1985) 128.

749. T. Greatorex and P. B. Smith, Flow Injection Analysis—A Review of Experiences in a Water Authority Laboratory. *J. Inst. Water Eng. Scient.*, **39** (1985) 81.

750. K. H. Kroner and M. R. Kula, On-Line Measurement of Extracellular Enzymes During Fermentation by Using Membrane Techniques. *Anal. Chim. Acta,* **163** (1984) 3.

751. Y. Baba, N. Yoza, and S. Ohashi, Simultaneous Determination of Phosphate and Phosphonate by Flow Injection Analysis and High Performance Liquid Chromatography with a Series Detection System. *J. Chromatogr.*, **318** (1985) 319.

752. T. Yao, M. Sato, and T. Wasa, Flow Injection Analysis for Uric Acid by the Combined Use of an Immobilized Uricase Reactor and a Peroxidase Electrode [in Japanese]. *Nippon Kagaku Kaishi,* **2** (1985) 189.

753. M. Koupparis and P. Anagnostopoulou, An Automated Microprocessor Based Spectrophotometric Flow Injection Analyser. *J. Aut. Chem.*, **6** (1984) 186.

754. M. A. Marshall and H. A. Mottola, Performance Studies under Flow Con-

ditions of Silica-Immobilized 8-Quinolinol and Its Application as a Preconcentration Tool in Flow Injection/Atomic Absorption Determinations. *Anal. Chem.*, **57** (1985) 729.

755. K. G. Miller, G. E. Pacey, and G. Gordon, Automated Iodometric Method for Determination of Trace Chlorate Ion Using Flow Injection Analysis. *Anal. Chem.*, **57** (1985) 734.
756. L. T. M. Prop, P. C. Thijssen, and L. G. G. van Dongen, A Software Package for Computer-Controlled Flow-Injection Analysis. *Talanta,* **32** (1985) 230.
757. A. R. Rios, M. D. Luque de Castro, and M. Valcárcel, New Approach to the Simultaneous Determination of Pollutants in Waste Waters by Flow Injection Analysis. *Analyst,* **110** (1985) 277.
758. B. Olsson, H. Lundbäck, and G. Johansson, Galactose Determinations in an Automated Flow-Injection System Containing Enzyme Reactors and an On-Line Dialyzer. *Anal. Chim. Acta,* **167** (1985) 123.
759. A. T. Faizullah and A. Townshend, Application of a Reducing Column for Metal Speciation by Flow Injection Analysis. Spectrophotometric Determination of Iron(III) and Simultaneous Determination of Iron(II) and Total Iron. *Anal. Chim. Acta,* **167** (1985) 225.
760. M. Maeda and A. Tsuji, Enzymatic Immunoassay of α-Fetoprotein, Insulin and 17-α-Hydroxyprogesterone Based on Chemiluminescence in a Flow-Injection System. *Anal. Chim. Acta,* **167** (1985) 231.
761. P. van Zoonen, D. A. Kamminga, C. Gooijer, N. H. Velthorst, and R. W. Frei, Flow Injection Determination of Hydrogen Peroxide by Means of a Solid-State Peroxyoxalate Chemiluminescence Reactor. *Anal. Chim. Acta,* **167** (1985) 249.
762. Y. Hirai and K. Tomokuni, Extraction-Spectrophotometric Determination of Anionic Surfactants with a Flow-Injection System. *Anal. Chim. Acta,* **167** (1985) 409.
763. A. G. Cox, I. G. Cook, and C. W. McLeod, Rapid Sequential Determination of Chromium(III)—Chromium(VI) by Flow Injection Analysis—Inductively Çoupled Plasma Atomic-Emission Spectrometry. *Analyst,* **110** (1985) 331.
764. A. G. Fogg, A. M. Summan, and M. A. Fernández-Arciniega, Flow Injection Amperometric Determination of Ascorbic Acid and Dopamine at a Sessile Mercury Drop Electrode without Deoxygenation. *Analyst,* **110** (1985) 341.
765. A. G. Fogg, M. A. Fernández-Arciniega, and R. M. Alonso, Flow Injection Amperometric Determination of Nitroprusside at a Glassy Carbon Electrode and at a Sessile Mercury Drop Electrode. *Analyst,* **110** (1985) 345.
766. T. Fukasawa, S. Kawakubo, T. Okabe, and A Mizuike, Catalytic Determination of Vanadium by Micro Ion Exchange Separation—Flow Injection Method and Its Application to Rain Water [in Japanese]. *Bunseki Kagaku,* **33** (1984) 609.

767. M. Miyazaki, N. Okubo, K. Hayakawa, and T. Umeda, Specific and Selective Determination Method for Halide Anions by a Flow Injection Technique. *Chem. Pharm. Bull.,* **32** (1984) 3702.

768. S. Tanabe, T. Shiori, K. Murakami, and T. Imanari, A New Method for Assay of Ferroxidase Activity and Its Application to Human and Rabbit Sera. *Chem. Pharm. Bull.,* **32** (1984) 4029.

769. H. Cui, L. Meng, and Z. Zhu, Automatic Flow Injection Colorimetric Determination of Available Aluminum in Soils [in Chinese]. *Fenxi Huaxue,* **12** (1984) 754.

770. M. Herrera, L. S. Kao, D. J. Curran, and E. W. Westhead, Flow-Injection Analysis of Catecholamine Secretion from Bovine Adrenal Medulla Cells on Microbeads. *Anal. Biochem.,* **144** (1985) 218.

771. E. Martins, M. Bengtsson, and G. Johansson, On-Line Dialysis of Some Metal Ions and Metal Complexes. *Anal. Chim. Acta,* **169** (1985) 31.

772. K. Bäckström, L.-G. Danielsson, and L. Nord, Design and Evaluation of a New Phase Separator for Liquid-Liquid Extraction in Flow Systems. *Anal. Chim. Acta,* **169** (1985) 43.

773. S. Storgaard Jørgensen, K. M. Petersen, and L. A. Hansen, A Simple Multifunctional Valve for Flow Injection Analysis. *Anal. Chim. Acta,* **169** (1985) 51.

774. R. C. Schothorst and G. den Boef, The Application of Strongly Oxidizing Agents in Flow Injection Analysis. Part 1. Silver(II). *Anal. Chim. Acta,* **169** (1985) 99.

775. E. A. Jones, Spectrophotometric Determination of Uranium(VI) with 2-(5-Bromo-2-Pyridylazo)-5-Diethylaminophenol in a Flow-Injection System. *Anal. Chim. Acta,* **169** (1985) 109.

776. F. Lazaro, M. D. Luque de Castro, and M. Valcarcel, Sequential and Differential Catalytic-Fluorimetric Determination of Manganese and Iron by Flow Injection Analysis. *Anal. Chim. Acta,* **169** (1985) 141.

777. M. Gallego and M. Valcárcel, Indirect Atomic Absorption Spectrometric Determination of Perchlorate by Liquid-Liquid Extraction in a Flow-Injection System. *Anal. Chim. Acta,* **169** (1985) 161.

778. F. Cui, E. H. Hansen, and J. Růžička, Evaluation of Critical Parameters for Measurement of pH by Flow Injection Analysis. Determination of pH in Soil Extracts. *Anal. Chim. Acta,* **169** (1985) 209.

779. R. Appelqvist, G. Marko-Varga, L. Gorton, A. Torstensson, and G. Johansson, Enzymatic Determination of Glucose in a Flow System by Catalytic Oxidation of the Nicotinamide Coenzyme at a Modified Electrode. *Anal. Chim. Acta,* **169** (1985) 237.

780. E. B. Milosavljevic, J. Růžička, and E. H. Hansen, Simultaneous Determination of Free and EDTA-Complexed Copper Ions by Flame Atomic Absorption Spectrometry with an Ion-Exchange Flow-Injection System. *Anal. Chim. Acta,* **169** (1985) 321.

781. Y. Hayashi, K. Zaitsu, and Y. Ohkura, Flow Injection Analysis of Hydrogen Peroxide Using Immobilized Horseradish Peroxidase and Its Fluorogenic Substrate 3-(*p*-Hydroxyphenyl)propionic Acid. *Anal. Sci.,* **1** (1985) 65.

782. L. Fossey and F. F. Cantwell, Determination of Acidity Constants by Solvent Extraction/Flow Injection Analysis Using a Dual-Membrane Phase Separator. *Anal. Chem.,* **57** (1985) 922.

783. A. MacDonald and T. A. Nieman, Flow Injection and Liquid Chromatography Detector for Amino Acids Based on a Postcolumn Reaction with Luminol. *Anal. Chem.,* **57** (1985) 936.

784. F. L. Boza, M. D. Luque de Castro, and M. Valcárcel Cases, Flow Injection Environmental Analysis. A Review. *Quim. Anal.,* **13** (1985) 147.

785. T. Toyoda, S. S. Kuan, and G. G. Guilbault, Determination of Lactate Dehydrogenase Isoenzyme (LD-1) Using Flow Injection Analysis with Electrochemical Detection after Immunochemical Separation. *Anal. Lett.,* **18** (1985) 345.

786. M. H. Ho and M. U. Asouzu, Use of Immobilized Enzymes in Flow Injection Analysis. *Ann. N.Y. Acad. Sci.,* **434** (1984) 526.

787. U. Jönsson, I. Rönnberg, and H. Malmqvist, Flow Injection Ellipsometry—An In Situ Method for the Study of Biomolecular Adsorption and Interaction at Solid Surfaces. *Colloid. Surface,* **13** (1985) 333.

788. T. Hara, M. Toriyama, T. Ebuchi, and M. Imaki, Flow Injection Analysis of α-Amino Acids by the Chemiluminescence Method. *Chem. Lett.,* **3** (1985) 341.

789. C. W. McLeod, I. G. Cook, P. J. Worsfold, J. E. Davies, and J. Queay, Analyte Enrichment and Matrix Removal in Flow Injection Analysis Inductively Coupled Plasma-Atomic Emission Spectrometry. Determination of Phosphorous in Steels. *Spectrochim. Acta,* **40** (1985) 57.

790. J. F. Tyson, Flow Injection Analysis Techniques for Atomic-Absorption Spectrometry. A Review. *Analyst,* **110** (1985) 419.

791. J. F. Tyson, C. E. Adeeyinwo, J. M. H. Appleton, S. R. Bysouth, A. B. Idris, and L. L. Sarkissian, Flow Injection Techniques of Method Development for Flame Atomic-Absorption Spectrometry. *Analyst,* **110** (1985) 487.

792. R. A. Sherwood, B. F. Rocks, and C. Riley, Controlled-Dispersion Flow Analysis with Atomic-Absorption Detection for the Determination of Clinically Relevant Elements. *Analyst,* **110** (1985) 493.

793. M. Bengtsson and G. Johansson, Preconcentration and Matrix Isolation of Heavy Metals through a Two-Stage Solvent Extraction in a Flow System. *Anal. Chim. Acta,* **158** (1984) 147.

794. R. Gnanasekaran and H. A. Mottola, Flow Injection Determination of Penicillins Using Immobilized Penicillinase in a Single Bead String Reactor. *Anal. Chem.,* **57** (1985) 1005.

795. P. K. Dasgupta and H. Hwang, Application of a Nested Loop System for the Flow Injection Analysis of Trace Aqueous Peroxides. *Anal. Chem.*, **57** (1985) 1009.

796. K. Fujiwara and K. Fuwa, Liquid Core Optical Fiber Total Reflection Cell as a Colorimetric Detector for Flow Injection Analysis. *Anal. Chem.*, **57** (1985) 1012.

797. M. Yamada, H. Kanai, and S. Suzuki, Flavin Mononucleotide Chemiluminescence for Determination of Traces of Copper(II) by Continuous Flow and Flow Injection Methods. *Bull. Chem. Soc. Jpn.*, **58** (1985) 1137.

798. J. B. Landis, Flow Injection Analysis (in Pharmaceutical Analysis). *Drugs Pharm. Sci.*, **11** (1984) 217.

799. Z. Niegreisz, L. Szucs, J. Fekete, G. Horvai, K. Toth, and E. Pungor, Modifications of the Wall-Jet Electrochemical Detector for Liquid Chromatography and Flow Analysis. *J. Chromatogr.*, **316** (1984) 451.

800. K. Hult, R. Fuchs, M. Peraica, R. Plestina, and S. Ceovic, Screening for Ochratoxin A in Blood by Flow Injection Analysis. *J. Appl. Toxicol.*, **4** (1984) 326.

801. C. B. Ranger, An Automated Ion Analyzer. *Am. Lab.*, **17** (1985) 92.

802. M. A. Gómez-Nieto, A. D. Luque de Castro, A. Martin, and M. Varcárcel, Prediction of the Behaviour of a Single Flow-Injection Manifold. *Talanta,* **32** (1985) 319.

803. J. T. Vanderslice and G. R. Beecher, Comments on the Paper by Gómez-Nieto, Luque de Castro, Martin, and Valcárcel. *Talanta,* **32** (1985) 334.

804. M. Valcárcel and M. D. Luque de Castro, Reply to the Comments by Vanderslice and Beecher. *Talanta,* **32** (1985) 339.

805. F. B. Bigley, Studies Using Flow Injection Analysis and Post-Column Reaction Detection Analysis with High Performance Liquid Chromatography. *Diss. Abstr. Int. B,* **45**(8) (1985) 2528.

806. J. Martinez Calatayud, Flow Injection Analysis (FIA). An Analytical Innovation with a Promise for the Future [in Spanish]. *Tec. Lab.*, **9** (1984) 358.

807. K. K. Stewart, Flow Injection Analysis: A New Tool for the Automation of the Determination of Food Components. In *Modern Methods of Food Analysis* (Kent K. Stewart and John R. Whitaker, Eds.), AVI Publ. Co., CT (1984) 369.

808. M. Gallego, M. D. Luque de Castro, and M. Valcárcel, Atomic Spectroscopic Techniques in Flow Injection Analysis. A Review. *At. Spectrosc.*, **6** (1985) 16.

809. J. L. Burguera, M. Burguera, and D. F. Flores, Determination of Some Phosphorous-Containing Compounds by Flow Injection with a Molecular Emission Cavity Detector. *Anal. Chim. Acta,* **170** (1985) 331.

810. H. Hwang and P. K. Dasgupta, Fluorimetric Determination of Trace Hydrogen Peroxide in Water with a Flow Injection Analysis. *Anal. Chim. Acta,* **170** (1985) 347.

811. A. Tanaka, M. Miyazaki, and T. Deguchi, New Simultaneous Catalytic Determination of Thiocyanate and Iodide by Flow Injection Analysis. *Anal. Lett.,* **18** (1985) 695.

812. J. Tyson, Flow Injection Techniques for Flame Atomic Absorption Spectrophotometry. *Trends Anal. Chem.,* **4** (1985) 124.

813. R. Kuroda, I. Ida, and H. Kimura, Spectrophotometric Determination of Silicon in Silicates by Flow Injection Analysis. *Talanta,* **32** (1985) 353.

814. M. A. Koupparis, P. Anagnostopoulou, and H. V. Malmstadt, Automated Flow-Injection Pseudotitration of Strong and Weak Acids, Ascorbic Acid and Calcium, and Catalytic Psudotitrations of Aminopolycarboxylic Acids bu Use of a Microcomputer-Controlled Analyser. *Talanta,* **32** (1985) 411.

815. A. Rios, M. D. Luque de Castro, and M. Valcárcel, Flow Injection Analysis: A New Approach to Pharmaceutical Determinations. *J. Pharm. Biomed. Anal.,* **3** (1985) 105.

816. J. Möller and B. Winter, Application of Flow Injection Techniques for the Analysis of Inorganic Anions. *Fresenius Z. Anal. Chem.,* **320** (1985) 451.

817. Z. Fang, J. M. Harris, J. Růžička, and E. H. Hansen, Simultaneous Flame Photometric Determination of Lithium, Sodium, Potassium and Calcium by Flow Injection Analysis with Gradient Scanning Standard Addition. *Anal. Chem.,* **57** (1985) 1457.

818. J. A. Polta and D. C. Johnson, Pulsed Amperometric Detection of Electroinactive Adsorbates at Platinum Electrodes in a Flow Injection System. *Anal. Chem.,* **57** (1985) 1373.

819. T. P. Tougas, J. M. Jannetti, and W. G. Collier, Theoretical and Experimental Response of a Biamperometric Detector for Flow Injection Analysis. *Anal. Chem.,* **57** (1985) 1377.

820. M. Yamamoto, M. Yasuda, and Y. Yamamoto, Hydride-Generation Atomic Absorption Spectrometry Coupled with Flow Injection Analysis. *Anal. Chem.,* **57** (1985) 1382.

821. C. C. Y. Chan, Semiautomated Method for Determination of Selenium in Geological Materials Using a Flow Injection Analysis Technique. *Anal. Chem.,* **57** (1985) 1482.

822. K. S. Johnson, R. L. Petty, and J. Thomsen, Flow Injection Analysis for Seawater Micronutrients. *Adv. Chem. Ser.,* **209** (1985) 7.

823. M. Maeda and A. Tsuji, Fluorescence and Chemiluminescence Determination of Steroid and Bile Acid Sulphates with Lucigenin by Flow Injection Analysis Based on Ion-Pair Solvent Extraction. *Analyst,* **110** (1985) 665.

824. N. Yoza, H. Hirano, Y. Baba, and S. Ohashi, Characterization of Enzymatic Hydrolysis of Inorganic Polyphosphates by Flow Injection Analysis and High-Performance Liquid Chromatography. *J. Chromatogr.,* **325** (1985) 385.

825. F. Lazaro, M. D. Luque de Castro, and M. Valcárcel, Catalytic-Fluorimetric Determination of EDTA and Iron(III) by Flow Injection Analysis. Inhibition Methods. *Fresenius Z. Anal. Chem.,* **321** (1985) 467.

826. A. Recktenwald, K. H. Kroner, and M. R. Kula, Rapid On-Line Protein Detection in Biotechnological Processes by Flow Injection Analysis (FIA). *Enzyme Microb. Technol.,* **7** (1985) 146.

827. Z. Gao and J. Chen, Determination of Total Nitrogen in Soils with an Ammonia Gas-Sensitive Electrode by the Flow Injection Merging Zone Method [in Chinese]. *Fenxi Huaxue,* **12** (1984) 1096.

828. H. Matschiner and H. H. Ruettinger, Wet Chemical Analysis—Economically Executed [in German]. *Wiss. Fortschr.,* **35** (1985) 53.

829. Y. Suzuki, Class Determination of Aliphatic Aldehydes and Ketones as Their 2,4-Dinitrophenylhydrazones by Flow Injection Analysis [in Japanese]. *Bunseki Kagaku,* **34** (1985) 414.

830. K. Oguma, Y. Kato, and R. Kuroda, Continuous Spectrophotometric Determination of Calcium in Silicates by Flow-Injection Analysis [in Japanese]. *Bunseki Kagaku,* **34** (1985) T98.

831. A. G. Fogg, M. A. Fernández-Arciniega, and R. M. Alonso, Oxidative Amperometric Flow Injection Determination of Sulphite at an Electrochemically Pretreated Glassy Carbon Electrode. *Analyst,* **110** (1985) 851.

832. P. W. Alexander, P. R. Haddad, and M. Trojanowicz, Potentiometric Flow-Injection Determination of Copper-Complexing Organic Ligands with a Copper-Wire Indicating Electrode. *Anal. Chim. Acta,* **171** (1985) 151.

833. M. Masoom and A. Townshend, Simultaneous Determination of Sucrose and Glucose in Mixtures by Flow Injection Analysis with Immobilized Enzymes. *Anal. Chim. Acta,* **171** (1985) 185.

834. A. Rios, M. D. Luque de Castro, and M. Valcárcel, Spectrophotometric Determination of Acidity-Constants of Unstable Compounds by Flow Injection Analysis. *Anal. Chim. Acta,* **171** (1985) 303.

835. P. J. Worsfold and A. Nabi, The Bioluminescent Determination of Adenosine Triphosphate with a Flow-Injection System. *Anal. Chim. Acta,* **171** (1985) 333.

836. M. C. U. Araujo, C. Pasquini, R. E. Bruns, and E. A. G. Zagatto, A Fast Procedure for Standard Additions in Flow Injection Analysis. *Anal. Chim. Acta,* **171** (1985) 337.

837. R. S. Schifreen, Flow Enthalpimetry (Isoperibol Flow-Injection Calorimetry). *Chem. Anal. (N.Y.),* **79** (1985) 97.

838. J. M. Hungerford, G. D. Christian, J. Růžička, and J. C. Giddings, Reaction Rate Measurement by Flow Injection Analysis Using the Gradient Stopped-Flow Method. *Anal. Chem.,* **57** (1985) 1794.

839. M. R. Straka, G. Gordon, and G. E. Pacey, Residual Aqueous Ozone Determination by Gas Diffusion Flow Injection Analysis. *Anal. Chem.,* **57** (1985) 1799.

840. A. Rios, M. D. Luque de Castro, and M. Valcárcel, Multidetection in Unsegmented Flow Systems with a Single Detector. *Anal. Chem.,* **57** (1985) 1803.

841. W. R. LaCourse, I. S. Krull, and K. Bratin, Photoelectrochemical Detector for High-Performance Liquid Chromatography and Flow Injection Analysis. *Anal. Chem.,* **57** (1985) 1810.

842. J. M. Harnly and G. R. Beecher, Two-Valve Injector to Minimize Nebulizer Memory for Flow Injection Atomic Absorption Spectrometry. *Anal. Chem.,* **57** (1985) 2015.

843. J. T. Dyke and Q. Fernando, Deconvolution Techniques for Rapid Flow-Injection Analysis. *Talanta,* **32** (1985) 807.

844. R. E. A. Escott and A. F. Taylor, Determination of Water by Flow Injection Analysis Using Karl Fischer Reagent with Electrochemical Detection. *Analyst,* **110** (1985) 847.

845. J. A. Wise, W. R. Heineman, and P. T. Kissinger, Flow Injection System for Stripping Voltammetry. *Anal. Chim. Acta,* **172** (1985) 1.

846. A. T. Faizullah and A. Townshend, Spectrophotometric Determination of Copper by Flow Injection Analysis with an On-Line Reduction Column. *Anal. Chim. Acta,* **172** (1985) 291.

847. H. Wada, H. Mori, and G. Nakagawa, Spectrophotometric Determination of Fluoride with Lanthanum/Alizarin Complexone by Flow Injection Analysis. *Anal. Chim. Acta,* **172** (1985) 297.

848. J. Růžička and E. H. Hansen, Optosensing at Active Surfaces—A New Detection Principle in Flow Injection Analysis. *Anal. Chim. Acta,* **173** (1985) 3.

849. E. A. G. Zagatto, M. F. Giné, E. A. N. Fernandes, B. F. Reis, and F. J. Krug, Sequential Injections in Flow Systems as an Alternative to Gradient Exploitation. *Anal. Chim. Acta,* **173** (1985) 289.

850. J. Kurzawa, Determination of Sulphur(II) Compounds by Flow Injection Analysis with Application of the Induced Iodine/Azide Reaction. *Anal. Chim. Acta,* **173** (1985) 343.

851. F. Lázaro, M. D. Luque de Castro, and M. Valcárcel, Catalytic-Fluorimetric Determination of Titanium(IV) by Flow Injection Analysis. *Anal. Lett.,* **18** (1985) 1209.

852. B. Bernhardsson, E. Martins, and G. Johansson, Solute Transfer in On-Line Analytical Flow-Through Dialyzers. *Anal. Chim. Acta,* **167** (1985) 111.

853. J. Wang and L. D. Hutchins, Thin-Layer Electrochemical Detector with a Glassy Carbon Electrode Coated with a Base-Hydrolyzed Cellulosic Film. *Anal. Chem.,* **57** (1985) 1536.

854. P. Linares, M. D. Luque de Castro, and M. Valcárcel, Simultaneous Determination of Pyridoxal and Pyridoxal 5-Phosphate in Human Serum by Flow Injection Analysis. *Anal. Chem.,* **57** (1985) 2101.

855. Olympus Optical Co. Ltd., Flow-Injection-Type Immunochemical Analysis [in Japanese]. Jpn. Kokai Tokkyo Koho, Pat. No. 85107568 (1985).

856. Olympus Optical Co. Ltd., Flow-Injection-Type Apparatus for Immuno-

chemical Analysis [in Japanese]. Jpn. Kokai Tokkyo Koho, Pat. No. 85089757 (1985).

857. R. A. Mowery, Jr., The Potential of Flow Injection Techniques for On Stream Process Applications. *ISA Trans.*, **24** (1985) 1.

858. P. Martinez-Jiménez, M. Gallego, and M. Valcárcel, Indirect Atomic Absorption Determination of Uranium by Flow Injection Analysis Using an Air-Acetylene Flame. *At. Spectrosc.*, **6** (1985) 65.

859. J. F. van Staden, Automated Prevalve Dilution in Flow-Injection Analysis. The Automated Determination of Chloride in Surface, Ground and Domestic Water. *Fresenius Z. Anal. Chem.*, **322** (1985) 36.

860. M. A. Marshall, Silica-Immobilized 8-Quinolinol: Synthesis, Characterization, and Applications in Flow Systems. *Diss. Abstr. Int. B*, **45** (1985) 3795.

861. A. J. Faske, K. R. Snable, A. W. Boorn, and R. F. Browner, Microlitre Sample Introduction for ICP-AES. *Appl. Spectrosc.*, **39** (1985) 542.

862. H. Huck, A. Schelter-Graf, and H. L. Schmidt, Measurement and Calculation of the Calibration Graphs for Flow Injection Analysis Using Enzyme Reactors with Immobilized Dehydrogenases and an Amperometric NADH Detector. *Bioelectrochem. Bioenerg.*, **13** (1984) 199.

863. K. Kurkijärvi, T. Heinonen, T. Lövgren, J. Lavi, and R. Raunio, Flow-Injection Analysis with Immobilized Bacterial Bioluminescence Enzymes. *Anal. Appl. Biolumin. Chemilumin.*, **3rd** (1984) 125.

864. J. A. Sweileh and F. F. Cantwell, Use of Peak Height for Quantification in Solvent Extraction/Flow Injection Analysis. *Can. J. Chem.*, **63** (1985) 2559.

865. S. Melnik and J. Fejes, Injection Devices in Flow Injection Analysis [in Czech]. *Chem. Listy*, **79** (1985) 910.

866. M. J. Whitaker and M. F. Bryant, The Determination of Pyritic Sulfur in Coal or Iron(III) in Aqueous Solutions by Flow Injection Analysis. *J. Coal Qual.*, **4** (1985) 68.

867. H. Lundbäck and B. Olsson, Amperometric Determination of Galactose, Lactose and Dihydroxyacetone Using Galactose Oxidase in a Flow Injection System with Immobilized Enzyme Reactors and On-Line Dialysis. *Anal. Lett.*, **18** (1985) 871.

868. M. A. Koupparis, E. P. Diamandis, and H. V. Malmstadt, Automated Stopped-Flow Analyzer in Clinical Chemistry: Determination of Albumin with Bromcresol Green and Purple. *Clin. Chim. Acta*, **149** (1985) 225.

869. T. Yao, M. Sato, Y. Kobayashi, and T. Wasa, Amperometric Assays of Total and Free Cholesterols in Serum by the Combined Use of Immobilized Cholesterol Esterase and Cholesterol Oxidase Reactors and Peroxidase Electrode in a Flow Injection System. *Anal. Biochem.*, **149** (1985) 387.

870. I. Nordin-Andersson and A. Cedergren, Spectrophotometric Determination of Water by Flow Injection Analysis Using Conventional and Pyridine-Free Two-Component Karl Fischer Reagents. *Anal. Chem.*, **57** (1985) 2571.

871. L. Sun, L. Li, and Z. Fang, Flow Injection Analysis of Soil Available Zinc by Solvent Extraction with Dithizone [in Chinese]. *Fenxi Huaxue,* **13** (1985) 447.

872. D. K. Cope and D. E. Tallman, Calculation of Convective Diffusion Current at a Strip Electrode in a Rectangular Flow Channel: Inviscid Flow. *J. Electroanal. Chem. Interfacial Electrochem.,* **188** (1985) 21.

873. K. L. Jansen and J. M. Harris, Double-Beam Thermal Lens Spectroscopy. *Anal. Chem.,* **57** (1985) 2434.

874. K. Matsumoto, O. Hamada, H. Ukeda, and Y. Osajima, Flow Injection Analysis of Lactose in Milk Using a Chemically Modified Lactose Electrode. *Agric. Biol. Chem.,* **49** (1985) 2131.

875. T. Yao, Y. Matsumoto, and T. Wasa, Determination of Ethanol in Serums by the Use of Immobilized Enzymes in a Flow Injection-Amperometry System [in Japanese]. *Bunseki Kagaku,* **34** (1985) 513.

876. Z. Fang, Discussion on Modernization of Soil and Plant Analysis [in Chinese]. *Turang Tongbao,* **15** (1984) 230.

877. G. Schulze, W. Bönigk, and W. Frenzel, Matrix Exchange Technique for the Simultaneous Determination of Several Elements in Flow Injection Potentiometric Stripping Analysis. *Fresenius Z. Anal. Chem.,* **322** (1985) 255.

878. L. Ilcheva and K. Cammann, Flow Injection Analysis of Chloride in Tap and Sewage Water Types by Adsorption Differential Pulse Voltammetry. *Fresenius Z. Anal. Chem.,* **322** (1985) 323.

879. K. Uchida, M. Tomoda, and S. Saito, Determination of Calcium by Flow Injection Analysis with Hydroxynaphthol Blue [in Japanese]. *Bunseki Kagaku,* **34** (1985) 568.

880. P. C. Thijssen, L. T. M. Prop, G. Kateman, and H. C. Smit, A Kalman Filter for Calibration. Evaluation of Unknown Samples and Quality Control in Drifting Systems. Part 4. Flow Injection Analysis. *Anal. Chim. Acta,* **174** (1985) 27.

881. J. J. Harrow and J. Janata, Heterogeneous Samples in Flow-Injection Systems. Part 1. Whole Blood. *Anal. Chim. Acta,* **174** (1985) 115.

882. J. J. Harrow and J. Janata, Heterogeneous Samples in Flow-Injection Systems. Part 2. Standard Addition. *Anal. Chim. Acta,* **174** (1985) 123.

883. M. Masoom and A. Townshend, Determination of Cholesterol by Flow Injection Analysis with Immobilized Cholesterol Oxidase. *Anal. Chim. Acta,* **174** (1985) 293.

884. K. Toei, T. Zaitsu, and C. Igarashi, Spectrophotometric Determination of Micro Amounts of Cationic Polymeric Flocculants by Flow Injection Analysis. *Anal. Chim. Acta,* **174** (1985) 369.

885. M. J. Whitaker, Determination of Total Chromium by Flow Injection Analysis. *Anal. Chim. Acta,* **174** (1985) 375.

886. M. Yamada, S. Kamiyami, and S. Suzuki, Eosin Y-Sensitized Chemiluminescence of 7,7,8,8-Tetracyanoquinodimethane in Surfactant Vesicles

for Determination of Manganese(II) at Sub-nanogram Levels by Flow Injection Method. *Chem. Lett.*, **10** (1985) 1597.

887. G. L. Abdullahi and J. N. Miller, Application of Gradient Flow Injection Analysis to Studies of Drug-Protein Binding. *Analyst*, **110** (1985) 1271.

888. A. G. Fogg, M. A. Fernandez-Arciniega, and R. M. Alonso, Amperometric Flow Injection Determination of Ethylenediaminetetraacetic Acid (EDTA) at an Electrochemically Pretreated Glassy Carbon Electrode. *Analyst*, **110** (1985) 1201.

889. E. H. Hansen, Recent Advances in Flow Injection Analysis. *Int. Lab.*, **15**(8) (1985) 14.

890. A. S. Attiyat and G. D. Christian, Discrete Microsample Injection into a Gaseous Carrier. *Talanta*, **31** (1984) 463.

891. Y. P. Wu and G. E. Pacey, Spectrophotometric Determination of Lithium Ion with the Chromogenic Crown Ether, 2″,4″-Dinitro-6″-Trifluoromethylphenyl-4′-Aminobenzo-14-Crown-4. *Anal. Chim. Acta*, **162** (1984) 285.

892. Y. Hirai, N. Koga, T. Hasegawa, and K. Tomokuni, Dynamics of the Anionic Surfactants in River and Creek Waters of Saga City [in Japanese]. *Jpn. J. Hyg.*, **39** (1984) 787.

893. J. Reuter and B. Neidhart, A New Chemical Reaction System for Flow Through Analysis and Post Column Chromatographic Detection of Manganese and Chromium Species. *Trace Elem. Anal. Chem. Med. Biol. Proc. Int. Workshop.*, **3rd** (1984) 285.

894. E. Pungor, K. Tóth, and A. Hrabéczy-Páll, Application of Ion-Selective Electrodes in Flow Analysis. *Trends Anal. Chem.*, **3** (1984) 28.

895. G. Lange, H. Holesch, and G. Kaltenborn, Microanalyzer with High Performance and Low Reagent Use [in German]. *Wiss. Z. (Martin Luther Univ., Halle Wittenberg, Math.-Naturwiss. Reihe)*, **33**(6) (1984) 5.

896. K. Toei, Flow Injection Analysis of Inorganic Contaminants in Environmental Waters [in Japanese]. *J. Flow Injection Anal.*, **1**(1) (1984) 2.

897. T. Kojima and F. Morishita, Flow Injection Analysis with Enzyme Immobilized Open Tubular Reactor [in Japanese]. *J. Flow Injection Anal.*, **1**(1) (1984) 9.

898. Y. Hirai, Flow Injection Analysis of Sulphite [in Japanese]. *J. Flow Injection Anal.*, **1**(1) (1984) 16.

899. T. Imato and N. Ishibashi, Experimental Studies of Dispersion in FIA-Solvent Extraction Process [in Japanese]. *J. Flow Injection Anal.*, **1**(1) (1984) 23.

900. N. Ishibashi and T. Imato, Flow Injection Analysis Based on Volumetric Analysis-Reaction with Buffer Solutions [in Japanese]. *J. Flow Injection Anal.*, **1**(2) (1984) 2.

901. M. Munemori, Flow Injection Analysis with Membrane Separation [in Japanese]. *J. Flow Injection Anal.*, **1**(2) (1984) 13.

902. T. Korenaga and H. Ikatsu, Development of Micro Spectrometric Detector

Using Light Emitting Diode and Photo Diode, and Its Application to Flow Injection Analysis [in Japanese]. *J. Flow Injection Anal.,* **1**(2) (1984) 21.

903. M. Kato, M. Yamada, and S. Suzuki, Flavin Mononucleotide Sensitized and Polyoxyethylene (20) Sorbitan Trioleate Micelle-Enhanced Chemiluminescence for Sulfite Determination by Flow Injection Method [in Japanese]. *J. Flow Injection Anal.,* **1**(2) (1984) 31.
904. Y. Baba, H. Hirano, N. Yoza, and S. Ohashi, Optimization of Flow Injection Analysis for the Determination of Inorganic Phosphates; Effect of Reaction Temperature [in Japanese]. *J. Flow Injection Anal.,* **1**(2) (1984) 40.
905. Y. T. Shih and P. W. Carr, Flow-Rate Dependence of Post-Column Reaction Chromatographic Detectors and Optimization of Reactor Length for Slow Chemical Reaction. *Anal. Chim. Acta,* **167** (1985) 144.
906. M. Ikeda, Determination of Arsenic at the Picogram Level by Atomic Absorption Spectrometry with Miniaturized Suction-Flow Hydride Generation. *Anal. Chim. Acta,* **167** (1985) 289.
907. M. Ikeda, Determination of Selenium by Atomic Absorption Spectrometry with Miniaturized Suction-Flow Hydride Generation and On-Line Removal of Interferences. *Anal. Chim. Acta,* **170** (1985) 217.
908. T. Kumamaru, Y. Nitta, F. Nakata, H. Matsuo, and M. Ikeda, Determination of Cadmium by Suction-Flow Liquid-liquid Extraction Combined with Inductively-Coupled Plasma Atomic Emission Spectrometry. *Anal. Chim. Acta,* **174** (1985) 183.
909. C. Maekoya, F. Shiratsuchi, and K. Yano, Trace Analysis of Heavy Metals in Water by Flow-Injection Analysis Using Catalytic Reactions [in Japanese]. *Hitachi Sci. Instum. News,* **28**(1) (1985) 13.
910. T. Yokoyama and T. Tarutani, Flow Injection Analysis of Silicic Acid in Geothermal Water [in Japanese]. *J. Jpn. Geotherm. Energy Assoc.,* **22** (1985) 9.
911. B. F. Rocks, S. M. Wartel, R. A. Sherwood, and C. Riley, Determination of Albumin with Bromocresol Purple Using Controlled-Dispersion Flow Analysis. *Analyst,* **110** (1985) 669.
912. T. Korenaga, Successive Automation on the Measurement of Chemical Oxygen Demand by Flow Injection Analysis [in Japanese]. *J. Flow Injection Anal.,* **2**(1) (1985) 5.
913. G. Nagakawa and H. Wada, Highly Sensitive Chromogenic Reagents for the Use in FIA [in Japanese]. *J. Flow Injection Anal.,* **2**(1) (1985) 15.
914. T. Yokoyama and T. Tarutani, Flow Injection Analysis of Silicic Acid: Application to Study of the Polymerization of Silicic Acid [in Japanese]. *J. Flow Injection Anal.,* **2**(1) (1985) 30.
915. T. Kawashima, T. Minami, M. Ata, M. Kamada, and S. Nakano, Catalytic Determination of Ultratrace Amounts of Cobalt by Flow Injection Analysis [in Japanese]. *J. Flow Injection Anal.,* **2**(1) (1985) 40.

916. K. Zaitsu, Y. Hayashi, and Y. Ohkura, Fluorimetric Flow Injection Analysis for Inosine in Human Plasma Using Multi-Connected Immobilized Enzyme Columns [in Japanese]. *J. Flow Injection Anal.,* **2**(1) (1985) 50.

917. F. Scheller, N. Siegbahn, B. Danielsson, and K. Mosbach, High-Sensitivity Enzyme Thermistor Determination of L-Lactate by Substrate Recycling. *Anal. Chem.,* **57** (1985) 1740.

918. S. D. Rothwell and A. A. Woolf, A Timed Solenoid Injector for Flow Analysis. *Talanta,* **32** (1985) 431.

919. A. Rios, M. D. Luque de Castro, and M. Valcarcel, Injection Analysis with Flow-Gradient Systems: A New Approach to Unsegmented Flow Techniques. *Talanta,* **32** (1985) 845.

920. H. Mikasa, S. Motomizu, and K. Toei, Fluorometric Determination of Ammonia in River Water by Flow Injection Analysis [in Japanese]. *Bunseki Kagaku,* **34** (1985) 518.

921. Y. Baba, Y. Yamamoto, N. Yoza, and S. Ohashi, Flow Injection Analysis for Nucleotides [in Japanese]. *Bunseki Kagaku,* **34** (1985) 692.

922. K. Oguma and R. Kuroda, Simultaneous Determinations by Flow-Injection Analysis [in Japanese]. *J. Flow Injection Anal.,* **2**(2) (1985) 98.

923. T. Yao, Flow Injection Analysis Using Immobilized Enzyme [in Japanese]. *J. Flow Injection Anal.,* **2**(2) (1985) 115.

924. M. Yamamoto, M. Yasuda, and Y. Yamamoto, Flow Injection-Hydride Generation–Atomic Absorption Spectrometry with Gas Diffusion Unit Using Microporous PFTF Tube [in Japanese]. *J. Flow Injection Anal.,* **2**(2) (1985) 134.

925. K. Uchida, M. Tomoda, and S. Saito, Comparison of Sample Injectors Used for Flow Injection Analysis [in Japanese]. *J. Flow Injection Anal.,* **2**(2) (1985) 143.

926. J. Toei and N. Baba, Flow Injection Analysis for Glucose Using Multi-Function Pump for HPLC [in Japanese]. *J. Flow Injection Anal.,* **2**(2) (1985) 151.

927. M. J. Milano and W. G. Liesegand, Laboratory Automation and Information Management for Continuous Flow Analysis. *Amer. Lab.,* **17**(2) (1985) 132.

928. C. Wechter, N. Sleszynski, J. J. O'Dea, and J. Osteryoung, Anodic Stripping Voltammetry with Flow Injection Analysis. *Anal. Chim. Acta,* **175** (1985) 45.

929. L. Nord, S. Johansson, and H. Brötell, Flow Injection Extraction and Gas-Chromatographic Determination of Terodiline in Blood Serum. *Anal. Chim. Acta,* **175** (1985) 281.

930. T. Yao and T. Wasa, Simultaneous Determination of L(+) and D(−)-Lactic Acid by Use of Immobilized Enzymes in a Flow-Injection System. *Anal. Chim. Acta,* **175** (1985) 301.

931. R. C. Schothorst and G. den Boef, The Application of Strongly Reducing

Agents in Flow Injection Analysis. Part 5. Chromium(II) and Vanadium(II) in Acidic Medium. *Anal. Chim. Acta,* **175** (1985) 305.

932. T. Kato, Flow-Injection Spectrophotometric Determination of Enalapril in Pharmaceuticals with Bromothymol Blue. *Anal. Chim. Acta,* **175** (1985) 339.

933. P. J. Worsfold, Selective Biochemical Reagents in Flow Injection Analysis. *Anal. Proc.,* **22** (1985) 357.

934. A. Townshend, Recent Advances in Chemiluminescence and Flow Injection Analysis. *Anal. Proc.,* **22** (1985) 370.

935. P. J. Worsfold, A. Hughes, and D. J. Mowthorpe, Determination of Human Serum Immunoglobulin G Using Flow Injection Analysis with Rate Turbidimetric Detections. *Analyst,* **110** (1985) 1303.

936. V. P. Y. Gadzekpo, G. J. Moody, and J. D. R. Thomas, Coated-Wire Lithium Ion-Selective Electrodes Based on Polyalkoxylate Complexes. *Analyst,* **110** (1985) 1381.

937. F. Belal and J. L. Anderson, Flow Injection Analysis of Three *N*-Substituted Phenothiazine Drugs with Amperometric Detection at a Carbon Fibre Array Electrode. *Analyst,* **110** (1985) 1493.

938. K. Ohshita, H. Wada, and G. Nagakawa, Synthesis of Bidentate Pyridylazo and Thiazolylazo Reagents and the Spectrophotometric Determination of Copper in a Flow-Injection System. *Anal. Chim. Acta,* **176** (1985) 41.

939. E. N. Chaney Jr. and R. P. Baldwin, Voltammetric Determination of Doxorubicin in Urine by Adsorptive Preconcentration and Flow Injection Analysis. *Anal. Chim. Acta,* **176** (1985) 105.

940. T. Imato and N. Ishibashi, Flow Injection Analysis of Concentrated Aqueous Solution of Strong Acids and Bases. *Anal. Sci.,* **1** (1985) 481.

941. P. W. Alexander, P. R. Haddad, and M. Trojanowicz, Potentiometric Flow-Injection Determination of Sugars Using a Metallic Copper Electrode. *Anal. Lett.,* **18** (1985) 1953.

942. D. K. Morgan, N. D. Danielson, and J. E. Katon, Aqueous Flow Injection Analysis with Fourier Transform Infrared Detection. *Anal. Lett.,* **18** (1985) 1979.

943. G. Johansson, Flow Injection Analysis: A Powerful and Versatile Tool for Chemical Analysis [in Swedish]. *Kemisk Tidskrift,* **9** (1985) 58.

944. Bo Karlberg, Liquid-Liquid Extractions in FIA. *FIAstar NEWSlett.,* **2** (1985) 1.

945. J. Möller, Determination of Aluminium in Water and Soils. *FIAstar NEWSlett.,* **2** (1985) 2.

946. H. Casey and S. Smith, Application of Flow Injection Analysis to Water Pollution Studies. *Trends Anal. Chem.,* **4** (1985) 256.

947. D. A. Hollowell, G. E. Pacey, and G. Gordon, Selective Determination of Chlorine Dioxide Using Gas Diffusion Flow Injection Analysis. *Anal. Chem.,* **57** (1985) 2851.

948. M. L. Hitchman and F. W. M. Nyasulu, Indirect Potentiometric Monitoring of Proteins with a Copper Electrode. *Anal. Chim. Acta,* **173** (1985) 337.

949. Toyo Soda Manufacturing Co., Ltd., Flow-Injection Analysis [in Japanese]. Jpn. Kokai Tokkyo Koho, Pat. No. 8582968 (1985).

950. Toshiba Corp., Flow-Injection Analysis Method [in Japanese]. Jpn. Kokai Tokkyo Koho, Pat. No. 8561646 (1985).

951. P. I. Anagnostopoulou and M. A. Koupparis, Automated Flow-Injection Pseudotitration of Boric Acid. *J. Pharm. Sci.,* **74** (1985) 886.

952. Y. Lu, Y. Zhu, and Z. Yan, A Simple Single-Channel Flow-Injection Spectrophotometer [in Chinese]. *Fenxi Huaxue,* **13** (1985) 545.

953. Olympus Optical Co., Ltd., Determination of Enzyme Activity by Flow Injection Method [in Japanese]. Jpn. Kokai Tokkyo Koho, Pat. No. 85126096 (1985).

954. Olympus Optical Co., Ltd., Flow Injection-Type Analytical Method [in Japanese]. Jpn. Kokai Tokkyo Koho, Pat. No. 85125562 (1985).

955. A. F. Gomez, J. R. Polonio, M. D. Luque de Castro, and M. Valcarcel Cases, Automatic Enzymatic-Fluorimetric Determination of Ethanol in Blood by Flow Injection Analysis. *Anal. Chim. Acta,* **148** (1985) 131.

956. P. W. Alexander, P. R. Haddad, and M. Trojanowicz, Response Characteristics of a Potentiometric Detector with a Copper Electrode for Flow-Injection and Chromatographic Determinations of Metal Ions. *Anal. Chim. Acta,* **177** (1985) 183.

957. T. J. Cardwell, R. W. Cattrall, P. J. Iles, and I. C. Hamilton, Photo-Cured Polymers in Ion-Selective Electrode Membranes. Part 2. A Calcium Electrode for Flow Injection Analysis. *Anal. Chim. Acta,* **177** (1985) 239.

958. D. J. Curran and W. G. Collier, Determination of Phenyl Isocyanate in a Flow-Injection System with Infrared Spectrometric Detection. *Anal. Chim. Acta,* **177** (1985) 259.

959. P. Linares, M. D. Luque de Castro, and M. Valcarcel, Spectrofluorimetric Determination of Silicon by Flow Injection Analysis. *Anal. Chim. Acta,* **177** (1985) 263.

960. A. Recktenwald, K.-H. Kroner, and M.-R. Kula, On-Line Monitoring of Enzymes in Downstream Processing by Flow Injection Analysis (FIA). *Enzyme Microb. Technol.,* **7** (1985) 607.

961. H. Hustedt, K.-H. Kroner, and M.-R. Kula, On-Line-Determination of Enzymes and Proteins for Process Control in Enzyme Production [in German]. *Biotech. Forum,* **2** (1985) 57.

962. S. Alegret, J. Alonso, J. Bartroli, J. L. F. C. Lima, A. A. S. C. Machado, and J. M. Paulis, Flow-Through Sandwich PVC Matrix Membrane Electrode for Flow Injection Analysis. *Anal. Lett.,* **A18** (1985) 2291.

963. H. Haraguchi and T. Hasegawa, ICP Emission Spectrometry [in Japanese]. *Kagaku,* **40** (1985) 476.

964. J. Martinez Calatayud, Applications of Flow Injection to Samples of Agricultural Interest [in Spanish]. *Tec. Lab.,* **9** (1985) 18.

965. Olympus Optical Co., Ltd., Automated Analyzer Based on Immunochemical Reactions [in Japanese]. Jpn. Kokai Tokkyo Koho, Pat. No. 85159648 (1985).

966. Olympus Optical Co., Ltd., Flow Injection-Type Analysis Based on Antigen-Antibody Reactions [in Japanese]. Jpn. Kokai Tokkyo Koho, Pat. No. 85159649 (1985).

967. P. Hernandez, L. Hernandez, J. Vicente, and M. T. Sevilla, Use of an FIA Ion Exchanger-Atomic Absorption System for Determining Manganese, Lead, and Copper [in Spanish]. *An. Quim. Ser. B,* **81** (1985) 117.

968. J. Ruz, A. Rios, M. D. Luque de Castro, and M. Valcarcel, Simultaneous and Sequential Determination of Chromium(VI) and Chromium(III) by Unsegmented Flow Methods. *Fresenius Z. Anal. Chem.,* **322** (1985) 499.

969. P. Martinez-Jiménez, M. Gallego, and M. Valcárcel, Indirect Atomic Absorption Determination of Cerium and Lanthanum by Flow Injection Analysis Using an Air-Acetylene Flame. *At. Spectrosc.,* **6** (1985) 137.

970. E. A. Jones, The Determination of Uranium(VI) by Flow-Injection Analysis, *Rep.-MINTEK (S. Africa),* **M206** (1985) 25.

971. V. V. Kuznetsov, Flow-Injection Photometric Analysis. Determination of Iron(III) [in Russian]. *Zh. Anal. Khim.,* **40** (1985) 1859.

972. D. Betteridge, A. P. Wade, and A. G. Howard, Reflections on the Modified Simplex—I. *Talanta,* **32** (1985) 709.

973. T. Nomura, H. Ukeda, K. Matsumoto, and Y. Osajima, Electrochemical Measurement of Sugar Content of Food. VII. Flow Injection Analysis System for Conductometric Measurement of Sugar Content [in Japanese]. *Nippon Shokuhin Kogyo Gakkaishi,* **32** (1985) 576.

974. R. A. Salerno, C. Odell, N. Cryanovich, B. Bubnis, W. Morges, and A. Gray, Lowry Protein Determination by Automated Flow Injection Analysis for Bovine Serum Albumin and Hepatitis B Surface Antigen. *Anal. Biochem.,* **151** (1985) 309.

975. F. Belal and J. L. Anderson, Flow Injection Analysis of Warfarin Sodium with Amperometric Detection. *Mikrochim. Acta,* **II** (1985) 145.

976. B. Olsson, A Flow-Injection System Using Immobilized Peroxidase and Chromogenic Reagents for Possible Determination of Hydrogen Peroxide. *Mikrochim. Acta,* **II** (1985) 211.

977. K. Kashiwabara, T. Hubo, E. Kobayashi, and S. Suzuki, Flow Injection Analysis for Traces of Zinc with Immobilized Carbonic Anhydrase. *Anal. Chim. Acta,* **178** (1985) 209.

978. O. Røyset, Determination of Phosphate Species in Nutrient Solutions and Phosphorous in Plant Material as Phosphovanadomolybdic Acid by Flow Injection Analysis. *Anal. Chim. Acta,* **178** (1985) 217.

979. O. Røyset, Comparison of Four Chromogenic Reagents for the Flow-Injection Determination of Aluminium in Water. *Anal. Chim. Acta,* **178** (1985) 223.

980. C. A. Scolari and S. D. Brown, Multicomponent Determination in Flow-

Injection Systems with Square-Wave Voltammetric Detection Using the Kalman Filter. *Anal. Chim. Acta,* **178** (1985) 239.

981. L. Gorton, A Carbon Electrode Sputtered with Palladium and Gold for the Amperometric Detection of Hydrogen Peroxide. *Anal. Chim. Acta,* **178** (1985) 247.
982. H. W. Shih and C. O. Huber, Amperometric Determination of Formate with a Nickel Electrode. *Anal. Chim. Acta,* **178** (1985) 313.
983. A. N. Tsaousis and C. O. Huber, Flow-Injection Amperometric Determination of Chlorine at a Gold Electrode. *Anal. Chim. Acta,* **178** (1985) 319.
984. H. Ma and H. Yan, Flow Injection Analysis [in Chinese]. *Huanjing Kexue Congkan,* **1** (1983) 44.
985. H. A. Mottola, C.-M. Wolff, A. Iob, and R. Gnanasekaran, Potentiometric and Amperometric Detection in Flow Injection Analysis. In *Modern Trends in Analytical Chemistry* (Proc. of Two Scientific Symps., Mátrafüred, Hungary, 1982. I. "Electrochemical Detection in Flow Analysis") (E. Pungor, I. Bizás and G. E. Veress, Eds.). *Anal. Chem. Symp. Series,* **18** (1984) 49.
986. H. Matschiner, H. H. Ruettinger, P. Sivers, and U. Mann, Quantitative Determination of Chlorate Ions [in German]. German (East) Pat. No. 216543 (1984).
987. H. Matschiner, P. Sivers, and H. H. Ruettinger, Photometric Flow Analyzer [in German]. German (East) Pat. No. 220883 (1985).
988. J. D. R. Thomas, Continuous Analysis with Ion-Selective Electrodes: Flow-Injection Analysis and Monitoring of Enzyme-Based Reactions. In *Ion-Selective Electrodes, 4* (Proc. of the 4th Symp., Mátrafüred, Hungary, 1984) (E. Pungor and I. Buzás, Eds.). *Anal. Chem. Symp. Series,* **22** (1985) 213.
989. Y. Yang, Thermal Lens Spectrometry Based on Single-Laser/Dual-Beam Configuration with Lock-in Detection. *Proc. Int. Conf. Lasers,* (1985) 445.
990. M. Koizumi, T. Ono, T. Tanaka, T. Yokoyama, and A. Kiyama, Evaluation of a Nitrogen Autoanalyzer for Wastewater. Proc. 4th IAWPRC Workshop (Instrum. Control Water Wastewater Treat. Transp. Syst.), Pergamon, New York, (1985) 533.
991. T. P. Ruiz, M. H. Cordoba, C. M. Lozano, and C. Sanchez-Pedreno, Spectrophotometric Determination of Mercury by Flow Injection Analysis [in Spanish]. *Quim. Anal.,* **4** (1985) 72.
992. A. Müller, H. Eulenberg, and W. Kopprasch, Automated Analyzer ADM300 with Microcomputer-Controlled Evaluator AE2-2 [in German]. *Wasserwirtsch.-Wassertech.,* **35** (1985) 98.
993. H. Hashimoto, I. Iwahashi, and M. Saito, Flow-Injection Analyzer [in Japanese]. Jpn. Kokai Tokkyo Koho, Pat. No. 85128365 (Hitachi, Ltd.) (1985).
994. W. R. Seitz and M. L. Grayeski, The Flow Injection Approach for Making Analytical Chemiluminescence Measurements. *Biolumin. Chemilumin.: Instrum. Appl.,* **2** (1985) 95.

995. M. Koupparis, P. Macheras, and C. Tsaprounis, Automated Flow-Injection Colorimetric Determination of Acetominophen for Assays and Dissolution Studies of Multicomponent Dosage Forms. *Int. J. Pharm.*, **27** (1985) 349.

996. M. Uejima, Flow Through-Type Analyzer [in Japanese]. Jpn. Kokai Tokkyo Koho, Pat. No. 85205346 (Shimadzu Corp.) (1985).

997. G. H. Heider, Jr., The Application of Chemically Modified Electrodes in Flow Injection Analysis. *Diss. Abstr. Int. B,* **46** (1985) 499.

998. J. Růžička, Flow Injection Analysis: From Serial Assay to a New Concept of Measurement in the Chemical Laboratory. Proc. Symp. New Dir. Chem. Anal., Texas A&M Univ., **3rd**. (1985) 199.

999. T. A. Kelly, Separation-Free Enzyme Fluorescence Immunoassay by Continuous Flow Injection Analysis. *Proc. Enzyme-Mediated Immunoassay,* Plenum, New York (1985) 191.

1000. E. H. Hansen, Flow Injection Analysis. Polyteknisk Forlag, Copenhagen, Denmark, (1986). (D.Sc. Diss.).

1001. C. A. Broyles, Creatine Kinase Isoenzymes Determination by Selective Inhibition and by Flow Injection Analysis. *Diss. Abstr. Int. B,* **46** (1985) 1144.

1002. S. Takitani, M. Suzuki, M. Takamatu, M. Murayama, and A. Sano, Studies on Fluorometric Determination of Cyclophosphamide with Nicotinamide and Acetophenone [in Japanese]. *Iyakuhin Kenkyo,* **16** (1985) 1136.

1003. E. A. Jones, The Determination, by Flow Injection Analysis, of Fluoride, Chloride, Phosphate, Ammonia, Nitrite, and Nitrate. *Rep.-MINTEK (S. Africa),* **M200** (1985) 65.

1004. K. Peisker and H. Matschiner, Some Test Parameters of Four Methods for the Determination of Protein with Respect to Their Application in Flow Injection Analysis [in German]. *Z. Med. Laboratoriumsdiagn.,* **26** (1985) 422.

1005. I. G. Cook, C. W. McLeod, and P. J. Worsefold, Use of Activated Alumina as a Column Packing Material for Adsorption of Oxyanions in Flow Injection Analysis with ICP-AES Detection. *Anal. Proc.,* **23** (1986) 5.

1006. D. C. Stone and J. F. Tyson, Effect of Flow Cell on Dispersion in Flow Injection Analysis. *Anal. Proc.,* **23** (1986) 23.

1007. R. W. Abbott and A. Townshend, The Chemiluminescence Determination of Drugs. *Anal. Proc.,* **23** (1986) 25.

1008. A. T. Haj-Hussein and G. D. Christian, Simultaneous Spectrophotometric Determination of Copper, Nickel and Palladium by Flow Injection Analysis. *Analyst,* **111** (1986) 65.

1009. S. Maspoch, M. Blanco, and V. Cerdá, Catalytic Determination of Manganese at Ultra-Trace Levels by Flow Injection Analysis. *Analyst,* **111** (1986) 69.

1010. A. T. Haj-Hussein, G. D. Christian, and J. Růžička, Determination of Cyanide by Atomic Absorption Using a Flow Injection Conversion Method. *Anal. Chem.,* **58** (1986) 38.

1011. C. A. Koerner and T. A. Nieman, Chemiluminescence Flow Injection Analysis Determination of Sucrose Using Enzymatic Conversion and a Microporous Membrane Flow Cell. *Anal. Chem.,* **58** (1986) 116.

1012. P. Linares, M. D. Luque de Castro, and M. Valcárcel, Flow Injection Analysis of Binary and Ternary Mixtures of Arsenite, Arsenate, and Phosphate. *Anal. Chem.,* **58** (1986) 120.

1013. K. W. Pratt and W. F. Koch, Determination of Trace-Level Chromium(VI) in the Presence of Chromium(III) and Iron(III) by Flow Injection Amperometry. *Anal. Chem.,* **58** (1986) 124.

1014. K. R. Wehmeyer, H. B. Halsall, W. R. Heineman, C. P. Volle, and I-W. Chen, Competitive Heterogeneous Enzyme Immunoassay for Digoxin with Electrochemical Detection. *Anal. Chem.,* **58** (1986) 135.

1015. F. Winquist, I. Lundström, and B. Danielsson, Determination of Creatinine by an Ammonia-Sensitive Semiconductor Structure and Immobilized Enzymes. *Anal. Chem.,* **58** (1986) 145.

1016. P. I. Anagnostopoulou and M. A. Koupparis, Automated Flow-Injection Phenol Red Method for Determination of Bromide and Bromide salts in Drugs. *Anal. Chem.,* **58** (1986) 322.

1017. R. S. Vithanage and P. K. Dasgupta, Quantitative Study of Chemical Equilibria by Flow Injection Analysis with Diode Array Detection. *Anal. Chem.,* **58** (1986) 326.

1018. G. E. Pacey, M. R. Straka, and J. R. Gord, Dual Phase Gas Diffusion Flow Injection Analysis/Hydride Generation Atomic Absorption Spectrometry. *Anal. Chem.,* **58** (1986) 502.

1019. P. K. Dasgupta and E. L. Loree, Versatile Instrument for Pulse Width Measurement. *Anal. Chem.,* **58** (1986) 507.

1020. A. F. Kapauan and M. C. Magno, Continuous-Flow Injector for Flow Injection Analysis. *Anal. Chem.,* **58** (1986) 509.

1021. M. D. Luque de Castro, Speciation Studies by Flow-Injection Analysis. *Talanta,* **33** (1986) 45.

1022. J. M. H. Appleton and J. F. Tyson, Flow Injection Atomic Absorption Spectrometry: The Kinetics of Instrument Response. *J. Anal. Atom. Spectrom.,* **1** (1986) 63.

1023. J. M. Harnly and G. R. Beecher, Signal to Noise Ratios for Flow Injection Atomic Absorption Spectrometry. *J. Anal. Atom. Spectrom.,* **1** (1986) 75.

1024. J. L. Burguera, M. Burguera, and O. M. Alarcón, Determination of Sodium, Potassium, Calcium, Magnesium, Iron, Copper and Zinc in Cerebrospinal Fluid by Flow Injection Atomic Absorption Spectrometry. *J. Anal. Atom. Spectrom.,* **1** (1986) 79.

1025. M. Gallego, M. Silva, and M. Valcárcel, Determination of Nitrate and Nitrite by Continuous Liquid-Liquid Extraction with a Flow-Injection Atomic-Absorption Detection System. *Fresenius Z. Anal. Chem.,* **323** (1986) 50.

1026. V. P. Y. Gadzekpo, G. J. Moody, and J. D. R. Thomas, Lithium Ion-Selective Electrodes in Flow Injection Analysis. *Anal. Proc.*, **23** (1986) 62.

1027. A. B. Ghawji and A. G. Fogg, Reduction in Size by Electrochemical Pre-treatment at High Negative Potentials of the Background Currents Obtained at Negative Potentials at Glassy Carbon Electrodes and its Application in the Reductive Flow Injection Amperometric Determination of Nitrofurantoin, *Analyst,* **111** (1986) 157.

1028. F. Lázaro, A. Rios, M. D. Luque de Castro, and M. Valcárcel, Determination of Vitamin C by Flow Injection Analysis. *Analyst,* **111** (1986) 163.

1029. F. Lázaro, A. Rios, M. D. Luque de Castro, and M. Valcárcel, Determination of Vitamin C in Urine by Flow Injection Analysis. *Analyst,* **111** (1986) 167.

1030. M. Burguera and J. L. Burguera, Flow Injection—Hydride Generation System for the Determination of Arsenic by Molecular Emission Cavity Analysis. *Analyst,* **111** (1986) 171.

1031. A. G. Fogg, R. M. Alonso, and M. A. Fernández-Arciniega, Oxidative Amperometric Flow Injection Determination of Oxalate at an Electrochemically Pre-treated Glassy Carbon Electrode. *Analyst,* **111** (1986) 249.

1032. N. Ishibashi and T. Imato, Flow Injection Analysis by Using Acid-Base or Metal-Ligand Buffers and Color-Indicators. *Fresenius Z. Anal. Chem.,* **323** (1986) 244.

1033. T. Deguchi, A. Higashi, and I. Sanemasa, Flow Injection Analysis of Cobalt(II) by Catalytic Oxidations of Stilbazo and Pyrocatechol Violet. *Bull. Chem. Soc. Jpn.,* **59** (1986) 295.

1034. H. Satake, Y. Kohri, and S. Ikeda, Flow Injection Analysis of Hydroquinone, Pyrocatechol, Resorcinol and Pyrogallol with Amperometric Detector [in Japanese]. *Nippon Kagaku Kaishi,* **1** (1986) 43.

1035. M. J. Whitaker, Laboratory Note: Industrial Applications of Flow Injection Analysis. *Amer. Lab.,* **18**(3) (1986) 154.

1036. J. F. Coetzee and C. W. Gardner, Jr., Determination of Sulfate, Orthophosphate, and Triphosphate Ions by Flow Injection Analysis with the Lead Ion Selective Electrode as Detector. *Anal. Chem.,* **58** (1986) 608.

1037. J. F. Coetzee and C. Gunaratna, Potentiometric Gas Sensor for the Determination of Free Chlorine in Static or Flow Injection Analysis Systems. *Anal. Chem.,* **58** (1986) 650.

1038. A. Rios, M. D. Luque de Castro, and M. Valcárcel, New Configuration for Construction of pH Gradients in Flow Injection Analysis. *Anal. Chem.,* **58** (1986) 663.

1039. M. Maeda and A. Tsuji, Chemiluminescence Flow Injection Analysis of Biological Compounds Based on the Reaction with Lucigenin. *Anal. Sci.,* **2** (1986) 183.

1040. T. Yamane, Determination of Manganese at Trace Levels in Natural Waters with Continuous Flow System Utilizing On-Line Cation-Exchange Separation and Catalytic Detection. *Anal. Sci.,* **2** (1986) 191.

1041. A. Tanaka, K. Obata, and T. Deguchi, Fluorimetric Catalytic Determination of Iodide by Flow Injection Analysis. *Anal. Sci.,* **2** (1986) 197.

1042. M. A. Koupparis and A. Barcuchová, Automated Flow Injection Spectrophotometric Determination of Some Phenothiazines Using Iron Perchlorate: Applications in Drug Assays, Content Uniformity and Dissolution Studies. *Analyst,* **111** (1986) 313.

1043. P. J. Worsfold, Flow Analysis III—A Report on the Third International Conference on Flow Analysis, Held in Birmingham, U.K., 5-8 September, 1985. *Trends Anal. Chem.,* **5** (1986) XI.

1044. M. D. Luque de Castro and M. Valcárcel, Flow Injection Methods Based on Multidetection. *Trends Anal. Chem.,* **5** (1986) 71.

1045. H. Nakajima and H. Ichiki, The Rapid Examination for Phosphate Fertility in Grazing and Farm Land Soil Derived from Neutral Volcanic Ash. 3. Application of Flow Injection Analysis to 10^{-2} H_2SO_4 Extraction [in Chinese]. *Tohoku Agric. Res.,* **37** (1985) 135.

1046. H. Nakajima and H. Ichiki, The Rapid Examination for Phosphate Fertility in Grazing and Farm Land Soil Derived from Neutral Volcanic Ash. 4. Application of Flow Injection Analysis to $NH_4F{\cdot}HCl$ Extraction [in Chinese]. *Tohoku Agric. Res.,* **37** (1985) 137.

1047. F. T. M. Dohmen and P. C. Thijssen, A FORTH Package for Computer-Controlled Flow-Injection Analysis. *Talanta,* **33** (1986) 107.

1048. S. Uchiyama, Y. Tohfuku, S. Suzuki, and G. Muto, Determination of High Concentrations of Glucose by Flow Injection Analysis with Immobilized Enzyme Tube Reactor [in Japanese]. *Bunseki Kagaku,* **35** (1986) 134.

1049. J. Ruz, A. Ríos, M. D. Luque de Castro, and M. Valcárcel, Flow-Injection Analysis with Multidetection as a Useful Technique for Metal Speciation. *Talanta,* **33** (1986) 199.

1050. A. Trojanek and S. Bruckenstein, Novel Flow-Through Pneumatoamperometric Detector for Determination of Nanogram and Subnanogram Amounts of Nitrite by Flow-Injection Analysis. *Anal. Chem.,* **58** (1986) 866.

1051. A. Trojanek and S. Bruckenstein, Flow-Injection Analysis of Volatile, Electroinactive Organic Compounds at a Platinum Gas Diffusion Membrane Electrode by Use of a Redox Mediator. *Anal. Chem.,* **58** (1986) 981.

1052. A. Trojanek and S. Bruckenstein, Flow-Injection Analysis of Phenols via Bromination and Detection of Unreacted Bromine at a Platinized Gas Diffusion Membrane Electrode. *Anal. Chem.,* **58** (1986) 983.

1053. E. W. Kristensen, R. L. Wilson, and R. M. Wightman, Dispersion in Flow Injection Analysis Measured with Microvoltammetric Electrodes. *Anal. Chem.,* **58** (1986) 986.

1054. F. Belal and J. L. Anderson, Flow Injection Determination of Ergonovine Maleate with Amperometric Detection at the Kel-F-Graphite Composite Electrode. *Talanta,* **33** (1986) 448.

1055. J. Růžička and E. H. Hansen, The First Decade of Flow Injection Analysis: From Serial Assay to Diagnostic Tool. *Anal. Chim. Acta,* **179** (1986) 1.

1056. K. K. Stewart, Time-Based Flow Injection Analysis. *Anal. Chim. Acta,* **179** (1986) 59.

1057. C. Riley, B. F. Rocks, and R. A. Sherwood, Controlled-Dispersion Flow Analysis. Flow-Injection Analysis Applied to Clinical Chemistry. *Anal. Chim. Acta,* **179** (1986) 69.

1058. J. N. Miller, G. L. Abdullahi, H. N. Sturley, V. Gossain, and P. L. McCluskey, Studies of Interacting Biochemical Systems by Flow Injection Analysis. *Anal. Chim. Acta,* **179** (1986) 81.

1059. W. E. van der Linden, Flow Injection Analysis in On-Line Process Control. *Anal. Chim. Acta,* **179** (1986) 91.

1060. F. J. Krug, H. Bergamin Fº, and E. A. G. Zagatto, Commutation in Flow Injection Analysis. *Anal. Chim. Acta,* **179** (1986) 103.

1061. J. T. Vanderslice, A. G. Rosenfeld, and G. A. Beecher, Laminar-Flow Bolus Shapes in Flow Injection Analysis. *Anal. Chim. Acta,* **179** (1986) 119.

1062. J. F. Tyson, Peak Width and Reagent Dispersion in Flow Injection Analysis. *Anal. Chim. Acta,* **179** (1986) 131.

1063. M. Gisin, C. Thommen, and K. F. Mansfield, Hydrodynamically Limited Precision of Gradient Techniques in Flow Injection Analysis. *Anal. Chim. Acta,* **179** (1986) 149.

1064. H. L. Pardue and P. Jager, Kinetic Treatment of Unsegmented Flow Systems. Part 3. Flow-Injection System with Gradient Chamber Evaluated with a Linearly Responding Detector. *Anal. Chim. Acta,* **179** (1986) 169.

1065. H. Wada, S. Hiraoka, A. Yuchi, and G. Nakagawa, Sample Dispersion with Chemical Reaction in a Flow-Injection System. *Anal. Chim. Acta,* **179** (1986) 181.

1066. C. J. Patton and S. R. Crouch, Experimental Comparison of Flow-Injection Analysis and Air-Segmented Continuous Flow Analysis. *Anal. Chim. Acta,* **179** (1986) 189.

1067. B. Olsson, B. Stålbom, and G. Johansson, Determination of Sucrose in the Presence of Glucose in a Flow-Injection System with Immobilized Multi-Enzyme Reactors. *Anal. Chim. Acta,* **179** (1986) 203.

1068. H. Hirano, Y. Baba, N. Yoza, and S. Ohashi, Measurements of Kinetic Parameters of Inorganic Pyrophosphatase by Flow-Injection Procedures. *Anal. Chim. Acta,* **179** (1986) 209.

1069. M. Masoom and P. J. Worsfold, The Kinetic Determination of Clinically Significant Enzymes in an Automated Flow-Injection System with Fluorescence Detection. *Anal. Chim. Acta,* **179** (1986) 217.

1070. B. F. Rocks, R. A. Sherwood, M. M. Hosseinmardi, and C. Riley, The Use of Holding Coils to Facilitate Long Incubation Times in Unsegmented Flow Analysis. Determination of Serum Prostatic Acid Phosphatase. *Anal. Chim. Acta,* **179** (1986) 225.

1071. A. T. Faizullah and A. Townshend, Applications of Ion-Exchange Minicolumns in a Flow-Injection System for the Spectrophotometric Determination of Anions. *Anal. Chim. Acta,* **179** (1986) 233.

1072. K. S. Johnson, C. L. Beehler, and C. M. Sakamoto-Arnold, A Submersible Flow Analysis System. *Anal. Chim. Acta,* **179** (1986) 245.

1073. G. E. Pacey, D. A. Hollowell, K. G. Miller, M. R. Straka, and G. Gordon, Selectivity Enhancement by Flow Injection Analysis. *Anal. Chim. Acta,* **179** (1986) 259.

1074. J. M. H. Appleton, J. F. Tyson, and R. P. Mounce, The Rapid Determination of Chemical Oxygen Demand in Waste Waters and Effluents by Flow Injection Analysis. *Anal. Chim. Acta,* **179** (1986) 269.

1075. F. Lázaro, A. Ríos, M. D. Luque de Castro, and M. Valcárcel, Simultaneous Multiwavelength Detection in Flow Injection Analysis. *Anal. Chim. Acta,* **179** (1986) 279.

1076. M. Milla, R. M. de Castro, M. Garcia-Vargas, and J. A. Muñoz-Leyva, Batch and Flow-Injection Determination of Ethylenediamine in Pharmaceutical Preparations. *Anal. Chim. Acta,* **179** (1986) 289.

1077. R. C. Schothorst, O. O. Schmitz, and G. den Boef, The Application of Strongly Oxidizing Agents in Flow Injection Analysis. Part 2. Manganese(III). *Anal. Chim. Acta,* **179** (1986) 299.

1078. P. J. Worsfold and A. Nabi, Bioluminescent Assays with Immobilized Firefly Luciferase Based on Flow Injection Analysis. *Anal. Chim. Acta,* **179** (1986) 307.

1079. Y. Sahleström and B. Karlberg, An Unsegmented Extraction System for Flow Injection Analysis. *Anal. Chim. Acta,* **179** (1986) 315.

1080. Z. Fang, S. Xu, X. Wang, and S. Zhang, Combination of Flow-Injection Techniques with Atomic Spectrometry in Agricultural and Environmental Analysis. *Anal. Chim. Acta,* **179** (1986) 325.

1081. M. Silva, M. Gallego, and M. Valcárcel, Sequential Atomic Absorption Spectrometric Determination of Nitrate and Nitrite in Meats by Liquid-Liquid Extraction in a Flow-Injection System. *Anal. Chim. Acta,* **179** (1986) 341.

1082. M. Burguera, J. L. Burguera, and O. M. Alarcón, Flow Injection and Microwave-Oven Sample Decomposition for Determination of Copper, Zinc and Iron in Whole Blood by Atomic Absorption Spectrometry. *Anal. Chim. Acta,* **179** (1986) 351.

1083. K. Tóth, J. Fucskó, E. Lindner, Zs. Fehér, and E. Pungor, Potentiometric Detection in Flow Analysis. *Anal. Chim. Acta,* **179** (1986) 359.

1084. G. Marko-Varga, R. Appelqvist, and L. Gorton, A Glucose Sensor Based on Glucose Dehydrogenase Adsorbed on a Modified Carbon Electrode. *Anal. Chim. Acta,* **179** (1986) 371.

1085. G. G. Neuburger and D. C. Johnson, Constant-Potential Pulse Polarographic Detection in Flow-Injection Analysis without Deaeration of Solvent or Sample. *Anal. Chim. Acta,* **179** (1986) 381.

1086. W. Frenzel and P. Brätter, Flow-Injection Potentiometric Stripping Analysis—A New Concept for Fast Trace Determinations. *Anal. Chim. Acta,* **179** (1986) 389.

1087. M. Masoom and A. Townshend, Flow-Injection Determination of Sulphite and Assay of Sulphite Oxidase. *Anal. Chim. Acta,* **179** (1986) 399.

1088. J. F. van Staden, A Coated Tubular Solid-State Chloride-Selective Electrode in Flow-Injection Analysis. *Anal. Chim. Acta,* **179** (1986) 407.

1089. D. C. Stone and J. F. Tyson, Flow Cell and Diffusion Coefficient Effects in Flow Injection Analysis. *Anal. Chim. Acta,* **179** (1986) 427.

1090. M. Wasberg and A. Ivaska, A Computer-Controlled Voltammetric Flow-Injection System. *Anal. Chim. Acta,* **179** (1986) 433.

1091. M. Granados, S. Maspoch, and M. Blanco, Determination of Sulphur Dioxide by Flow Injection Analysis with Amperometric Detection. *Anal. Chim. Acta,* **179** (1986) 445.

1092. T. Anfält and S. Twengström, The Determination of Bromide in Natural Waters by Flow Injection Analysis. *Anal. Chim. Acta,* **179** (1986) 453.

1093. M. J. Whitaker, Spectrophotometric Determination of Nonionic Surfactants by Flow Injection Analysis Utilizing Ion-Pair Extraction and an Improved Phase Separator. *Anal. Chim. Acta,* **179** (1986) 459.

1094. A. Ríos, M. D. Luque de Castro, and M. Valcárcel, Simultaneous Determination by Iterative Spectrophotometric Detection in a Closed Flow System. *Anal. Chim. Acta,* **179** (1986) 463.

1095. K. H. Al-Sowdani and A. Townshend, Simultaneous Spectrofluorimetric Determination of Cerium (III) and Cerium (IV) by Flow Injection Analysis. *Anal. Chim. Acta,* **179** (1986) 469.

1096. O. F. Kamson, Spectrophotometric Determination of Iodate, Iodide and Acids by Flow Injection Analysis. *Anal. Chim. Acta,* **179** (1986) 475.

1097. S. R. Bysouth and J. F. Tyson, A Microcomputer-Based Peak-Width Method of Extended Calibration for Flow-Injection Atomic Absorption Spectrometry. *Anal. Chim. Acta,* **179** (1986) 481.

1098. A. G. Cox and C. W. McLeod, Preconcentration and Determination of Trace Chromium(III) by Flow Injection/Inductively-Coupled Plasma/Atomic Emission Spectrometry. *Anal. Chim. Acta,* **179** (1986) 487.

1099. C. G. Taylor and J. M. Trevaskis, Determination of Lead in Gasolin by a Flow-Injection Technique with Atomic Absorption Spectrometric Detection. *Anal. Chim. Acta,* **179** (1986) 491.

1100. J. L. Burguera and M. Burguera, Determination of Some Organophosphorous Insecticides by Flow-Injection with a Molecular Emission Cavity Detector. *Anal. Chim. Acta,* **179** (1986) 497.

1101. J. Alonso, J. Bartroli, J. L. F. C. Lima, and A. A. S. C. Machado, Sequential Flow-Injection Determination of Calcium and Magnesium in Waters. *Anal. Chim. Acta,* **179** (1986) 503.

1102. L. Risinger, Preconcentration of Copper(II) on Immobilized 8-Quinolinol in a Flow-Injection System with an Ion-Selective Electrode Detector. *Anal. Chim. Acta,* **179** (1986) 509.

1103. Bo Olsson, H. Lundbäck, G. Johansson, F. Scheller, and J. Nentwig, Theory and Application of Diffusion-Limited Amperometric Enzyme Electrode Detection in Flow Injection Analysis of Glucose. *Anal. Chem.,* **58** (1986) 1046.

1104. P. van Zoonen, D. A. Kamminga, C. Gooijer, N. H. Velthorst, and R. W. Frei, Quenched Peroxyoxalate Chemiluminescence as a New Detection Principle in Flow Injection Analysis and Liquid Chromatography. *Anal. Chem.,* **58** (1986) 1245.

1105. K. E. LaFreniere, G. W. Rice, and V. A. Fassel, Flow Injection Analysis with Inductively Coupled Plasma-Atomic Emission Spectroscopy: Critical Comparison of Conventional Pneumatic, Ultrasonic and Direct Injection Nebulization. *Spectrochim. Acta Pt. B—At. Spec.,* **40** (1985) 1495.

1106. T. Nomura, H. Ukeda, K. Matsumoto, and Y. Osajima, Electrochemical Measurement of Sugar Content of Food. VIII. Conductometric Measurement of Sugar Content with Flow Injection Analysis System [in Japanese]. *Nippon Shokuhin Kogyo Gakkaishi,* **32** (1985) 916.

1107. Y. Niiyama, J. Mori, and K. Sugawara, Enzyme Electrode and Its Use in Flow-Injection System [in Japanese]. Jpn. Kokai Tokkyo Koho, Pat. No. 85188839 (Hitachi, Ltd.), (1985).

1108. I. S. Krull, Recent Advances in New and Potentially Novel Detectors in High-Performance Liquid Chromatography and Flow Injection Analysis. *Amer. Chem. Soc. Symp. Ser.,* **297** (1986) 137.

1109. H. Matschiner, P. Sivers, and H. H. Ruettinger, Flow Injection Analysis—New Method in Wet Analysis [in German]. *Wiss. Z. Martin-Luther Univ. Halle-Wittenberg, Math.-Naturwiss. Reihe,* **34** (1985) 3.

1110. R. Oltner, S. Bengtsson, and K. Larsson, Flow Injection Analysis for the Determination of Urea in Cow Milk. *Acta Vet. Scand.,* **26** (1985) 396.

1111. P. Linares, M. D. Luque de Castro, and M. Valcárcel, Flow Injection Analysis in Clinical Chemistry. *Rev. Anal. Chem.,* **8** (1985) 229.

1112. G. den Boef and R. C. Schothorst, Unstable Reagents in Quantitative Analysis? Why Not Apply FIA! *Anal. Chim. Acta,* **180** (1986) 1.

1113. J. Emnéus, R. Appelqvist, G. Marko-Varga, L. Gorton, and G. Johansson, Determination of Starch in a Flow Injection System Using Immobilized Enzymes and a Modified Electrode. *Anal. Chim. Acta,* **180** (1986) 3.

1114. Z. Fang, Potential of the Flow Injection Technique in Soil Chemical Analysis. *Anal. Chim. Acta,* **180** (1986) 8.

1115. J. M. Harris, Flow Injection Analysis for Chemistry under Non-Ambient Conditions. *Anal. Chim. Acta,* **180** (1986) 11.

1116. J. Janata and J. J. Harrow, Heterogeneous Samples in Flow Injection Analysis. *Anal. Chim. Acta,* **180** (1986) 14.

1117. B. Karlberg, Flow Injection Analysis—Or the Art of Controlling Sample Dispersion in a Narrow Tube. *Anal. Chim. Acta,* **180** (1986) 16.

1118. W. E. van der Linden, Flow Injection Analysis—The Intermarriage of Analytical Chemistry and Chemical Engineering Science. *Anal. Chim. Acta,* **180** (1986) 20.

1119. J. N. Miller, V. K. Mahant, and H. Thakrar, Flow Injection Biochemical Analysis with Luminescence Detection: Towards the Simple Determination of Single Molecules. *Anal. Chim. Acta,* **180** (1986) 23.

1120. H. A. Mottola, Flow Injection Analysis: Sample and Reagent Processing in the Chemical Laboratory. *Anal. Chim. Acta,* **180** (1986) 26.

1121. H. Müller and V. Müller, An FIA System Using an Ion-Selective Electrode as Detector for the Catalimetric Determination of Molybdenum(VI) and Tungsten(VI). *Anal. Chim. Acta,* **180** (1986) 30.

1122. G. E. Pacey, Selectivity Enhancement Obtained by Using Flow Injection Analysis. *Anal. Chim. Acta,* **180** (1986) 36.

1123. C. Riley, What Flow-Injection Analysis Offers the Clinical Chemist. *Anal. Chim. Acta,* **180** (1986) 39.

1124. J. Růžička and E. H. Hansen, Peak Dimensions and Flow Injection Titrations. *Anal. Chim. Acta,* **180** (1986) 41.

1125. K. K. Stewart, G. R. Beecher, J. T. Vanderslice, and P. E. Hare, Flow Injection Analysis—A Personal View from the United States. *Anal. Chim. Acta,* **180** (1986) 46.

1126. A. Townshend, Ion-Exchange Minicolumns. *Anal. Chim. Acta,* **180** (1986) 49.

1127. J. F. Tyson, Flow Injection Techniques for Extending the Working Range of Atomic Absorption Spectrometry and U.V.-Visible Spectrophotometry. *Anal. Chim. Acta,* **180** (1986) 51.

1128. M. Valcárcel and M. D. Luque de Castro, Flow Injection Analysis: An Available Tool for Solving Analytical Problems. *Anal. Chim. Acta,* **180** (1986) 54.

1129. P. Worsfold, Flow Injection Techniques for Monitoring Biochemically Selective Interactions. *Anal. Chim. Acta,* **180** (1986) 56.

1130. N. Yoza, N. Ishibashi, and K. Ueno, Flow Injection Analysis in Japan. *Anal. Chim. Acta,* **180** (1986) 58.

1131. F. J. Krug, E. A. G. Zagatto, and H. Bergamin Fº, Ten Years of Flow Injection Analysis in a Brazilian Laboratory. *Anal. Chim. Acta,* **180** (1986) 59.

1132. K. Toei, S. Motomizu, M. Oshima, and M. Onoda, Spectrophotometric Determination of Boron by Flow Injection Analysis [in Japanese]. *Bunseki Kagaku,* **35** (1985) 344.

1133. H. Ohura, T. Imato, S. Yamasaki, and N. Ishibashi, Potentiometric Flow Injection Analysis of Trace Bromate Based on Its Redox-Reaction with the

Iron(III)-Iron(II) Buffer Solution Containing Bromide [in Japanese]. *Bunseki Kagaku,* **35** (1985) 349.

1134. M. Ishii, M. Yamada, and S. Suzuki, Determination of Copper(II) at Subpicogram Level by Flow Injection Method Using Micellar Enhanced Chemiluminescence of 1,10-Phenanthroline [in Japanese]. *Bunseki Kagaku,* **35** (1985) 373.

1135. M. Ishii, M. Yamada, and S. Suzuki, Determination of Copper(II) at Sub-Picogram Level by Flow Injection Method Using Micellar Enhanced Chemiluminescence of 1,10-Phenanthroline—Application to Rabbit Lens [in Japanese]. *Bunseki Kagaku,* **35** (1985) 379.

1136. J. Toei and N. Baba, New Type Valve for Flow Injection Analysis. 1. Application to Stopped-Flow Spectrophotometric Determination of Glucose [in Japanese]. *Bunseki Kagaku,* **35** (1985) 411.

1137. R. C. Schothorst and G. den Boef, The Application of Strongly Oxidizing Agents in Flow Injection Analysis. Part 3. Cobalt(III). *Anal. Chim. Acta,* **181** (1986) 235.

1138. F. P. Bigley, R. L. Grob, and G. S. Brenner, Pharmaceutical Applications of a High-Performance Flow Injection System. *Anal. Chim. Acta,* **181** (1986) 241.

1139. E. A. G. Zagatto, O. Bahia Fº, M. F. Giné, and H. Bergamin Fº, A Simple Procedure for Hydrodynamic Injection in Flow Injection Analysis Applied to the Atomic Absorption Spectrometry of Chromium in Steels, *Anal. Chim. Acta,* **181** (1986) 265.

1140. T. Kumamaru, H. Matsuo, Y. Okamoto, and M. Ikeda, Sensitivity Enhancement for Inductively-Coupled Plasma Atomic Emission Spectrometry of Cadmium by Suction-Flow On-Line Ion-Exchange Preconcentration. *Anal. Chim. Acta,* **181** (1986) 271.

1141. J. R. Chipperfield and P. J. Worsfold, A Simple Injection Valve for Flow Injection Analysis. *Anal. Chim. Acta,* **181** (1986) 283.

1142. N. Leon, J. L. Burguera, M. Burguera, and O. M. Alarcon, Determination of Cobalt and Manganese in Serum by Flow Injection Analysis and Atomic Absorption Spectroscopy. *Rev. Roum. Chim.,* **31** (1986) 353.

1143. I. N. Andersson, Development of Methods for Determination of Water Based on Flow Injection Analysis and Karl Fischer Reaction (Ph.D. Thesis). Dept. Anal. Chem, Univ. of Umeå, Sweden, 1986.

1144. V. P. Y. Gadzekpo, G. J. Moody, and J. D. R. Thomas, Problems in the Application of Ion-Selective Electrodes to Serum Lithium Analysis. *Analyst,* **111** (1986) 567.

1145. J.-S. Rhee and P. K. Dasgupta, Studies on Peak Width Measurement-Based FIA Acid-Base Determinations. *Mikrochim. Acta,* **III** (1985) 49.

1146. H. Hwang and P. K. Dasgupta, Flow Injection Analysis of Trace Hydrogen Peroxide Using an Immobilized Enzyme Reactor. *Mikrochim. Acta,* **III** (1985) 77.

1147. J.-S. Rhee and P. K. Dasgupta, Determination of Acids, Bases, Metal Ions and Redox Species by Peak Width Measurement Flow Injection Analysis with Potentiometric, Conductometric, Fluorometric and Spectrophotometric Detection. *Mikrochim. Acta,* **III** (1985) 107.

1148. J. Kirie and T. Ikue, Baseline Stabilization for Liquid Chromatography and Flow-Injection Analysis [in Japanese]. Jpn. Kokai Tokkyo Koho, Pat. No. 85168047 (Toyo Soda Mfg. Co., Ltd.) (1985).

1149. M. Nakamura, T. Kano, A. Tamagawa, and K. Doi, Flow-Injection Analysis Using Liquid Chromatography [in Japanese]. Jpn. Kokai Tokkyo Koho, Pat. No. 85161558 (Olympus Optical Co., Ltd.) (1985).

1150. Olympus Optical Co., Ltd., Apparatus for monitoring a Flowline Containing a Reaction Column [in Japanese]. Jpn. Kokai Tokkyo Koho, Pat. No. 85115852 (1985).

1151. J. T. Currie, Jr., A Response Surface Methodology Approach to Optimization in Flow Injection Analysis. *Diss. Abstr. Int. B,* **46** (1985) 498.

1152. Y. Hirai, Development of a High-Pressure Flow Injection System for the Determination of Inorganic Oxoacids of Phosphorous [in Japanese]. *Bunseki,* **12** (1985) 891.

1153. A. Miyawaki, H. Date, and Y. Kobayashi, Biosensor with Immobilized Physiologically Active Substances [in Japanese]. Jpn. Kokai Tokkyo Koho, Pat. No. 8603047 (1986).

1154. R. Klinkner and H. Engelhardt, Impact of Reaction Detectors on Flow Injection Analysis. *Proc. Int. Symp. Capillary Chromatogr.,* **6th** (1985) 857.

1155. F. Malecki, New Possibilities for Automation in Drug Control Using Flow Injection Analysis [in Polish]. *Farm. Pol.,* **41** (1985) 513.

1156. P. K. Dasgupta and V. K. Gupta, Membrane-Based Flow Injection System for Determination of Sulphur(IV) in Atmospheric Water. *Environ. Sci. Technol.,* **20** (1986) 524.

1157. D. A. Davies and C. O. Huber, Anodic Voltammetric Determination of Monochloramine in Water. *Proc. Conf. Water Chlorination: Environ. Impact Health Eff.,* **5th** (1985) 1091.

1158. G. Andersson, L. Andersson, and G. Carlström, Determination of Milk Urea by Flow Injection Analysis. *Zentralbl. Veterinärmed., Reihe A,* **33** (1986) 53.

1159. T. D. Yerian, T. P. Hadjioannou, and G. D. Christian, Flow-Injection Analysis with the Iron-Induced Perbromate-Iodide Reaction: Spectrophotometric Determination of Iron. *Talanta,* **33** (1986) 547.

1160. B. A. Petersson, E. H. Hansen, and J. Růžička, Enzymatic Assay by Flow Injection Analysis with Detection by Chemiluminescence: Determination of Glucose, Creatinine, Free Cholesterol and Lactic Acid Using an Integrated FIA Microconduit. *Anal. Lett.,* **19** (1986) 649.

1161. G. J. Moody, G. S. Sanghera, and J. D. R. Thomas, Amperometric Enzyme Electrode System for the Flow Injection Analysis of Glucose. *Analyst,* **111** (1986) 605.

1162. R. W. Abbott, A. Townshend, and R. Gill, Determination of Morphine by Flow Injection Analysis with Chemiluminescence Detection. *Analyst,* **111** (1986) 635.

1163. M. Aihara, M. Arai, and T. Taketatsu, Flow Injection Spectrofluorimetric Determination of Europium(III) Based on Solubilizing Its Ternary Complex with Thenoyltrifluoroacetone and Trioctylphosphine Oxide in Micellar Solution. *Analyst,* **111** (1986) 641.

1164. K. Ohshita, H. Wada, and G. Nakagawa, Spectrophotometric Determination of Silver with 4-(3,5-Dibromo-2-Pyridylazo)-*N*,*N*-Diethylaniline in the Presence of Sodium Dodecylsulfate. *Anal. Chim. Acta,* **182** (1986) 157.

1165. J. Georges and M. Khalil, A Potentiometric Detector for Following pH Shifts in Liquid Chromatography. *Anal. Chim. Acta,* **182** (1986) 281.

1166. D. Betteridge, A. P. Wade, and A. G. Howard, Reflections on the Modified Simplex—II. *Talanta,* **32** (1985) 723.

1167. E. H. Hansen and J. Růžička, The Growth of Published Papers on Flow Injection Analysis. *J. Flow Injection Anal.,* **3**(1) (1986) 2.

1168. K. K. Stewart, Some Thoughts About FIA in 1986. *J. Flow Injection Anal.,* **3**(1) (1986) 5.

1169. S. Hirata, Y. Umezaki, and M. Ikeda, Determination of Cadmium of ppb Level by Column Preconcentration-Atomic Absorption Spectrometry [in Japanese]. *J. Flow Injection Anal.,* **3**(1) (1986) 8.

1170. K. Uchida, M. Tomoda, and S. Saito, The Development of a Manually Operated Sample Injector Employing a Pair of Six-Valve Systems [in Japanese]. *J. Flow Injection Anal.,* **3**(1) (1986) 18.

1171. J. Toei, Y. Saigusa, and N. Baba, FIA Titration Using Volume-Variable Mixer (1) [in Japanese]. *J. Flow Injection Anal.,* **3**(1) (1986) 27.

1172. N. Yoza, Y. Sagara, H. Morioka, T. Handa, H. Hirano, Y. Baba, and S. Ohashi, Fate Analysis of Phosporous Compounds in Environmental Waters by Flow Injection Analysis and High-Performance Liquid Chromatography [in Japanese]. *J. Flow Injection Anal.,* **3**(1) (1986) 37.

1173. H. Hwang and P. K. Dasgupta, Fluorometric Flow Injection Determination of Aqueous Peroxides at Nanomolar Level Using Membrane Reactors. *Anal. Chem.,* **58** (1986) 1521.

1174. D. A. Hollowell, J. R. Gord, G. Gordon, and G. E. Pacey, Selective Chlorine Dioxide Determination Using Gas-Diffusion Flow Injection Analysis with Chemiluminescence Detection. *Anal. Chem.,* **58** (1986) 1524.

1175. H. Gunasingham, B. T. Tay, and K. P. Ang, Automated Mercury Film Electrode for Flow Injection Analysis and High-Performance Liquid Chromatography Detection. *Anal. Chem.,* **58** (1986) 1578.

1176. M. C. Gosnell, R. E. Snelling, and H. A. Mottola, Construction and Performance of Plastic-Embedded Controlled-Pore Glass Open Tubular Reactors for Use in Continuous-Flow Systems. *Anal. Chem.,* **58** (1986) 1585.

1177. E. H. Hansen, Flow Injection Analysis [in Danish]. *Dansk Kemi,* **67** (1986) 100.

1178. J. Chen and W. Zhu, Flow Injection Analysis—New Technique in Wet Analysis. (1) [in Chinese]. *Huaxue Shijie,* **27** (1986) 91.

1179. J. Martinez Calatayud, Flow Injection Analysis of Pharmaceuticals. *Pharmazie,* **41** (1986) 92.

1180. T. Korenaga, T. Moriwake, and T. Takahashi, Evaluation of Three Flow Injection Analysis Methods for the Determination of Chemical Oxygen Demand. *Mem. Sch. Eng., Okayama Univ.,* **19** (1984) 53.

1181. K. Matsumuto, K. Ishida, and Y. Osajima, Approach for Conductometric Flow Injection Analysis of the Salt Content in Food. *Nippon Shokuhin Kogyo Gakkaishi,* **33** (1986) 61.

1182. M. Yamamoto, Flow Injection Analyzer [in Japanese]. Jpn. Kokai Tokkyo Koho, Pat. No. 85181656 (Toshiba Corp.) (1985).

1183. J. F. van Staden, Flow-Injection Turbidimetric Analysis of Sulfate in Water [in Afrikaans]. *Water SA,* **12** (1986) 43.

1184. D. List, I. Ruwisch, and P. Langhans, Application of Flow Injection Analysis in Fruit Juice Analysis [in German]. *Fluess. Obst,* **53** (1986) 10.

1185. E. Schwerdtfeger, Automated Nitrate Assay in Plant Material Following the FIA (Flow Injection Analysis) Principle [in German]. *Landwirtsch. Forsch., Kongressband* **1984** (1985) 99.

1186. J. M. Hwang, J. C. Tsung, and Y. M. Chen, Catalytic Spectrophotometric Determination of Vanadium by Flow Injection Analysis. *J. Chin. Chem. Soc.,* **32** (1985) 405.

1187. J. M. Hwang and Y. M. Chen, Amperometric Determination of Cr(III) and Cr(VI) by Flow Injection Analysis. *J. Chin. Chem. Soc.,* **32** (1985) 411.

1188. L. Sun and Z. Fang, Spectrophotometry of Soil Organic Matter by Flow Injection Analysis [in Chinese] *Nongye Ceshi Fenxi,* **2** (1985) 23

1189. X. Wang and Z. Fang, Determination of Trace Amounts of Selenium in Environmental Samples by Hydride Generation Atomic Absorption Spectrometry Combined with Flow Injection Analysis [in Chinese]. *Kexue Tongbao,* **30** (1985) 1598.

1190. X. Wang and Z. Fang, Determination of Trace Amounts of Selenium in Environmental Samples by Hydride Generation—Atomic Absorption Spectrometry Combined with Flow Injection Analysis Technique. *Kexue Tongbao,* **31** (1986) 791.

1191. Z. Fang, The Automation of Soil Chemical Analysis by Flow Injection Analysis. *Proc. Multinat. Instrum. Conf.,* Beijing, China, (1986) 555.

1192. Z. Zhu and Z. Fang, Gas Diffusion Separation Flow Injection Analysis of Cyanide in Waste Waters [in Chinese]. *Kexue Tongbao,* **31** (1986) 800.

1193. A. T. Haj-Hussein and G. D. Christian, Novel Flow Injection/Potentiometric Measurement of Slightly Soluble Salts in Small Volumes: Determination of Solubility Product Constants of Some Silver Salts. *Anal. Lett.,* **19** (1986) 825.

1194. F. Lázaro, M. D. Luque de Castro, and M. Valcárcel, Photometric Determination of Tartaric Acid in Wine by Flow Injection Analysis. *Analyst,* **111** (1986) 729.
1195. T. Uchida, I. Kojima, C. Iida, and K. Goto, Discrete Nebulisation in Inductively Coupled Plasma Atomic Emission Spectrometry of Manganese. *Analyst,* **111** (1986) 791.
1196. J. Wang, P. A. M. Farias, and J. S. Mahmoud, Measurements of Low Concentrations of Diltiazem by Adsorptive Stripping Voltammetry and Flow Amperometry. *Analyst,* **111** (1986) 837.
1197. C. Pasquini and W. A. de Oliveira, Determination of Iron in Iron Ores Using Enthalpimetric Flow Injection Analysis. *Analyst,* **111** (1986) 857.
1198. A. G. Fogg, Reverse Flow Injection Formation of Monitorands in Direct and Indirect On-Line Flow Injection Analysis Methods. *Analyst,* **111** (1986) 859.
1199. Y. Ikariyama and W. R. Heineman, Polypyrrole Electrode as a Detector for Electroinactive Anions by Flow Injection Analysis. *Anal. Chem.,* **58** (1986) 1803.
1200. R. Y. Xie and G. D. Christian, Serum Lithium Analysis by Coated Wire Lithium Ion Selective Electrodes in a Flow Injection Analysis Dialysis System. *Anal. Chem.,* **58** (1986) 1806.
1201. M. H. Memon and P. J. Worsfold, The Use of Microemulsions in Flow Injection Analysis. Spectrofluorimetric Determination of Primary Amines. *Anal. Chim. Acta,* **183** (1986) 179.
1202. M. L. Grayeski, E. J. Woolf, and P. J. Helly, Quantitation of Enzymatically Generated Hydrogen Peroxide Based on Aqueous Peroxylate Chemiluminescence. *Anal. Chim. Acta,* **183** (1986) 207.
1203. T. Fukasawa, S. Kawakubo, and A. Unno, Flow-Injection Spectrophotometric Determination of Trace Vanadium Based on Catalysis of the Gallic Acid Bromate Reaction. *Anal. Chim. Acta,* **183** (1986) 269.
1204. J. A. Osborn, A. M. Yacynych, and D. C. Roberts, A Flow-Injection System for Assay of the Activity of an Immobilized Enzyme Chemically-Modified Electrode. *Anal. Chim. Acta,* **183** (1986) 287.
1205. M. D. Luque de Castro, Flow-Injection Analysis: A New Tool to Automate Extraction Processes. *J. Autom. Chem.,* **8** (1986) 56.
1206. J. Ruz, A. Torres, A. Rios, M. D. Luque de Castro, and M. Valcárcel, Automation of a Flow-Injection System for Multispeciation. *J. Autom. Chem.,* **8** (1986) 70.
1207. K. Yasuda, F. Takahata, T. Kuroishi, and H. Hachino, Improvement of Analytical Precision of Flow-Injection Analysis [in Chinese]. *Guangpuxue Yu Guangpu Fenxi,* **6** (1986) 20.

1208. C. F. Mascone, Flow Injection Analysis Moves into CPI Plants. *Chem. Eng.*, **93**(13) (1986) 14.

1209. M. Ishii, M. Yamada, and S. Suzuki, Determination of Cyanide Ion by Flow Injection Method Using Surfactant Bilayer Vesicle-Enhanced Chemiluminescence of Brilliant Sulfoflavine [in Japanese]. *Bunseki Kagaku,* **35** (1986) 542.

1210. D. C. Johnson, J. A. Polta, T. Z. Polta, G. G. Neuburger, J. Johnson, A. P.-C. Tang, I.-H. Yeo, and J. Baur, Anodic Detection in Flow-Through Cells. *J. Chem. Soc. Faraday Trans. I,* **82** (1986) 1081.

1211. C. E. Lomen, U. de Alwis, and G. S. Wilson, Use of a Reversibly Immobilized Enzyme in the Flow Injection Amperometric Determination of Picomole Glucose Levels. *J. Chem. Soc. Faraday Trans. I,* **82** (1986) 1265.

1212. J. Martinez Calatayud and C. Falco, Determination of Levamisole Hydrochloride with Tetraiodomercurate(II) by a Turbidimetric Method and Flow-Injection Analysis. *Talanta,* **33** (1986) 685.

1213. T. Hara, T. Ebuchi, A. Arai, and M. Imaki, The Determination of a Small Amount of a Biological Constituent by the Use of Chemiluminescence. VI. The Flow-Injection Analysis of Protein Using 1,10-Phenanthroline-Hydrogen Peroxide System. *Bull. Chem. Soc. Jpn.,* **59** (1986) 1833.

1214. T. Komatsu, M. Ohira, M. Yamada, and S. Suzuki, Analytical Use of Luminescence Induced Ultrasonically in Solution. I. Sonic Chemiluminescence of Luminol for Determination of Cobalt(II) at Sub-pg Levels in Flow Injection and Continuous Flow Metods. *Bull. Chem. Soc. Jpn.,* **59** (1986) 1849.

1215. J. Fejes and S. Melnik, Pumping Techniques in Flow Injection Analysis [in Slovak]. *Chem. Listy,* **80** (1986) 586.

1216. H. Müller and V. Müller, Kinetic Determination of Traces of Iron by Means of Flow Injection Analysis Based on the Catalytic Oxidation of Malachitgreen Leuco Base by Hydrogenperoxide. *Z. Chem.,* **26** (1986) 142.

1217. M. Bengtsson, F. Malamas, A. Torstensson, O. Regnell, and G. Johansson, Trace Metal Ion Preconcentration for Flame Atomic Absorption by an Immobilized *N,N,N'*-Tri-(2-Pyridylmethyl)ethylene Diamine (TriPEN) Chelate Ion Exchanger in a Flow Injection Analysis. *Mikrochim. Acta,* **III** (1985) 209.

1218. J. Kinkeldei, A Novel Spectrophotometer as a Detector for an FIA (Flow Injection Analysis) [in German]. *Gewässerschutz, Wasser, Abwasser,* **79** (1985) 115.

1219. C. A. Pohl, Flat Membrane Reactor Device with Structure in Flow Channel. Eur. Pat. No. 177265 (Dionex Corp., USA) (1986).

1220. L. Sun, L. Li, and Z. Fang, Determination of Total Nitrogen in Soil Digests by Gas Diffusion Flow Injection Analysis [in Chinese]. *Turang Tongbao,* **17** (1986) 37.

1221. E. Gaetani, C. F. Laureri, M. Vitto, L. Borghi, G. F. Elia, and A. Novarini, Determination of Oxalate in Urine by Flow Injection Analysis. *Clin. Chim. Acta,* **156** (1986) 71.

1222. F. A. Shaver and D. J. O'Conner, Method and Apparatus for Dosing a Flowing Liquid Stream. Eur. Pat. No. 153530 (Sybron Corp., USA) (1985).

1223. J. Růžička and E. H. Hansen, *Flow Injection Analysis* [In Chinese. Translated by Fang Zhaolun]. Science Press, Beijing, People's Republic of China (1986).

1224. A. Rios, M. D. Luque de Castro, and M. Valcárcel, Determination of Reaction Stoichiometrics by Flow Injection Analysis. *J. Chem. Educ.,* **63** (1986) 552.

1225. C. P. Parker, M. G. Gardell, and D. Di Biasio, A Complete System for Fermentation Monitoring. *Int. Biotech. Lab.,* **4**(2) (1986) 33.

1226. E. H. R. von Barsewisch, Flow Injection Analysis. Univ. Witwatersrand, Johannesburg, South Africa (1985). (Ph.D. Thesis).

1227. A. T. Haj-Hussein and G. D. Christian, Multicomponent Flow Injection Analysis Using Spectrophotometric Detection with Reagent Spectral Overlap: Application to Determination of Calcium and Magnesium in Blood Serum Using Eriochrome Black T. *Microchem. J.,* **34** (1986) 67.

1228. S. D. Hartenstein, G. D. Christian, and J. Růžička, Applications of an On-Line Preconcentrating Flow Injection Analysis System for Inductively Coupled Plasma Atomic Emission Spectrometry. *Can. J. Spectrosc.,* **30** (1986) 144.

1229. H. Ishii, M. Aoki, T. Aita, and T. Odashima, Flow Injection-Spectrophotometric Determination of Trace Amounts of Iron with 2-Pyridyl-3′-Sulfophenylmethanone 2-Pyrimidylhydrazone and Possibility of Sensitization by Analogue-Derivative Spectrophotometric Monitoring. *Anal. Sci.,* **2** (1986) 125.

1230. J. J. Sullivan, T. A. Hollingworth, M. M. Wekell, R. T. Newton, and J. E. Larose, Determination of Sulfite in Food by Flow Injection Analysis. *J. Assoc. Off. Anal. Chem.,* **69** (1986) 542.

1231. J. Růžička, Flow Injection Analysis—A Survey of Its Potential for Spectroscopy. *Fresenius Z. Anal. Chem.,* **324** (1986) 745.

1232. T. D. Yerian, G. D. Christian, and J. Růžička, Enzymatic Determination of Urea in Water and Serum by Optosensing Flow Injection Analysis. *Analyst,* **111** (1986) 865.

1233. A. Y. Chamsi and A. G. Fogg, Application of the Reductive Flow Injection Amperometric Determination of Iodine at a Glassy Carbon Electrode to the Iodometric Determination of Hypochlorite and Hydrogen Peroxide. *Analyst,* **111** (1986) 879.

1234. K. A. Holm, Automatic Spectrophotometric Determination of Amyloglucosidase Activity Using *p*-Nitrophenyl-α-D-Glucopyranoside and a Flow Injection Analyser. *Analyst,* **111** (1986) 927.

1235. G. J. Moody, J. M. Slater, and J. D. R. Thomas, Some Parameters of Ion-Sensitive Field Effect Transistor (ISFET) Sensors. *Anal. Proc.,* **23** (1986) 287.

1236. B. Bouzid and A. M. G. Macdonald, Flow Injection Determination of 5-Fluorouracil with Voltammetric Detection. *Anal. Proc.,* **23** (1986) 295.

1237. J. F. Tyson, Extended Range Calibrations by Flow Injection Analysis. *Anal. Proc.,* **23** (1986) 304.

1238. W. A. McCrum, Determination of Dissolved Humic Substances in River Waters Using Flow Injection Analysis with Fluorimetric Detection. *Anal. Proc.,* **23** (1986) 307.

1239. D. C. Johnson, S. G. Weber, A. M. Bond, R. M. Wightman, R. E. Shoup, and I. S. Krull, Electroanalytical Voltammetry in Flowing Solutions. *Anal. Chim. Acta,* **180** (1986) 187.

1240. P. Macheras, M. Koupparis, and C. Tsaprounis, An Automated Flow Injection-Serial Dynamic Dialysis Technique for Drug-Protein Binding Studies. *Int. J. Pharm.,* **30** (1986) 123.

1241. J. Chen and W. Zhu, Flow Injection Analysis—New Technique in Wet Analysis. (2) [in Chinese]. *Huaxue Shijie,* **27** (1986) 137.

1242. M. Straka, Selectivity Enhancement Using Flow Injection Analysis. *Diss. Abstr. Int. B,* **46** (1986) 3421. (Ph.D. Thesis, Miami Univ., Oxford, OH, USA.)

1243. D. K. Morgan, Prederivatization Liquid Chromatography for the Determination of Inorganic Phosphate and of Methylmalonic Acid and Flow Injection Analysis with Infrared Detection. *Diss. Abstr. Int. B,* **46** (1986) 3421. (Ph.D. Thesis, Miami Univ., Oxford, OH, USA).

1244. E. A. Jones, Determination, by Flow Injection Analysis, of Fluoride, Chloride, Phosphate, Ammonia, Nitrite, and Nitrate. *Rep.-MINTEK (S. Africa),* **M200** (1985) 68.

1245. J. F. Tyson, J. R. Mariara, and J. M. H. Appleton, A Variable Dispersion Flow Injection Manifold for Calibration and Sample Dilution in Flame Atomic Absorption Spectrometry. *J. Anal. Atom. Spectrom.,* **1** (1986) 273.

1246. K. Oguma, T. Nara, and R. Kuroda, Atomic Absorption Spectrophotometric Determination of Magnesium in Silicates by Flow Injection Method [in Japanese]. *Bunseki Kagaku,* **35** (1986) 690.

1247. B. A. Petersson, Z. Fang, J. Růžička, and E. H. Hansen, Conversion Techniques in Flow Injection Analysis. Determination of Sulphide by Precipitation with Cadmium Ions and Detection by Atomic Absorption Spectrometry. *Anal. Chim. Acta,* **184** (1986) 165.

1248. D. F. Leclerc, P. A. Bloxham, and E. C. Toren, Jr., Axial Dispersion in Coiled Tubular Reactors. *Anal. Chim. Acta,* **184** (1986) 173.

1249. N.-H. Tioh, Y. Israel, and R. M. Barnes, Determination of Arsenic in Glycerine by Flow Injection, Hydride Generation and Inductively-Coupled Plasma/Atomic Emission Spectrometry. *Anal. Chim. Acta,* **184** (1986) 205.

1250. J. Hernández-Méndez, A. A. Mateos, M. J. A. Parra, and C. G. de María, Spectrophotometric Flow-Injection Determination of Ascorbic Acid by Generation of Triiodide. *Anal. Chim. Acta,* **184** (1986) 243.

1251. R. Y. Xie, V. P. Y. Gadzekpo, A. M. Kadry, Y. A. Ibrahim, J. Růžička, and G. D. Christian, Use of a Flow-Injection System in the Evaluation of the Characteristic Behaviour of Neutral Carriers in Lithium Ion-Selective Electrodes. *Anal. Chim. Acta,* **184** (1986) 259.

1252. E. G. Sarantonis and A. Townshend, Flow-Injection Determination of Iron(II), Iron(III) and Total Iron with Chemiluminescence Detection. *Anal. Chim. Acta,* **184** (1986) 311.

1253. A. Ivaska, Flow Injection Analysis. A Dynamic Method of Analysis [in Finnish]. *Kemia-Kemi,* **13**(5) (1986) 494.

1254. J. T. Dobbins, Jr. and J. M. Martin, Flow Injection Analysis: Twelve Years Old and Growing. *Spectroscopy,* **1** (1986) 20.

1255. S. Motomizu, H. Mikasa, and K. Toei, Fluorometric Determination of Nitrite in Natural Waters with 3-Aminonaphthalene-1,5-Disulphonic Acid by Flow Injection Analysis. *Talanta,* **33** (1986) 729.

1256. B. Olsson, Immobilized Enzymes in Flow Injection Analysis. Methods with Hydrogen Peroxide Detection. Univ. of Lund, Sweden (1985). (Ph.D. Thesis).

1257. Z. Fang, Flow Injection Techniques in Atomic Spectroscopy [in Chinese]. *Fenxi Huaxue,* **14** (1986) 549.

1258. L. Ilcheva and K. Cammann, A Simple, Selective and Sensitive Liquid-Chromatographic or Flow-Injection Detector for Chloro-Organic Compounds Based on Ion-Selective Electrodes. *Fresenius Z. Anal. Chem.,* **325** (1986) 11.

1259. S. V. Olesik, S. B. French, and M. Novotny, Reaction Monitoring in Supercritical Fluids by Flow Injection Analysis with Fourier Transform Infrared Spectrometric Detection. *Anal. Chem.,* **58** (1986) 2256.

1260. D. Betteridge, T. J. Sly, and A. P. Wade, Versatile Automatic Development System for Flow Injection Analysis. *Anal. Chem.,* **58** (1986) 2258.

1261. M. Gallego, M. Silva, and M. Valcárcel, Indirect Atomic Absorption Determination of Anionic Surfactants in Wastewaters by Flow Injection Continuous Liquid-Liquid Extraction. *Anal. Chem.,* **58** (1986) 2265.

1262. J. Toei and N. Baba, Multifunction Valve for Flow Injection Analysis. *Anal. Chem.,* **58** (1986) 2346.

1263. G. Schulze, O. Elsholz, and W. Frenzel, Double-Detection Technique for Flow-Injection Analysis [in German]. *Fresenius Z. Anal. Chem.,* **320** (1985) 650.

1264. G. Schulze and O. Elsholz, Fast On-Line Preconcentration for Atomic Absorption Spectrometry [in German]. *Fortschr. Atomspek. Spurenanal.,* **2** (1986) 261.

1265. J. W. Keller, T. F. Gould, and K. T. Aubert, The ASYST Language for the IBM PC: An Application to Flow Injection Analysis. *J. Chem. Educ.*, **63** (1986) 709.

1266. M. Masoom and A. Townshend, Cyclic Regeneration of Nicotinamide Adenine Dinucleotide with Immobilized Enzymes. Flow-Injection Spectrofluorimetric Determination of Ethanol. *Anal. Chim. Acta,* **185** (1986) 49.
1267. F. Lázaro, M. D. Luque de Castro, and M. Valcárcel, Individual and Simultaneous Enzymatic Determination of Ethanol and Acetaldehyde in Wines by Flow Injection Analysis. *Anal. Chim. Acta,* **185** (1986) 57.
1268. P. E. Macheras and M. A. Koupparis, An Automated Flow-Injection Serial Dynamic Dialysis Technique for Drug-Protein Binding Studies. *Anal. Chim. Acta,* **185** (1986) 65.
1269. O. Røyset, Flow-Injection Spectrophotometric Determination of Aluminium in Water with Pyrocatechol Violet. *Anal. Chim. Acta,* **185** (1986) 75.
1270. W. Frenzel and P. Brätter, The Fluoride Ion-Selective Electrode in Flow Injection Analysis. Part 1. General Methodology. *Anal. Chim. Acta,* **185** (1986) 127.
1271. P. Peták and K. Stulík, Use of Ion-Selective Electrodes in Industrial Flow Analysis. *Anal. Chim. Acta,* **185** (1986) 171.
1272. Y. Sahleström and B. Karlberg, Flow-Injection Extraction with a Microvolume Module Based on Integrated Conduits. *Anal. Chim. Acta,* **185** (1986) 259.
1273. S. D. Kolev and E. Pungor, Description of an Axially-Dispersed Plug Flow Model for the Flow Pattern in Elements of Fluid Systems. *Anal. Chim. Acta,* **185** (1986) 315.
1274. A. A. Alwarthan and A. Townshend, Chemiluminescence Determination of Buprenorphine Hydrochloride by Flow Injection Analysis. *Anal. Chim. Acta,* **185** (1986) 329.
1275. S. D. Brown, Studies of the Analyte-Carrier Interface in Multicomponent Flow Injection Analysis. *Energy Res. Abstr.,* **10**(16) (1985) Abstr. No. 31594.
1276. E. Pungor and M. Gratzl, Examples of the Application of Chemometrics in Electroanalysis [in Hungarian]. *Nagy. Kem. Foly.,* **92** (1986) 242.
1277. Y. Nagaishi and K. Yokoyama, Speciation Determination Using Flow-Injection Analysis [in Japanese]. Jpn. Kokai Tokkyo Koho, Pat. No. 85236063 (Mitsubishi Electric Corp.) (1985)
1278. H. Yan, F. Li, and Y. Li, Determination of Phenols in Surface Water with Flow-Injection Analysis [in Chinese]. *Fenxi Huaxue,* **14** (1986) 359.
1279. M. D. Luque de Castro, Flow Injection Analysis: Its Possibilities and Applications in Food Analysis [in Spanish]. *An. Bromatol.,* **37** (1986) 197.
1280. J. Toei and N. Baba, Flow Injection Titration Analysis for the Determination of Brines Using an Ion Exchange Column [in Japanese]. *Bunseki Kagaku,* **35** (1986) 832.
1281. P. Martinez-Jimenez, M. Gallego, and M. Valcárcel, Indirect Atomic Ab-

sorption Determination of Aluminum by Flow Injection Analysis. *Microchem. J.,* **34** (1986) 190.

1282. Y. Bao, Flow Injection Analysis. A Review [in Chinese]. *Fenxi Ceshi Tongbao,* **2** (1983) 1.

1283. L. Wang, S. Zhang, Q. Wang, and Y. Zhu, A Study on the Environmental Redox Reaction Kinetics of Cr(VI) and Sulfides in Natural Waters [in Chinese]. *Acta Scientiae Circumstantiae,* **3** (1983) 247.

1284. H. Ma, Development of Automated Analytical Instruments in Wet Chemical Analysis [in Chinese]. *Fenxi Yiqi,* **5** (1983) 57.

1285. Z. Fang, Flow Injection Analysis. In *Instrumental Analysis of Soils and Biological Materials* (Z. Fang, Ed.) [in Chinese]. Science Press, Beijing, China (1983).

1286. C. Wang, Flow Injection Analysis: A New Approach in Analytical Chemistry [in Chinese]. *Shanghai Youse Jinshu,* **2** (1984) 23.

1287. H. Cui, L. Li, L. Sun, and Z. Fang, Modifications in the Flow Injection Analysis of Total Nitrogen in Soils [in Chinese]. *Nongye Ceshi Fenxi,* **1** (1984) 63.

1288. Y. Li, A Brief Introduction to Flow Injection Analysis. A Review [in Chinese]. *Dianzi Sheji Shuichuli Dongtai,* **1** (1984) 1.

1289. H. Cui and Y. Li, Determination of Fluoride in Waters by Flow Injection Analysis [in Chinese]. *Lihua Jianyan,* **22** (1985) 21.

1290. H. Ma, Applications of Flow Injection Analysis in Metallurgy and Geology. A Review [in Chinese]. *Lihua Jianyan,* **22** (1985) 374.

1291. H. Ma, Flow Injection Analysis and the Automation of Wet Chemical Analysis [in Chinese]. *Fenxie Shiyanshi,* **4** (1985) 43.

1292. L. Xu, X. Liu, Y. Yang, and F. Yan, Determination of Total Phosphate in Plant Samples by Flow Injection Analysis [in Chinese]. *Dongbei Linxueyuan Xuebao,* **13** (1985) 33.

1293. W. Cheng and Y. Li, Development of the FIA-T-721 Model Flow Injection Spectrophotometric Analyzer [in Chinese]. *Fenxi Yiqi,* **3** (1986).

1294. Y. Guo, The Determination of Calcium in Siliceous Earths by Flow Injection Atomic Absorption Spectrometry [in Chinese]. *Jilin Daxue Ziran Kexue Xuebao,* **4** (1986).

1295. W. Han and L. Fan, Ion Selective Electrode—Flow Injection Analysis [in Chinese]. *Fenxie Huaxue,* **14** (1986) 387.

1296. Z. Fang, S. Xu, X. Wang, and S. Zhang, Combination of Flow Injection Analysis with Atomic Absorption Spectrophotometry [in Chinese]. *Guangpuxue yu Guangpu Fenxi,* **6** (1986) 31.

1297. W. Cheng and Y. Li, The Flow Injection Spectrophotometric Determination of Phosphate in Power Plant Boilers [in Chinese]. *Gongye Shui Chuli,* **3** (1986).

1298. B. Xia and P. Mo, Flow Injection Analysis and Its Application in Clinical Chemistry [in Chinese]. *Zhonghua Yixue Jianyan Zazhi,* **9** (1986) 117.

1299. G. Wei and H. Ma, Determination of Total Phosphate in Samples from Trees by Flow Injection Analysis—Molybdenum Yellow Photometric Method [in Chinese]. *Linye Daxue Xuebao,* **3** (1986).

1300. J. F. van Staden, A. E. Joubert, and H. R. van Vliet, Flow-Injection Determination of Nitrate in Natural Waters with Copper- and Copperised Cadmium Tubes in the Reaction Manifold System. *Fresenius Z. Anal. Chem.,* **325** (1986) 150.

1301. J. A. Cox and K. R. Kulkarni, Flow Injection Determination of Nitrite by Amperometric Detection at a Modified Electrode. *Analyst,* **111** (1986) 1219.

1302. B. A. Woods, J. Růžička, G. D. Christian, and R. J. Charlson, Measurement of pH in Solutions of Low Buffering Capacity and Low Ionic Strength by Optosensing Flow Injection Analysis. *Anal. Chem.,* **58** (1986) 2496.

1303. J. J. Thompson and R. S. Houk, Inductively Coupled Plasma Mass Spectrometric Detection for Multielement Flow Injection Analysis and Elemental Speciation by Reversed-Phase Liquid Chromatography. *Anal. Chem.,* **58** (1986) 2541.

1304. J. A. Koropchak and D. H. Winn, Thermospray Interfacing for Flow Injection Analysis with Inductively Coupled Plasma Atomic Emission Spectrometry. *Anal. Chem.,* **58** (1986) 2558.

1305. K. Wolf and P. J. Worsfold, Flow Injection Analysis as a Sample Handling Technique for Diode Array Spectroscopy. *Anal. Proc.,* **23** (1986) 365.

1306. J. van Staden, Incorporation of a Coated Tubular Solid-State Iodide-Selective Electrode into the Conduits of a Flow-Injection System. *Fresenius Z. Anal. Chem.,* **325** (1986) 247.

1307. P. Hernandez, L. Hernandez, and J. Losada, Determination of Aluminium in Hemodialysis Fluids by a Flow Injection System with Preconcentration on a Synthetic Chelate-Forming Resin and Flame Atomic Absorption Spectrophotometry. *Fresenius Z. Anal. Chem.,* **325** (1986) 300.

1308. F. Morishita, Y. Nishikawa, and T. Kojima, Simultaneous Determination of Three Species by Flow Injection Analytical Method Using Enzyme-Immobilized Open-Tubular Reactors. *Anal. Sci.,* **2** (1986) 411.

1309. T. Iwachido, M. Onoda, and S. Motomizu, Determination of Potassium in River Water by Solvent Extraction-Flow Injection Analysis. *Anal. Sci.,* **2** (1986) 493.

1310. E. B. Milosavljevic, L. Solujic, J. H. Nelson, and J. L. Hendrix, Simultaneous Determination of Chromium(VI) and Chromium(III) by Flame Atomic Spectrometry with a Chelating Ion-Exchange Flow Injection System. *Mikrochim. Acta,* **III** (1985) 353.

1311. H. Wada, Y. Deguchi, and G. Nagakawa, Determination of Calcium in Chlor-Alkali Brine by Flow Injection Analysis. *Mikrochim. Acta,* **III** (1985) 393.

1312. D. E. Davey, D. E. Mulcahy, and G. R. O'Connell, Rapid Determination of Fluoride in Potable Waters by Potentiometric Flow Injection Analysis. *Anal. Lett.,* **19** (1986) 1387.

1313. J. F. van Staden, Flow-Injection Analysis of Chloride in Milk with a Dialyzer and a Coated Tubular Inorganic Chloride-Selective Electrode. *Anal. Lett.,* **19** (1986) 1407.

1314. D. E. Davey, The Analysis of Galvanizing Preflux Solutions for Zinc by Flow Injection Analysis. *Anal. Lett.,* **19** (1986) 1573.

1315. M. Ishii, M. Yamada, and S. Suzuki, Didodecyldimethylammonium Bromide Bilayer Vesicle-Catalyzed and Uranine-Sensitized Chemiluminescense for Determination of Free Cyanide at Picogram Levels by Flow Injection Analysis. *Anal. Lett.,* **19** (1986) 1591.

1316. F. Scheller, F. Schubert, B. Olsson, L. Gorton, and G. Johansson, Flow Injection Analysis of Lactate and Lactate Dehydrogenase Using an Enzyme Membrane in Conjunction with a Modified Electrode. *Anal. Lett.,* **19** (1986) 1691.

1317. G. B. Martin and M. E. Meyerhoff, Membrane-Dialyzer Injection Loop for Enhancing the Selectivity of Anion-Responsive Liquid-Membrane Electrodes in Flow Systems. Part I. A Sensing System for NO_x and Nitrite. *Anal. Chim. Acta,* **186** (1986) 71.

1318. Q. Chang and M. E. Meyerhoff, Membrane-Dialyzer Injection Loop for Enhancing the Selectivity of Anion-Responsive Liquid-Membrane Electrodes in Flow Systems. Part II. A Selective Sensing System for Salicylate. *Anal. Chim. Acta,* **186** (1986) 81.

1319. D. Taylor and T. A. Nieman, Bipolar Pulse Conductometric Detection of Enzyme Reactions in Flow-Injection Systems. Urea in Serum and Urine. *Anal. Chim. Acta,* **186** (1986) 91.

1320. Y. Hayashi, K. Zaitsu, and Y. Ohkura, Flow-Injection Determination of Adenosine and Inosine in Blood Plasma with Immobilized Enzyme Columns Connected in Series and Fluorimetric Detection. *Anal. Chim. Acta,* **186** (1986) 131.

1321. J. Ruz, A. Rios, M. D. Luque de Castro, and M. Valcárcel, Flow-Injection Configurations for Chromium Speciation with a Single Spectrophotometric Detector. *Anal. Chim. Acta,* **186** (1986) 139.

1322. J. L. Burguera, M. Burguera, L. La Cruz, and O. R. Naranjo, Determination of Lead in the Urine of Exposed and Unexposed Adults by Extraction and Flow-Injection/Atomic Absorption Spectrometry. *Anal. Chim. Acta,* **186** (1986) 273.

1323. A. Al-Wehaid and A. Townshend, Spectrophotometric Flow-Injection Determinaion of Nitrate Based on Reduction with Titanium(III) Chloride. *Anal. Chim. Acta,* **186** (1986) 289.

1324. J. Fejes and S. Melnik, Optical Methods of Detection in Flow Injection Analysis [in Czech]. *Chem. Listy,* **80** (1986) 1009.

1325. P. Linares, M. D. Luque de Castro, and M. Valcárcel, Fluorimetric Differential-Kinetic Determination of Silicate and Phosphate in Waters by Flow-Injection Analysis. *Talanta,* **33** (1986) 889.

1326. H. Matschiner and P. Sivers, Flow Injection Analysis—Special Solutions for General Problems [in German]. *Wiss. Fortschr.*, **35** (1985) 142.

1327. Z.-N. Zhou, R. Y. Xie, and G. D. Christian, Relation of Plasticizers and Ionophore Structures on Selectivities of Lithium Ion-Selective Electrodes. *Anal. Lett.*, **19** (1986) 1747.

1328. S. Hirata, Y. Umezaki, and M. Ikeda, Determination of Chromium(III), Titanium, Vanadium, Iron(III), and Aluminum by Inductively Coupled Plasma Atomic Emission Spectrometry with an On-Line Preconcentrating Ion-Exchange Column. *Anal. Chem.*, **58** (1986) 2602.

1329. G. Audunsson, Aqueous/Aqueous Extraction by Means of a Liquid Membrane for Sample Cleanup and Preconcentration of Amines in a Flow System. *Anal. Chem.*, **58** (1986) 2714.

1330. C. A. Lucy and F. F. Cantwell, Simultaneous Determination of Phenylephrine Hydrochloride and Pheniramine Maleate in Nasal Spray by Solvent Extraction-Flow Injection Analysis Using Two Porous-Membrane Phase Separators and One Photometric Detector. *Anal. Chem.*, **58** (1986) 2727.

1331. K. Matsumoto, O. Hamada, H. Ukeda, and Y. Osajima, Amperometric Flow Injection Determination of Fructose with an Immobilized Fructose 5-Dehydrogenase Reactor. *Anal. Chem.*, **58** (1986) 2732.

1332. P. K. Dasgupta and H.-C. Yang, Trace Determination of Aqueous Sulfite, Sulfide, and Methanethiol by Fluorometric Flow Injection Analysis. *Anal. Chem.*, **58** (1986) 2839.

1333. J. W. Bixler and A. M. Bond, Amperometric Detection of Picomole Samples in a Microdisk Electrochemical Flow-Jet Cell with Dilute Supporting Electrolyte. *Anal. Chem.*, **58** (1986) 2859.

1334. J. F. van Staden, Electrodes in Series. Simultaneous Flow Injection Determination of Chloride and pH with Ion-Selective Electrodes. *Analyst,* **111** (1986) 1231.

1335. G. J. Moody, G. S. Sanghera, and J. D. R. Thomas, Modified Platinum Wire Glucose Oxidase Amperometric Electrode. *Analyst,* **111** (1986) 1235.

1336. M. A. Koupparis and P. I. Anagnostopoulou, Automated Flow Injection Spectrophotometric Determination of Zinc Using Zincon: Applications to Analysis of Waters, Alloys and Insulin Formations. *Analyst,* **111** (1986) 1311.

1337. A. Nabi and P. J. Worsold, Bioluminescence Assays with Immobilized Bacterial Luciferase Using Flow Injection Analysis. *Analyst,* **111** (1986) 1321.

1338. J.-S. Rhee, Peak Width Measurement Based Flow Injection Analysis. Texas Tech. Univ., Lubbock, Texas (1986). (Ph.D. Thesis).

1339. L. K. Shpigun, I. Y. Kolotyrkina, and Y. A. Zolotov, Flow-Injection Analysis. Spectrophotometric Determination of Nickel [in Russian]. *Zh. Anal. Khim.*, **41** (1986) 1224.

1340. C. M. Sakamoto-Arnold, K. S. Johnson, and C. L. Beehler, Determination of Hydrogen Sulfide in Seawater Using Flow Injection Analysis and Flow Analysis. *Limnol. Oceanogr.,* **31** (1986) 894.

1341. K. Matsumuto, K. Ishida, and Y. Osajima, Correction for Conductometric Flow Injection Analysis of the Salt Content in Food. *Nippon Shokuhin Kogyo Gakkaishi,* **33** (1986) 345.

1342. M. Burguera, J. L. Burguera, P. C. Rivas, and O. M. Alarcon, Determination of Copper, Zinc, and Iron in Parotid Saliva by Flow Injection with Flame Atomic Absorption Spectrophotometry. *At. Spectrosc.,* **7** (1986) 79.

1343. J. Yamamoto, M. Yamamoto, and M. Yasuda, Flow-Injection Analysis [in Japanese]. Jpn. Kokai Tokkyo Koho, Pat. No. 8607467 (Mitsui Toatsu Chemicals, Inc.) (1986).

1344. K. Uchida, T. Shibata, M. Tomoda, S. Saito, and S. Inayama, A New Convenient Method for Microdetermination of Hydroxyproline by Flow Injection Analysis. *Chem. Pharm. Bull.,* **34** (1986) 2649.

1345. K. Matsumuto, K. Ishida, and Y. Osajima, Application of Conductometric Flow Injection Analysis to the Salt Contents in Soy Sauce and Worcester Sauce. *Nippon Shokuhin Kogyo Gakkaishi,* **33** (1986) 450.

1346. Y. Nishikawa, F. Morishita, and T. Kojima, Determination of Urate by a FIA Method Using a Uricase-Immobilized Open-Tubular Reactor [in Japanese]. *Bunseki Kagaku,* **35** (1986) 575.

1347. H. Ohura, T. Imato, Y. Asano, S. Yamasaki, and N. Ishibashi, Potentiometric FIA of Disaccharides Using the Hexacyanoferrate(III)-Hexacyanoferrate(II) Potential Buffer Solution [in Japanese]. *Bunseki Kagaku,* **35** (1986) 807.

1348. J. Olafsson, The Semi-Automated Determination of Manganese in Sea Water with Leuco-Malachite Green. *Sci. Total Environ.,* **49** (1986) 101.

1349. M. Aihara, M. Arai, and T. Tomitsugu, Flow Injection Spectrofluorimetric Determination of Terbium(III) Based on Solubilizing its Ternary Complex in Micellar Solution. *Anal. Lett.,* **19** (1986) 1907.

1350. P. van Zoonen, I. de Herder, C. Gooijer, N. H. Velthorst, R. W. Frei, E. Küntzberg, and G. Gübitz, Detection of Oxidase Generated Hydrogenperoxide by a Solid State Peroxylate Chemiluminescence Detector. *Anal. Lett.,* **19** (1986) 1949.

1351. J. Martinez Calatayud, M. C. Pascual Marti, and S. Sagrado Vives, Determination of Paracetamol by a Flow Injection-Spectrophotometric Method. *Anal. Lett.,* **19** (1986) 2023.

1352. E. Pungor, Message for the Journal of the Japanese Association of Flow Injection Analysis. *J. Flow Injection Anal.,* **3**(2) (1986) 74.

1353. T. Yamane, Catalytic Kinetic Methods of Analysis with Flow Injection System [in Japanese]. *J. Flow Injection Anal.,* **3**(2) (1986) 77.

1354. T. Korenaga, H. Yoshida, Y. Yokota, S. Kaseno, and T. Takahashi, Simulation of FIA Peak without Molecular Diffusion in Laminar Capillary Flow [in Japanese]. *J. Flow Injection Anal.,* **3**(2) (1986) 91.

1355. T. Imato, C. Azemori, Y. Asano, and N. Ishibashi, Flow Injection Analysis of Organic Acids and Amino Acids in Sake (Japanese Rice Wine) by Using pH-Sensitive Glass Electrode and Acid-Base Buffer Solution [in Japanese]. *J. Flow Injection Anal.*, **3**(2) (1986) 103.

1356. M. Sato and T. Yamada, Electrogenerated Chemiluminescence Detector for Flow Injection Analysis. *Anal. Sci.*, **2** (1986) 529.

1357. P. Linares, M. D. Luque de Castro, and M. Valcárcel, Spectrophotometric Determination of Selenium(IV) and Selenium(VI) with Flow Injection. *Analyst*, **111** (1986) 1405.

1358. L. K. Shpigun, I. D. Eremina, and Y. A. Zolotov, Flow-Injection Analysis. Determination of Lead and Sulfate Ions by Using a Lead-Selective Electrode [in Russian]. *Zh. Anal. Khim.*, **41** (1986) 1557.

1359. M. H. Ho, Automated Flow Injection Analysis System for Formaldehyde Determination. *ACS Symp. Ser.*, **316** (1986) 107.

1360. M. E. Herrera, Analytical and Neurochemical Studies with Flow Injection Analysis and Electrochemical Detection. Part I. Pulsed Flow Electrolysis with the Reticulated Vitreous Carbon Electrode. Part II. On-Line Determination of Secreted Catecholamines from Cultured Adrenal Cells. *Diss. Abstr. Int. B*, **46** (1986) 1894.

1361. C. S. I. Lai, G. J. Moody, and J. D. R. Thomas, Piezoelectric Quartz Crystal Detection of Ammonia Using Pyridoxine Hydrochloride Supported on a Polyethoxylate Matrix. *Analyst*, **111** (1986) 511.

1362. S. R. Bysouth and J. F. Tyson, On-Line Sample and Standard Manipulation for Flame Atomic Absorption Spectrometry. *Anal. Proc.*, **23** (1986) 412.

1363. A. Nabi and P. J. Worsfold, Indirect Assays with Immobilized Firefly Luciferase Based on Flow Injection Analysis. *Anal. Proc.*, **23** (1986) 415.

1364. M. H. Memon and P. J. Worsfold, Analytical Applications of Microemulsions. *Anal. Proc.*, **23** (1986) 418.

1365. D. Eadington and J. A. W. Dalziel, Reductive Determination of Metal Ions by Flow Injection Analysis Using Amperometric Amalgam Detectors. *Anal. Proc.*, **23** (1986) 434.

1366. G. J. Moody, G. S. Sanghera, and J. D. R. Thomas, Enzyme Electrode Systems for Glucose Analysis. *Anal. Proc.*, **23** (1986) 446.

1367. J. Pawliszyn, Properties and Applications of the Concentration Gradient Sensor to Detection of Flowing Samples. *Anal. Chem.*, **58** (1986) 3207.

1368. J. A. Cox and K. R. Kulkarni, Comparison of Inorganic Films and Poly(4-Vinyl-pyridine) Coatings as Electrode Modifiers for Flow-Injection Systems. *Talanta*, **33** (1986) 911.

1369. K. W. Simonsen, B. Nielsen, A. Jensen, and J. R. Andersen, Direct Microcomputer Controlled Determination of Zinc in Human Serum by Flow Injection Atomic Absorption Spectrometry. *J. Anal. Atom. Spectrom.*, **1** (1986) 453.

1370. W. Frenzel and P. Brätter, Fluoride Ion-Selective Electrode in Flow Injection Analysis. Part 2. Interference Studies. *Anal. Chim. Acta,* **187** (1986) 1.

1371. A. Ríos, M. D. Luque de Castro, and M. Valcárcel, Simultaneous Flow-Injection Fluorimetric Determination of Ammonia and Hydrazine with a Novel Mode of Forming pH Gradients. *Anal. Chim. Acta,* **187** (1986) 139.
1372. K. Bäckström, L.-G. Danielsson, and L. Nord, Dispersion in Phase Separators for Flow-Injection Extraction Systems. *Anal. Chim. Acta,* **187** (1986) 255.
1373. Y. Sahleström, S. Twengström, and B. Karlberg, An Automated Flow-Injection Extraction Method for Determination of Bittering Compounds in Beer. *Anal. Chim. Acta,* **187** (1986) 339.
1374. P. Jager and H. L. Pardue, Kinetic Treatment of Unsegmented Flow Systems. Part 4. Equations for a System with Gradient Chamber Corrected to Account for Detector with Finite Sensitivities. *Anal. Chim. Acta,* **187** (1986) 343.
1375. K. E. LaFreniere, Evaluation of a Direct Injection Nebulizer Interface for Flow Injection Analysis and High Performance Liquid Chromatography with Inductively Coupled Plasma-Atomic Emission Spectroscopic Detection. *Diss. Abstr. Int. B,* **47**(4) (1986) 1519.
1376. L. Gorton and T. Svensson, An Investigation of the Influences of the Background Material and Layer Thickness of Sputtered Palladium/Gold on Carbon Electrodes for the Amperometric Determination of Hydrogen Peroxide. *J. Mol. Catal.,* **38** (1986) 49.
1377. K. Stulik and V. Pacakova, Some Aspects of Design, Performance and Applications of Electrochemical Detectors in HPLC and FIA. *Ann. Chim.,* **76** (1986) 315.
1378. K. M. Pedersen, P. Haagensen, M. Kümmel, and H. Søeberg, Automated Measuring System for Biological Treatment Plant with Biological Removal of Phosphate. *IFAC Proc. Ser.,* **10** (Modell. Control Biotechnol. Processes) (1986) 131.
1379. F. Holz, Automatic, Continuous-Flow, Photometric Determination of Soil Enzyme Activities by the Use of (Enzyme) Oxidative Coupling Reactions. Part 1. Determination of Catalase Activity [in German]. *Landwirtsch. Forsch.,* **39** (1986) 139.
1380. J. D. Trent and S. Christiansen, Determination of Total Nonstructural Carbohydrates in Forage Tissue by p-Hydroxybenzoic Acid Hydrazide Flow Injection Analysis. *J. Agric. Food Chem.,* **34** (1986) 1033.
1381. T. Z. Polta and D. C. Johnson, Pulsed Amperometric Detection of Sulfur Compounds. Part 1. Initial Studies at Platinum Electrodes in Alkaline Solutions. *J. Electroanal. Chem. Interfacial Electrochem.,* **209** (1986) 159.
1382. F. Lazaro, A. Rios, M. D. Luque de Castro, and M. Valcárcel, Diode Array Detectors in Hydrodynamic Analytical Systems. *Analusis,* **14** (1986) 378.
1383. S. Motomizu, Determination of Nitrogen, Phosphorus and Anionic Sur-

factant in Water by Flow Injection Analysis [in Japanese]. *Kankyo Gijutsu,* **15** (1986) 525.

1384. J. M. Hwang, T. S. Wei, and Y. M. Chen, Catalytic Photometric Determination of Selenium by Flow Injection Analysis. *J. Chin. Chem. Soc.,* **33** (1986) 109.

1385. W. Frenzel and P. Brätter, The Fluoride Ion-Selective Electrode in Flow Injection Analysis. Part 3. Applications. *Anal. Chim. Acta,* **188** (1986) 151.

1386. M. Trojanowicz, A. Hulanicki, W. Matuszewski, M. Palys, A. Fuksiewicz, T. Hulanicka-Michalak, S. Raszewski, J. Szyller, and W. Augustyniak, Flow-Injection Catalytic Determination of Molybdenum with Biamperometric Detection in a Microprocessor-Controlled System. *Anal. Chim. Acta,* **188** (1986) 165.

1387. C. E. Lunte, S. Wong, T. H. Ridgway, W. R. Heineman, and K. W. Chan, Voltammetric/Amperometric Detection for Flow-Injection Systems. *Anal. Chim. Acta,* **188** (1986) 263.

1388. M. Masoom and P. J. Worsfold, Automated Flow-Injection Procedures for Glycerol and Triglycerides. *Anal. Chim. Acta,* **188** (1986) 281.

1389. K. A. Holm, Spectrophotometric Flow-Injection Determination of Cellobiose Dehydrogenase Activity in Fermentation Samples with 2,6-Dichlorophenolindophenol. *Anal. Chim. Acta,* **188** (1986) 285.

1390. F. Shirato, N. Kobayashi, and K. Yatsuno, Determination of Fluoride Ions at the pppm Level by a Flow-Injection Method [in Japanese]. Jpn. Kokai Tokkyo Koho, Pat. No. 86100641 (Hitachi, Ltd.) (1986).

1391. M. M. Cooper and S. R. Spurlin, Chemiluminescent Method for Flow Injection Analysis of Anions. *Anal. Lett.,* **19** (1986) 2221.

1392. J. M. Calatayud and P. C. Falcó, Spectrophotometric Determination of Chlorhexidine with Bromocresol Green by Flow-Injection and Manual Methods. *Anal. Chim. Acta,* **189** (1986) 323.

1393. J. Růžička, Flow Injection Analysis—A Survey of Its Potential for Continuous Monitoring of Industrial Processes. *Anal. Chim. Acta,* **190** (1986) 155.

1394. M. Gisin and C. Thommen, Industrial Process Control by Flow Injection Analysis. *Anal. Chim. Acta,* **190** (1986) 165.

1395. H. Bergamin F$^{\underline{o}}$, F. J. Krug, E. A. G. Zagatto, E. C. Arruda, and C. A. Coutinho, On-Line Electrolytic Dissolution of Alloys in Flow-Injection Analysis. Part 1. Principles and Application in the Determination of Soluble Aluminium in Steels. *Anal. Chim. Acta,* **190** (1986) 177.

1396. N. Ishibashi, T. Imato, and K. Tsukiji, Potentiometric and Spectrophotometric Flow-Injection Determinations of Metal Ions with Use of Metal Ion Buffers. *Anal. Chim. Acta,* **190** (1986) 185.

1397. Zs. Niegreisz, G. Horvai, K. Toth, and E. Pungor, Silicone Rubber Wall-Jet Electrode in Hydrodynamic Voltammetry. (Comparison of Various Carbon Electrodes). *Symp. Biol. Hung.,* **31** (1986) 83.

1398. T. S. Stevens, N. N. Frawley, D. E. Diedering, W. C. Harris, and D. J. Swart, Apparatus and Method for Addition of a Reagent to a Flowing Stream of a Mobile Phase [in German]. Ger. Pat. No. 3606938 (Dow Chemical Co) (1986).

1399. C. Pohlandt-Watson, A Simplified Flow-Injection Method for Determination of Free Cyanide in Process Solutions. *Rep-MINTEK (S. Africa),* **M275** (1986) 6.

1400. M. Ishii, M. Yamada, and S. Suzuki, Determination of Ultratraces of Cyanide Ion by Flow-Injection Analysis with Surfactant Bilayer Vesicle-Catalyzed and Uranine-Sensitized Chemiluminescence [in Japanese]. *Bunseki Kagaku,* **35** (1986) 955.

1401. F. Malamas, M. Bengtsson, and G. Johansson, Trace Metal Enrichment by Immobilized Chelate Ion Exchangers in a Flow Injection System. *Proc. 5th Heavy Metal Environ. Int. Conf.,* **2** (1985) 525.

1402. D. A. Hollowell, The Utilization of Flow Injection Analysis for Selective Determination. *Diss. Abstr. Int. B.,* **46** (1986) 3417. (Ph.D. Thesis).

1403. B. Hoeyer and L. Kryger, Evaluation of Single-Point Calibration in Flow Potentiometric Stripping Analysis. *Talanta,* **33** (1986) 883.

1404. K. G. Miller, Selective Methods of Analysis Utilizing Flow Injection Analysis. *Diss. Abstr. Int. B,* **47** (1986) 598. (Ph.D. Thesis).

1405. P. Macheras, M. Koupparis, and C. Tsaprounis, Drug Dissociation Studies in Milk Using the Automated Flow Injection Serial Dynamic Dialysis Technique. *Int. J. Pharm.,* **33** (1986) 125.

1406. H. Ikeda, T. Matsumoto, T. Kitamoto, M. Shimizu, T. Nakakuki, M. Yoshida, and M. Okada, Apparatus and Method for Quantitative Analysis of Aldoses [in Japanese]. Jpn. Kokai Tokkyo Koho, Pat. No. 86175559 (Shokuhin Sangyo Bioreactor System Gijutsu Kenkyu Kumiai) (1986).

AUTHOR INDEX TO FIA REFERENCES*

* These numbers correspond to reference numbers in the list of FIA References on pages 383–476.

SUBJECT INDEX

Bold Face denotes main source

(*continued from front*)

Vol. 66. **Solid Phase Biochemistry: Analytical and Synthetic Aspects.** Edited by William H. Scouten

Vol. 67. **An Introduction to Photoelectron Spectroscopy.** By Pradip K. Ghosh

Vol. 68. **Room Temperature Phosphorimetry for Chemical Analysis.** By Tuan Vo-Dinh

Vol. 69. **Potentiometry and Potentiometric Titrations.** By E. P. Serjeant

Vol. 70. **Design and Application of Process Analyzer Systems.** By Paul E. Mix

Vol. 71. **Analysis of Organic and Biological Surfaces.** Edited by Patrick Echlin

Vol. 72. **Small Bore Liquid Chromatography Columns: Their Properties and Uses.** Edited by Raymond P. W. Scott

Vol. 73. **Modern Methods of Particle Size Analysis.** Edited by Howard G. Barth

Vol. 74. **Auger Electron Spectroscopy.** By Michael Thompson, M. D. Baker, Alec Christie, and J. F. Tyson

Vol. 75. **Spot Test Analysis: Clinical, Environmental, Forensic and Geochemical Applications.** By Ervin Jungreis

Vol. 76. **Receptor Modeling in Environmental Chemistry.** By Philip K. Hopke

Vol. 77. **Molecular Luminescence Spectroscopy—Parts I and II: Methods and Applications.** Edited by Stephen G. Schulman

Vol. 78. **Inorganic Chromatographic Analysis.** Edited by John C. MacDonald

Vol. 79. **Analytical Solution Calorimetry.** Edited by J. K. Grime

Vol. 80. **Selected Methods of Trace Metal Analysis: Biological and Environmental Samples.** By Jon C. VanLoon

Vol. 81. **The Analysis of Extraterrestrial Materials.** By Isidore Adler

Vol. 82. **Chemometrics.** By Muhammad A. Sharaf, Deborah L. Illman, and Bruce R. Kowalski

Vol. 83. **Fourier Transform Infrared Spectrometry.** By Peter R. Griffiths and James A. de Haseth

Vol. 84. **Trace Analysis: Spectroscopic Methods for Molecules.** Edited by Gary Christian and James B. Callis

Vol. 85. **Ultratrace Analysis of Pharmaceuticals and Other Compounds of Interest.** Edited by S. Ahuja

Vol. 86. **Secondary Ion Mass Spectrometry: Basic Concepts, Instrumental Aspects, Applications and Trends.** By A. Benninghoven, F. G. Rüdenauer, and H. W. Werner

Vol. 87. **Analytical Applications of Lasers.** Edited by Edward H. Piepmeier

Vol. 88. **Applied Geochemical Analysis.** by C. O. Ingamells and F. F. Pitard

Vol. 89. **Detectors for Liquid Chromatography.** Edited by Edward S. Yeung

Vol. 90. **Inductively Coupled Plasma Emission Spectroscopy: Part I: Methodology, Instrumentation, and Performance; Part II: Applications and Fundamentals.** Edited by J. M. Boumans

Vol. 91. **Applications of New Mass Spectrometry Techniques in Pesticide Chemistry.** Edited by Joseph Rosen

Vol. 92. **X-Ray Absorption: Principles, Applications, Techniques of EXAFS, SEXAFS, and XANES.** Edited by D. C. Konnigsberger

Vol. 93. **Quantitative Structure–Chromatographic Retention Relationships.** By Roman Kaliszan

Vol. 94. **Laser Remote Chemical Analysis.** Edited by Raymond M. Measures

Vol. 95. **Inorganic Mass Spectrometry.** Edited by F. Adams, R. Gijbels, and R. Van Grieken